International Technical Conference on Diamond, Cubic Boron Nitride and Their Applications (INTERTECH 2019)

New Orleans, Louisiana, USA
29 - 31 May 2019

ISBN: 978-1-7138-0350-8

Printed from e-media with permission by:

Curran Associates, Inc.
57 Morehouse Lane
Red Hook, NY 12571

Some format issues inherent in the e-media version may also appear in this print version.

Industrial Diamond Association of America Inc. (IDA)
P.O. Box 29460
Columbus, Ohio 43229
USA

Phone: 614.797.2265
Fax: 614.797.2264

tkane-ida@insight.rr.com

Additional copies of this publication are available from:

Curran Associates, Inc.
57 Morehouse Lane
Red Hook, NY 12571 USA
Phone: 845-758-0400
Fax: 845-758-2633
Email: curran@proceedings.com
Web: www.proceedings.com

TABLE OF CONTENTS

Welcome to...

INTERTECH 2019!

INDUSTRIAL DIAMOND ASSOCIATION OF AMERICA, INC. (IDA)

- Mailing Address -

P.O. Box 29460
Columbus, Ohio 43229

- Shipping Address -

6081 Central Park Drive
Columbus, Ohio 43231
Phone: 614-797-2265 • Fax: 614-797-2264
E-Mail: tkane@superabrasives.org

From the President and Executive Director of the Industrial Diamond Association of America...

Welcome to the Hyatt Regency New Orleans and INTERTECH 2019! New Orleans is known as "The Big Easy" thanks to the late Betty Guillaud, a gossip columnist from the Times-Picayune. It is said that Guillaud used the term first to compare life in New York City, "The Big Apple" to life in New Orleans, The Big Easy… Enthralling, captivating and always a good time, that's New Orleans through the ages. In New Orleans the past isn't dead; it lives day to day throughout the city. No other city in America keeps its history as alive as New Orleans. Neighborhoods, architecture, cemeteries with crypts called "cities of the dead"; cobblestone streets and tree lined streets of southern mansions serve as permanent reminders of days gone by! Look around, in New Orleans traditions and history can be a band led funeral or the sights and sounds of Bourbon Street and the French Quarter with live jazz pouring from the doorways! Enter INTERTECH 2019 with a history all its own after being established almost 20 years ago to be the premier International Conference on diamond, cubic boron nitride (cBN) and their applications. INTERTECH has always provided a venue where attendees can experience outstanding technical presentations highlighting new innovations along with proven application successes across all manufacturing industries. INTERTECH has heralded major innovations and accomplishments in our industry. This is where we have heard of revolutionary prosthetic applications for diamond and nanodiamond applications to carry drugs to kill cancerous tumors. We have heard of laser windows, composite drilling and blood filters made of diamond… Of course the new developments in traditional machining and grinding applications are always of great interest. This year we have focused on the materials used throughout all industries and how these materials can only be cost-effectively machined and ground with superabrasives. INTERTECH has held true to its mission of holding a conference dedicated not only to the new technology being developed on superabrasives, but also show how that new technology can be applied in the real world to increase productivity and improve the quality of our lives. With a technically stimulating program, the companionship of your peers in the industry, and exciting social possibilities both organized and spontaneous INTERTECH 2019 will be a memorable event. If you have any questions related to INTERTECH 2019 during your stay, don't hesitate to call on any of the Officers or Directors from the Industrial Diamond Association of America (IDA) or the INTERTECH 2019 staff. We will be present during the Conference to make your INTERTECH 2019 worthwhile and your stay enjoyable.... Welcome!

Shane Hollenbaugh, President
Industrial Diamond Association of America

Terry M. Kane, Executive Director
Industrial Diamond Association of America

GENERAL INFORMATION AND GUIDELINES FOR *INTERTECH 2019*

YOUR COURTESY IS GREATLY APPRECIATED

PLEASE MAKE EVERY EFFORT TO RESPECT PRESENTERS AND TURN OFF ALL CELL PHONES AND PAGERS

WE ASK THAT YOU STAY FOR THE ENTIRE CONFERENCE AND TREAT THE LAST SPEAKER WITH THE SAME COURTESY AND ENTHUSIASM AS THE FIRST

DRESSCODE

Dress for the daytime sessions will be business casual, which means no tie, unless you prefer to wear one. For those signed up on the Thursday "TOUR OF NEW ORLEANS" it will be casual... a light jacket or sweater is recommended, enjoy!

The Welcome Reception on Wednesday evening is casual.

We encourage men speakers to wear a coat and tie during their presentations.

Thank you for your cooperation!

NAME BADGES

You are asked to wear your Name Badge to all INTERTECH functions. This will not only help us all remember names it will help the INTERTECH and Hyatt staff easily identify you as a paid attendee for admittance to the technical sessions, meals and functions.

IF YOU LOSE YOUR NAMEBADGE PLEASE GO TO THE INTERTECH REGISTRATION DESK FOR A REPLACEMENT.

SMOKING

There is no smoking allowed in any of the conference areas or the hotel.

PLEASE CHECK WITH HOTEL STAFF OR ADHERE TO SIGNS INDICATING SMOKING AREAS

AUDIO/VIDEO RECORDINGS ARE NOT PERMITTED!

No recording is permitted in the technical sessions. All sessions and content as presented is copyrighted to the Industrial Diamond Association of America...

THIS MEANS NO CELL PHONE PHOTOS OR VIDEOS AS WELL!

SPECIAL OR EMERGENCY REQUIREMENTS

Please contact any one of the INTERTECH 2019 staff or see someone at the reception desk. We will do our best to make your stay enjoyable.

International Cosponsoring Organizations
IDA Board of Directors and INTERTECH 2019 Organizing Committees
INTERTECH Corporate Sponsors.

Wednesday Morning Keynote - DR. K. RENEE HORTON, SPACE LAUNCH SYSTEM (SLS) LEAD METALLIC/ WELD ENGINEER NASA RESIDENTIAL MANAGEMENT OFFICE -
Dr. Horton is a graduate of Louisiana State University with a B.S. of Electrical Engineering with a minor in Math in 2002 and a Ph.D. in Material Science with a concentration in Physics, becoming the first African American to graduate from the University of Alabama in 2011 in this area. Dr. Horton currently serves as the Space Launch System (SLS) Lead Metallic/Weld Engineer in the NASA Residential Management Office at Michoud Assembly Facility (MAF) in New Orleans.

Wednesday Afternoon Keynote - JIM DELAND, TECHNICAL SUPPORT ENGINEER
APACHE AEROSPACE - Jim's responsibility at **APACHE AEROSPACE** involves providing technical support for drilling solutions to major aerospace companies and their suppliers. His background in aerospace includes over 28 years at **THE BOEING COMPANY** as a Research & Technology Engineer where he worked in the Assembly Technology area supporting technology development for production.

Wednesday Technical Sessions - AEROSPACE DEVELOPMENTS
SUPERABRASIVE GRINDING DEVELOPMENT
AUTOMOTIVE IMPROVEMENTS

Thursday Morning Keynote - MIKE FELIX, GLOBAL CHIEF ENGINEER
FORD POWERTRAIN MANUFACTURING ENGINEERING - Mike's responsibilities include Global Machining and Assembly Processes for Engines, Transmissions, Axles and Electrification components; including Hybrid Batteries, eTransaxles and eMotor manufacturing. Mike has been with Ford for over 30 Years. He started his career at the Cleveland Engine Plant Facility in 1988 as an Engineering Graduate. Mike's career has progressed through a number of engineering and operations positions within Powertrain. He has worked in nine different Ford Plants throughout his career and has spent 11 Years as Plant Manager in 5 different Plants across Ford Powertrain, as well as international and Staff Engineering assignments.

Thursday Technical Sessions - SUPERABRASIVE APPLICATION ADVANCEMENTS
SUPERABRASIVE RESEARCH AND DEVELOPMENT

Friday Technical Sessions - SUPERABRASIVE SCIENCE & TECHNOLOGY
Keynote Speaker Biographies
Biographies of Presenters & Coauthors
Diamond/CBN Application Chart
Note Pages with Presentations
Additional Note Pages

INTERTECH 2019 COSPONSORS

Abrasive Engineering Society

ASM International

Assodiamante

Concrete Sawing & Drilling Association

F.E.P.A.

Industrial Diamond Association of America

The Association for Manufacturing Technology

The American Ceramics Society

Elsevier Diamond Conference

International Association Of Concrete Drillers and Sawers

Industrial Diamond Association of China

Industrial Diamond Association of India

Industrial Diamond Association of Japan

Japan Society for Abrasive Technology

Society of Manufacturing Engineers (SME)

Unified Abrasive Manufacturers Association

INTERTECH 2019 is an internationally sponsored event and these leading Global Organizations represent the far reaching memberships that support the ongoing technological research and worldwide cooperation for superabrasive use and development.

Industrial Diamond Association of America (IDA) Board of Directors & Organizing Committees

IDA BOARD and STEERING COMMITTEE

OFFICERS

President - Shane Hollenbaugh, YG-1 USA
Vice President - Kevin Stiles, RADIAC ABRASIVES, A TYROLIT COMPANY
Secretary/Treasurer - Aaron Nolan, WORLDWIDE SUPERABRASIVES SOLIDS DIVISION
Past President - Ed Francis, CRYSTALLUME ENGINEERED DIAMOND PRODUCTS

DIRECTORS

Renee Nord, 3M ABRASIVES SYSTEMS DIVISION
Eric Bieberich, FORT WAYNE WIRE DIE, INC.
Julie Griffin, PRECISION EFORMING
Charles "Chip" Rarey, HYPERION MATERIALS & TECHNOLOGIES
Michael Lew, MWI ELOQUENCE CORPORATIONS
William Fisher, NATIONAL RESEARCH COMPANY
Marc Modersitzki, US SYNTHETIC CORPORATION

INTERTECH 2019 Chairman

IDA Executive Director - Terry M. Kane, INDUSTRIAL DIAMOND ASSOCIATION OF AMERICA

INTERTECH 2019 Meeting Planner

IDA Business Manager - Kathy A. Kane, INDUSTRIAL DIAMOND ASSOCIATION OF AMERICA

The INTERTECH 2019 Technical Advisory Committee (TAC) provides technical guidance and recommendations for the technical portion of INTERTECH 2019. They insured the quality and content of each paper is relevant and representative of the technology for the superabrasives industry so each conference session will consist of the best research papers and state-of-the-art topics for our industry.

INTERTECH 2019 – Technical Advisory Committee

Founding Chairman - Ion C. Benea, PhD Vice President Operations - Hyprez Products Engis Corporation USA
Prof. John Angus, PhD, Honorary Member Case Western Reserve University USA
Jim Boland, PhD, Honorary Research Fellow CSIRO AUSTRALIA
James Butler, PhD, Honorary Member Research Associate - Smithsonian National Museum of Natural History USA
Boris Feigelson, PhD, Naval Research Laboratory Washington, DC USA
Richard Hall, PhD, Independent Consultant United Kingdom
Prof. František Holešovský, MSc., PhD, University of West Bohemia Czech Republic
Yukata Kobayashi, General Manager, Super Hard Materials Research Department Sumitomo Electric Industries Japan
Prof. Syuhei Kurokawa, Kyushu University Japan
Prof. Zhengxin Li, PhD, Henan University of Technology China
Abds-Sami (Sami) Malik, PhD, Technology Manager Hyperion Materials & Technology USA
Prof. Ioan Marinescu, PhD, Director of Precision Micro-Machining Center University of Toledo USA
Christoph Nebel, PhD, Head of Micro and Nanosensor Department Fraunhofer-Institute Germany
Jim Pickles, PhD, Commercial Strategy Manager Element Six Group UK
Prof. Luiz Eduardo de Angelo Sanchez, PhD Sao Paulo State University Brazil
Gerald Shin, R&D Center ILJIN Diamond Co., Ltd. Korea
John Webster, PhD, Senior Research Engineer Worldwide Superabrasives USA
Prof. Liangchi Zhang, PhD, The University of New South Wales, Australia

INTERTECH 2019 CORPORATE SPONSORS

PREMIUM - ALL EVENTS

◆Worldwide Superabrasives, *USA*

PRESENTERS & A/V

◆Ohio Carbon Blank, *USA*

◆Worldwide Superabrasives, *USA*

WEDNESDAY NIGHT WELCOME RECEPTION

◆American Superabrasives & ANCO Industrial Diamond Corp. *USA*

◆National Research Company, *USA*

◆Ohio Carbon Blank, *USA*

◆Protech Diamond, *USA & Canada*

◆Worldwide Superabrasives, *USA*

CONFERENCE PROGRAM

◆**3M Abrasive System Division**, *USA*

◆Crystallume Engineered Diamond Products, *USA*

◆Fort Wayne Wire Die, *USA*

◆MegaDiamond, *USA*

A Schlumberger Company

◆MWI Eloquence Corporation *USA*

◆National Research Company, *USA*

◆Ohio Carbon Blank, *USA*

◆Precision eForming, *USA*

◆Protech Diamond, *USA & Canada*

◆Radiac Abrasives, A Tyrolit Company *USA*

◆US Synthetic, *USA*

◆Worldwide Superabrasives, *USA*

◆YG-1, *USA & Korea*

WEDNESDAY MAY 29, 2019*

7:00 AM - 4:00 PM - REGISTRATION DESK OPEN
7:00 - 7:45 - CONTINENTAL BREAKFAST
7:45 - 8:00 - CELESTIN BALLROOM - WELCOME - **Shane Hollenbaugh, *President IDA* & Terry Kane, *INTERTECH Chalrman***
8:00 - 8:45 - CELESTIN BALLROOM "E" - KEYNOTE SPEAKER - KEYNOTE SPEAKER - **DR. K. RENEE HORTON,** NASA SPACE LAUNCH SYSTEM LEAD METALLIC/WELD ENGINEERING MANAGER

AEROSPACE DEVELOPMENTS

CELESTIN BALLROOM "E"

Chairperson - SHANE HOLLENBAUGH, YG-1 USA

9:00- 9:45 — **#1941 GRINDING OF AEROSPACE MATERIALS USED FOR HIGH-TEMPERATURE APPLICATIONS**
Mr. Philip K. Varghese, PhD, Group Leader-Advanced Application Engineering, Norton/Saint-Gobain Abrasives USA

9:45-10:30 — **#1929 INNOVATED CUTTING TOOLS AND METHOD FOR EFFICIENT COMPOSITE MATERIALS MACHINING**
Mr. Abdelatif Atarsia, PhD, Global Composites Industry Manager, YG1 USA

10:30 - 10:45 — **— BREAK –**

10:45-11:30 — **#1930 GRIT SIZE, THERMAL DAMAGE, SURFACE FINISH, DRESSING PARAMETERS AND COMMON MISTAKES WHEN GRINDING HARDENED STEEL AND NICKEL ALLOYS WITH CBN GRITS.**
Mr. Jeffrey Badger, PhD, President, The Grinding Doc USA

11:30-12:30 — **—LUNCH—**

SUPERABRASIVE GRINDING DEVELOPMENT

CELESTIN BALLROOM "H&G"

Chairperson - WILLIAM FISHER, National Research Company

9:00- 9:45 — **#1924 FIXED ABRASIVE THREE-DIMENSIONAL PLATE FOR LAPPING AND POLISHING OF ADVANCED MATERIALS**
Mr. Clement Onyenemezu, PhD - Engis Corporation USA

9:45 - 10:30 — **#1927 THEORETICAL EXPERIMENTAL STUDY OF SILICON CARBIDE GRINDING**
Mr. Mohammed Tharwan, PhD, Engineer, University of Toledo USA

10:30-10:45 — **— BREAK —**

10:45-11:30 — **#1928 MICROSCOPIC CHARACTERIZATION AND MODELING OF OXIDE LAYER FOR ELECTROLYTIC IN-PROCESS DRESSING (ELID) GRINDING**
Mr. Bader Alqahtani, PhD, Engineer Mechanical Engineering Department, The University of Toledo USA

11:30-12:30 — **—LUNCH—**

12:45 - 1:45 CELESTIN BALLROOM "E" - JIM DELAND, TECHNICAL SUPPORT ENGINEER APACHE AEROSPACE, INC.

AEROSPACE DEVELOPMENTS

CELESTIN BALLROOM "E"

Chairperson - MICHAEL LEW, MWI Eloquence Corporation

1:45-2:30 — **#1926 TRIBOLOGY OF DIAMOND BEARINGS FOR SPACE APPLICATIONS**
Mr. Kenneth Street, PhD, Analytical Chemist - Diamond Materials Inc. USA

2:30-3:15 — **#1936 INFLUENCE OF ABRASIVE TYPE ON GRINDING INDUCED RESIDUAL STRESSES AND SURFACE INTEGRITY**
Mr. John Webster, PhD, Senior Research Engineer, Worldwide Superabrasives USA

3:15-3:30 — **— BREAK —**

3:30-4:15 — **#1940 GRINDABILITY OF ADDITIVELY MANUFACTURED AEROSPACE ALLOYS**
Mr. Philip K. Varghese, PhD, Group Leader-Advanced Application Engineering, Norton/Saint-Gobain Abrasives USA

AUTOMOTIVE IMPROVEMENTS

CELESTIN BALLROOM "H&G"

Chairperson - KEVIN STILES, RadiacAbrasives, A Tyrolit Company

1:45 - 2:30 — **#1922 THE NEXT GENERATION OF VITRIFIED CBN FOR CYLINDRICAL GRINDING IN THE AUTOMOTIVE INDUSTRY**
Mr. Florian Rechfelden, Senior Application Engineer - Radiac Abrasives, A Tyrolit Company

2:30 - 3:15 — **#1931 INVESTIGATING AND MODELING THE BALL-BURNISHING FACTORS IMPACT ON RESIDUAL STRESS OF AISI 8620 STEEL**
Mr. Abdulaziz Alshareef, PhD Candidate, Senior Researcher at PMMC-MIME, University of Toledo USA

PROGRAM AND SCHEDULE SUBJECT TO REVISION

6:00 PM - 8:00 PM WELCOME RECEPTION - HORS D' OEUVRES & COCKTAILS

THURSDAY MAY30, 2019*

7:30 AM - 4:00 PM - REGISTRATION DESK OPEN
7:30 - 8:30 - CONTINENTAL BREAKFAST
8:30 - 8:45 - CELESTIN BALLROOM - WELCOME - **Shane Hollenbaugh, *President IDA*** **& Terry Kane, *INTERTECH Chalrman***
9:00 - 9:45 - CELESTIN BALLROOM "E" - KEYNOTE SPEAKER - **MIKE FELIX,** FORD MOTOR COMPANY, GLOBAL CHIEF ENGINEER POWERTRAIN MANUFACTURING

SUPERABRASIVE APPLICATION ADVANCEMENTS

CELESTIN BALLROOM "E"

Chairperson - AARON NOLAN, Worldwide Superabrasives Solids Division

10:00 - 10:45 **#1933 THE USE OF ELECTRON-MICROSCOPY IN THE STUDY OF SUPERABRASIVES IN GRINDING**
Mr. Jeffrey Badger, PhD, President, The Grinding Doc USA

10:45 - 11:00 **— BREAK —**

11:00 - 11:45 **#1934 SINGLE-CRYSTAL DIAMOND GRIT IN CONSTRUCTION: APPLICATIONS AND CHARACTERIZATION**
Mr. Stuart Nailer, PhD, Research & Development Expert, Hilti Corporation, Schaan, Liechtenstein

11:45 -12:45 **—LUNCH—**

SUPERABRASIVE RESEARCH AND DEVELOPMENT

CELESTIN BALLROOM "H&G"

Chairperson - ERIC BIEBERICH, Fort Wayne Wire Die

10:00- 10:45 **#1939 CHARACTERIZATION OF THE RESIDUAL STRESSES INTRODUCED BY A NEW JOINING METHOD IN DIAMOND AND TUNGSTEN CARBIDE COMPOSITES**
Mr. Timothy Dumm, PhD, Technology Manager, Abrasives - Hyperion Materials & Technologies USA

10:45 - 11:00 **— BREAK —**

11:00 - 11:45 **#1938 EVALUATION OF THE INFLUENCE OF IMPURITIES AND CRYSTALLINITY IN SINGLE CRYSTAL DIAMOND AND NANO-POLYCRYSTALLINE DIAMOND (NPD) ON WEAR CHARACTERISTICS**
Ms. Minori Teramoto, Manager, Sumitomo Electric Industries Ltd. Japan

11:45-12:45 **—LUNCH—**

1:00 - 1:45 CELESTIN BALLROOM "E" - FEATURED SPEAKER **- TARIFF ACTIONS AFFECTING THE SUPERABRASIVES INDUSTRY- Mr. Daniel Pickard, Attorney At Law - Wiley Rein USA**

SUPERABRASIVE APPLICATION ADVANCEMENTS

CELESTIN BALLROOM "E"

Chairperson - JULIE GRIFFIN, Precision eForming

1:45 - 2:30 **#1937 DETERMINING THE FLOWRATE, PRESSURE AND NOZZLE APERTURE IN A GRINDING PROCESS WITH CONSIDERATION OF ABRASIVE TYPE, COOLANT TYPE AND WHEEL STRUCTURE**
Mr. John Webster, PhD, Senior Research Engineer, Worldwide Superabrasives USA

2:30 - 3:15 **#1944 ADVANCES IN THE PROCESSING OF SEMICONDUCTOR MATERIALS USING DIAMOND TECHNOLOGIES**
Mr. Timothy Dumm, PhD, Technology Manager, Abrasives - Hyperion Materials & Technologies USA

3:15 - 3:30 **— BREAK —**

3:30 - 4:15 **#1943 LASER TECHNOLOGY AND MACHINING BRITTLE MATERIALS WITH PCD APPLICATIONS**
Mr. Stephen Kelly, PhD, Regional Technical Manager the Americas - Element Six US Corporation USA

SUPERABRASIVE RESEARCH AND DEVELOPMENT

CELESTIN BALLROOM "H&G"

Chairperson - ED FRANCIS, Crystallume Engineered Diamond Products

1:45 - 2:30 **#1945 UNDERSTANDING SOURCES OF RESIDUAL STRESS IN PDC BY X-RAY DIFFRACTION**
Mr. Deb K. Mukhopadhyay, PhD, Senior Scientist - US Synthetic Corporation USA

2:30 - 3:15 **#1925 INSPECTING ULTRA-HARD MATERIALS AND PROCESSES USING ACOUSTIC MICRO IMAGING (AMI)**
Mr. Jack Richtsmeier, Business Development Manager, Nordson Sonoscan

3:15 - 3:30 **— BREAK —**

3:30 - 4:15 **#1932 THEORETICAL RESEARCH ON THE UNIFORMITY OF PLANETARY DOUBLEï¼•SIDE LAPPING TRAJECTORY**
Mr. Yaoyu Wu, PhD, Professor, Zhengzhou Institute of Technology, Henan China

PROGRAM AND SCHEDULE SUBJECT TO REVISION

6:30 PM - 10:30 PM NEW ORLEANS TOUR AND DINNER - BUS LEAVES FROM HYATT MAIN ENTRANCE

FRIDAY MAY 31, 2019*

7:30 - 4:00 - REGISTRATION DESK OPEN
7:00 - 8:00 - CONTINENTAL BREAKFAST
7:30 - 8:30 - IDA Board Meeting - **CELESTIN BALLROOM "H"**
8:30 - 8:45 - CELESTIN BALLROOM "E" - WELCOME - **Kevin Stiles, *President IDA* & Terry Kane, *INTERTECH Chalrman***

SUPERABRASIVE SCIENCE & TECHNOLOGY

CELESTIN BALLROOM "E"

Chairperson - MARC MODERSITZKI, US SyntheticCorporation

8:45 - 9:30 **#1923 VOLUME PRODUCTION OF HIGH PRECISION TOOLS WITH NEW 5-AXIS LASER MACHINING CENTRE**
Mr. Jacques Coderre, US Business Manager - Synova USA

9:30 - 10:15 **#1935 PERFORMANCE IMPROVEMENT WITH NOVEL NANO BINDER TECHNOLOGY IN POLYCRYSTALLINE CUBIC BORON NITRIDE**
Mr. Stephen Kelly, PhD, Regional Technical Manager - the Americas, Element Six US Corporation USA

10:15 - 10:30 **— BREAK —**

10:30 - 11:30 **#1942 DIAMOND SYNTHESIS "THE SCIENCE"**
Mr. Ion Benea, PhD, Vice President Operation - Hyprez Products, Engis Corporation USA

12:00 **Conference Close**

PROGRAM AND SCHEDULE SUBJECT TO REVISION

AFTERNOON & EVENING - SIGHTSEEING AND NETWORKING ON YOUR OWN

INTERTECH 2019 KEYNOTE SPEAKERS

INTERTECH 2019 IS PROUD TO PRESENT AN IMPRESSIVE LIST OF KEYNOTE SPEAKERS REPRESENTING LEADING INDUSTRY BUSINESSES AND INNOVATIVE DIAMOND RESEARCH

DR. K. RENEE HORTON, SPACE LAUNCH SYSTEM (SLS) LEAD METALLIC/WELD ENGINEER
NASA RESIDENTIAL MANAGEMENT OFFICE

Dr. Horton is a native of Baton Rouge, Louisiana and lifelong lover of science and NASA. A graduate of Louisiana State University with a B.S. of Electrical Engineering with a minor in Math in 2002 and a Ph.D. in Material Science with a concentration in Physics, becoming the first African American to graduate from the University of Alabama in 2011 in this area. Dr. Horton currently serves as the Space Launch System (SLS) Lead Metallic/Weld Engineer in the NASA Residential Management Office at Michoud Assembly Facility (MAF) in New Orleans. She worked for NASA, first as a student from 2009 to 2011, and then started her career as a mechanical test engineer in 2012. In 2014 she was promoted to her current position. In 2016, Renee was elected President of the National Society of Black Physicists (NSBP) as the second woman to hold the office. She has served the physics community abroad as a member of the International Union of Pure and Applied Physics (IUPAP) Women in Physics Working Group and currently serves on several advisory boards dedicated to a more diverse inclusion in physics. In 2017, she was elevated to a Fellow in the NSBP, which is the highest honor bestowed upon a member. Renee has been an invited speaker for the first International Women and Girls Day at the United Nations, Essence Power Stage, March for Science – New Orleans and recently the LSU Engineering Spring Commencement. She has spoken all over the world including South Africa, Brazil, South Korea, Canada, Mexico and the beautiful Virgin Islands. She is featured in NASA's I am Building SLS, Physics Today, and the LSU Alumni Magazine. She recently graced the cover of the Hearing Health Foundation Magazine, winter edition that can be found online. She is the author of Dr. H Explores the Universe, Dr. H and her Friends, and Dr. H Explores the ABCs.

MIKE FELIX, GLOBAL CHIEF ENGINEER
FORD POWERTRAIN MANUFACTURING ENGINEERING.

Mike's responsibilities include Global Machining and Assembly Processes for Engines, Transmissions, Axles and Electrification components; including Hybrid Batteries, eTransaxles and eMotor manufacturing. Mike has been with Ford for over 30 Years. He started his career at the Cleveland Engine Plant Facility in 1988 as an Engineering Graduate. Mike's career has progressed through a number of engineering and operations positions within Powertrain. He has worked in nine different Ford Plants throughout his career and has spent 11 Years as Plant Manager in 5 different Plants across Ford Powertrain, as well as international and Staff Engineering assignments. Mike has a BSME in Mechanical Engineering from Akron University and an MBA from Baldwin Wallace College both in Ohio. Mike assumed his current role as Chief Engineer in 2017.

JIM DELAND, TECHNICAL SUPPORT ENGINEER
APACHE AEROSPACE

Jim's responsibility at **APACHE AEROSPACE** involves providing technical support for drilling solutions to major aerospace companies and their suppliers. His background in aerospace includes over 28 years at **THE BOEING COMPANY** as a Research & Technology Engineer where he worked in the Assembly Technology area supporting technology development for production. Jim was frequently called on to develop manufacturing solutions for new airplane programs and was instrumental in successfully developing and implementing new drilling technology on both the 787 & 777X Program. During his career he has earned 9 patents both in assembly technology and new cutting tool development. Based on his engineering achievements he was recognized as a Boeing Associate Technical Fellow in 2011. Recently Blue Origin reached out to Jim to provide his drilling and assembly expertise as a consultant on the New Glen rocket program. Jim has published multiple papers for SAE on the topic of cutting tool technology and has presented at multiple conferences about drilling technology. Jim is a graduate of the Oregon Institute of Technology with a B.S. in Manufacturing Engineering Technology.

INTERTECH 2019 SPEAKERS & COAUTHORS BIOGRAPHIES

(Received by press time)

ABDULAZIZ ALSHAREEF is a PhD Candidate in the Department of Mechanical, Industrial, and Manufacturing Engineering at the University of Toledo, Ohio. He is Senior Researcher at the Precision Micro-Machining Center (PMMC). He has a master's degree in Mechanical Engineering (M.S.M.E.) from Wilkes University, Pennsylvania, USA and has a bachelor's degree in Mechanical Engineering from Umm Al-Qura University, Saudi Arabia. Mr. Alshareef is a member of the American Society of Mechanical Engineers (ASME). Mr. Alshareef's areas of research interest include tribological investigation of mechanical properties of metals, burnishing, metal working, and design of experiment.

ABDELATIF ATARSIA, PhD Sales & Marketing Director, YG-1 USA - Dr. Atarsia has led and participated in multiple projects involving the development of new machining strategies for Aerospace parts and structure for higher performance and lower cost. Dr. Atarsia worked as an Engineering Specialist at Bombardier Aerospace leading the drilling project for the C Series Aft Fuse work package. Dr. Atarsia holds a Bachelor of Mechanical Engineering in 1992 from Annaba University then M.S. in 1994 and Ph.D. degrees (1999) in Mechanical Engineering, from École Polytechnique of Montreal – Canada.

JEFFREY BADGER, PhD, The Grinding Doc - Dr. Jeffrey Badger is an expert in the industrial process of grinding. He works independently as a consultant, helping companies around the world improve their grinding operations and troubleshooting grinding problems. He has worked in grinding facilities in over 30 countries.

Dr. Badger is the author of the regular question/answer column "Ask The Grinding Doc" in Cutting Tool Engineering. He is well known for his practical, down-to-earth, yet high-tech approach to the subject.

ION C. BENEA, PhD - Dr. Benea is currently the Vice President Operations – Hyprez Products, with Engis Corporation in Wheeling, Illinois (www.engis.com). He received his Bachelor of Science in Solid State Physics from the University of Iasi, Romania. In 1991 he received his Ph.D. Degree in Physics from the University of Cluj-Napoca, Romania, for his research work in high pressure-high temperature synthesis of diamond and cubic boron nitride, and characterization of their physical and chemical properties. Dr. Benea's career in superabrasives industry encompasses forty-five years: twenty years with "Dacia" Synthetic Diamond Factory – Bucharest, Romania as Manager of HPHT R&D Lab. (1974 – 1992) and as Vice President (1992-1994); six years (1995 – 2001) with Warren Diamond/Saint-Gobain – Olyphant, Pennsylvania as Director of R&D and eighteen years (2001 – present) with Engis Corporation – Wheeling, Illinois as Director of R&D (2001-2004) and Vice President Operations (2004 – present).

His research work has been focused on science and technology of superabrasive materials synthesis and the characterization of physical and chemical properties of mesh and micron diamond and CBN powders. He holds four patents and published a number of technical and scientific papers on synthesis and characterization of diamond and CBN, as well as, a number of unpublished internal R&D reports.

Dr. Benea has served on the Board of Directors of the Industrial Diamond Association (IDA) of America for eleven years (2000 and 2011) and was the President of IDA in 2009. He is currently the founding Chairman of the Technical Advisory Committee of the Industrial Diamond Association of America (2010 – present).

KEN BERTAGNOLLI, PhD is the Vice President of Research and Development at US Synthetic Corp. in Orem, Utah. Dr. Bertagnolli has been directing the Company's research efforts since October 2001. Prior to this position, he worked as a Senior Research Engineer developing new products for the fixed cutter market. He and his team have achieved industry-leading performance by incorporating lean manufacturing principles into the product development process. Dr. Bertagnolli has a background in diamond chemical vapor deposition (CVD), laser diagnostics, and numerical modeling. He received his PhD in Mechanical Engineering from the University of Illinois in Urbana-Champaign and BS and MS degrees in Mechanical Engineering from Utah State University.

ANDREW BIRO, Applications Engineer_Norton/Saint-Gobain Abrasives - Andrew is an applications engineer with Saint-Gobain Abrasives in Worcester, MA. His work focuses on delivering innovative grinding solutions to customers in the precision aerospace markets through application, product and process improvements. Andrew received his B.Sc. in Materials Science from Rensselaer Polytechnic Institute and his M.Sc. in Materials Engineering from Worcester Polytechnic Institute. At WPI, his thesis research focused on friction stir welding of wrought and cast aluminum and magnesium alloys, as well as investigation of novel applications for the technology.

MARIA CANN graduated with a BSc in Physics from the University of Leicester in 2008. She then went on to manufacture lasers, working as a Production Engineer at Laser Quantum. She then went on to work in the R&D department at M-Solv Ltd, developing laser processes and systems for manufacture. She then spent six years at M-Solv, during which she competed an industrial doctorate in Nanomaterials at the University of Surrey. This was sponsored by M-Solv Ltd and focused on laser processing of alternative transparent conductors suitable for touch sensor manufacture. She joined Element Six in 2018 as a Laser Ablation Specialist, exploring the emerging opportunities in laser processing of super hard materials for tool manufacture.

JACQUES CODERRE, US Business Manager Synova USA, Inc

Jacques Coderre has been Synova USA's business manager since 2012. Jacques leverages a career at IBM, having occupied various engineering and managerial positions in Bromont, QC. In 1999, he moved to Binghamton NY to work for Universal Instruments Corporation, a supplier of capital equipment. There, he focused on marketing, product management and strategic business development, with international responsibilities.

Out of the Synova offices in Secaucus, NJ, Jacques manages all US operations and personnel. In addition to overall management, responsibilities include strategic business development, customer care operations, and pre-sales applications engineering management.

TIMOTHY F. DUMM, PhD, Technology Manager, Abrasives Hyperion Materials & Technology

Tim Dumm graduated from the Pennsylvania State University in 1983 with a B.S. in Mining Engineering. After working for one year as an engineer in an underground coal mine in Utah, he returned to Penn State and obtained his Masters and Ph.D. in Mineral Processing. Tim's thesis research focused on characterization of respirable coal dust and modeling of the milling of cement clinker. In 1989 he was hired by GE Superabrasives as a Senior Engineer where he developed new processes and characterization methods for diamond and cubic boron nitride powders. Tim has been a working contributor on both the FEPA and UAMA abrasives standards committees. In1997 Tim received the GE Wilson Award for his contribution to development of a new mesh diamond process. In 2002, Tim introduced composite diamond wear products to Sandvik and has qualified and grown that product line in areas of pump wear, mineral exploration and a number of industrial wear applications. He is co-inventor of *Hyperion*® diamond that has revolutionized the lapping and polishing processes for sapphire, silicon carbide, ceramic and glass products used in LED's, consumer products and power supply semiconductors. Tim has been awarded 12

US patents for various wear resistant coatings and diamond powder products and tools. Tim has published 18 technical papers in international journals and presented at numerous conferences. He is currently the R&D Manager for the Abrasives products at Hyperion Materials & Technologies.

LUIZ FRANCA is a Principal Application Research Engineer at Element Six Global Innovation Centre

in the UK. He has a solid scientific background that includes a Ph.D., a M.Sc. and a B.Sc. in Mechanical Engineering, and a vast knowledge in drilling mechanics built over 10+years, managing R&D projects with in the Oil/Gas sector. He is the author of more than 50 publications, consisting of peer-review journals, conference papers, book chapters, and 4 patents. Recently, his research has been directed to tool/workpiece interaction modelling and in setting up test programmes to evaluate superabrasive materials in different applications.

CHRIS GRAHAM – Element 6, Senior Advanced Manufacturing Research Engineer

6 years' experience at Element Six with extensive experience in the application of super hard advanced materials for precision applications in both Automotive and Aerospace. Tortured Newcastle United fan that lives in the outback of England, recently got electricity.

JOHN HAGAN, Senior Application Engineer Norton/Saint-Gobain Abrasives

John Hagan has been a Senior Application Engineer for Norton I Saint-Gobain for 8 years where he is tasked with designing and completing grinding projects focused on new process development and the optimization of existing processes. In addition to this John is heavily involved training related to grinding technology for both internal personnel and customers. With more than a total of 29 years of experience with Norton/Saint-Gobain, John's past experiences include 12 years as a Manufacturing/Process Engineer and several years as a R&D Engineer.

STEPHAN HUTH, PhD Group Manager, Combi Tools stephan.huth@hilti.com

Stephan holds a degree in Mechanical Engineering, with a special focus on materials, from Ruhr-Universität Bochum, Germany. He finished his PhD in 2009 in the field of PM steel alloy design and processing. Afterwards he worked as Post-Doctoral Officer and from 2011 to 2014 as Assistant Professor (Junior Professor) at Ruhr-Universität where he focused on alloy design and production technology for high alloyed metallic materials for service under harsh conditions.

At the end of 2014 he joined Hilti in Liechtenstein to lead research and development projects for diamond inserts for construction applications. In October 2018 he relocated to Hilti in Germany, where he leads the development group for large combi hammers for the drilling and chiseling of construction materials.

TERRY M. KANE, Executive Director - Industrial Diamond Association of America - P.O. Box

29460, Columbus, Ohio 43229 USA

Phone: 614-797-2265 - Fax: 614-797-2264 - Email: tkane@superabrasives.org

Mr. Kane has 49 years of professional experience in all aspects of the superabrasives industry. The past 19 years he has been Executive Director for the Industrial Diamond Association of America and responsible for managing this international association including publishing the Association's Trade Magazine. Mr. Kane has developed educational literature, books and Education Sessions at International Technical Conferences. Mr. Kane's first 31 years were with the General Electric Company's diamond business, where he acquired extensive experience in superabrasive technology including manufacturing operations & training, engineering, R&D, application development and marketing communications. Mr. Kane was instrumental in developing the first Website launch for General Electric Company… He also produced the first Superabrasives Education Course for the News and Trade Media plus customer and end user training and educational programs on superabrasive materials, principles and applications. Mr. Kane was part of original R&D teams to develop processes for Polycrystalline Diamond, Cubic Boron Nitride (cBN) and diamond products, as well as various bond, coating and plating systems for the Industrial Diamond business. He also developed a successful in-house coating & nickel plating operation and authored a detailed operational procedure for base-line coating operations. Mr. Kane established testing criteria for grinding products and evaluated customer products for analysis and recommended improvements to GE's customer base. His work included metallurgical analysis, crystal wear evaluation and written assessment of grinding wheels and their interaction with various workpiece materials in diverse grinding modes and bond systems. Mr. Kane was instrumental in organizing the original Manufacturing Training Programs department that developed detailed training materials and testing criteria for the businesses' entire International diamond and cBN manufacturing operations.

BERNARD H. KEAR, Materials Science and Engineering Rutgers University - Bernard H. Kear is

Distinguished Professor Emeritus, Department of Materials Science and Engineering, Rutgers University. His most recent research has been focused on synthesis and processing of nanostructured materials by solution, flame and plasma methods; surface treatments by thermal spraying of nanophase powders; liquid phase sintering of nanophase cermets; pressure-assisted sintering of nanophase oxides, diamond and cubic-boron nitride; and high-shear mixing/injection molding of high performance graphitic-nanoflake reinforced polymer matrix composites. Kear has published 350 papers, edited 10 books, and been granted 45 patents. He was awarded the Howe Medal of ASM in 1970 and the Mathewson Gold Medal of TMS-AIME in 1971. He was elected to the National Academy of Engineering in 1979. From 1983, he served as a member of the National Materials Advisory Board, and was Chairman from 1986-1989.

STEPHEN KELLY PhD, BE (Hons.) Mechanical Engineer, MSc Management, CEng

Stephen Kelly is Regional Technical Manager for Element Six responsible for technical business development in the US, Canada and Latin America markets. Stephen graduated in 2002 with an Honor's Degree in Mechanical Engineering from University College Dublin, Ireland and has an MSc in Management. Stephen joined Element Six in 2005, becoming Senior Applications Engineer before being promoted to Project Manager – Innovative Product Development where he managed a highly trained innovation team of qualified engineers, experienced production operators, senior technicians and materials experts developing and establishing new production processes of innovative products. Stephen joined the Element Six US team in 2010 where he works closely with strategic customers and end users in the Americas' region ensuring applications focused revenue growth. Stephen is a Chartered Engineer.

THORSTEN KLEIN, Dipl.-Ing Development Expert, Diamond - Thorsten holds a degree in Materials

Engineering from the University of Applied Sciences in Osnabrück, Germany. In the earlier stages of his career in industrial diamond products, Thorsten performed various roles within the Boart Longyear group, rising to the position of Head of Development and Production in its subsidiary Longdia GmbH in Celle, Germany.
Since joining Hilti in 2010, Thorsten holds the position of development expert for diamond inserts for construction applications, with responsibility for driving technology concepts through to final products. He has expertise in all facets of diamond insert technology, from segment design and diamond arrangement through to insert behaviour in coring and sawing.

OLIVIER LAVIGNE is a R&D Engineer – Materials Scientist at Hyperion Materials and Technologies.

He received his B.S. Degree in Physics and M.S. Degree in Materials Science from University of Montpellier, France, his M.E. from Polytech Montpellier, France, and his Ph.D. from INSA of Lyon, France. He previously worked as a Research Fellow at the Fracture and Reliability Research Institute, Tohoku University, Japan, and at the School of Mechanical Engineering, University of Adelaide, Australia, on the environmental assisted degradation of metals and alloys for nuclear power plant and gas pipeline industries.

IOAN D. MARINESCU, PHD - Dr. Marinescu is a professor in the Department of Mechanical, Industrial, and Manufacturing Engineering at the University of Toledo, Ohio. He is the director of the Precision Micro-Machining Center (PMMC). He has a doctorate from the Department of Manufacturing Engineering at the University of Galatzi, Romania. Marinescu is a member of the American Society of Mechanical Engineers (ASME), Society of Manufacturing Engineers (SME), American Ceramic Society (ACerS), and American Society for Abrasive Technology (ASAT). Marinescu's areas of research interest include tribological investigation of abrasive processes, minimum lubrication in metal cutting, investigation of friction and wear mechanisms, and molecular dynamics simulation of friction and wear. PhD - Manufacturing Engineering, University of Galatzi, Romania, 1991, MS - Mechanical Engineering, Polytechnic Institute of Bucharest, Romania, 1976, BS - Manufacturing Engineering/Industrial Engineering, Polytechnic Institute of Bucharest, Romania, 1974, Post-Doc. - Tribology Program, MIT, 1994 (Rabinowicz, Nam Suh)

RICHARD J MONDRY II was a mechanical engineering technician IV for Gillcrest Electric under contract to NASA, John Glenn Research Center at Lewis field from 2001 through 2014. While at NASA, Rick specialized in acquiring tribometry data using various pin-on-disc rigs, high temperature fretting rigs and performed postmortem analysis of sample wear by interferometric profilometry. The experimental samples evaluated in these rigs ranged from diamond and diamond-like materials to specialty ceramics and surface coatings of nanocarbon materials in ultrahigh vacuum, cryoscopic conditions and in inert atmospheres like CO_2. Retired, Rick is enjoying life and going for walks with the family. He is currently keeping busy in real estate as well as assisting in the startup of several small businesses.

STEVEN MOSELEY, PhD Chief Scientist and Key Expert - Steve holds a bachelor's degree and Ph.D. in Metallurgy from the University of Sheffield, UK, and has worked in the field of powder metallurgy since 1987. Following his research work on HIPing of ceramic-metal powder composites and dissimilar material diffusion bonding at HIP Ltd (a Bodycote Company), he spent 3 years in the steel industry with Davy McKee before returning to PM as Technical/Quality Manager for Edgar Allen Danite, a UK-based hardmetal manufacturer. He has worked at Hilti in Liechtenstein since 1996 in R&D for cutting materials, joining technologies and drilling systems, covering the range of hard and superhard materials from steels through cemented tungsten carbide hardmetals and on to industrial diamond. His current positions are Chief Scientist – Hard Materials and Key Expert – Inserts within the Business Area Electric Tools and Accessories. He is also active internationally within the European powder metallurgy community as Co-Chairman of EuroHM (the European Hard Materials Group); an EPMA Council Member; a steering board member of the IAPK in Germany (Institute for the Application of Powder Metallurgy and Ceramics, RWTH-Aachen); and a member of the Particulate Engineering Committee of IOM3 (the UK Institute of Materials, Minerals and Mining).

DEB MUKHOPADHYAY, PhD is a senior scientist at US Synthetic Corporation, Orem, Utah. Dr. Mukhopadhyay has a Ph.D. degree in Metallurgy from University of Idaho. Prior to US Synthetic, Dr. Mukhopadhyay worked as a Chief Metallurgist at General Carbide Corporation at Greensburg, Pennsylvania and at Vista Metals Inc., McKeesport, Pennsylvania. He was a visiting faculty at Oregon State University, Corvallis, Oregon before joining Vista Metals Inc. Most of his early work was on nanocrystalline material and material for fusion reactor. He had published more than twenty five papers in various referred journals and in the conference proceedings. He also has more than ninety patents. His recent work includes development of cemented carbide and polycrystalline diamond compact.

STUART NAILER, PhD - A British national, Stuart has specialised in diamond science and technology since the final year of his Bachelor's degree in Physics from the University of London in 1994. Following a PhD in x-ray and optical studies of diamond growth and perfection from the same institute in 1998, Stuart started his industrial career at De Beers Industrial Diamonds (the future Element Six) in Ireland. As technical product manager, Stuart led a team with responsibility for the product development and market support activities for the full range of diamond and cBN grit products. Since 2010, Stuart has held the positions of project manager and R&D expert at Hilti in Liechtenstein, driving strategies for testing and sourcing diamond products for the drilling of construction materials, and project managing research and development into next-generation core bits with advanced diamond materials.

CLEMENT ONYENEMEZU, PhD is the R&D Manager, Chemistry at Engis Corp. He received his Ph.D. degree from University of Minnesota in Polymer/Analytical Chemistry. His expertise is in particle dispersions, emulsions and polymers. Since joining Engis in 2009 Dr. Onyenemezu has been involved in all aspects of diamond slurry development, lapping plate development, fixed abrasive lapping plate development, and applications.

ANDREAS PLANAKIS, Superabrasives Process Optimization Center (SPOC) Director and Director of Process Analytics at Worldwide Superabrasives, LLC (WWSA) Boynton Beach, Florida and Kaifeng, China. Andreas has been serving the Superabrasives Industry for over 10 years:

- Utilizing his extensive background in Chemistry and Biology along with individually grinding +450 wheels over the past 4 years at SPOC from +25 global companies with a variety of grinding kinematics, Andreas has developed a distinct and diverse knowledge to optimize superabrasive raw materials for industry
- Andreas manages the testing for many industries at SPOC, including, but not limited to, aerospace, automotive, semiconductor, round tool, valve processing, polycrystalline diamond and CBN products, super finishing, as well as serving many industries looking to optimize their grinding processes
- Andreas's knowledge of utilizing multiple grinding platforms along with world-class analytical equipment provides WWSA/SPOC with industry leading information to improve manufacturing processes for industry
- Andreas has hosted over 200 visits to WWSA/SPOC over the past 4 years to work directly with customers to optimize their superabrasive raw material choices and as a result, produce the optimal tools for industry
- By combining an extensive grinding knowledge with mastering the ability to analyze and process the data/results, Andreas continues to add value to industry for process optimization

HYESEUNG RACHEL PARK, R&D Engineer Engis Corporation - Hyeseung joined Engis Corporation in 2017 upon graduating from University of Illinois at Urbana-Champaign with B.S. degree in Material Science and Engineering. At Engis, she is currently engaged in work with diamond characterization as well assisting in the development/production of fixed abrasive lapping techniques.

DAN PICKARD is a Partner in the International Trade practice of Wiley Rein LLP in Washington DC. He counsels U.S. and international clients on the laws and regulations governing international trade, with particular emphasis on antidumping investigations. He has successfully represented domestic companies in obtaining relief from unfairly priced imports involving diamond sawblades, plastic ribbons, stainless steel products, wind towers, and many others. He regularly appears before the U.S. International Trade Commission (ITC), the U.S. Department of Justice (DOJ), the U.S. Department of Commerce, and the Federal Courts. Additionally, he is listed by Chambers Global as one of the "Leaders in Their Field" for his work in Trade Remedies & Trade Policy.

FLORIAN RECHFELDEN is a Senior Application Engineer for the Austrian grinding wheel manufacturer TYROLIT. After attending the Technical Collage for Electrical Engineering in Innsbruck he started his career as a trainee with TYROLIT in 2001. After a year as trainee he became an Application Engineer for grinding wheels within the Aerospace and Turbine Industry division of TYROLIT. A few years later Florian took the opportunity to switch to the Automotive division within TYROLIT. Since 2008 he´s working as a Senior Application Engineer with international customers of the Automotive sector. His main tasks are to specify and provide high end grinding tools and solutions to assure a cost effective grinding process for the customers.

JACK RICHTSMEIER holds a Bachelor's degree in Physical Science from The University of St. Thomas in St. Paul, MN (1981). His professional background includes over 35 years of engineering, sales and marketing experience within the industrial and scientific marketplace. He has over 25 years of expertise in acoustics and ultrasonics working directly within the industrial, petrochemical, biomedical and microelectronics sectors. Mr. Richtsmeier is presently employed as Business Development Manager for Nordson *SONOSCAN* headquartered in Elk Grove Village, IL.

BENJAMIN R. ROSCZYK, R&D Manager – Superabrasives Engis Corporation

Benjamin received his M.S. degree in Materials Science and Engineering from the University of Illinois at Urbana-Champaign after studying mixed oxide powders synthesized using novel organic methods. He joined Engis Corporation in 2003 as a R&D Engineer and was promoted to Senior R&D Engineer in 2012 and R&D Manager of Superabrasives in 2019. Benjamin has 16 years' experience in the characterization of diamond and cBN powders, specializing in particle size, shape, intrinsic defects and surface chemistry.

GUNTER STECKEL, Tyrolit manager product development of vitrified super abrasives

1977-1982 study at Bergakademie Freiberg development of inorganic non-metallic materials
1982 Schleifkörper-Union Dresden R&D department
1990 Firma Ernst Winter &Sohn Norderstedt technology/production vitrified super abrasive tools
1995 Saint-Gobain Diamantwerkzeuge R&D vitrified super abrasives tools
2006 DVS Präwema honing technology

KENNETH W. STREET, PhD - Dr. Street, an Analytical Chemist, joined the staff of the National Aeronautics and Space Administration's John H. Glenn Research Center in 1991 where he ran an environmental analytical laboratory. He later joined the Tribology Branch where he evaluated space lubricants and ran a vibrational micro-spectroscopy laboratory. In addition, he was involved with development of Lunar soil simulants prior to his retirement in 2011. Since NASA he has been a private consultant to Diamond Materials, Inc. Prior to NASA he was an assistant professor at Loyola University and at Kent State University. A native of Waterford, Connecticut, he earned his Bachelor's degree in chemistry from the University of Connecticut in 1970, and a Master's degree in chemistry from Cleveland State University in 1973. He received his Doctorate in 1977 from Wayne State University. Ken has authored over 100 technical papers and has presented over 200 papers at international, national and local scientific meetings. In 1996 and 1997, he was nominated for the NASA Inventor of the Year Award. He is a member of several professional societies: American Chemical Society Fellow, Society for Applied Spectroscopy, and Sigma Xi The Scientific Research society.

MINORI TERAMOTO, Sumitomo Electric Industries - I graduated and received a Master's Degree in Physical Science from Nara Women's University in 2005. Then I have been working for Sumitomo Electric Industries. For about 10 years I have experienced the development of PCBN and PVD coating for cutting tool materials, and technological development for the production of PCBN, PCD and HPHT synthetic single crystal diamond. I moved to laboratory in 2016, and I am in charge of research and development of diamond materials as manager

MOHAMMED Y. THARWAN - Born in 26 February 1984 in Adarb city, Jazan, Saudi Arabia. PhD student in Mechanical, Industrial & Manufacturing Engineering Department at the University of Toledo, United States BS & MS degrees in Mechanical Engineering at Wilkes University (2008 to 2013), United States. Lecturer in Mechanical Engineering Department at Jazan University, Saudi Arabia. Interested in non-traditional grinding of ceramic materials, machining of composite materials and its applications.

K. PHILIP VARGHESE, PhD - Group Leader, Advanced Application Engineering, Saint-Gobain Abrasives, Worcester, MA, USA - Dr. K. Philip Varghese has worked for Saint-Gobain Abrasives since 2011, leading the Advanced Application Engineering Group focusing on providing grinding system solutions to customers, and education and training in grinding science & technology. Prior to this, he worked as Chief Scientific Officer for CR Foundation leading a team of scientists and engineers focusing on dental materials evaluation. He has a PhD in Mechanical Engineering from University of Utah and an M.S. in Mechanical Engineering from University of Michigan. He has authored over 25 technical papers and articles in the field of machining, dental products/materials and grinding technology and has presented at numerous conferences and tradeshows hosted by AMT, ASME, SME, ASM International and others

OLEG A. VORONOV, PhD Dr. Voronov is co-founder and President of Diamond Materials, Inc. (DMI). While working for DMI, he invented a High Pressure-High Temperature (HPHT) apparatus for generating static high pressure and high temperature in relatively large volume. He discovered and developed a new line of carbon materials, diamond-based composites and hard metals. DMI was originally incorporated in New Jersey (1995). It operates now in Pennsylvania under the name Diamond Materials of Monroe. It is a small, 100% USA private company, for profit, to perform R&D in the area of physical sciences, including high pressure and high temperature physics. Several contracts have been performed for government and industry. He graduated from Moscow Federal University (Moskovskiy Gosudarstvenniy Universitet) with diploma of Physicist. He received his doctor-I degree in physics and mathematics (kandidat physiko-matematicheskih nauk) and doctor-II degree (doctor physiko-matematicheskih nauk) from the Certification Committee of the U.S.S.R. Dr. Voronov worked for the Academy of Sciences in the Vereshchagin Institute. Managing the contracts of Academy with industry, he was advanced to the position of Senior Staff Scientist and later promoted to the position of Chief of Laboratory. For problem solutions in the field of carbon-carbon (fibers-graphite) composites and diamond materials, he was elected to the Head position of the Department for Carbon Materials. Oleg has more than 100 scientific publications (in Russian and English), contract final reports, author's certificates of invention (U.S.S.R.-[Russian Federation]) and US patents.

JOHN A. WEBSTER, PhD, a.k.a. Dr. Cool, Senior Research Engineer at Worldwide Superabrasives

John Webster is a specialist in grinding process theory, with over 40 years of university-based research, graduate level teaching/mentoring of more than 50 PhD and MS students at the University of Connecticut, and hands-on experience in all areas of the grinding process. In 2004 Webster launched Cool-Grind Technologies, Connecticut, a global supplier of unique coolant application products, spear-heading coherent-jet nozzle designs into the grinding market. In early 2017 Webster joined Worldwide Superabrasives to help develop a world-class grinding application test and analysis center in Boynton Beach, to benchmark CBN and diamond grinding wheel performance. The test center also develops end-user grinding applications, with emphasis on optimizing speeds and feeds, residual stress creation, surface finish characterization, and wheel wear/life analysis. Seventeen journal papers and 60 conference papers in grinding from 1979 - present

CHRIS WINKEL, President and CEO of Worldwide Superabrasives, LLC (WWSA) Boynton Beach, Florida, Columbus, Ohio and Kaifeng, China. Chris has been serving the Superabrasives industry for over 22 years in the following capacities:

- Business Manager for Saint-Gobain at ABC Superabrasives 2001-2003
- Operations Manager for General Industrial Diamond (now 3M) 2003-2004
- Founder of WWSA in 2004
- Co-founder of diamond and CBN manufacturer BESCO Superabrasives 2007
- Founder of the Superabrasives Process Optimization Center (SPOC) in 2014
- Former President of the Industrial Diamond Association of America (IDA) in 2011 and Board member for 10 years.

Chris's education and experience over the past 22 years as both a manufacturer of superabrasive raw materials as well as superabrasive grinding tools and related products provide a unique perspective in the superabrasives industry. Chris's goal over the years has been to continuously pursue the next level of superabrasives knowledge and bring this understanding to industry.

WU YAOYU is a professor and senior engineer at Zhengzhou institute of technology. In 1985, he

graduated from Henan university of science and technology with a bachelor of engineering in fluid transmission and control. In 1988, he obtained a master of engineering in grinding direction in machinery manufacturing major of Guangxi university. He was engaged in the research and development of superhard material(CBN) technology in Zhengzhou Research institute for Abrasive & Grinding Co.,Ltd. In 2008, he entered Zhengzhou institute of technology to teach. Presided over the completion of 4 provincial-level or above scientific research projects, and obtained 2 provincial-level or above achievements; He has obtained 9 national patents, including 3 invention patents. Published more than 30 papers and 3 books; Presided over the formulation of national industry standards 1. He mainly teaches courses such as" mechanical engineering testing technology "and " Hydraulic and pneumatic transmission". The research direction is precision machining of superhard materials.

AMÉDÉE ZRYD, PhD Dir. Application and R&D, Synova SA

Dr Zryd graduated in Physics from the Swiss Federal Institute of Technology Lausanne (EPFL) and completed a PhD thesis at the Swiss Federal Institute of Technology in Zürich (ETHZ). After postdoctoral research at the EPFL and Osaka University, he joined the industry where he spent more than 25 years in several technical leadership positions, before joining Synova S.A. in 2018. He has also been Professor in Material Sciences at the University of Applied Science Western Switzerland.

DIAMOND/CBN APPLICATION CHART*

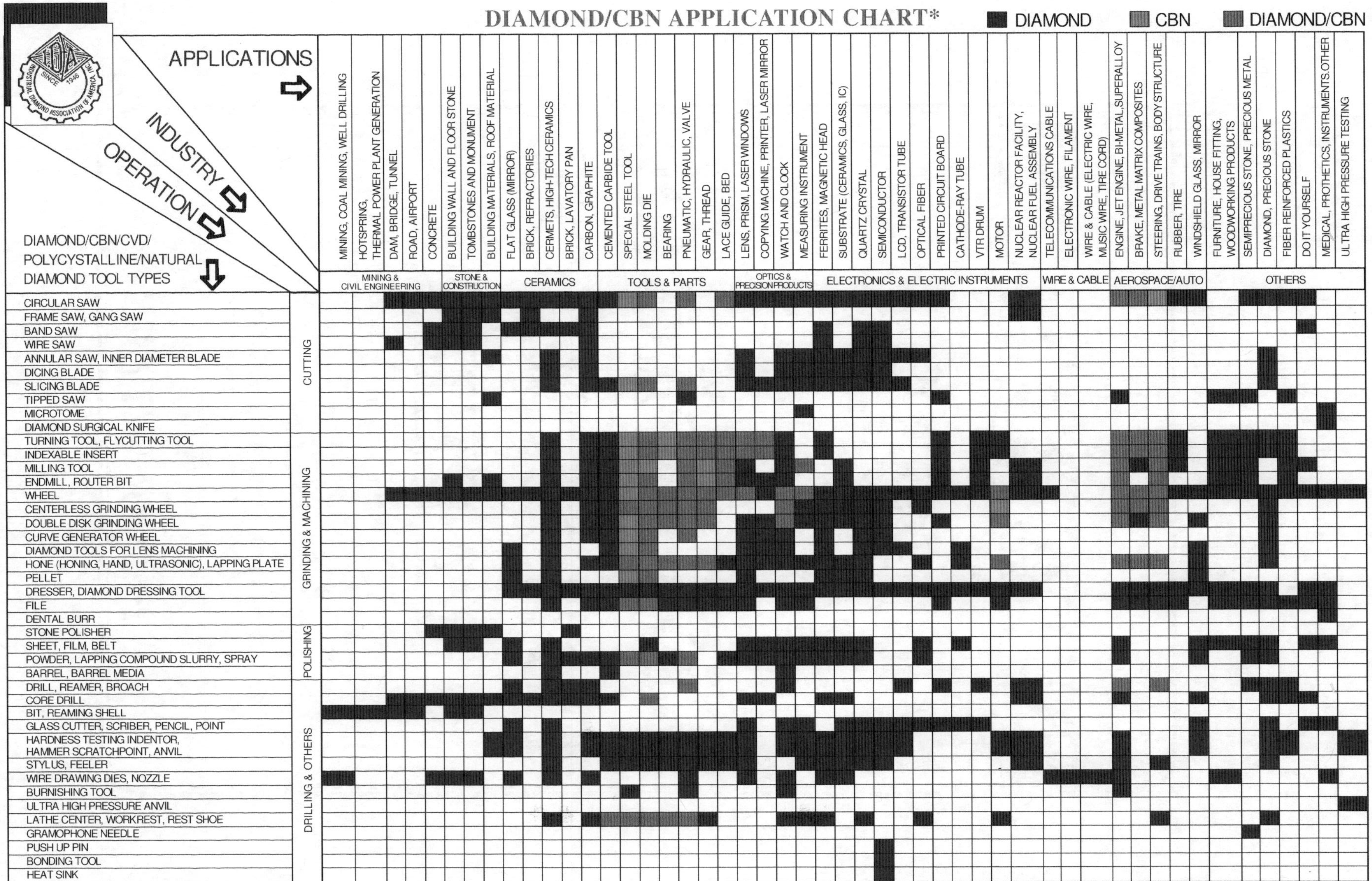

* *Typical Applications - Not to be Used for Specific Tool Selection*

INTERTECH 2019

NOTE PAGES FOR SESSION PAPERS

WITH PRESENTERS AND PAPER TITLES

WEDNESDAY AM KEYNOTE DR. K. RENEE HORTON, SPACE LAUNCH SYSTEM (SLS) LEAD METALLIC/WELD ENGINEER NASA RESIDENTIAL MANAGEMENT OFFICE - Dr. Horton is a graduate of Louisiana State University with a B.S. of Electrical Engineering with a minor in Math in 2002 and a Ph.D. in Material Science with a concentration in Physics, becoming the first African American to graduate from the University of Alabama in this area. She worked for NASA, first as a student from 2009 to 2011, and then started her career as a mechanical test engineer in 2012. In 2014 she was promoted to her current position. In 2016, Renee was elected President of the National Society of Black Physicists (NSBP) as the second woman to hold the office. She has served the physics community abroad as a member of the International Union of Pure and Applied Physics (IUPAP) Women in Physics Working Group and currently serves on several advisory boards dedicated to a more diverse inclusion in physics. In 2017, she was elevated to a Fellow in the NSBP, which is the highest honor bestowed upon a member. She has spoken all over the world including South Africa, Brazil, South Korea, Canada, Mexico and the Virgin Islands. She is featured in NASA's I am Building SLS, Physics Today, and the LSU Alumni Magazine. She recently graced the cover of the Hearing Health Foundation Magazine, winter edition that can be found online. She is the author of Dr. H Explores the Universe, Dr. H and her Friends, and Dr. H Explores the ABCs.

#1941 GRINDING OF AEROSPACE MATERIALS USED FOR HIGH-TEMPERATURE APPLICATIONS

Mr. Philip K. Varghese, PhD, Group Leader-Advanced Application Engineering - Norton/Saint-Gobain Abrasives USA

#1929 INNOVATED CUTTING TOOLS AND METHOD FOR EFFICIENT COMPOSITE MATERIALS MACHINING

Mr. Abdelatif Atarsia, PhD, Global Composites Industry Manager - YG1 USA

#1930 GRIT SIZE, THERMAL DAMAGE, SURFACE FINISH, DRESSING PARAMETERS AND COMMON MISTAKES WHEN GRINDING HARDENED STEEL AND NICKEL ALLOYS WITH CBN GRITS.

Mr. Jeffrey Badger, PhD, The Grinding Doc USA

WEDNESDAY PM KEYNOTE - JIM DELAND, TECHNICAL SUPPORT ENGINEER
APACHE AEROSPACE - Jim's responsibility at **APACHE AEROSPACE** involves providing technical support for drilling solutions to major aerospace companies and their suppliers. His background in aerospace includes over 28 years at **THE BOEING COMPANY** as a Research & Technology Engineer where he worked in the Assembly Technology area supporting technology development for production. Jim was frequently called on to develop manufacturing solutions for new airplane programs and was instrumental in successfully developing and implementing new drilling technology on both the 787 & 777X Program. During his career he has earned 9 patents both in assembly technology and new cutting tool development. Based on his engineering achievements he was recognized as a Boeing Associate Technical Fellow in 2011. Recently Blue Origin reached out to Jim to provide his drilling and assembly expertise as a consultant on the New Glen rocket program. Jim has published multiple papers for SAE on the topic of cutting tool technology and has presented at multiple conferences about drilling technology. Jim is a graduate of the Oregon Institute of Technology with a B.S. in Manufacturing Engineering Technology.

#1926 TRIBOLOGY OF DIAMOND BEARINGS FOR SPACE APPLICATIONS

Mr. Kenneth Street, PhD, Analytical Chemist - Diamond Materials Inc. USA

#1936 INFLUENCE OF ABRASIVE TYPE ON GRINDING INDUCED RESIDUAL STRESSES AND SURFACE INTEGRITY

Mr. John Webster, PhD, Senior Research Engineer - Worldwide Superabrasives USA

#1940 GRINDABILITY OF ADDITIVELY MANUFACTURED AEROSPACE ALLOYS

Mr. Philip K. Varghese, PhD, Group Leader-Advanced Application Engineering - Norton/Saint-Gobain Abrasives USA

#1924 FIXED ABRASIVE THREE-DIMENSIONAL PLATE FOR LAPPING AND POLISHING OF ADVANCED MATERIALS

Mr. Clement Onyenemezu, PhD - Engis Corporation USA

#1927 THEORETICAL EXPERIMENTAL STUDY OF SILICON CARBIDE GRINDING

Mr. Mohammed Tharwan, PhD, Engineer - University of Toledo USA

#1928 MICROSCOPIC CHARACTERIZATION AND MODELING OF OXIDE LAYER FOR ELECTROLYTIC IN-PROCESS DRESSING (ELID) GRINDING

Mr. Bader Alqahtani, PhD, Engineer Mechanical Engineering Department The University of Toledo USA

#1922 GENIS 2- THE NEXT GENERATION OF VITRIFIED CBN FOR CYLINDRICAL GRINDING IN THE AUTOMOTIVE INDUSTRY

Mr. Florian Rechfelden, Senior Application Engineer - Radiac Abrasives A Tyrolit Company USA

#1931 INVESTIGATING AND MODELING THE BALL-BURNISHING FACTORS IMPACT ON RESIDUAL STRESS OF AISI 8620 STEEL

Mr. Abdulaziz Alshareef, PhD Candidate, Senior Researcher at University of Toledo USA

<u>THURSDAY AM KEYNOTE SPEAKER</u> - MIKE FELIX, GLOBAL CHIEF ENGINEER FORD POWERTRAIN MANUFACTURING ENGINEERING - Mike's responsibilities include Global Machining and Assembly Processes for Engines, Transmissions, Axles and Electrification components; including Hybrid Batteries, eTransaxles and eMotor manufacturing. Mike has been with Ford for over 30 Years. He started his career at the Cleveland Engine Plant Facility in 1988 as an Engineering Graduate. Mike's career has progressed through a number of engineering and operations positions within Powertrain. He has worked in nine different Ford Plants throughout his career and has spent 11 Years as Plant Manager in 5 different Plants across Ford Powertrain, as well as international and Staff Engineering assignments. Mike has a BSME in Mechanical Engineering from Akron University and an MBA from Baldwin Wallace College both in Ohio. Mike assumed his current role as Chief Engineer in 2017.

#1933 THE USE OF ELECTRON-MICROSCOPY IN THE STUDY OF SUPERABRASIVES IN GRINDING

Mr. Jeffrey Badger, PhD, The Grinding Doc USA

#1934 SINGLE-CRYSTAL DIAMOND GRIT IN CONSTRUCTION: APPLICATIONS AND CHARACTERIZATION

Mr. Stuart Nailer, PhD, Research & Development Expert - Hilti Corporation - Liechtenstein

#1937 DETERMINING THE FLOWRATE, PRESSURE AND NOZZLE APERTURE IN A GRINDING PROCESS WITH CONSIDERATION OF ABRASIVE TYPE, COOLANT TYPE AND WHEEL STRUCTURE

Mr. John Webster, PhD, Senior Research Engineer - Worldwide Superabrasives USA

THURSDAY PM FEATURED SPEAKER - DAN PICKARD, Esq.

#1921 TARIFFS AND TRADE ACTIONS AFFECTING THE SUPERABRASIVES INDUSTRY

Mr. Daniel Pickard, Attorney At Law - Partner Wiley Rein USA

#1944 ADVANCES IN THE PROCESSING OF SEMICONDUCTOR MATERIALS USING DIAMOND TECHNOLOGIES

Mr. Timothy Dumm, PhD, Technology Manager, Abrasives - Hyperion Materials & Technologies USA

#1943 LASER TECHNOLOGY AND MACHINING BRITTLE MATERIALS WITH PCD APPLICATIONS

Mr. Stephen Kelly, PhD, Regional Technical Manager the Americas - Element Six US Corporation USA

#1939 CHARACTERIZATION OF THE RESIDUAL STRESSES INTRODUCED BY A NEW JOINING METHOD IN DIAMOND AND TUNGSTEN CARBIDE COMPOSITES

Mr. Timothy Dumm, PhD, Technology Manager, Abrasives - Hyperion Materials & Technologies USA

#1938 EVALUATION OF THE INFLUENCE OF IMPURITIES AND CRYSTALLINITY IN SINGLE CRYSTAL DIAMOND AND NANO-POLYCRYSTALLINE DIAMOND (NPD) ON WEAR CHARACTERISTICS

Ms. Minori Teramoto, Manager - Sumitomo Electric Industries Ltd. Japan

#1945 UNDERSTANDING SOURCES OF RESIDUAL STRESS IN PDC BY X-RAY DIFFRACTION

Mr. Deb K. Mukhopadhyay, PhD, Senior Scientist - US Synthetic Corporation USA

#1925 INSPECTING ULTRA-HARD MATERIALS AND PROCESSES USING ACOUSTIC MICRO IMAGING (AMI)

Mr. Jack Richtsmeier, Business Development Manager - Nordson Sonoscan USA

#1932 THEORETICAL RESEARCH ON THE UNIFORMITY OF PLANETARY DOUBLEÏ¼•SIDE LAPPING TRAJECTORY

Mr. Yaoyu Wu, PhD, Professor - Zhengzhou Institute of Technology
China

#1923 VOLUME PRODUCTION OF HIGH PRECISION TOOLS WITH SYNOVA'S NEW 5-AXIS LASER MACHINING CENTRE LCS 305

Mr. Jacques Coderre, US Business Manager - Synova USA

#1935 PERFORMANCE IMPROVEMENT WITH NOVEL NANO BINDER TECHNOLOGY IN POLYCRYSTALLINE CUBIC BORON NITRIDE

Mr. Stephen Kelly, PhD, Regional Technical Manager - the Americas - Element Six US Corporation USA

#1942 DIAMOND SYNTHESIS "THE SCIENCE"

Mr. Ion Benea, PhD, Vice President Operation - Hyprez Products - Engis Corporation USA

National Aeronautics and
Space Administration
NASA
SPACE LAUNCH SYSTEM
FRICTION STIR WELDING
SLS Ambassador Program
K Renee Horton
05/29/2019

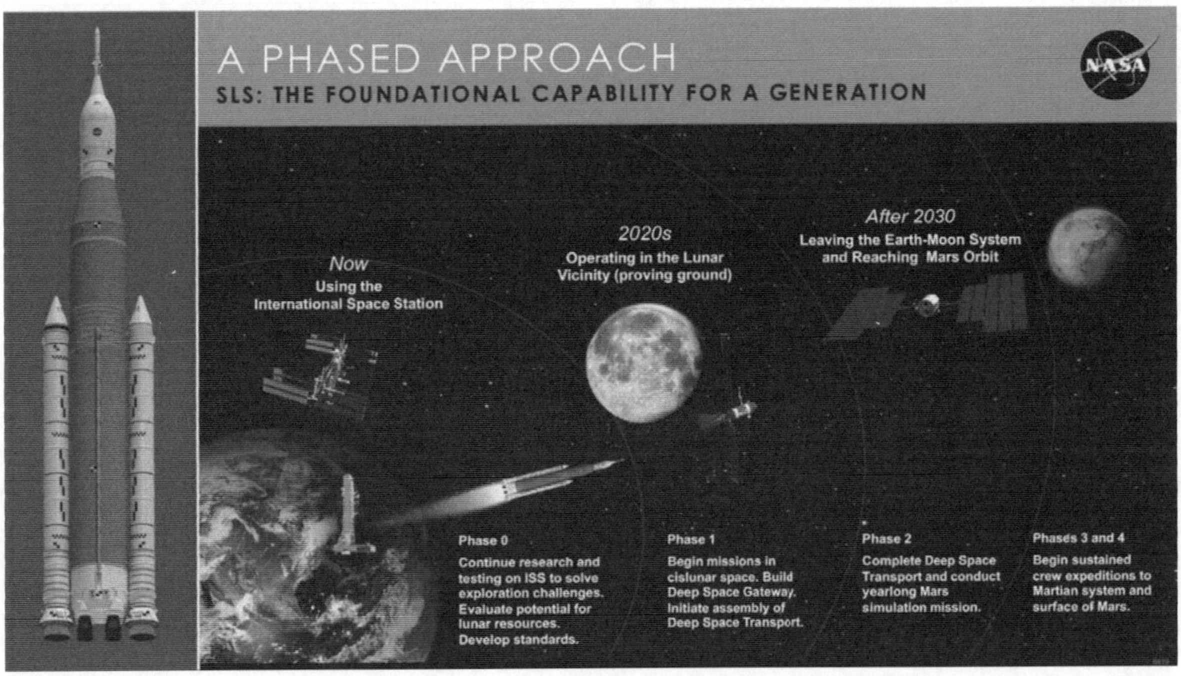
A PHASED APPROACH
SLS: THE FOUNDATIONAL CAPABILITY FOR A GENERATION
NASA
Now
Using the
International Space Station
2020s
Operating in the Lunar
Vicinity (proving ground)
After 2030
Leaving the Earth-Moon System
and Reaching Mars Orbit
Phase 0
Continue research and testing on ISS to solve exploration challenges. Evaluate potential for lunar resources. Develop standards.
Phase 1
Begin missions in cislunar space. Build Deep Space Gateway. Initiate assembly of Deep Space Transport.
Phase 2
Complete Deep Space Transport and conduct yearlong Mars simulation mission.
Phases 3 and 4
Begin sustained crew expeditions to Martian system and surface of Mars.

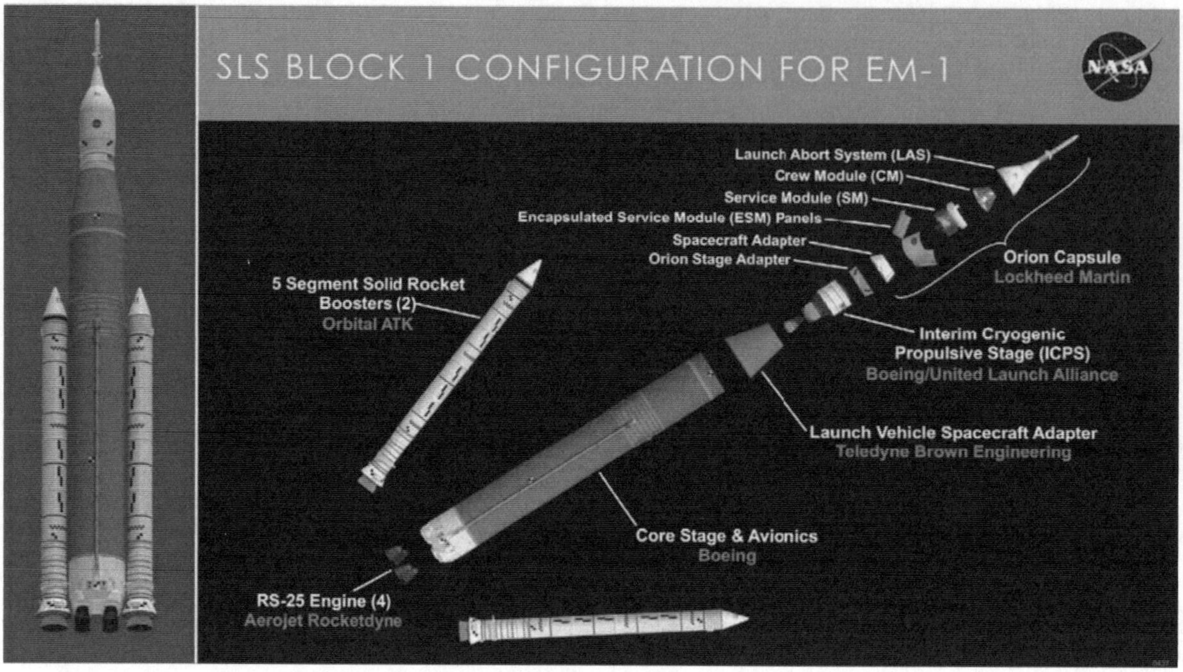

SLS BLOCK 1 CONFIGURATION FOR EM-1
Launch Abort System (LAS)
Crew Module (CM)
Service Module (SM)
Encapsulated Service Module (ESM) Panels
Spacecraft Adapter
Orion Stage Adapter
Orion Capsule
Lockheed Martin
5 Segment Solid Rocket Boosters (2)
Orbital ATK
Interim Cryogenic Propulsive Stage (ICPS)
Boeing/United Launch Alliance
Launch Vehicle Spacecraft Adapter
Teledyne Brown Engineering
Core Stage & Avionics
Boeing
RS-25 Engine (4)
Aerojet Rocketdyne

CORE STAGE

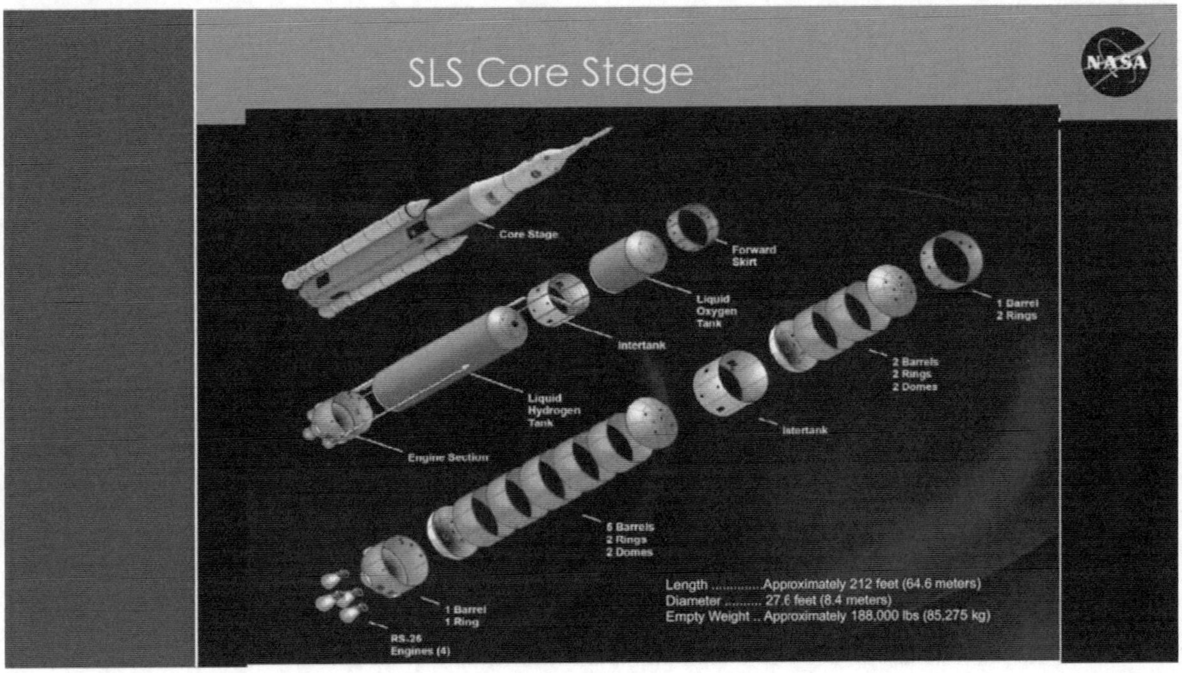
SLS Core Stage
NASA
Core Stage
Forward Skirt
Liquid Oxygen Tank
1 Barrel 2 Rings
Intertank
2 Barrels 2 Rings 2 Domes
Liquid Hydrogen Tank
Intertank
Engine Section
5 Barrels 2 Rings 2 Domes
1 Barrel 1 Ring
RS-25 Engines (4)
Length ....Approximately 212 feet (64.6 meters)
Diameter .... 27.6 feet (8.4 meters)
Empty Weight .. Approximately 188,000 lbs (85,275 kg)

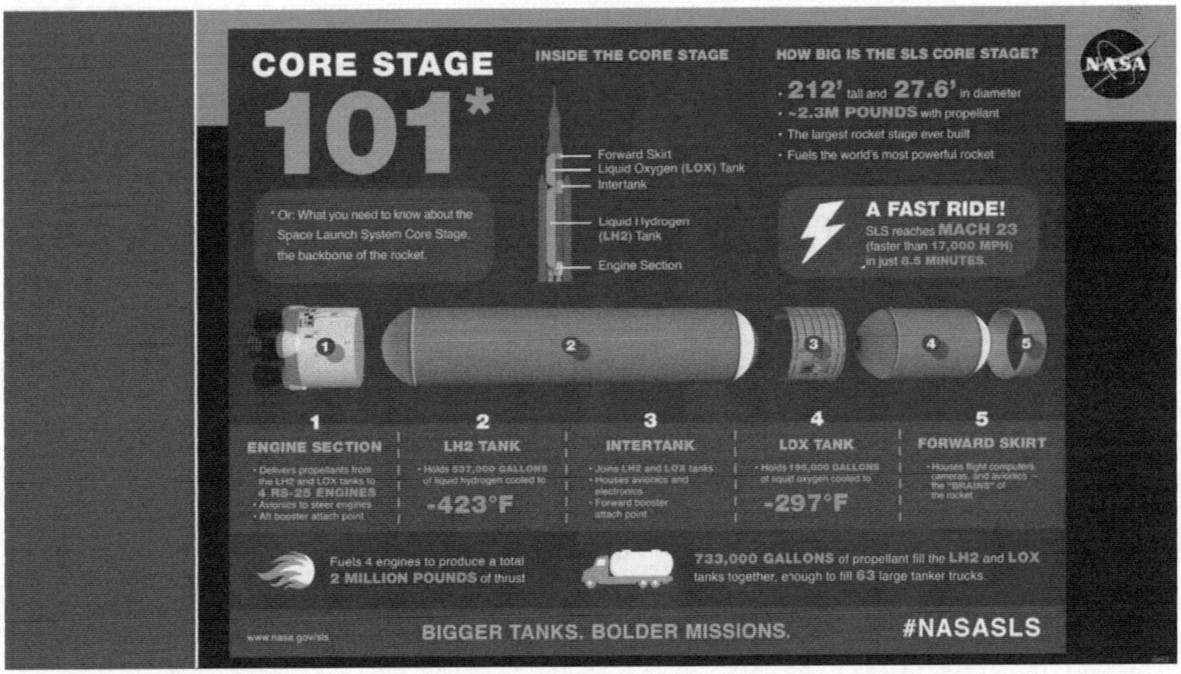
CORE STAGE 101*
* Or: What you need to know about the Space Launch System Core Stage, the backbone of the rocket.
INSIDE THE CORE STAGE
Forward Skirt
Liquid Oxygen (LOX) Tank
Intertank
Liquid Hydrogen (LH2) Tank
Engine Section
HOW BIG IS THE SLS CORE STAGE?
· 212' tall and 27.6' in diameter
· ~2.3M POUNDS with propellant
· The largest rocket stage ever built
· Fuels the world's most powerful rocket
A FAST RIDE!
SLS reaches MACH 23 (faster than 17,000 MPH) in just 8.5 MINUTES.
NASA
1 ENGINE SECTION
· Delivers propellants from the LH2 and LOX tanks to 4 RS-25 ENGINES
· Avionics to steer engines
· Aft booster attach point
2 LH2 TANK
· Holds 537,000 GALLONS of liquid hydrogen cooled to -423°F
3 INTERTANK
· Joins LH2 and LOX tanks
· Houses avionics and electronics
· Forward booster attach point
4 LOX TANK
· Holds 196,000 GALLONS of liquid oxygen cooled to -297°F
5 FORWARD SKIRT
· Houses flight computers, cameras, and avionics — the "BRAINS" of the rocket
Fuels 4 engines to produce a total 2 MILLION POUNDS of thrust
733,000 GALLONS of propellant fill the LH2 and LOX tanks together, enough to fill 63 large tanker trucks.
www.nasa.gov/sls
BIGGER TANKS. BOLDER MISSIONS.
#NASASLS

Vertical Assembly Center (VAC)

ENGINES
NASA

IT'S HAPPENING NOW!
www.nasa.gov
@NASA_SLS
NASASLS
google.com/+nasa
youtube.com/nasa
@explorenasa

Ford Motor Company

Mike Felix – Chief Engineer

Powertrain Manufacturing Engineering

Index

Business Fitness Through CapEx Reuse

Mike Felix

Manufacturing Strategic Priorities

FITNESS	FITNESS
Reset Revenue And Attack Costs	Redesign Business Operations
WINNING ASPIRATION	
STRATEGIC CHOICES	
CULTURAL IMPLICATIONS	

Manufacturing Business Fitness

- Capital/Expense (CapEx) and Hours Per Unit (HPU) Reduction
- Manufacturing Footprint Optimization
- Next Generation Manufacturing

Leverage Ford's core competency in Manufacturing to achieve step change improvement in operating costs and CapEx Reduction

SECRET

Fitness Redesign Of Our Product Development System – Co-Creation

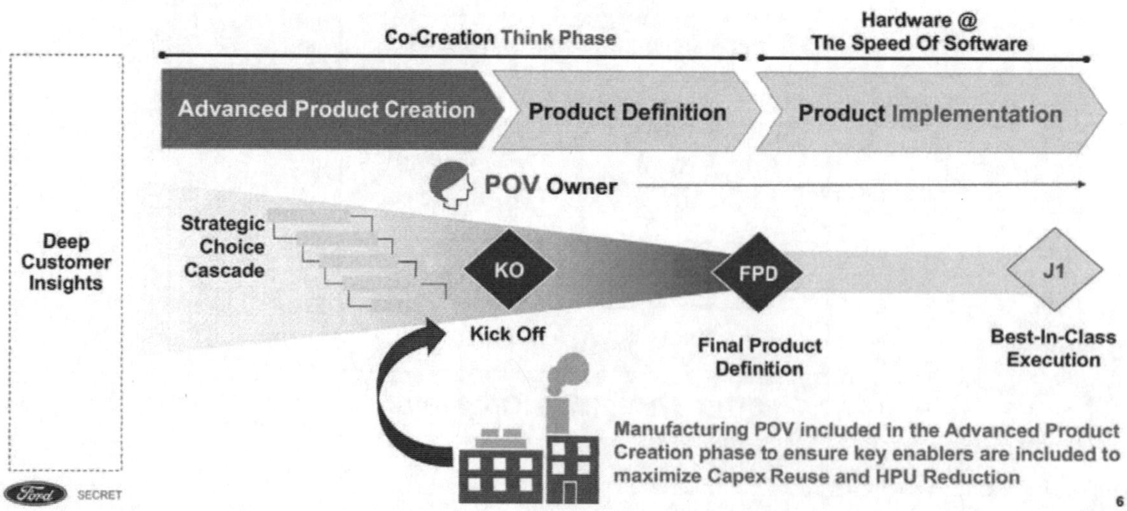

SECRET

CapEx Reduction Through Reuse Process

Principles → **Key Enablers** → **CapEx Reuse**

- **Maximum Asset Re-use:**
  - Tools
  - Equipment
  - Facilities
- **Minimum CapEx Investment**
- **Allow Product Design freedom within established:**
  - Bill of Process (BoP)
  - Hardpoints
  - Locators and attachments

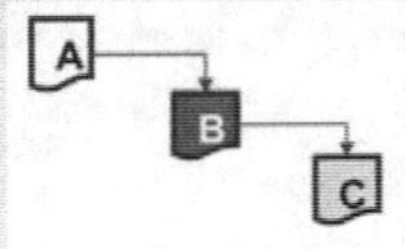

Bill of Process (BoP)

- Adhere to common sequence of assembly
- Use carry-over build sequence

Hardpoints

- Maintain existing facility and equipment hard points
- Contain within envelope dimensions (L x H x W)

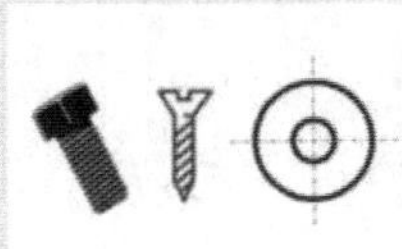

Locators and Attachment Strategy

- Adhere to carry-over locator strategy
- Use common fastener and attachment methods

SECRET 7

Ford Powertrain Electrification Overview

Executive Summary

Purpose

Develop a comprehensive Electrified Vehicle (EV) strategy for the Global Manufacturing, Purchasing, and Product Development organizations to support Electrification Strategy Prioritization

Strategic Goals

- Drive investment and variable cost efficiency for low-cost producer leadership
- Create Intellectual Property in support of Design and Manufacturing Leadership; leverage internal designs for key EV components
- Support efficient Global Powertrain Manufacturing footprint – Facilities & Labor Resources
- Develop EV component BoD/BoP jointly between PD and Global Manufacturing

Strategic Considerations

- Forecasted volume growth including volume impact to current ICE line-up and regional xEV requirements
- Manufacturing Synergies through the transition from ICE to xEV
- 2nd Cycle efficiencies of BoD/BoP Powertrain systems
- Support EV product differentiation (Brand Differentiated vs. Commodity)
- Value stream and scale for efficient manufacturing within PTO and Supply Base
- Rate of EV component technology change and opportunities for Alliances / Partnerships
- How our Global competitors are positioning themselves for EV growth

Powertrain – Major EV Components

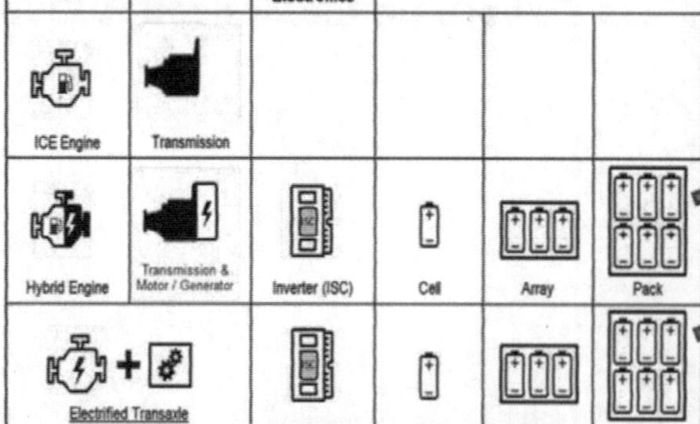

Vehicle Type	PT Cost %	Engine	Transmission	Power Electronics	Battery		
ICE	~20%	ICE Engine	Transmission				
HEV	~35%	Hybrid Engine	Transmission & Motor / Generator	Inverter (ISC)	Cell	Array	Pack
BEV	~60%	Electrified Transaxle E-Motor & Gearbox		Inverter (ISC)	Cell	Array	Pack

P/HEV

BEV

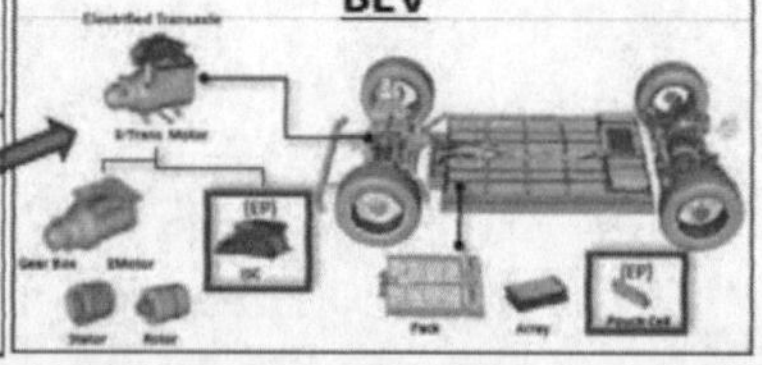

Major P/HEV Components Include A Hybrid Engine, Hybrid Transmission, Power Electronics, And A P/HEV Battery System; Major BEV Components Include An Electrified Transaxle (E-motor & Gearbox) Power Electronics, And A BEV Battery System

MBO MANUFACTURING BUSINESS OFFICE

Electrification Subsystem Products - Commodity POV (BEV, FHEV, PHEV)

	Product	Key Vehicle Impacts				Industry Commodity (Y/N)
		Vehicle Pkg	Crash	FE/Range	NVH	
1	Battery Pack	●	●	●	◑	No
2	Electric Motors	◕	◑	●	●	No
3	Electrified Transmission	◕	◕	●	●	No
4	Battery Array	◑	◔	○	○	Possible Commoditization
5	Cell	◔	◔	●	○	Commodity
6	Power Electronics	◔	◔	◑	◔	Trending toward Commoditization

Importance ◕

Note: The above ***Does Not*** apply to mHEV's

Battery Pack, Electric Motor, And Electrified Transmission Are Key For Platform Compatibility And Specific Vehicle Level Attributes/DNA And Therefore Not Expected To Be Commoditized

Originating Author: Name

Revision <#>

MBO MANUFACTURING BUSINESS OFFICE

Powertrain – EV Component Sourcing Criteria

Total Cost
(Efficiency)

- Investment & Variable cost efficiency for low-cost producer leadership
- Synergies with existing BoD / BoP footprint

Strategic Advantage
(Flexibility / Efficiency)

- Vehicle Integration – High value BoD/BoP for attributes & packaging
- Supply base footprint – optimize value stream
- Supply base risk – Intellectual property protection, production disruption

Design Leadership
(Technology)

- Research / D&R – Must be in-house for vertical integration
- Ability to develop intellectual property for competitive advantage
- Rate of technology change

Manufacturing Synergy
(Flexibility)

- Manufacturing Engineering competence exists or is easily acquired
- Design rate of technology change impacting BoD / BoP
- Scalable for volume

There Are (4) Criteria Categories That PT EV Component Sourcing Assessment Is Based On: Total Cost, Strategic Advantage, Design Leadership, And Manufacturing Synergy

Originating Author: Name

Revision <#>

MBO MANUFACTURING BUSINESS OFFICE — Ford Go Further

Powertrain – EV Component Sourcing Assessment

	Cross Functional Strategic Assessment: Total Cost	Strategic Advantage	Design Leadership	Mfg Synergy	Sourcing Decision	Rationale
Battery Pack	●	●	●	●	Make	• Strong synergy to existing processes • High value sub-system • Significant Impact on Pkg'g & Attributes
E-Motor	●	●	●	◕	Make	• Strong synergy to existing processes • No established supply base • Significant Impact on Pkg'g & Attributes
Gearbox	●	●	●	●	Make	• Existing commodity – leverage existing capacity/assets and design leadership
Battery Array	◔	◕	◑	◕	Make	• Synergy to existing processes • Ability to leverage state of technology, volume, in region cell prod availability
Battery Cell	◕	◑	○	○	Buy / Partner	• Specialized manufacturing requirements • High potential for technology change • Maintain access to external IP • Consider potential alliance opportunity
Power Electronics (ISC)	○	◔	◔	○	Buy	• No Mfg synergy or design integration • Evolving Technology

Note: Assessment of Battery Tray/Cover structures in-sourcing is planned for VI Phase 2 study

Importance | Make | Buy | Partner

Make In-house – E-motor, Gearbox, Battery Array, And Battery Pack
Purchase – Power Electronics (ISC) And Battery Cell
Future Consideration – OEM And Supplier JV / Alliances including Battery Cell

Originating Author: Name

Revision <#>

MBO MANUFACTURING BUSINESS OFFICE — Ford Go Further

Powertrain -- EV Component Sourcing Competitive Alignment

Proposed Vertical Integration	xEV	E-Transaxle: Electric Motor	E-Transaxle: Gearbox	Power Electronics	Cell	Battery: Array	Battery: Pack
TOYOTA	P/HEV	Partner a/ Make	Partner a/ Make	Partner a/ Make	Partner	Make	
VOLKSWAGEN GROUP	BEV	Make	Make	Make	Buy / Make		Make
RENAULT NISSAN	BEV	Make	Make	Buy / Make	Make / Buy	Make	
DAIMLER	BEV	Partner	Make	Buy	Buy	Make	
BMW GROUP	BEV	Make	Make	Buy	Buy		Make
TESLA	BEV	Make	Buy / Make	Make	Partner	Make	
GM	BEV	Buy / Make	Buy	Buy	Buy	Buy	Buy / Make
Ford	P/HEV / BEV	Make	Make	Buy	Buy / Partner	Make	

Make - | Buy | Alliance Partner /JV

Ford Proposed Strategy Will Align With The Majority Of Major OEM's

Originating Author: Name

Revision <#>

MANUFACTURING BUSINESS OFFICE

Powertrain Electrification Basics – EV Global Electrification Trends

NOW... NEAR...FAR....

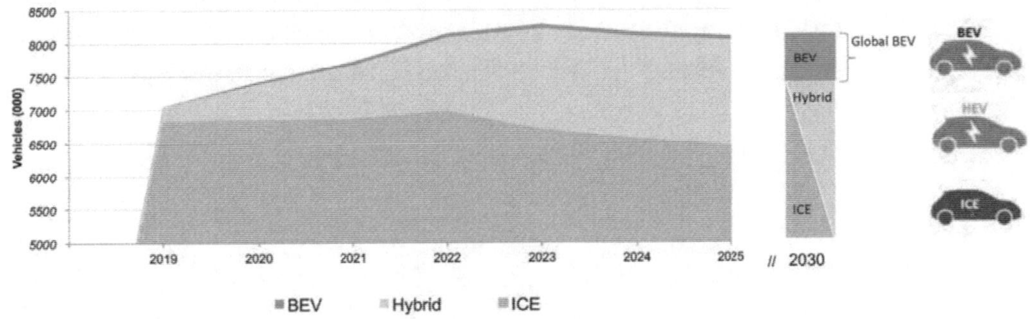

Manufacturing EV Strategy Should Support All Scenarios For Growth In P/HEV And BEV
P/HEV Volume Grows Significantly Through 2025, While BEV Volume Is Relatively Minor
Significantly Increases BEV Volumes By 2030

Originating Author: Name

Revision <#>

Ford Advanced Mfg Project

Gear Tooth Form Surface Finish

Goal

FORD PTME
Global Powertrain Manufacturing Engineering

- Improved fuel economy / emissions via improved transmission efficiency.
  - Improve gear tooth surface finish to reduce frictional losses.
- Downside
  - Higher operating cost to traditional gear finishing processes.

17

Patents Pending

Gear Tooth Profile Surface Finishes

FORD PTME
Global Powertrain Manufacturing Engineering

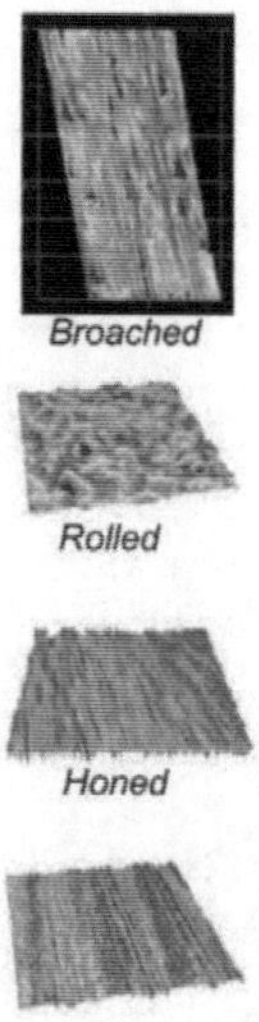

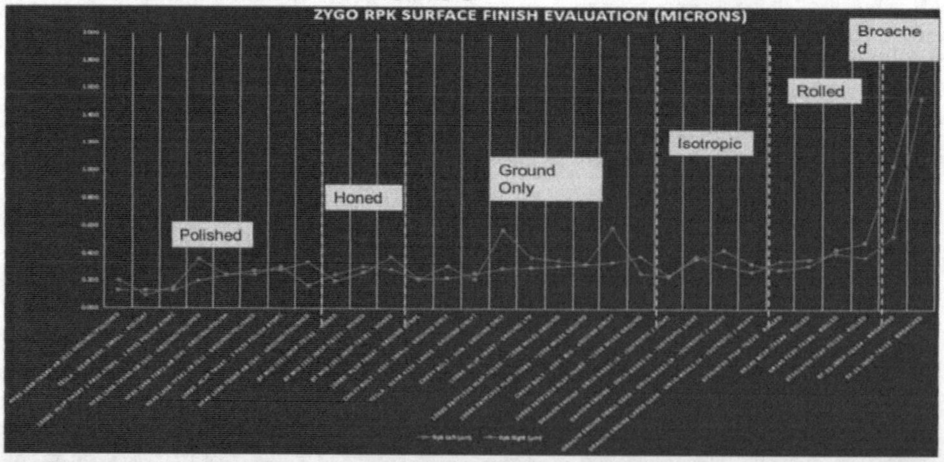

Various Manufacturing processes produce unique surface finish characteristics:

- Unique patterns, unique directions with respect to profile, and unique process capabilities (targeting and dispersion).

Tooth Profile Form Error (ffa 3 max typical) vs surface

FORD PTME
Global Powertrain Manufacturing Engineering

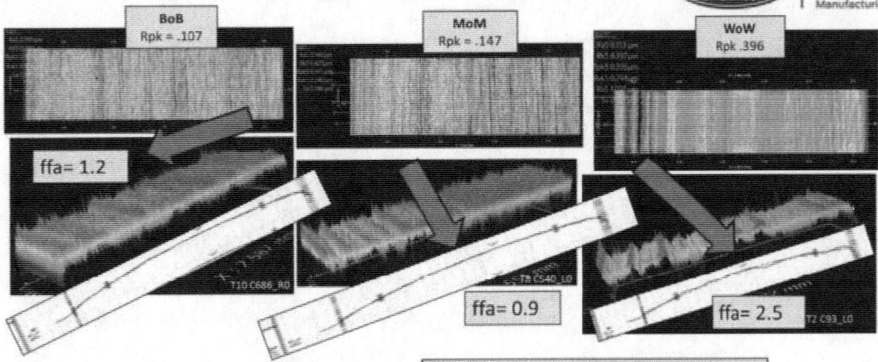

- Dresser life significantly reduced based on degradation of Rpk vs form error
- Surface finish can be easily confounded to tooth profile form error (ffa).

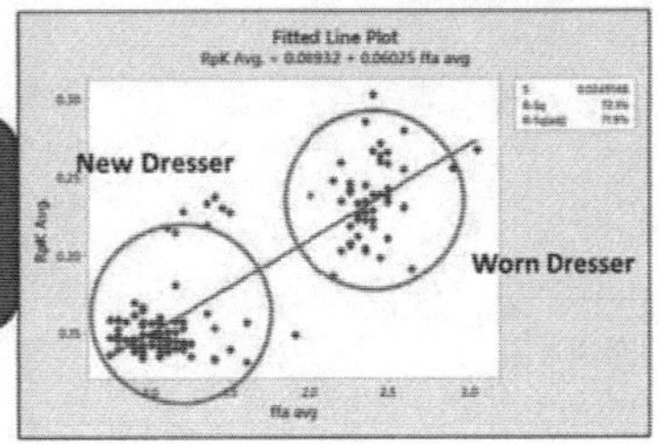

19

Patents Pending

Surface Finish after Transformation

FORD PTME

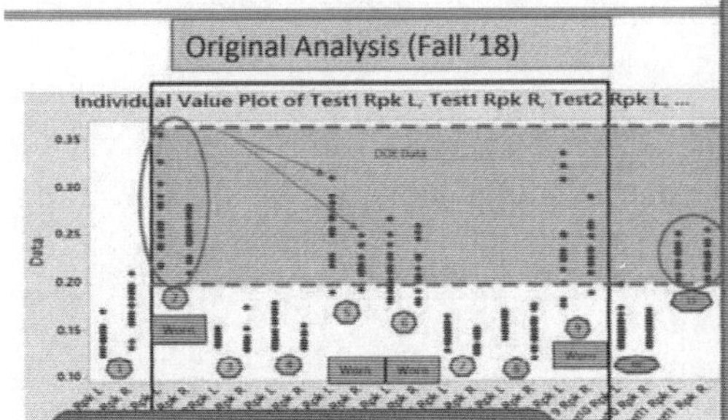

Baseline conventional grinding with new dresser is (almost) the same as polishing with a worn dresser.

1 Pass Polish (New Dresser / Old Dresser)					Grind only	
ZYGO 1um wide	T10 C686_R0	T8 C540_L0	T2 C93_L0	Average	C698R_0	ATO_GRIND_2L_0
Mean 5 slices	BoB	MoM	WoW		ATNPC Grind	LTP Grind
Ra	0.101	0.103	0.118	0.107	0.200	0.239
[illegible]	1.290	1.188	1.014	1.164	2.614	1.082
Rpk	0.204	0.185	0.150	0.180	0.359	0.412
Rvk	0.185	0.164	0.213	0.187	0.468	0.248

Applying a 2um areal filter, then a high pass 80um filter for getting roughness

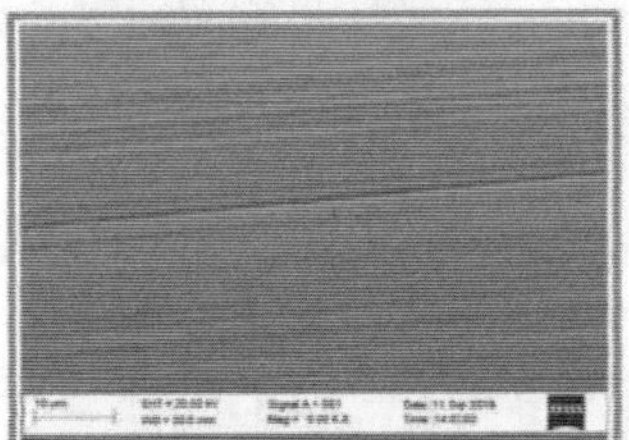

Sample 674: 1- Pass Polish at 500X

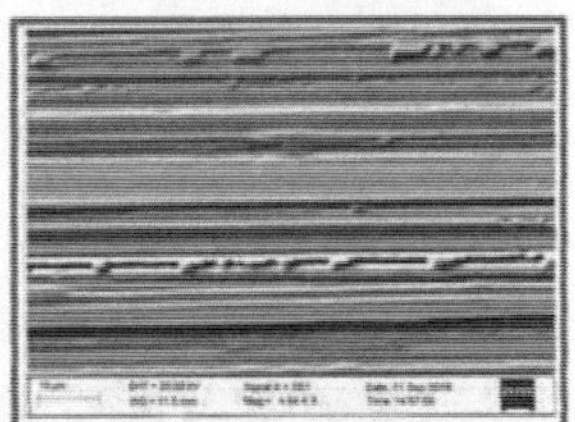

Ground Only ATNPC

By Decoupling ffa:
- Better separation between polish sample and conventional grind
- Minimal change across dresser life

Grinding Wheel composition

FORD PTME Global Powertrain Manufacturing Engineering

~ 600+ grit

Silicon carbide	40-70

Polyurethane	30-60
Aluminum oxide (Al2O3)	20-70
Chromium oxide (Cr2O3)	5-10

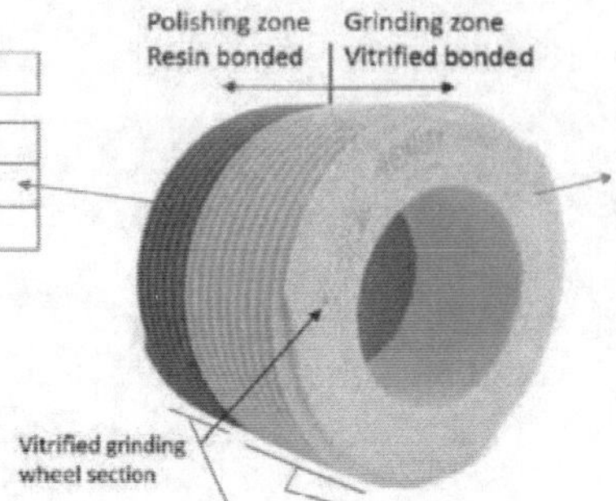

~ 80 grit

Information on ingredients:
Abrasive: Synthetic Aluminum Oxide (40-45%)
Vitrified: Glass (7-10%)
Bond constituents: SiO_2 (50-65%)
Bond constituents: Al_2O_3(10-17%)
Bond constituents: B_2O_3 (5-15%)
Bond constituents: Na_2O (2-7%)
Bond constituents: CaO (1-5%)

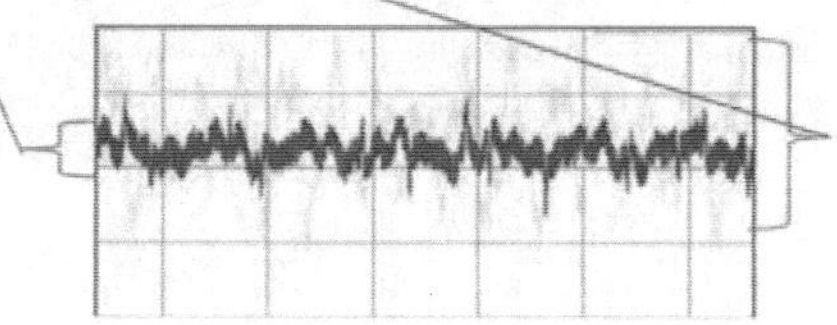

Goal: Keep the valleys for oil retention. Remove the peaks.

Challenges

FORD PTME Global Powertrain Manufacturing Engineering

- Correlating gear surface finish to transmission efficiency with enough discrimination to ensure desired results.
  - Doing so allows Manufacturing to select the "best" process to ensure those results are consistently delivered in a cost effective process.

"Typical" ground tooth profile, 5000X

"Polish ground" tooth profile, 5000X

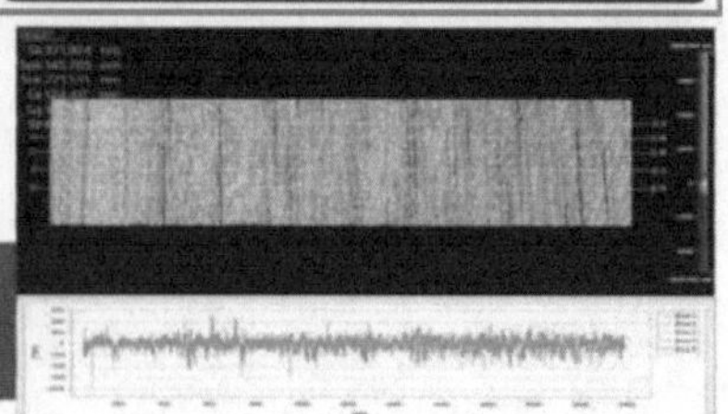

What Process Delivers the Key Attributes to Higher Efficiency?

22

Ford Advanced Mfg Project

Gear Tooth Form Periodicity Inspections and Abatements

Goals /Studies

FORD PTME
Global Powertrain
Manufacturing Engineering

Goals

A. Refine understanding of various periodicity inspection methods.

B. Refine understanding of how to change or target the periodicities.
   1. Feeds/Speeds.
   2. Machine/tool/part kinematics.
   3. Machine stiffness.

C. Understand possible system modeling tools to predict what periodicities to avoid or target.

Case Studies

1) Skived vs broach internal helical ring gear
2) 4.4th order gear noise.
3) Idler gear noise.

(1) Skived vs Broached Helical Ring Gears

FORD PTME
Global Powertrain
Manufacturing Engineering

Transmission Frequency Response

Component level /process (Ring Gear/skiving)

Form error Direction

PRELIMINARY Analysis:

- High Order response was very similar to broached baseline.
  - Skived ring feed pattern/periodicity does not appear to be impactful to performance.
  - Feed pattern and design line of contact are not parallel. Parallel alignment promotes higher order attenuation.

As customer NVH requirements tighten, and overall vehicle noise drop, transmission gear generated noise must be understood at a much refined level.

Conventional inspection to Frequency domain:

FORD PTME
Global Powertrain
Manufacturing Engineering

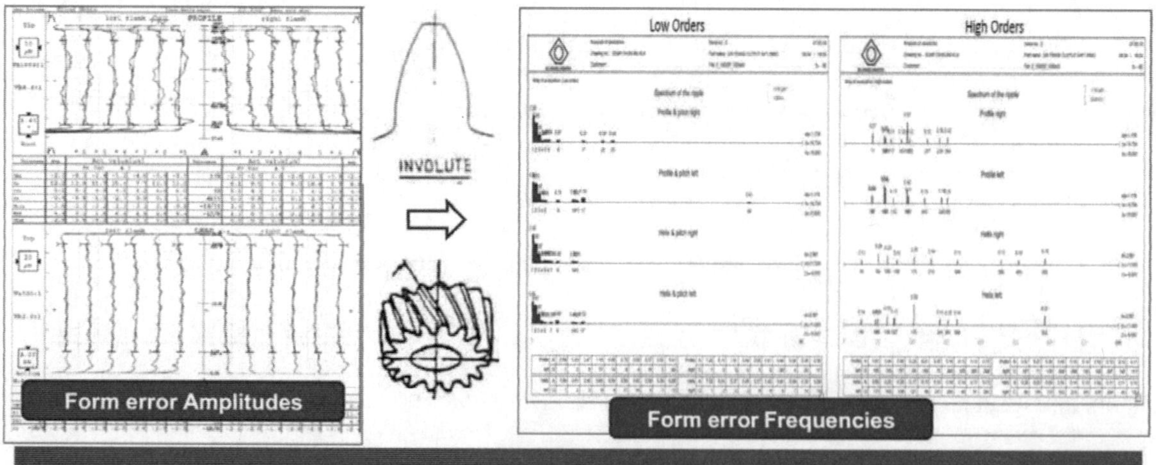

Form error Amplitudes

Form error Frequencies

Understanding a gear's tooth form "periodicity" is becoming more and more important. (Amplitude + direction + frequency)

26 Patents Pending

(2) 4.4th order gear noise :

FORD PTME
Global Powertrain
Manufacturing Engineering

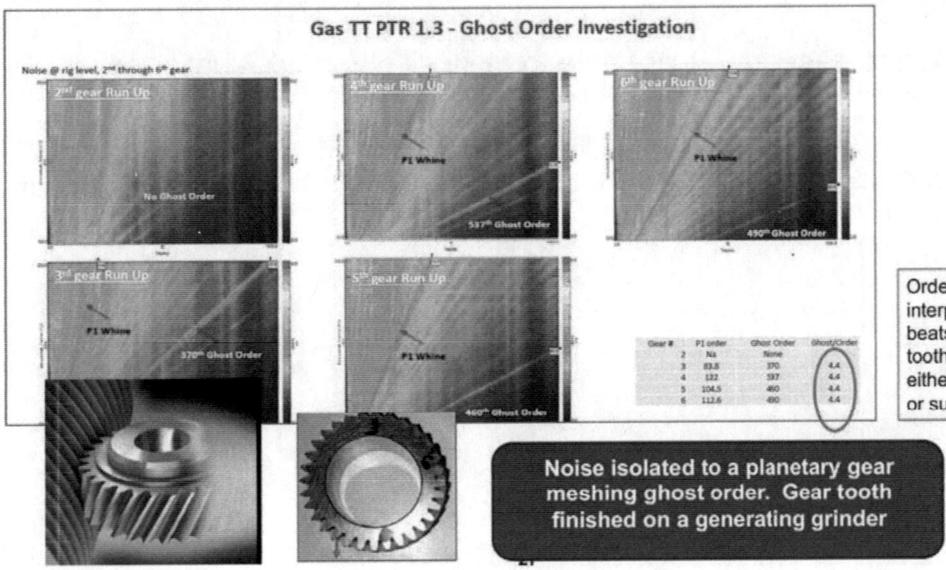

Order 4.4 interpretation: 4.4 beats per pinion tooth rolling thru either mating ring or sun tooth.

Noise isolated to a planetary gear meshing ghost order. Gear tooth finished on a generating grinder

Patents Pending

(3) Idler Gear Noise

FORD PTME
Global Powertrain
Manufacturing Engineering

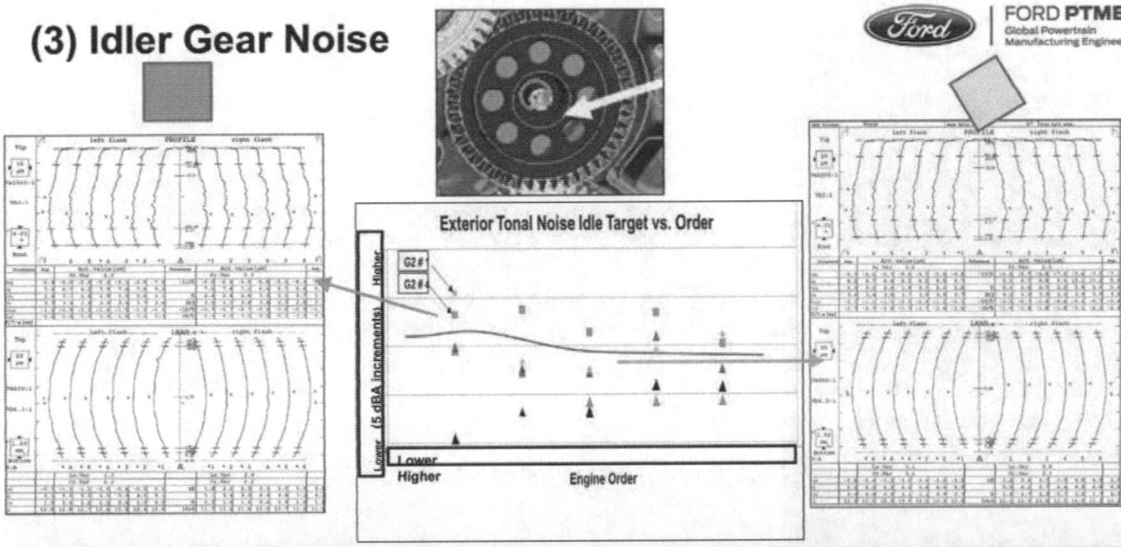

Gears appear nearly IDENTICAL as measured conventionally, yet perform quite differently in some transmission systems. Goal: Use FFT response, and machine parameters/responses to control performance when overly responsive systems cannot be designed out..

28

The Near Term

FORD PTME
Global Powertrain
Manufacturing Engineering

- Dimensional Gear Geometry is not able to fulfill all of today's challenging NVH challenges
- Sub micron errors and patterns require correlating real world system response to sub micron gear frequencies to control non-dimensional machine and machining characteristics.

29

Thank You

GRINDING OF AEROSPACE MATERIALS USED FOR HIGH-TEMPERATURE APPLICATIONS

Intertech 2019 – Paper #1941

K. Philip Varghese, PhD, Group Leader, Saint-Gobain Abrasives

John Hagan, Sr. Application Engineer, Saint-Gobain Abrasives

Andrew Biro, Application Engineer, Saint-Gobain Abrasives

Address: 9 Goddard Road, Northboro, MA, 01532

Keywords

Aerospace, Gamma Titanium Aluminide, Ceramic Matrix Composites, Nickel based superalloys, Grinding

Abstract

Many aerospace components used in the high and low pressure sections of a turbine engine are made from materials such as high nickel-based super alloys, intermetallics such as gamma titanium aluminide and ceramic matrix composites. These components that may be cast, forged or sintered using powder metallurgy techniques, are notorious for being difficult to machine due to their high strength, toughness, corrosion and fatigue resistance, and low thermal conductivity. Many of these parts have as much as two-thirds of their original weight removed to produce the finished component. Traditionally turning, milling and broaching processes are employed to remove most or all of the material from these components. However, with the use of modern grinding wheels composed of engineered grains and high strength bond technology, grinding can now remove the material more quickly and economically than turning or milling. This paper will show data/results of surface grinding and creep feed grinding tests conducted on advanced aerospace engine materials specifically gamma titanium aluminide, IN718 using both conventional and superabrasive grinding wheels. It will also compare the grinding of ceramic matrix composites with that of monolithic ceramics.

Background

Nickel-based superalloys, which constitute greater than 40% of the total weight of an aircraft engine, find applications in the combustor and turbine sections of the engine [1]. Their use is primarily due to the unique combination of high temperature strength, toughness and resistance to degradation in extreme environment. These properties also make them difficult to machine, and render high-speed machining of components made from these superalloys very costly. Only abrasive wheels designed specifically for these types of materials can survive the conflicting goals of higher productivity, higher wheel life and can grind at low threshold power/forces and low specific energy.

Titanium aluminides are an intermetallic compound consisting of nearly equal parts titanium and aluminum. Their high specific strength, high-temperature stability and oxidation resistance compared to conventional titanium and nickel alloys make them attractive for use in aerospace engines , Both the new Pratt and Whitney GTF engines and CFM Leap engines make use of γ-TiAl blades in their low pressure turbines [2]. The heat resistance typically associated with heavy nickel alloys, alongside the lightweight characteristics of titanium, make γ-TiAl attractive for these engine makers who are driving increased fuel efficiency and lower emissions. Titanium aluminide components are generally very expensive to process and can be 2 to 3 times the cost of nickel-based superalloys [3]. Near-gamma alloys are processed through ingot metallurgy, powder metallurgy or casting. Obtaining the final net shape in the component is usually done through some form of machining, grinding or by non-traditional methods such as electro-discharge machining, electro chemical machining or some combination of these processes. When compared with single-point machining, like with any intermetallic, grinding is often recognized as the best method for achieving final part dimensions and surface characteristics on γ-TiAl components [3].

CMCs are generally a system of materials made up of ceramic fibers or particles that lie in a ceramic matrix phase. They have many unique characteristics that lend them for use in high-temperature applications. They weigh a third of the weight of traditional aircraft engine metal alloys and can withstand temperatures of up to 2,400°F, much higher than the metal alloys. CMCs can be silicon carbide based or aluminum oxide based and their properties differ based on the degree of densification achieved after sintering and polymer or metal melt infiltration.

At the state –of-the-art Higgins Grinding Technology Center in Northboro, Massachusetts, engineers from Norton/Saint-Gobain have been studying the manufacturing of these innovative materials using grinding technology for many years. This has resulted in the development of grinding products and processes which enable more effective manufacture of components made from these materials. This article shows a sampling of data and results from grinding of high nickel-based super alloys, gamma titanium aluminide and ceramic matrix composites.

Grinding of high-Ni based superalloys

Testing consisted of creep-feed grinding of IN-718 and Waspaloy bars with two grades of TG2 wheels. Bars of IN-718 and Waspaloy measuring approximately 12.7 mm width by 76.2

mm thickness by 279 mm length were used. Instead of grinding slots into the material, the full width of the bars was ground. In this configuration coolant was not trapped by the sidewalls of the slot, but was able to escape along the sides of the part. A Campbell model 930 five Axis Grinder with straight oil coolant (Castrol Variocut B27) was used for this test. The large 30 kW. spindle is suitable for wheels up to 400mm in diameter and a 75,000 rpm high speed spindle is available for using with mounted points. A Field Instrumentation System (FIS) developed by Saint-Gobain Engineers was used to collect grinding power. Just two wheels (TG280-H20VTX2, TG280-G20VTX2) were tested and they were dressed with a BPR dress roll (BPR-2227). Truing was done at a wheel speed of 33 m/sec, at a speed ratio of 0.85. Depth per pass was 0.0254 mm and a feed of 0.508 mm per wheel revolution used for an overlap ratio of two. Figure 1 shows a picture of the test set-up used. Each wheel was tested at 2.54 mm depth per pass and at four feed rates. Under each condition approximately 17.8 mm of material was removed from the 279 mm long parts. Wheel speed was 43 m/sec and using a graphite coupon, wheel wear was determined by the step left between the used and unused portions of the wheel.

Figure 2 is a graph of the specific power for each condition tested. As expected for both materials the softer G grade wheel had a lower specific cutting energy (slope of the specific power) than the harder H grade wheel. The material type did not appear to have a significant influence on power. Although the specific power increased with increasing removal rate the specific grinding energy dropped as seen in the graph on Figure 3. As expected wheel wear increased as removal rate increased. This time separation due to the material type was more pronounced as seen in the G-Ratio graph on Figure 4. Figure 5 and 6 show the variation of surface finish at increasing removal rates in IN-718 and Waspalloy respectively. The harder H grade wheels show consistent surface finish values even at significantly high removal rates in case of both workpiece materials

At the completion of the testing, samples of the bars ground at the highest removal rates were cut on an abrasive water jet machine and sent for metallurgical analysis. The analysis done found no occurrence of white layer on any of the samples and surface stress and damage never exceeded 0.05 mm depth. Table 1 shows a summary of the results of metallurgical analysis on the ground IN-718 and Waspalloy workpieces.

The results show that both Ni-based superalloys were capable of being ground at very high removal rates without significant impact on surface finish or any white layer formation on the tested materials. Waspaloy was not much more difficult to grind than the IN-718 material from a grinding power perspective, however it did cause higher wheel wear rates particularly with the softer grade wheel where the G-Ratio dropped below five at a removal rate of 150 mm^3/sec/mm. These conditions being representative of a roughing process any surface damage caused during this process could easily be removed with a quick finish pass.

Grinding of gamma Titanium Aluminide

Recent product testing on gamma titanium aluminide focused around evaluating three abrasive types: diamond, cubic boron nitride (cBN) and conventional silicon carbide (SiC). The diamond and cBN products included electroplated and dressable multi-layer metal bonded wheels, while the SiC abrasives were used with a conventional vitrified bond system. For each superabrasive type (i.e. diamond & cBN), two grit sizes were evaluated to better understand the effect of grit size on force, power, and surface roughness. A summary of the wheels is shown in Table 2.

In order to evaluate the products in both mild and aggressive grinding conditions, slots were ground, as shown in Figure 7, into blocks of gamma titanium aluminide at two material removal rates. Only one set of grinding conditions was used per slot in order to preserve the ground surfaces for subsequent analysis. Testing was all in non-continuous dress mode and wheels were dressed in between slots (i.e. for each new set of grinding conditions). A summary of the grinding conditions is shown in Table 3.

Products were evaluated based on power, normal force, part surface finish, and part damage (cracking and/or burn). The results of this study showed that for the conventional SiC abrasives, the power (and forces) increased as a function of the amount of stock removed (shown in **Error! Reference source not found.**). Similar trends were also observed for cBN wheels; however, the power increase was much lower than for SiC. The diamond wheels consistently drew the lowest power, and had the most stable power as a function of the stock removed. The same trends were observed with the coarser superabrasive wheels (250 um) and the multi-layer superabrasive products.

The increase in power (and forces) with the SiC wheels resulted in cracking and mild burn on the bottom of the ground TiAl slots (shown in **Error! Reference source not found.**). No cracking was observed with the cBN or the diamond wheels after removing 1800 mm^3 of material. In fact, the diamond wheels removed ~48000 mm^3 of material without damaging the parts. The cBN and SiC wheels reached only about 6 & 10 % (respectively) of that before damage was observed.

The damage to the ground parts was attributed to metal adhesion on the tips of the grains. This adhesion is shown in **Error! Reference source not found.**, and was most prevalent in the SiC wheels. Metal adhesion was also observed on cBN grains, but to a much lower degree while very little metal adhesion was observed on the tips of the diamond grains. Adhesion of metal to the abrasive grains is known to lead to lower grinding efficiency (i.e. higher frictional contributions and a reduced ability for the grains to remove material).

While the diamond products were clearly superior in terms of cumulative stock removal without component damage, individual grinding processes may still dictate the use of SiC grinding wheels. Processes that require frequent wheel dressing in order to maintain strict form requirements could benefit from using SiC, as the cost/wheel is substantially lower and the form can be easily maintained through dressing.

Electroplated diamond wheels, while capable of removing substantial amounts of material, are limited in their tolerance capability. Once used to the point where the parts lose tolerance, no redressing can be done and the wheel must be replaced. With Saint-Gobain Abrasives's recent developments in new, multi-layer dressable metal bond diamond products, new possibilities have opened up to use these products in applications where form and finish have very strict tolerances that must be maintained over time [5].

Grinding of Ceramic Matrix Composites (CMC)

Higgins Grinding Technology Center carried out a grinding study where a Ceramic Matrix Composite material (CMC) and a Monolithic Ceramic material (MC) were evaluated in a plunge surface grinding mode. Figure 11 shows the test setup. Polymer infiltrated CMC plates (silicon carbide-in-silicon carbide matrix) were used to compare with monolithic ceramic (silicon carbide) made by Saint-Gobain Ceramic Materials. Table 4 shows a brief comparison of the properties of the tested materials. The study involved grinding multiple slots in each material at different Specific Material Removal Rates (MRR'). Each slot was 0.89mm wide and was completed by making 20 passes at 0.254mm depth/pass, resulting in an overall slot depth of 5.08mm. Both materials were evaluated at five MRR' ranging from 5.4 to 32.4 $mm^3/mm/sec$ (see Table 5). The range of MRR' was achieved by adjusting only the feedrate. Both the width of cut and the depth of cut/pass were maintained constant throughout the study at values of 0.89mm (wheel width) and at 0.254mm/pass, respectively.

The test wheel was a resin bonded, 150 grit, diamond, slicing wheel (1A1R shape). The test machine was an Elb Brillant surface-creepfeed grinder. The grinding wheel was trued and dressed prior to starting each condition. The workpieces were fixtured by waxing them onto a substrate plate. After securing the workpiece to the sub-plate, the assembly was bolted to a raiser block that was held in a machinist's vice. The wheel speed was 30m/s and the coolant was water-soluble oil. Table 6 shows the characteristics of the test wheel and some of the grinding conditions. Both materials were evaluated in terms of Specific Grinding Power, Specific Grinding Energy and Grinding Ratio. The grinding power was measured using a power monitoring system developed by Saint-Gobain engineers and built into the test machine. The Specific Grinding Energy (SGE) was calculated using the power measurements and the material removal rate (SGE = Power' ÷ MRR').The wheel diameter was measured before and after each grind. Using the wheel wear and the stock removed, the Grinding Ratio was calculated (G-Ratio = Volume of Work Material Removed ÷ Volume of Wheel Wear).

Figure 12 shows the Specific Power (W/mm) as a function of the cumulative number of grinds. The results show that the grinding power was significantly lower for the CMC material when compared to the MC material. Also the grinding power level was relatively flat throughout the test for the CMC material while the power for the MC material increased steadily as the number of cuts increased. Figure 13 shows the Specific Grinding Energy (J/mm^3) as a function of the material removal rate. The results showed that the SGE for the CMC material was significantly lower when compared to the MC material. Figure 14 shows the Grinding Ratio as a function of Material Removal Rate. The results showed that the Grinding Ratio is similar for both materials at lower MRR's. However, as the MRR' was increased, the G-Ratio for the MC material started to decrease while the CMC material remained relatively steady. This showed that at higher Material Removal Rates, the wheel wear will increase when grinding the MC material. Whereas, regardless of the MRR' the wheel wear remained relatively steady when grinding the CMC material. The results of this study showed that the CMC material, when compared to the MC material, required less Grinding Power, less Grinding Energy and the G-Ratio was higher (less wheel wear). Therefore, based on the results of this study, when comparing these materials, the CMC material is less difficult to grind than the MC material.

Conclusion

Materials such as CMCs, γ-TiAl, and high Ni-based superalloys used in newer aerospace engines are expanding in the aerospace sector primarily due to their unique combination of high temperature strength, toughness and resistance to degradation in extreme environment. These properties also make them difficult to machine and expensive. Higher productivity in machining is hampered by reduction in tool life and quality of machined components. However, grinding these materials with modern engineered conventional abrasive wheels and superabrasive wheels show that higher stock removal rates are achievable without part damage. Only highly engineered abrasive wheels that can remove stock at low threshold power/forces and low specific energy can achieve higher productivity as well higher wheel life compared to conventional milling / turning tools. The data presented here shows that newly developed abrasive products can meet the challenge to grind these difficult-to-machine materials economically.

References

1. R.R. Boyer, J.D.Cotton, M. Mohaghegh, R.E. Schafrik, "Materials Considerations for Aerospace Applications," MRS Bulletin, 40(12), pp. 1055-1066, December 2015
2. Aviation Week & Space Technology, "Pratt & Whitney's Geared Turbofan Growth Plan", July, 2013
3. D. Margolis, "Inter-metallic Wonder", Cutting Tool Engineering, 57(12), 2005
4. M. Hitchiner, J. Besse, K. P. Varghese, "Grinding Innovation in a Growing Defense Market," MFG4The Future Conference, Hartford, 2012

5. R. Bright, A. Biro, M. Hitchiner, "Grinding of γ-Titanium aluminide with superabrasives", Intertech Conference, Indianapolis, May 2015

Figures

Figure 1: Test Setup for grinding IN718 and Waspalloy with oil coolant

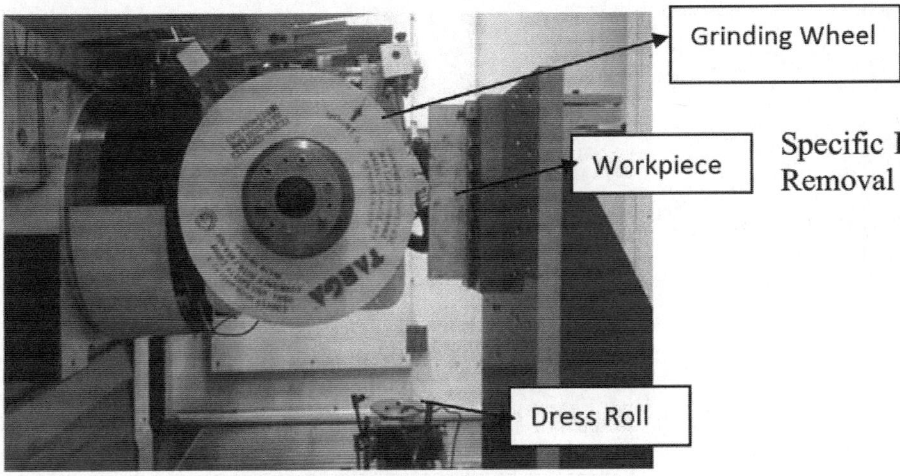

Figure 2: Variation of Specific Power vs. Specific Material Removal Rate

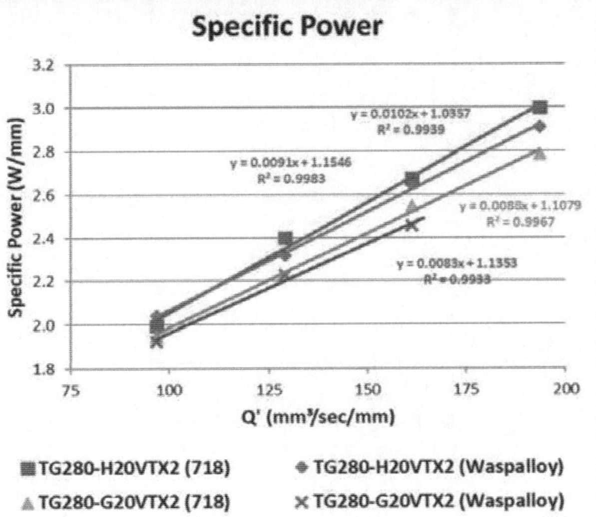

Figure 3: Variation of Specific Energy vs. Specific Material Removal Rate (shaded region represents specific energy in machining operations)

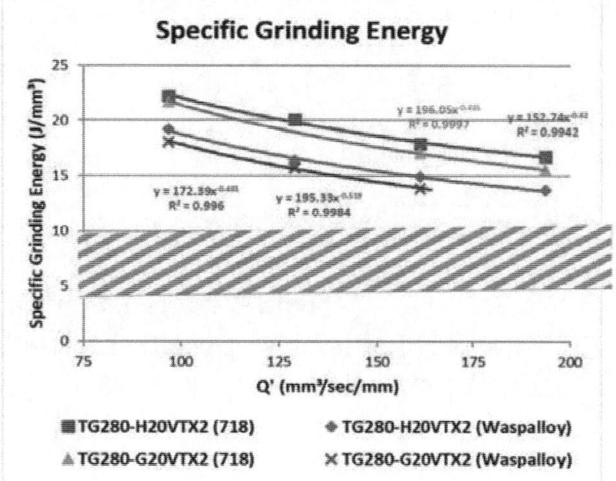

Figure 4: Variation of G-ratio vs. Specific Material Removal Rate

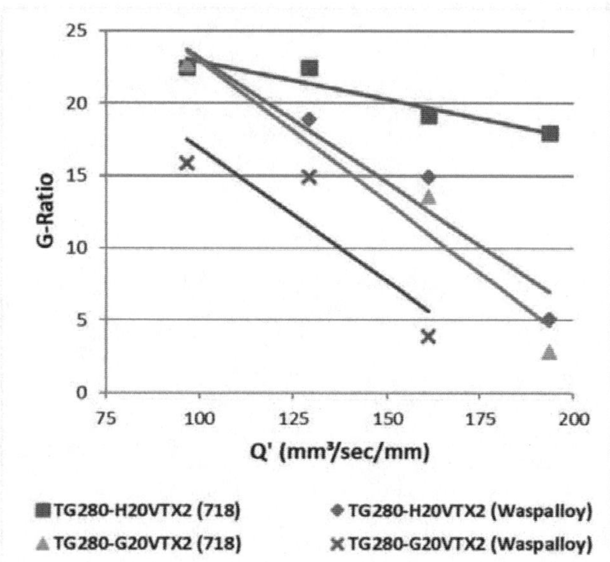

Figure 5: Variation of surface finish with Material Removal Rate in IN-718

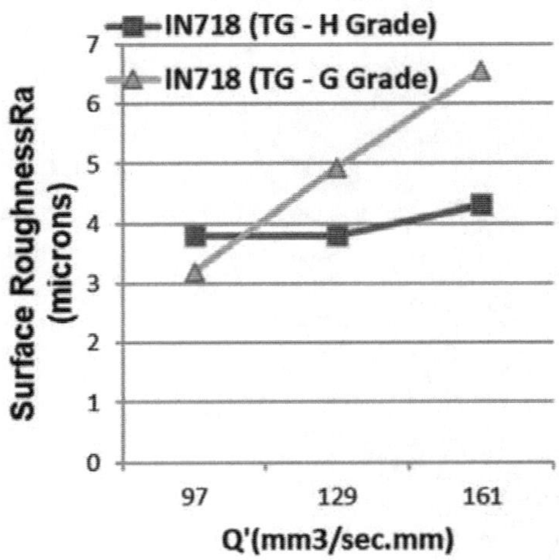

Figure 6: Variation of surface finish with Material Removal Rate in Waspalloy

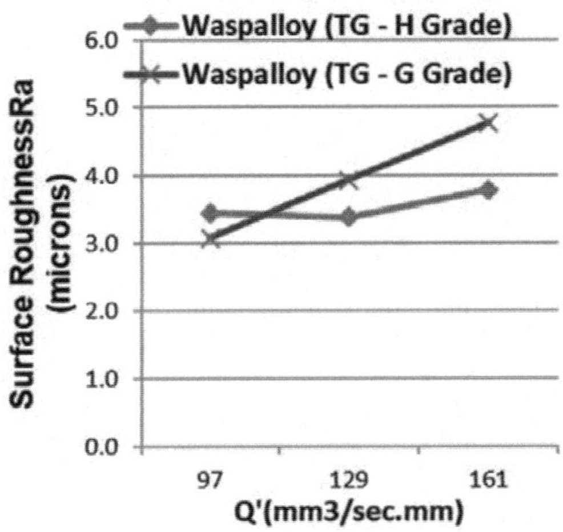

Figure 7: Setup on a creepfeed grinder used for grinding gamma Titanium Aluminide

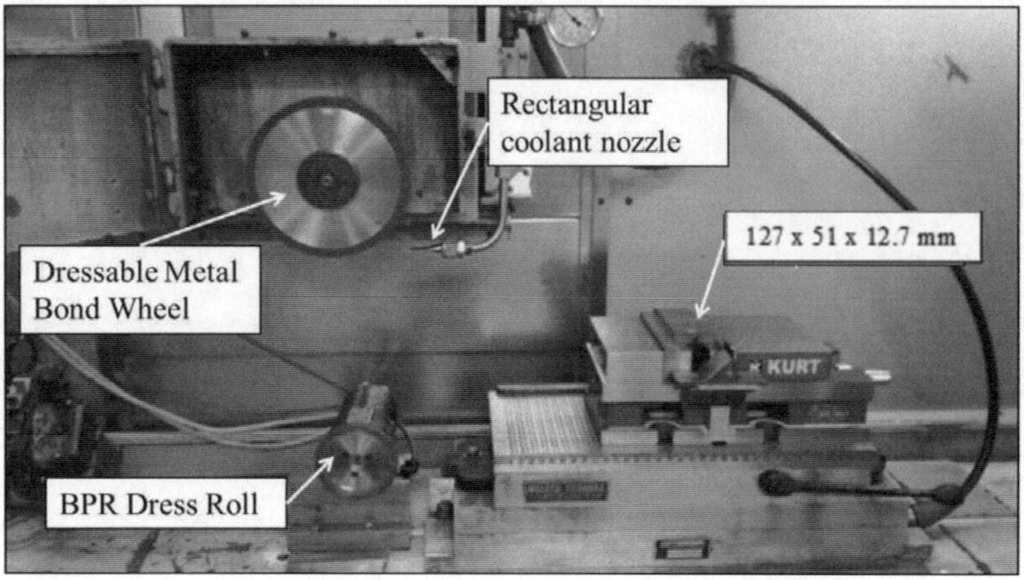

Figure 8: Plot of specific grinding power for each of the abrasives tested for grinding gamma titanium aluminide

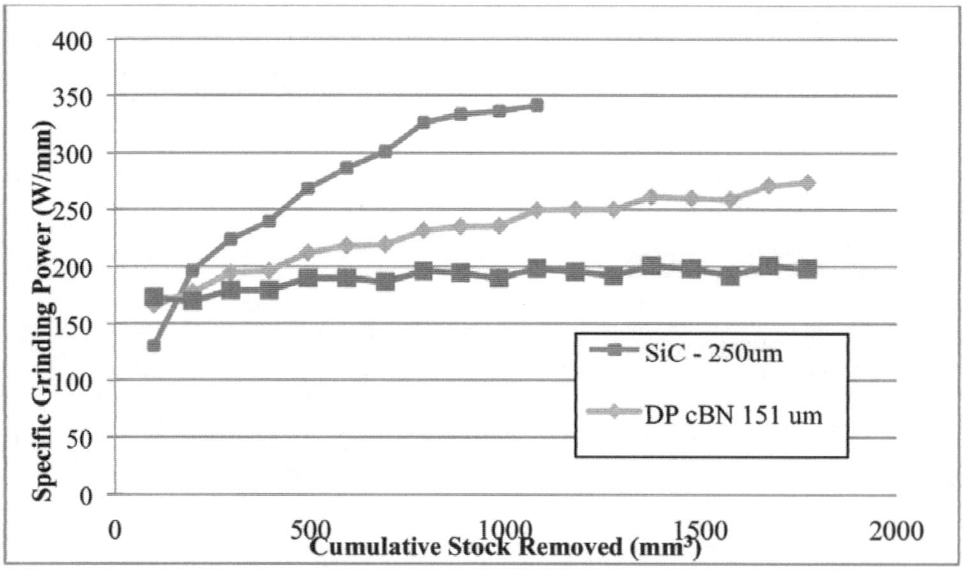

Figure 9: Images of cracking in a component a) as ground and b) using liquid dye penetrant to highlight the cracking

Figure 10: SEM of a used SiC wheel face showing metal adhesion on the tips of the active grains

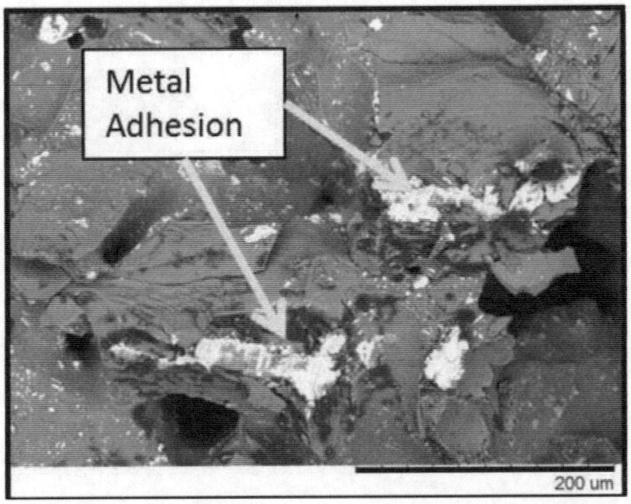

Figure 11: Test set-up for grinding a Ceramic Matrix Composite (CMC) plate

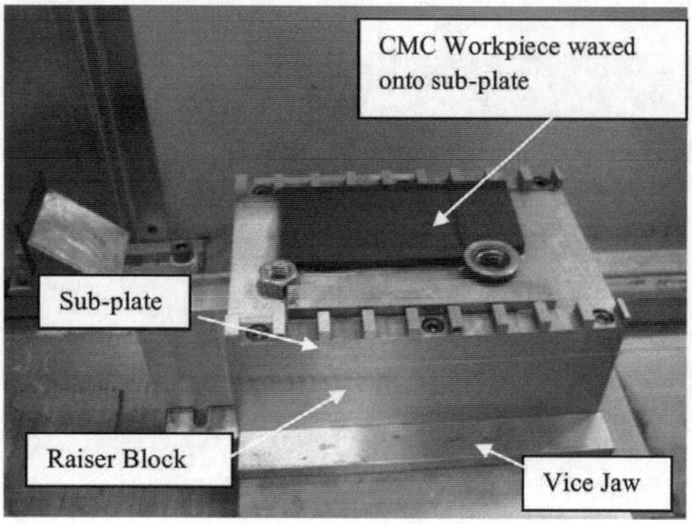

Figure 12: Variation of specific power with number of cuts when grinding MC and CMC materials

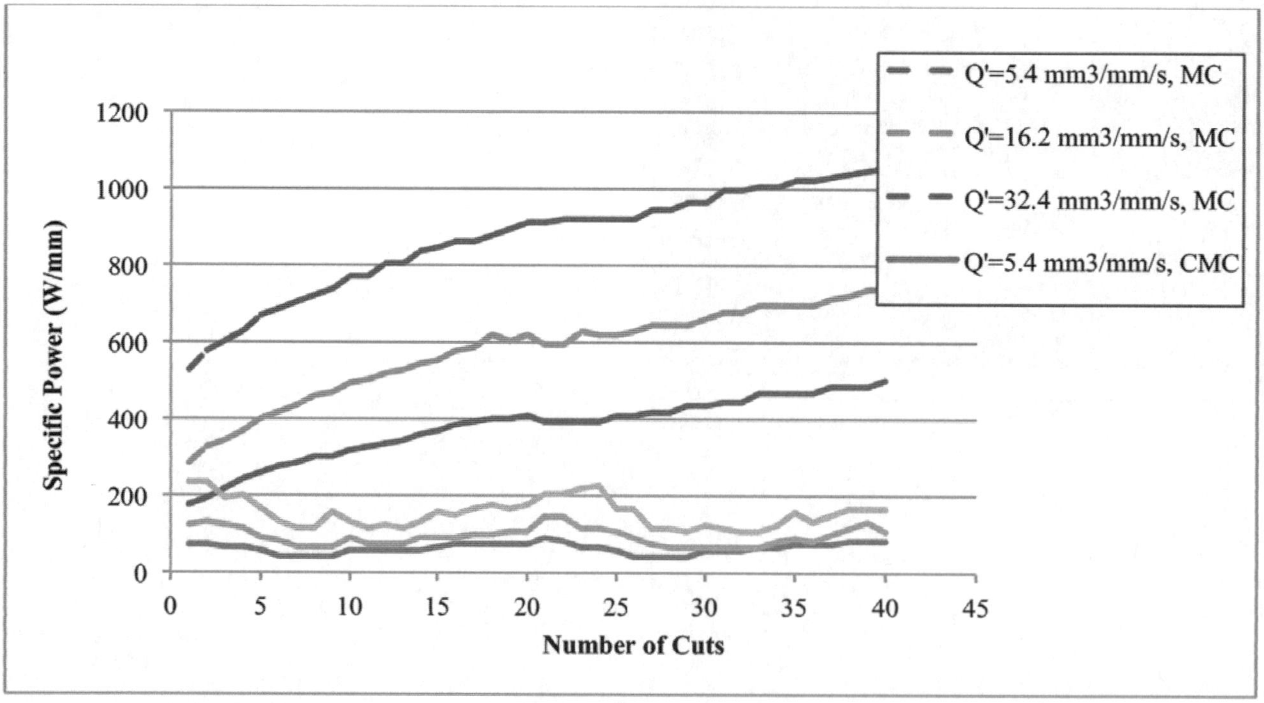

Figure 13: Variation of specific grinding energy with specific material removal rate

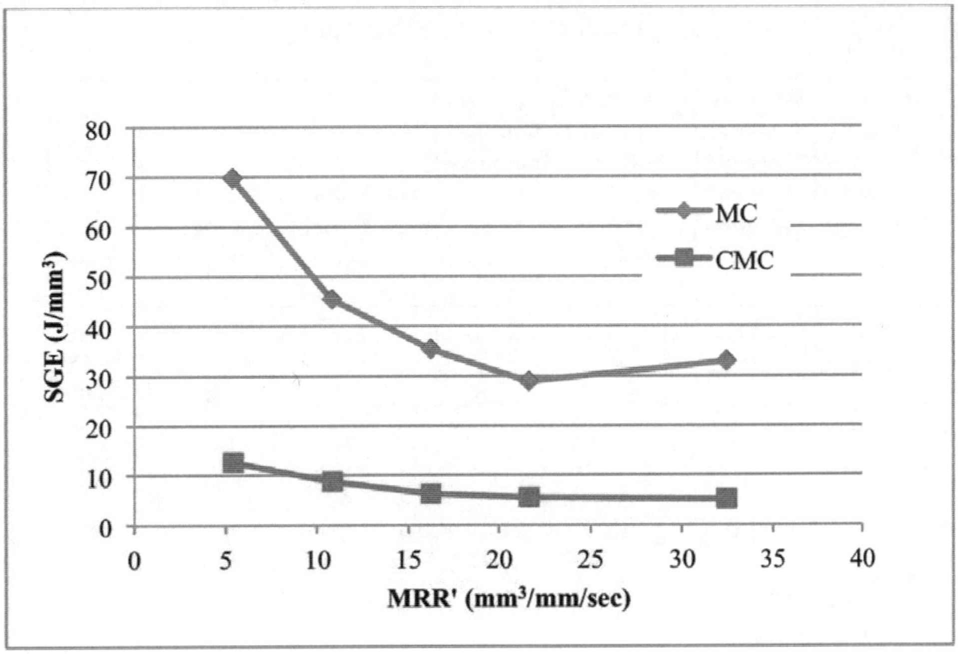

Figure 14: Variation of G-ratio with specific material removal rate

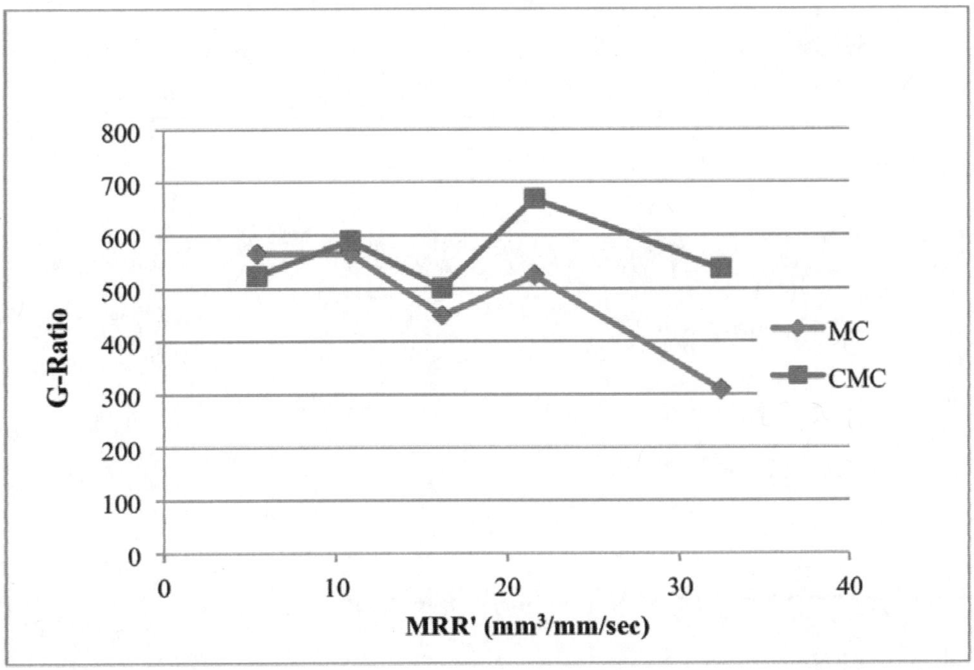

Tables

Table1: Summary of metallurgical analysis of ground IN-718 and Waspalloy

Sample	Q' (mm^3/sec/mm)	Maximum Depth of Longitudinal Surface Distortion (µm)	Maximum Depth of Transverse Surface Distortion (µm)	Maximum Depth of Longitudinal Strain Lines (µm)	Maximum Depth of Transverse Strain Lines (µm)	Longitudinal Imperfections (µm)	Transverse Imperfections (µm)	White Layer
IN718 - 3	193.5	15.2	5.1	30.5	15.2	Tearing 10.2	Tearing 5.1	none
IN718 - 6	193.5	10.2	10.2	25.4	45.7	Tearing 10.2	Tearing 7.6	none
Waspaloy - 4	161.3	5.1	5.1	15.2	20.3	none	Tearing 15.2	none
Waspaloy - 5	161.3	10.2	5.1	15.2	25.4	none	Gouge 15.2	none

Table 2: Summary of wheels and corresponding abrasive grit sizes tested

Wheel Type	**Abrasive Grit Size (avg. diameter)**
Vitrified-bonded SiC (hardness L)	60 grit (250 um)
Electroplated (EP) cBN	60/80 (250 um)
Dressable metal-bonded cBN	100/120 (151 um)
Electroplated diamond	60/80 & 100/120 (250 & 151 um)
Dressable metal-bonded diamond	100/120 (151 um)

Table 3: Summary of testing conditions

Machine Tool	
Machine	Elb Brilliant Surface/Creepfeed Grinder (7.5 kW spindle)
Grinding Mode	Slot grinding, non-continuous dress
Coolant	Trim E-812 (5%) water soluble 15.2 mm x 1.27 mm rectangular nozzle 45 liter per minute flow rate @ 0.413 MPa pressure
Dressing Conditions	
Dress Roll	Norton BPR 2203-20 (101 mm OD, 0.5 mm width)
Dress Speed Ratio	0.8
Dress Overlap Ratio	6
Dress Depth per Pass	0.013 mm
Workpiece Material	
Material	48-2-2 γ-Titanium Aluminide
Geometry	127 x 51 x 12.7 mm
Grinding Length	51 mm
Operational Conditions	
Wheel Speed	30 m/s
Material Removal Rates (MRR')	6.5 and 13.0 $mm^3/s/mm$
Depth of Cut	0.15 & 0.30 mm
Table Speed	42 mm/s

Table 4: Properties of the tested CMC and MC material.

Property	CMC Material	MC Material
Flexural Strength MPa (20°C)	379	395
Flexural Strength MPa (1000°C)	414	416
Tensile Strength MPa (20°C)	262	234
Tensile Strength MPa (1000°C)	300	240

Table 5: Material removal rate tested for the two materials

	Specific Material Removal Rate (mm^3/mm/sec)				
	5.4	10.8	16.2	21.6	32.4
Monolithic Ceramic (MC)	X	X	X	X	X
SiC-SiC CMC (CMC)	X	X	X	X	X

Table 6: Tested Wheel Characteristics

Wheel Specification	SD150R100B99-9/32
Wheel Shape	1A1R
Wheel Diameter (mm)	152.4
Wheel Width (mm)	0.89
Wheel Bore Size (mm)	31.75
Abrasive Depth (mm)	7.14
Usable Abrasive Approx. (mm)	5.1
Wheel Speed (m/s)	30
Wheel Speed (rpm)	3820
Depth per pass (mm)	0.254
Traverse Feedrate	Various rates tested

INNOVATED CUTTING TOOLS AND METHOD FOR EFFICIENT COMPOSITE MATERIALS MACHINING

Dr. Abdelatif Atarsia
Global Composite Industry Manager
YG-1 AMERICA

&

Shane Hollenbaugh
Executive Director of Sales
YG-1 USA

COMPOSITES IN AEROSPACE APPLICATION

COMPOSITES MACHINING CHALENGES

AUTOMATED MACHINING FOR ASSEMBLY

RELATED CUTTING TOOLS

CONCLUSION

INTERTECH 2019

COMPOSITES IN AEROSPACE APPLICATION

INTERTECH 2019

COMPOSITES IN AEROSPACE APPLICATION

From the Aluminum Design To the Composite New Technology

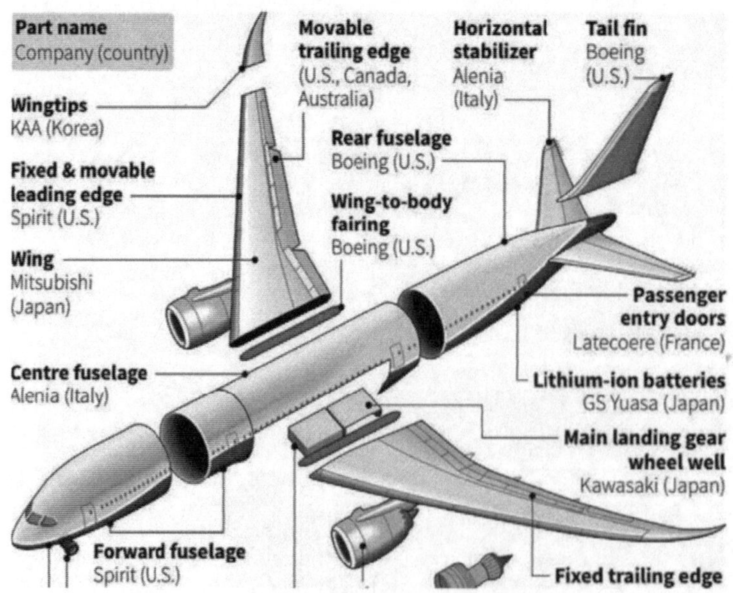

INTERTECH 2019

COMPOSITES IN AEROSPACE APPLICATION

WHY THE CHANGE FROM METAL COMPONENTS TO CFRP ONES ?

- WEIGHT REDUCTION
- PHYSICAL AND MECHANICAL BEHAVIORS
- REDUCE THE NUMBER OF PARTS
- REDUCE MAINTENANCE WORKS

INTERTECH 2019

COMPOSITES MACHINING CHALENGES

COMPOSITES MACHINING CHALENGES

DIFFERENT SPECS

MACHINING METALS

MACHINING COMPOSITES

- DIFFERENT MATERIAL
- DIFFERENT CUTTING DATA
- DIFFERENT SPECS

COMPOSITES MACHINING CHALENGES

SPECIFICATIONS OF COMPOSITE MACHINED PARTS THAT DO NOT EXIST IN METAL MACHINED ONES

- **Delamination**
- **Uncut fibers**
- **Surface finish based on the lay up sequence**
- **Resin burn**
- **Porosity**
- **Dry machined parts**

AUTOMATED MACHINING FOR ASSEMBLY

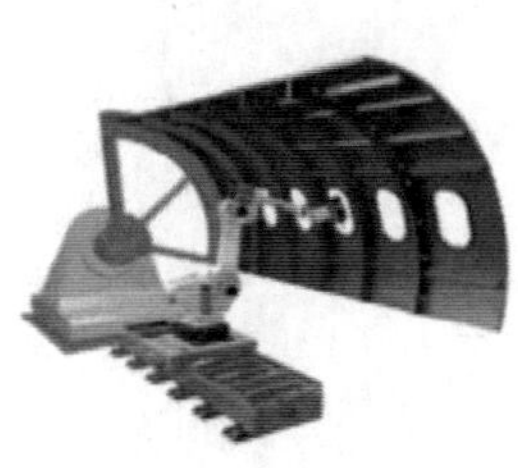

AUTOMATED MACHINING FOR ASSEMBLY

MACHINE TYPE & CHARACTERISTICS

- 3, 4 & 5 axis CNC Gantries
- High Spindles speed and high table feed
- Fixtures based on vacuum holding (POGOS) or bolting/ Toggles.
- Dry Cut with vacuum dust or coolant spray if necessary
- High accuracy machining

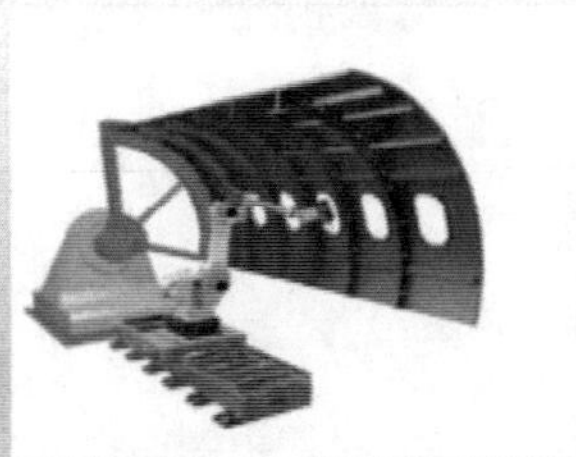

- Robot with Mill head or End Effector
- Lower accuracy compare to Gantries
- Limited in part size / need secondary equipment

AUTOMATED MACHINING FOR ASSEMBLY

FIXTURING SYSTEM

	- Designed for multipurpose applications - Cost effective and engineered to be friendly used - Risk of vibration induced during machining process
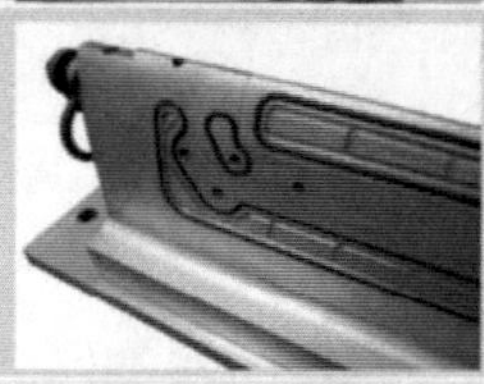	- Designed for single part shape - High cost per part investment - Better part holding and low vibration risk during process
	- Designed for single part shape - Time consuming process

RELATED CUTTING TOOLS

RELATED CUTTING TOOLS

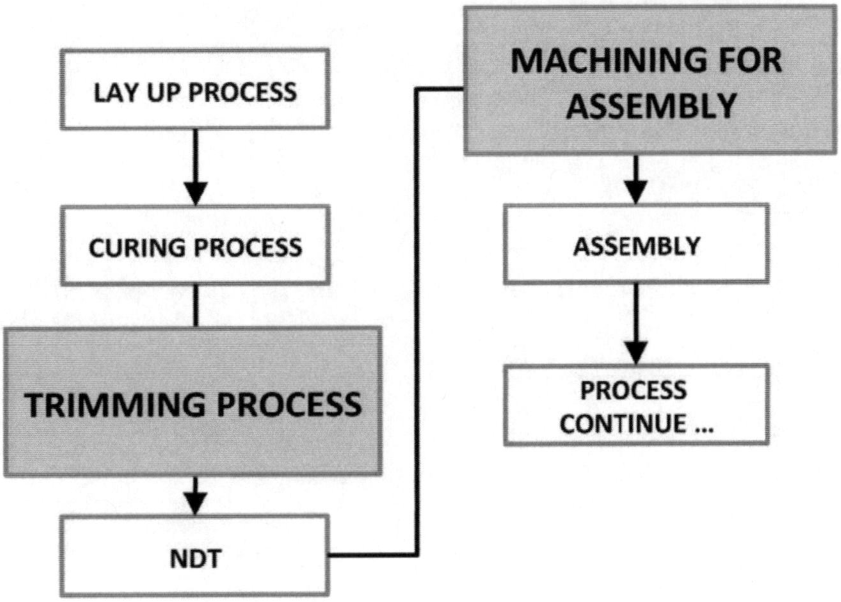

RELATED CUTTING TOOLS

WHY CFRP PARTS TRIMMING/ CUTTING IS IMPORTANT ?

- AEROSPACE COMPOSITE NET SHAPE PARTS DO NOT EXIST YET
- PARTS ASSEMBLY REQUIREMENTS (PROFILE AND QUALITY)

RELATED CUTTING TOOLS

CUTTING TOOLS FAMILIES

- CARBIDE TOOLS
- DIAMOND COATED TOOLS
- PCD TOOLS

RELATED CUTTING TOOLS

PROCESS TARGET (COST PER PART, QUALITY, TIME, ..)

- CUTTING TOOL
  - RAW MATERIAL
  - DESIGN
  - COATING
- MACHINING PROCESS
  - FEED & SPPED
  - COOLANT / NO

RELATED CUTTING TOOLS

CUTTING TOOL – RAW MATERIAL EFFECT

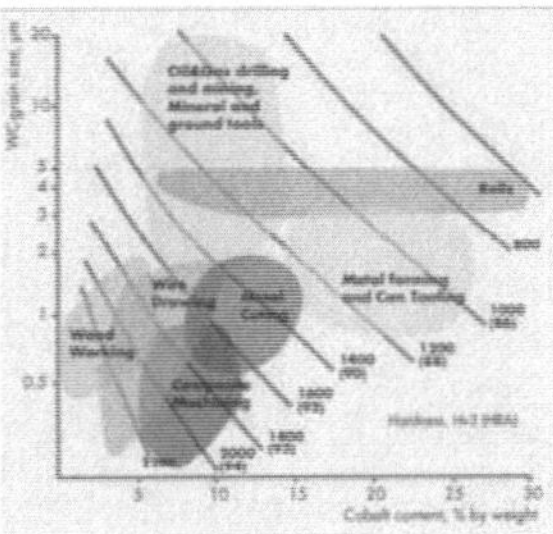

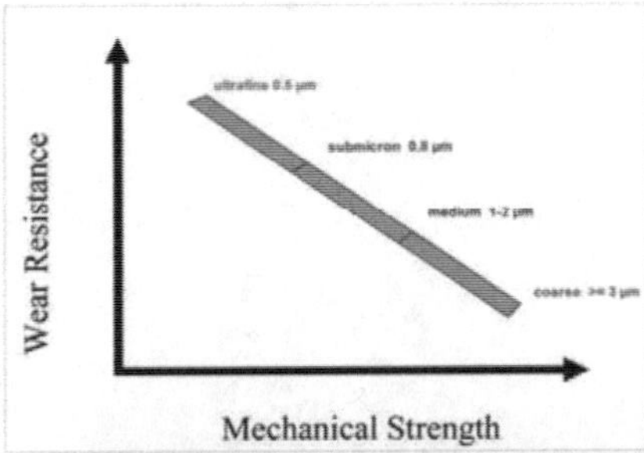

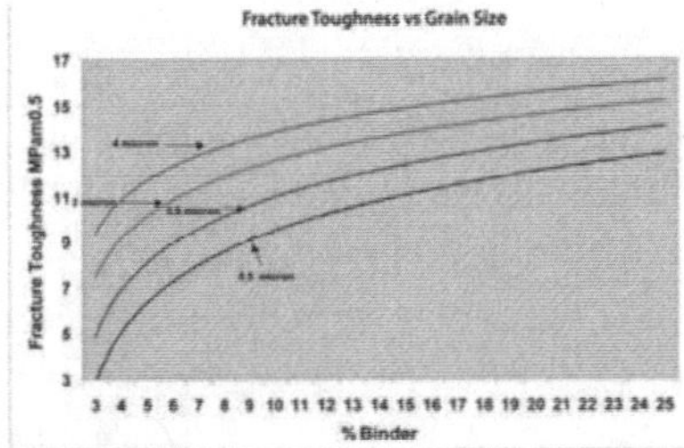

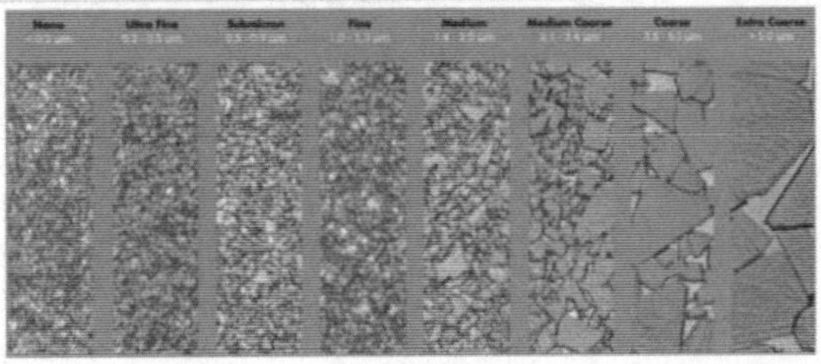

RELATED CUTTING TOOLS

CUTTING TOOL - DESIGN

- FLUTE HELIX PROFILE & ANGLE
- FLUTE DEPTH & TOOL CORE DEFINITION
- CUTTING EDGE DESIGN & RELIEF
- TOOL END DESIGN

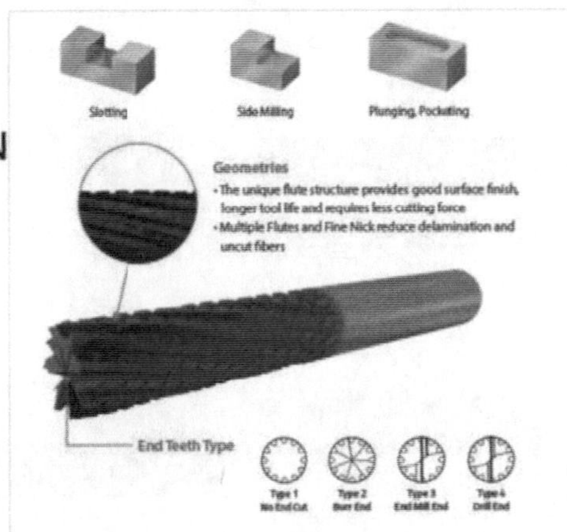

INTERTECH 2019

RELATED CUTTING TOOLS

DRILLING CFRP

If it's not perfect - it's not YG-1

INTERTECH 2019

RELATED CUTTING TOOLS

AEROSPACE STACK DRILLING (MOST USED STACK CONFIGURATIONS)

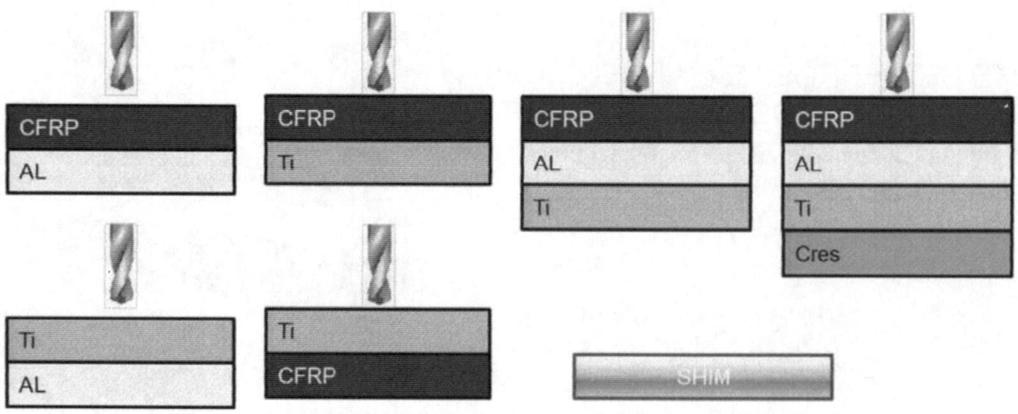

RELATED CUTTING TOOLS

TRIMMING CFRP

If it's not perfect - it's not YG-1

RELATED CUTTING TOOLS

TRIMMING CFRP

INTERTECH 2019

RELATED CUTTING TOOLS

TRIMMING CFRP

INTERTECH 2019

RELATED CUTTING TOOLS

INDEXABLE TOOLS

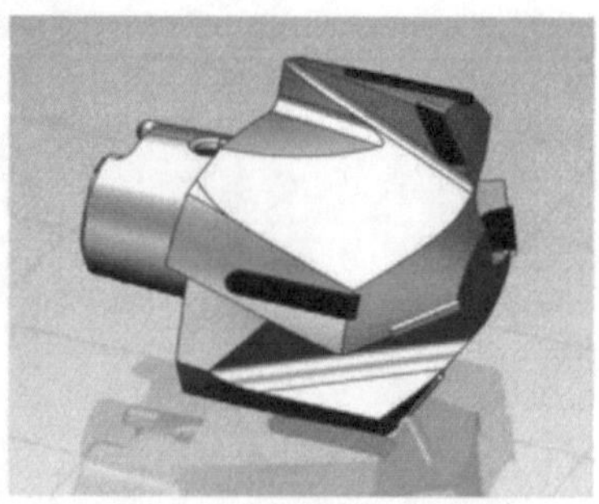

INTERTECH 2019

RELATED CUTTING TOOLS

INDEXABLE TOOLS

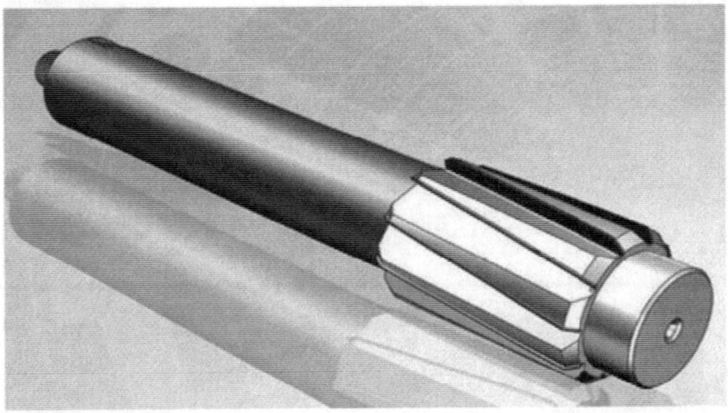

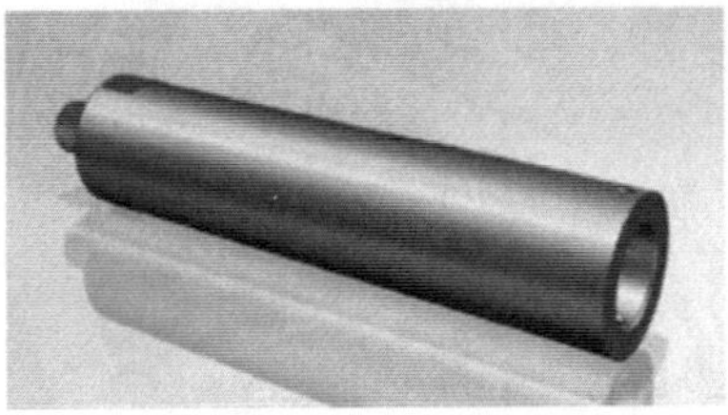

INTERTECH 2019

RELATED CUTTING TOOLS

MACHINING PROCESS

- FEED & SPPED
  - High Speed
  - High Feed

- COOLANT / DRY
  - Coolant is favorable for thick material to reduce overheat
  - Coolant needs evacuation managed process
  - Dry process is suitable for thin to medium thickness parts
  - Dry process induce heating which can be controlled by cutting data

CONCLUSION

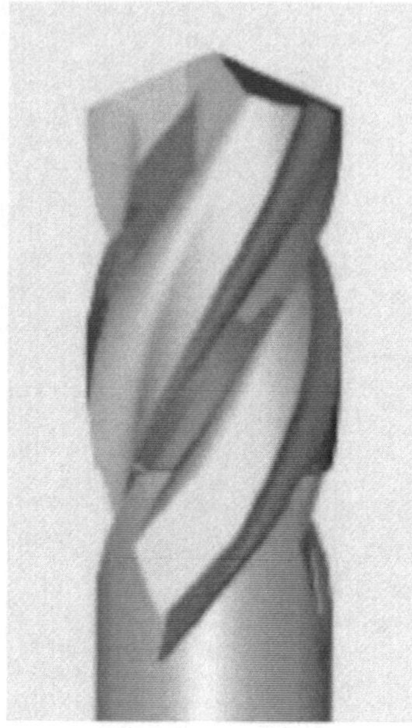

CONCLUSION

KNOWLEDGE IN CNC PROGRAMMING

KNOWLEDGE OF COMPOSITE MATERIALS

MACHINING CFRP

KNOWLEDGE OF CUTTING TOOLS

KNOWLEDGE OF COMPOSITE MATERIALS APPLICATIONS

THANK YOU

Dr. ABDELATIF ATARSIA
GLOBAL COMPOSITE INDUSTRY MANAGER
YG-1 AMERICA

Abdelatif.atarsia@yg1usa.com
+1 704 805 0193

INTERTECH 2019

Grit size, thermal damage, surface finish, dressing parameters and common mistakes when grinding hardened steel and nickel alloys with CBN grits.

Jeffrey A. Badger, Ph.D.
The Grinding Doc

HELPING COMPANIES ACHIEVE EXCELLENCE IN PRECISION GRINDING

THE GRINDING DOC

INTERTECH 2019

Abstract: An investigation was made into the relationship between grit size, specific energy and surface finish when grinding hardened steel and nickel alloys with CBN grits. It was found that, for a measured surface finish, there is a narrow general range of grit sizes that will achieve that finish with efficient grinding. Values below that can be achieved, but can only do so with dull dressing, resulting in higher specific energies (increasing the risk of thermal damage), higher grinding forces (increasing the risk of chatter and size-holding problems) and shorter dressing intervals (as a result of the CBN break-in period). Examples are given in industry of the common practice of choosing the wrong grit size and suffering the consequences. Practical recommendations are given.

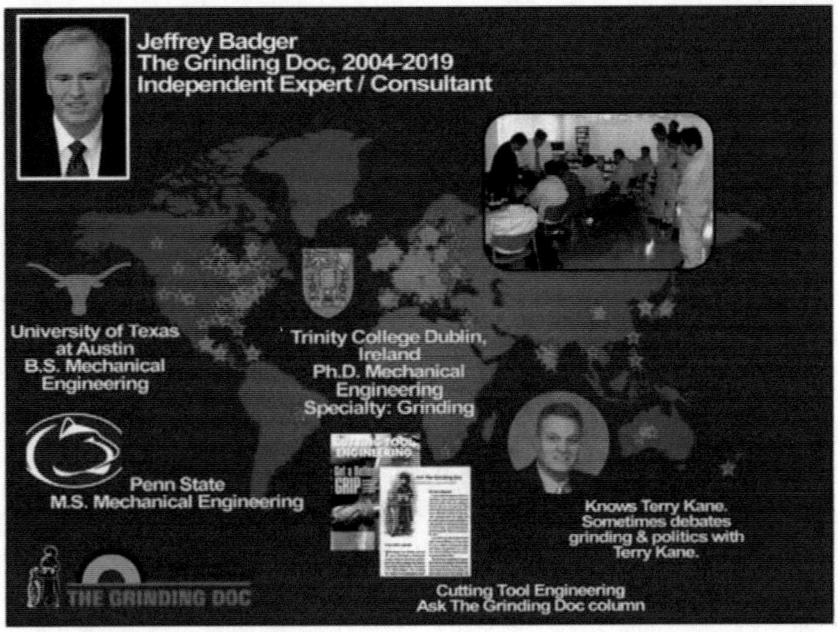

A little bit about The Grinding Doc.

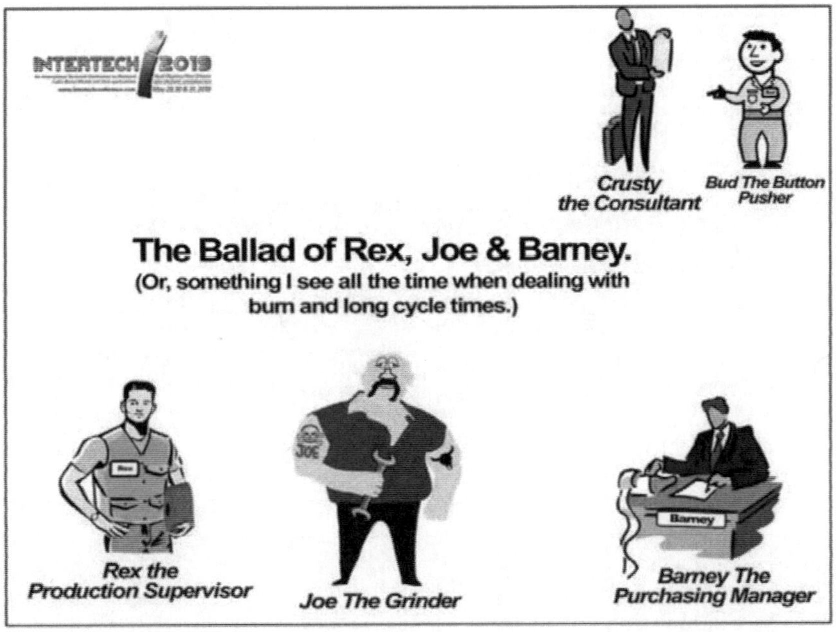

Here's something I see a lot when working with companies doing grinding, especially companies with grinding burn.

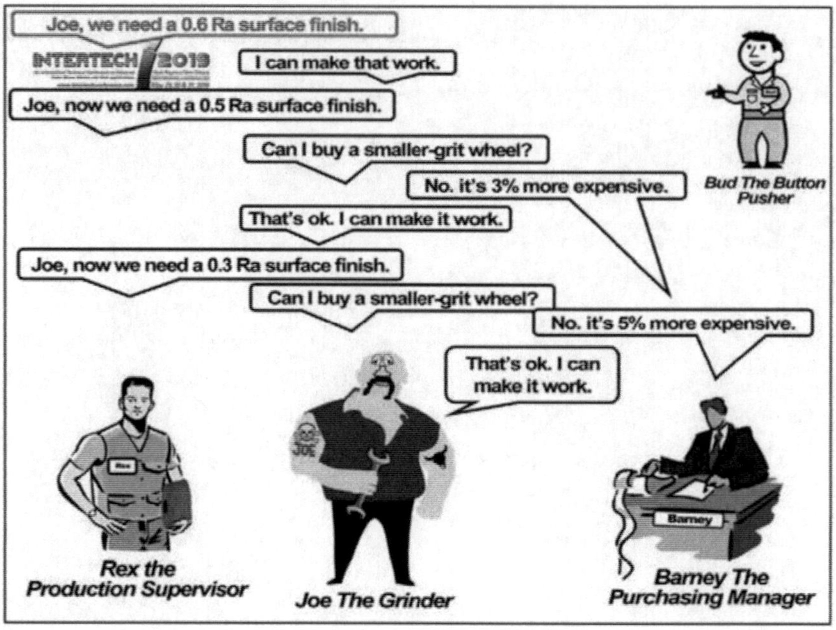

The company needs to get a certain surface finish. Then, over time, that surface finish requirement gets getting tighter and tighter (a lower Ra).

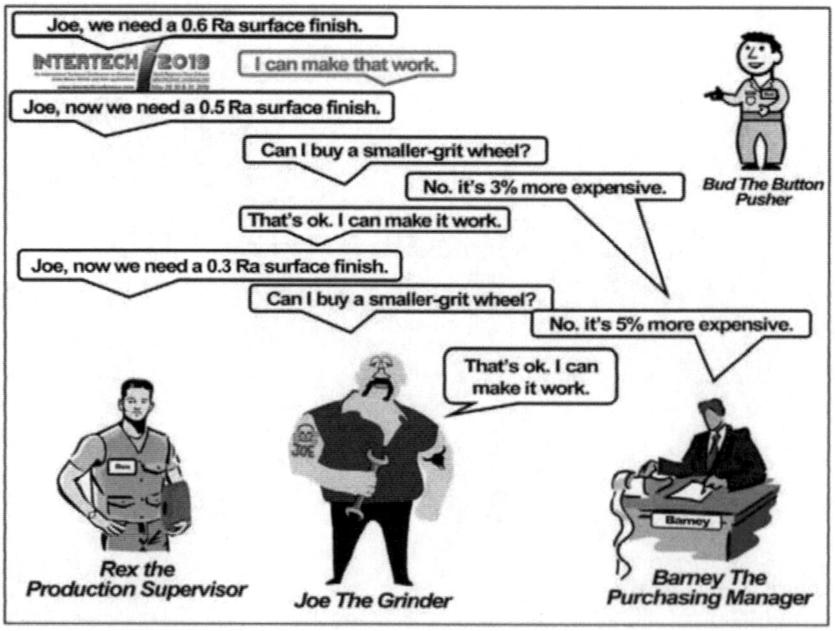

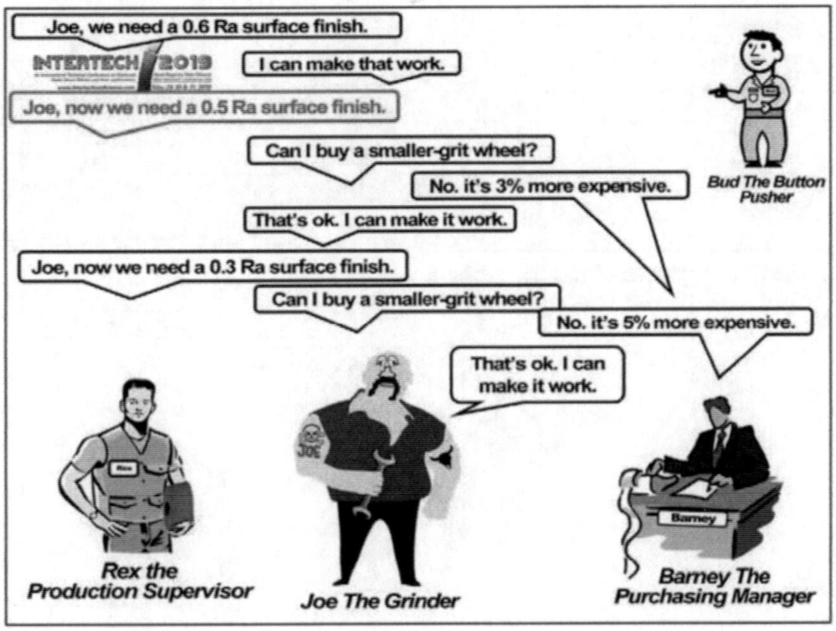
Joe, we need a 0.6 Ra surface finish.
INTERTECH 2019
I can make that work.
Joe, now we need a 0.5 Ra surface finish.
Can I buy a smaller-grit wheel?
No. it's 3% more expensive.
That's ok. I can make it work.
Joe, now we need a 0.3 Ra surface finish.
Can I buy a smaller-grit wheel?
No. it's 5% more expensive.
That's ok. I can make it work.
Bud The Button Pusher
JOE
Rex
Barney
Rex the Production Supervisor
Joe The Grinder
Barney The Purchasing Manager

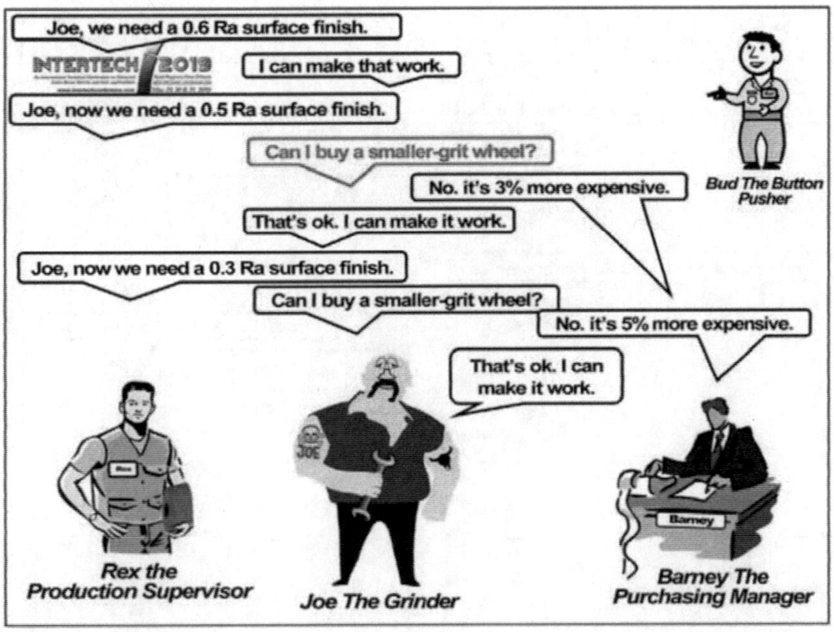
Joe, we need a 0.6 Ra surface finish.
INTERTECH 2019
I can make that work.
Joe, now we need a 0.5 Ra surface finish.
Can I buy a smaller-grit wheel?
No. it's 3% more expensive.
That's ok. I can make it work.
Joe, now we need a 0.3 Ra surface finish.
Can I buy a smaller-grit wheel?
No. it's 5% more expensive.
That's ok. I can make it work.
Bud The Button Pusher
JOE
Rex
Barney
Rex the Production Supervisor
Joe The Grinder
Barney The Purchasing Manager

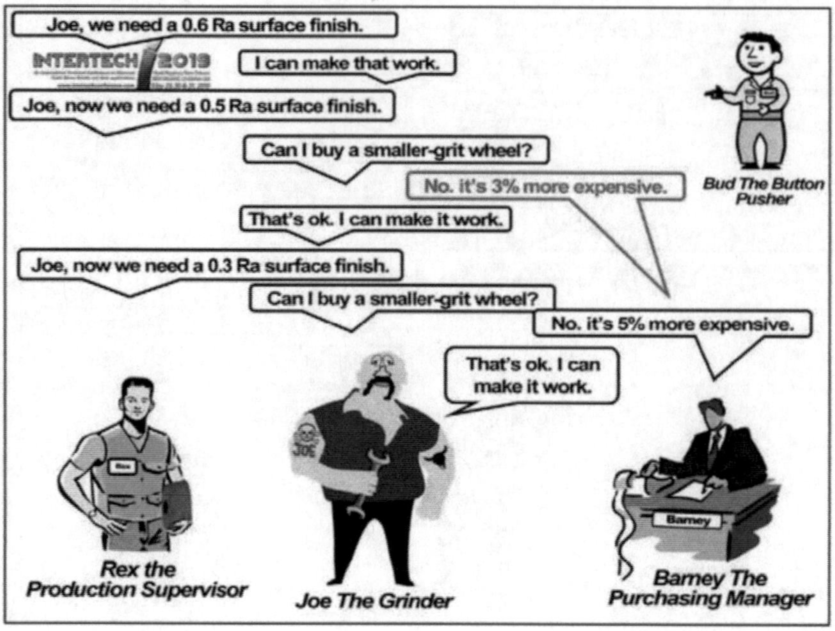
Joe, we need a 0.6 Ra surface finish.
INTERTECH 2019
I can make that work.
Joe, now we need a 0.5 Ra surface finish.
Can I buy a smaller-grit wheel?
No. it's 3% more expensive.
That's ok. I can make it work.
Joe, now we need a 0.3 Ra surface finish.
Can I buy a smaller-grit wheel?
No. it's 5% more expensive.
That's ok. I can make it work.
Bud The Button Pusher
Rex the Production Supervisor
Joe The Grinder
Barney The Purchasing Manager

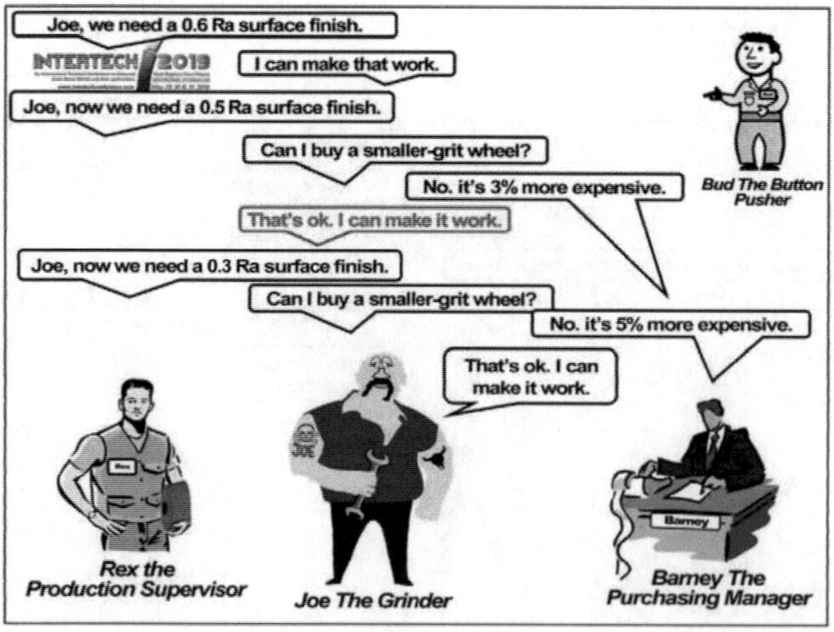
Joe, we need a 0.6 Ra surface finish.
INTERTECH 2019
I can make that work.
Joe, now we need a 0.5 Ra surface finish.
Can I buy a smaller-grit wheel?
No. it's 3% more expensive.
That's ok. I can make it work.
Joe, now we need a 0.3 Ra surface finish.
Can I buy a smaller-grit wheel?
No. it's 5% more expensive.
That's ok. I can make it work.
Bud The Button Pusher
Rex the Production Supervisor
Joe The Grinder
Barney The Purchasing Manager

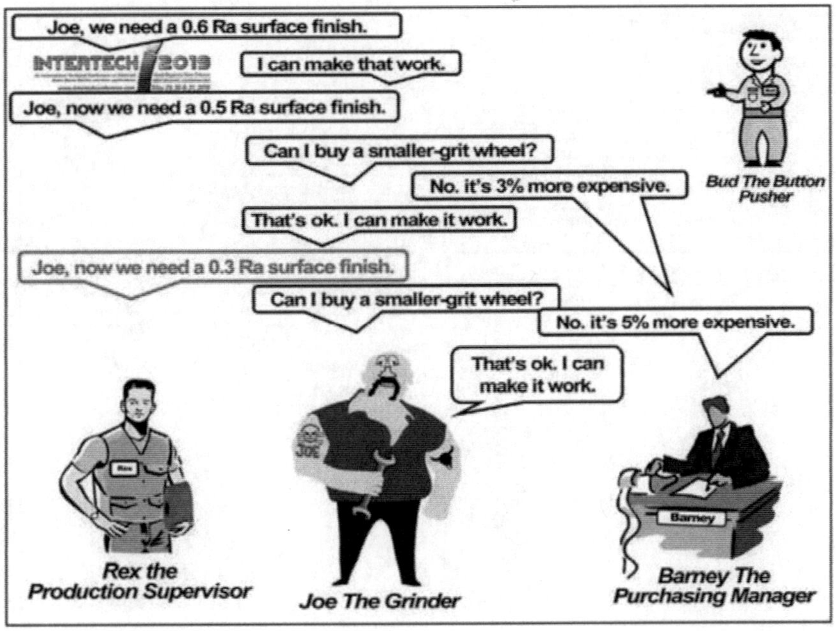
Joe, we need a 0.6 Ra surface finish.
INTERTECH 2019
I can make that work.
Joe, now we need a 0.5 Ra surface finish.
Can I buy a smaller-grit wheel?
No. it's 3% more expensive.
That's ok. I can make it work.
Joe, now we need a 0.3 Ra surface finish.
Can I buy a smaller-grit wheel?
No. it's 5% more expensive.
That's ok. I can make it work.
Bud The Button Pusher
Rex the Production Supervisor
Joe The Grinder
Barney The Purchasing Manager

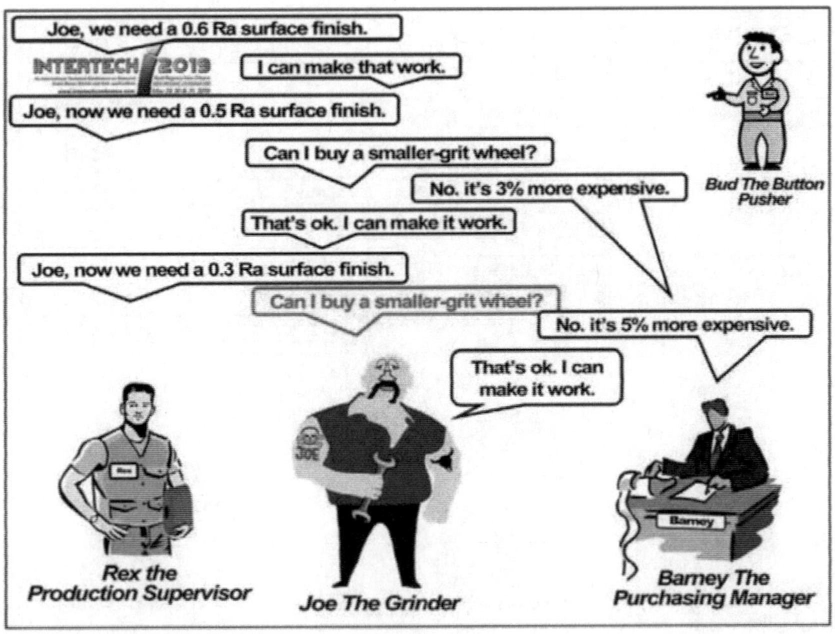
Joe, we need a 0.6 Ra surface finish.
INTERTECH 2019
I can make that work.
Joe, now we need a 0.5 Ra surface finish.
Can I buy a smaller-grit wheel?
No. it's 3% more expensive.
That's ok. I can make it work.
Joe, now we need a 0.3 Ra surface finish.
Can I buy a smaller-grit wheel?
No. it's 5% more expensive.
That's ok. I can make it work.
Bud The Button Pusher
Rex the Production Supervisor
Joe The Grinder
Barney The Purchasing Manager

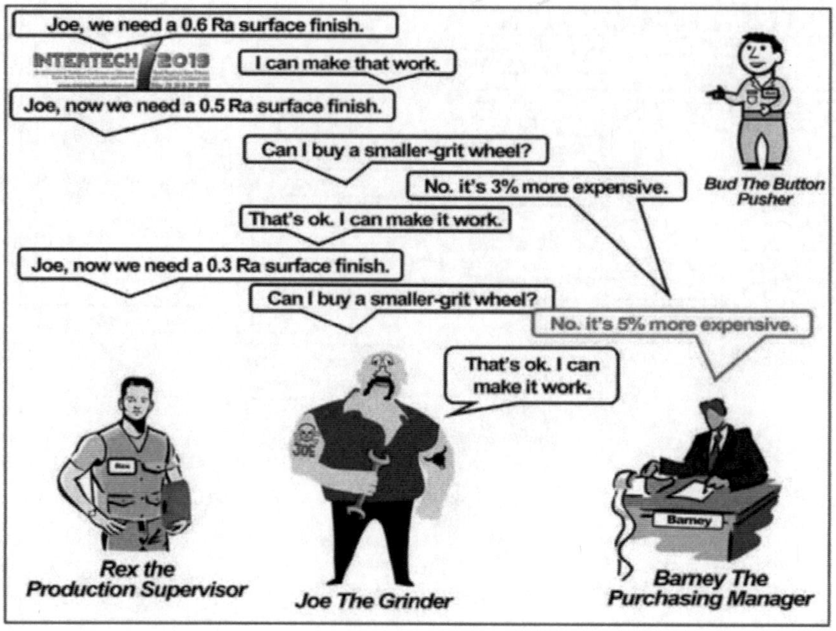

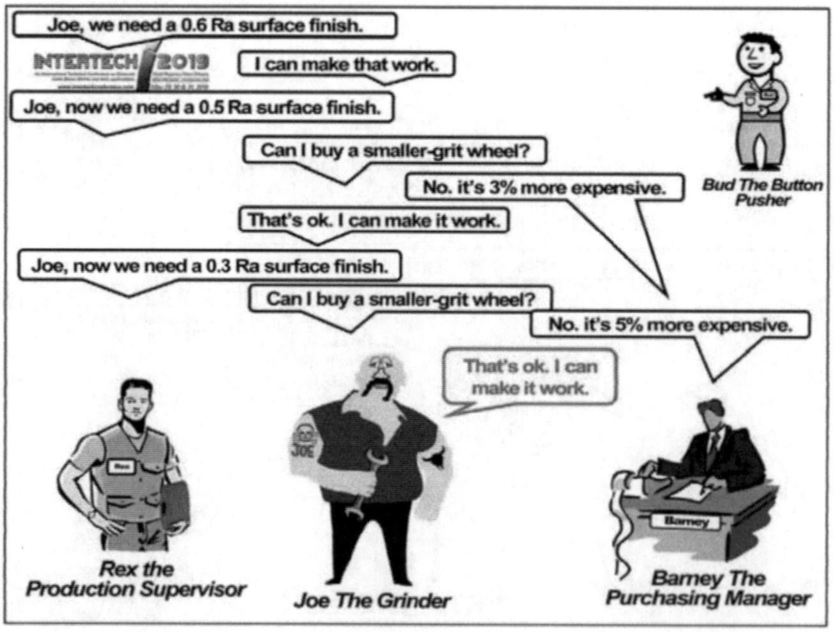

They should use a smaller grit size in the wheel. But that's a hassle: You have to order a special, smaller-grit wheel. So they just go with the wheel they have, and then dress it dull.

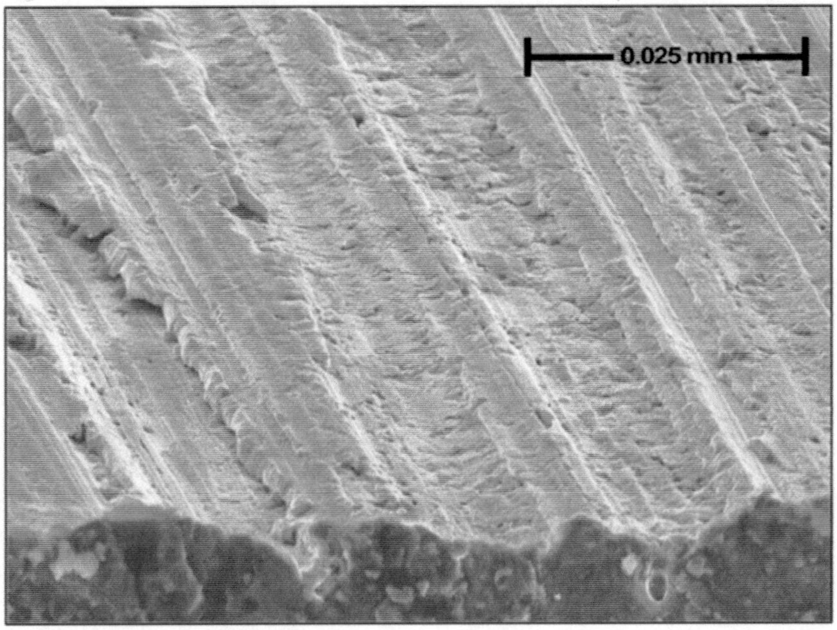

Here's a surface.

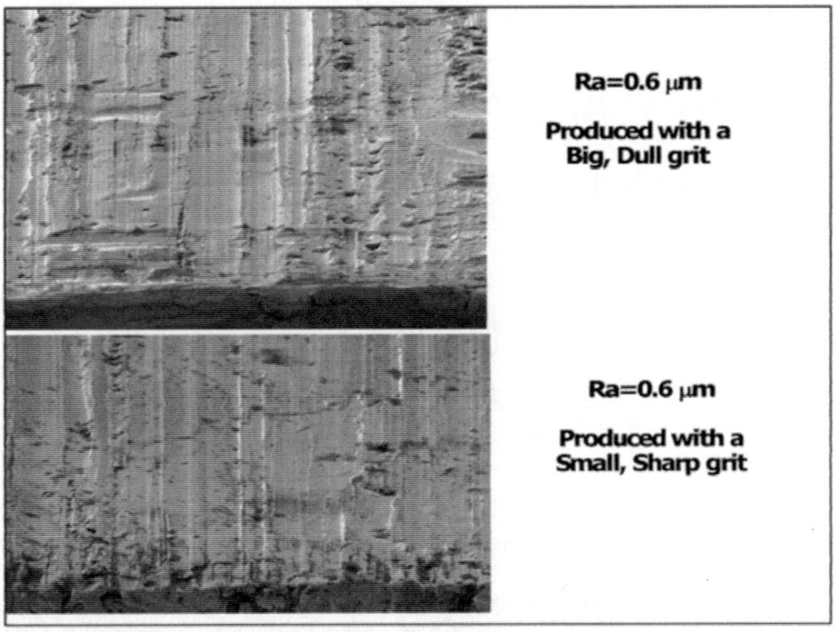

A given surface finish can be produced in two ways: a small-grit wheel dressed sharp; or a large-grit wheel dressed dull. Is there a difference?

First, conventional Abrasives

First let's look at conventional abrasives.

Grit Size to Achieve a Given Surface Finish

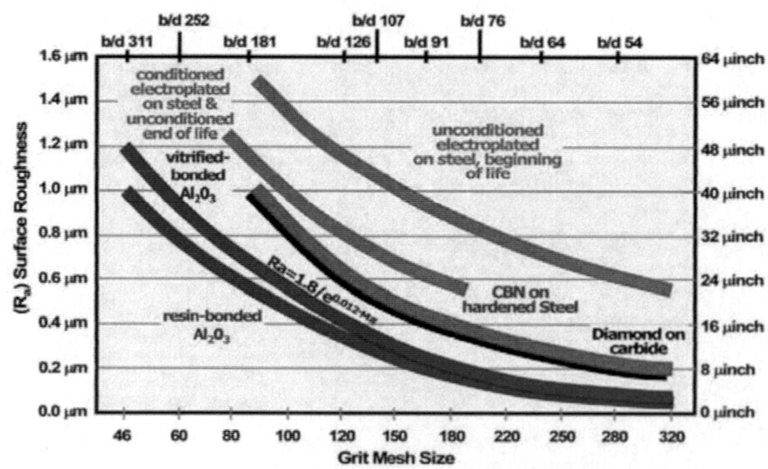

Excerpt from THE BOOK OF GRINDING

Here's a graph pulled from The Book of Grinding.

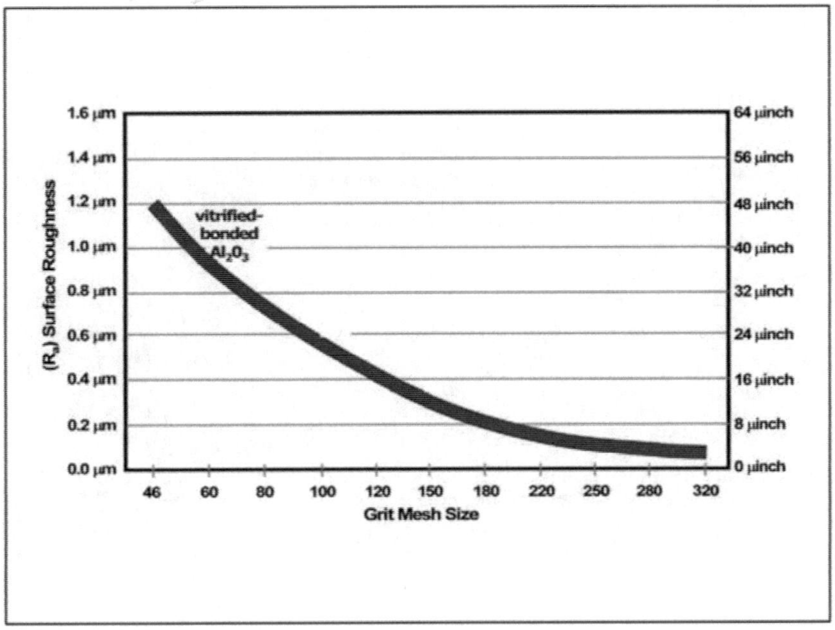

If we take aluminum-oxide, here are the mesh sizes required to produce a given surface finish assuming you have a sharp-dressed wheel. Of course, it depends on numerous factors – speeds and feeds, sparkout, coolant type, etc. – but this is still a ballpark curve for choosing the right grit size.

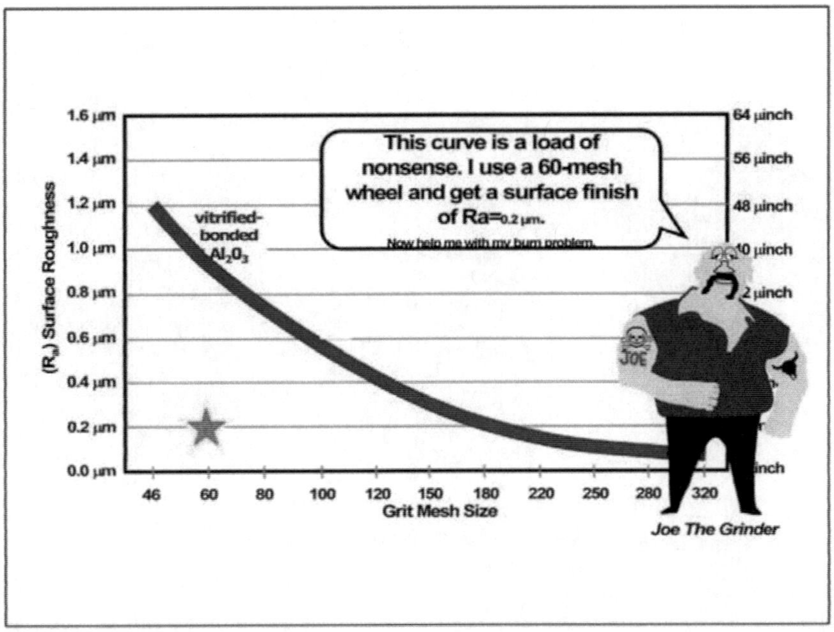

If I needed a 0.2 microns RA, I could do it. And I'd have to dress the wheel dull to achieve it.

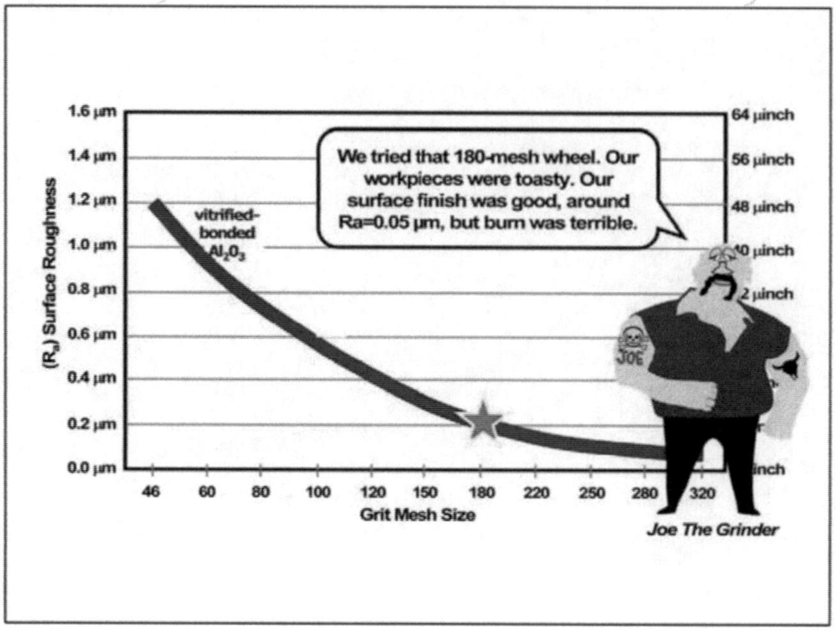

I could also achieve that 0.2 Ra with a small grit wheel dressed sharp. Is there a difference? Yes. The small-grit wheel will generate less heat.

Companies do this all the time: They choose too-large a grit size and dress dull.

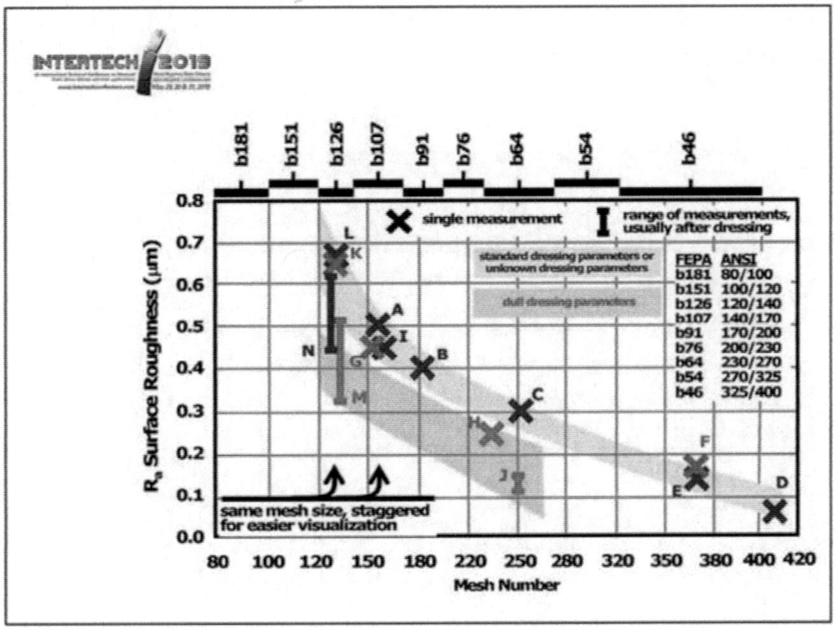

Over the years I've visited hundreds of companies doing grinding. And sometimes we measure surface finish. Here I have plotted measured values of Ra for different CBN grit sizes. I have also noted when the wheel was dressed dull.

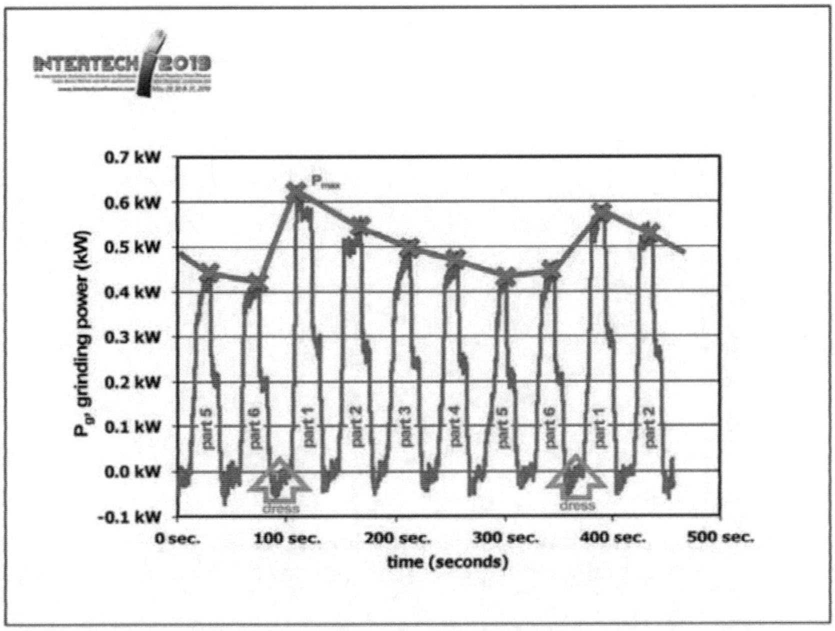

Here's a typical example. When you start grinding with a CBN wheel, the grinding power (here taken from ***The Grindometer***) is high but surface finish is good. Then, the wheel "breaks in" and power decreases to a near steady-state value. But what if you choose too-large a grit size. Then you have to dress your wheel before that steady-state period is achieved in order to get a low Ra. This is a bad situation.

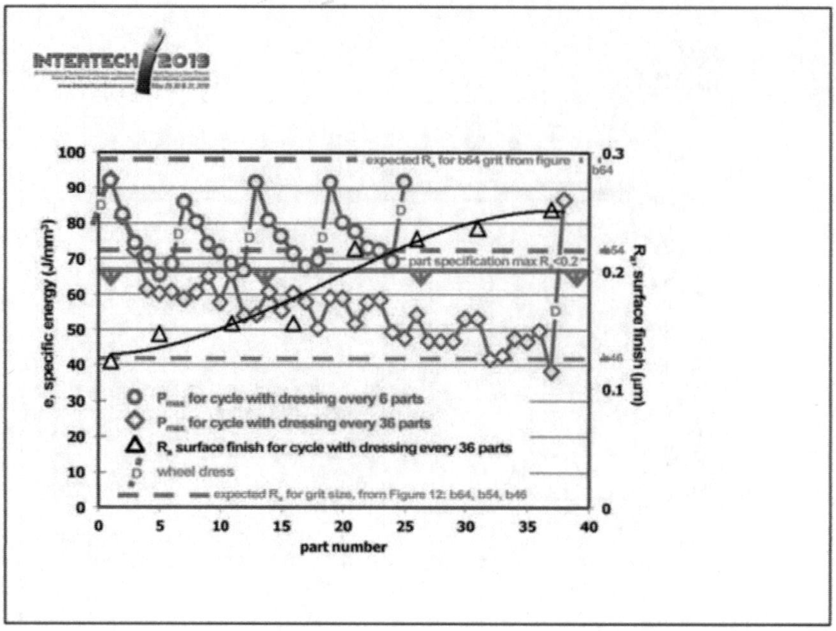

Here we've measured power and Ra. You can see how they needed to frequently dress to keep the RA below the specified value.

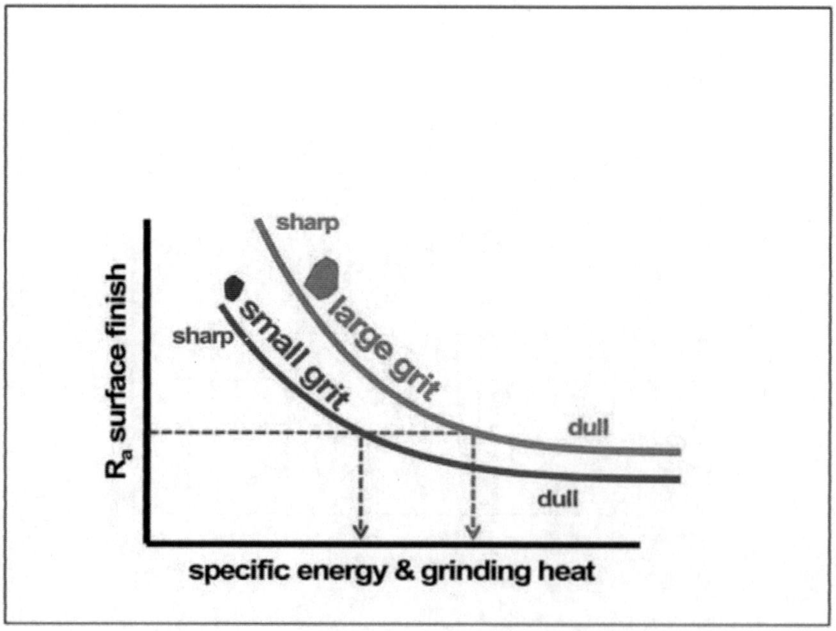

Here's a chart that illustrates the phenomenon well. In general: big-&-dull bad, small-&-sharp good.

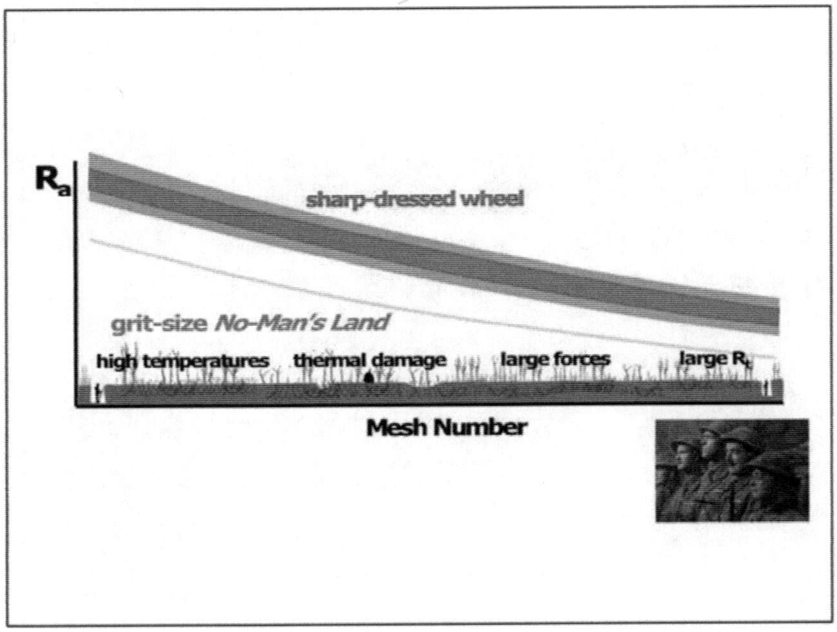

Whatever happens, you don't want to be in grit size No-Man's Land. It's a terrible place. Baldrick knows.

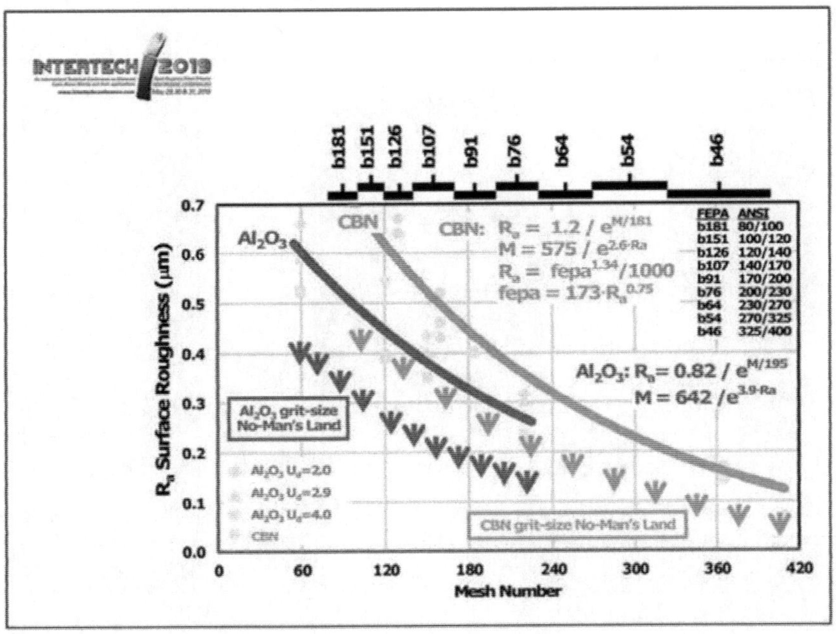

Here's a complete graph to tell you if you're in grit size No Man's Land (and probably dressing the wheel dull).

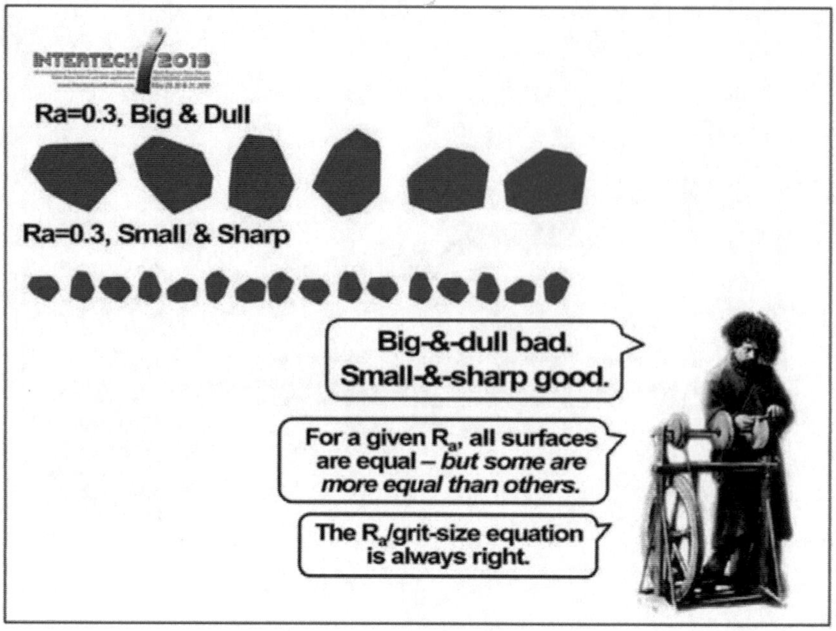

Repeat this mantra over and over again. Mantra's are good. Four legs good, two legs better!

It's all given in The Book of Grinding.

The Book of Grinding

- 4000 electronic pages
- Description with each page
- 20+ years, 20,000 hours
- A serious Investment for Serious Grinders

  $ US 4350

  € 3790

  £ 3

High Intensity Grinding Course

Upcoming Courses

October 8-10, 2019 **Grand Rapids, Michigan**
Hosted by SL Munson & Co. and IGA Abrasives

September 11-13, 2019 **Biberach, Germany**
Hosted by Vollmer **Focus on grinding of tungsten-carbide**

High Intensity Grinding Course, 2009, Hasle-Rüegsau, Switzerland. Hosted by Blaser Swisslube.

INTERTECH 2019

Grit size, thermal damage, surface finish, dressing parameters and common mistakes when grinding hardened steel and nickel alloys with CBN grits.

Jeffrey A. Badger, Ph.D.
The Grinding Doc

Thank you for your time!
Questions?

HELPING COMPANIES ACHIEVE EXCELLENCE IN PRECISION GRINDING
THE GRINDING DOC

Thanks for your time!

Tribology of Diamond Composite Bearings for Aerospace Applications

Kenneth W. Street, Jr.[1], Oleg A. Voronov[2], Bernard H. Kear[2] and Richard J. Mondry, II[3]

1. 140 Crystal Circle, Mooresville, NC 28117; kws1948@gmail.com
2. Diamond Materials, Inc., (Diamond Materials of Monroe), 23 Brahms Court, East Stroudsburg, PA 18301; oavoronov@aol.com
3. Gillcrest Electric, 570 Ternes Lawn, Elyria, OH 44053; rickrocks311@gmail.com

Abstract

Recent advances in high pressure-high temperature technology have led to improved Diamond and diamond-like materials for bearing applications. Diamond-composite discs made using this technology display exceptionally low friction and wear properties even under the dusty conditions encountered in a lunar environment. Diamond components can be synthesized near-net shape using high pressure-high temperature techniques, requiring minimal post processing to make the final component with toughness superior to Diamond itself and similar super-hardness.

Introduction

Carbon atoms can be well ordered into crystalline structures, or disordered in amorphous structures. Bulk carbon has two naturally occurring thermodynamically stable crystalline forms: graphite and Diamond. Graphite is stable at pressures <1 GPa and Diamond is stable at pressures >10 GPa. The range 1-10 GPa is the pressure range in which the standard thermodynamic potentials of graphite and Diamond are almost equal, with the equilibrium pressure depending on temperature.

Bulk carbon can consist of graphite-like or diamond-like crystallites. In some amorphous forms of carbon materials, or in regions of crystalline forms, carbon atoms may be disordered. Both crystalline forms exist as large single crystals or large polycrystals, as well as powders: sp^2-hybridization defines graphite-like material and sp^3-hybridization defines diamond-like material. It is not unusual for amorphous carbon to be represented by mixed sp^3, sp^2 and sp- hybridization.

We have developed a new type of bulk (monolithic) carbon material called "Diamonite". Diamonite is fullerene/nanotube-derived carbon, which is neither graphite-like nor diamond-like. It is almost as hard and tough as Diamond, and as thermally stable as graphite. When mixed fullerenes/nanotubes powders are subjected to pressure assisted polymerization, we call the product "Diamonite-B" to distinguish it from "Diamonite-A" where only fullerene, C_{60}, molecules are polymerized. "Diamonite-C" is a new superhard composite, where Diamond particles are uniformly dispersed in a Diamonite-B matrix. Diamonite-C is polishable to a mirror quality finish with unusual properties that are yet to be fully explored. The high hardness, low density and high-surface finish make this material suitable for aerospace (cosmic space) applications, e.g. lightweight precision bearings. High hardness is usually associated with low friction coefficient and low wear in bearings.

The Lunar environment consists of ultrahigh vacuum, modest to very low temperature (100 to -150°C, and as low as -185°C in the bottom of Shackleton Crater, a permanently shadowed area in the polar region) and a large quantity of very fine dust. Under these conditions unlubricated bearings are preferred as traditional liquid lubricants evaporate more quickly under vacuum and attract dust to the tribological contact area. In unprotected bearings, dust (grit) entering the contact area will cause additional wear when the hardness of the grit only slightly exceeds the hardness value of the bearing construction material.

In this paper we evaluate the tribological properties of Diamond and Diamonite-C materials in a number of relevant terrestrial and aerospace conditions. Coefficient of Friction (CoF) at several temperatures and atmospheric conditions is measured and wear is reported.

Experimental procedure

Diamond samples were made by transformation of graphite (powder or bulk) into Diamond at high pressure and high temperature (HPHT) [1] in the presence of Ni-base alloy catalyst. Single-crystalline Diamond powder were synthesized from graphite powder at P = 5.5-7 GPa, T = 1500-1600 °C, and t = 5-15 min. Polycrystalline monolithic Diamonds were made from monolithic pieces of graphite at P = 7-9 GPa, T = 1400-1500 °C, and t = 1-5 min. After grinding into powder, Diamond grains (~50-μm size) were separated using sieves. Admixtures of graphite and metallic alloy were removed by boiling in perchloric acid. The resulting high purity Diamond powder was consolidated into fully dense shaped samples of polycrystalline Diamond at P = 8-9 GPa, T = 1600-1800 °C, and t = 1-3 min.

In order to fabricate disc-shaped samples of Diamonite-C composite, Diamond grains were mixed with fullerene/nanotube powder, and subjected to pressure assisted treatment at P = 1-2 GPa, T = 500-900 °C, and t = 5-60 min. Under these conditions, the fullerene/nanotube mixture polymerizes, forming a very hard matrix that incorporates superhard Diamond grains. Thus, monolithic pins and discs of Diamonite-C were produced.

The discs and pins were ground and polished to a mirror surface with Diamond tools for further investigation. A tribological study was performed using an ultrahigh vacuum (UHV) pin-on-disc tribometer (PoD), a tilting plane tribometer and a dust PoD tribometer.

Diamond making procedure

A schematic of the HPHT apparatus [2] used to fabricate Diamond and its composites is depicted in Figure 1. The apparatus can be operated in the very high pressure range (P = 1-10 GPa), and in the high pressure range (P = 0.1-1 GPa).

Figure 1a. Apparatus for HPHT research and samples fabrication

I-hydraulic press (10-10,000 tf), II-HP unit, III- electrical unit (10-100 kW); 1-frame of press, 2-ram, 3-oil cylinder, 4-oil reservoir, 5-oil pump, 6, 7-oil pump valves (delivery and exhaust valves (flaps) of oil pump); 8-release valve, 9-oil pressure gauge, 10-silica fabric electrical insulation layers; 11-steel plates; 12-concrete base, 13-copper cables, 14-shunt, 15-mV-meter for electrical current value; 16-V-meter for voltage, 17-cooling device for steel plates 11 and HP unit II.

Figure 1b. HP Unit (position II on Figure 1a) for very high-pressure range P = 1-10 GPa

---

[1] O.A. Voronov, and B.H. Kear "Method for Rapidly Synthesizing Monolithic Polycrystalline Diamond Articles", Patent US 8,021,639 B1, 2011

[2] O.A. Voronov "High Pressure and High Temperature Apparatus", Patent US 6,942,729 B2, 2005

1-heater with sample; 2-lava stone container; 3-rubber ring; 4-profiled anvils; 5, 6-bearing inserts; 7-13-supporting steel rings; 14-16-plate levels; 17-rotational axis of symmetry; 18-mirror plane of symmetry.

The design of the HP Unit (position II on Figure 1a) for the very high-pressure range P = 1-10 GPa is shown in Figure 1b. This is called a "profiled anvils" design.

Profiled Anvils Design

In this design, the bearing plates on levels 14 and 15 of Figure 1b are joined to plates (11) of Figure 1a and are electrically insulated by silica fabric (10) from the frame (1) of the hydraulic press. The level 16 plates with profiled anvils (4) of Figure 1b, supporting rings (7-9), graphite heater (1), lava stone container (2) and rubber ring (3) (optional used for safety purposes) can be easily changed and replaced by another. This laboratory design, can also be used for industrial applications, where rapid and semi-automatic manufacturing of items is required in fast turnaround for HPHT treatments.

Profiled anvils (4) and inserts (5) of Figure 1b are made from tungsten carbide-cobalt (WC/Co) hard alloy, and insert (6) and pre-stressed supporting rings are made from hardened steel. Outer rings (9, 11, 13) are softer than inner rings (8, 10, 12). Inserts (6) and rings (7) are hardened to the highest achievable hardness. Round surfaces between rings (not shown in simplified figures) are conically shaped. Anvils (4), inserts (5, 6) and rings (7-13) are pressed into each other, so they are in pre-stressed condition. Such a design allows achievement of a pressure range that far exceeds the compressive strength of hardened steel, hard alloys and all known materials, and is required for transformation of graphite into Diamond. Limiting pressure is determined by the hardness of the anvil material, not by its strength. Dependence of pressure (P) on loading force (F) can now be approximated by the simple formula:

$$P = (F-F_0)/(S_0+F/H_V) \quad (1)$$

where S_0-is the area of the anvil's cavity, H_V-is the hardness of the anvils' material on the Vickers scale, F_0-is a friction force and also the force required to deform the lava stone (or clay) container causing its flow into a slit between the anvils. The F_0 is small in comparison with F and can be neglected.

The dependence of pressure and force can be calibrated using the established phase transitions in cerium (Ce) and bismuth (Bi). The dependence of temperature on electrical power can be calibrated using thermocouples and known melting points of certain metals under high pressure.

The following transformations can be performed in the P = 1-10 GPa pressure range with the previously described unit design:

1. Transformation of graphite powder into Diamond powder in the presence of Ni-base alloys. It requires approximately 15 min of holding at P = 5-7 GPa, and T ~ 1500 oC for Diamond crystals to grow up to 1 mm size from multiple points of recrystallization;

2. Rapid transformation of graphite ceramic pieces into monolithic polycrystalline Diamonds in the presence of Ni-base alloys. It requires approximately 1-3 min of holding at P = 7-10 GPa, and T = 1200-1500 oC to transform graphite phase carbon into monolithic Diamond phase carbon of about 3-5 mm size from one point of recrystallization;

3. Sintering of Diamond powder into monolithic polycrystalline Diamond;

4. Pressure Assisted Sintering of powders, or mixtures of powders into monolithic pellets.

The advantages of this design include the possibility of achieving pressure close to the value of hardness in a relatively large volume and the possibility of rapid turnaround in industrial applications.

Figure 2 illustrates moderate pressure (I), high pressure (II) and very high pressure (III) ranges for which the equipment was designed. Pressure is represented on a logarithmic scale (log P/P_0, where P_0 is ambient pressure). Regions 1 and 2 on this T vs. log P/P_0 diagram are the regions where Diamond samples were made. Region 1 is where single-crystalline Diamond materials are made from graphite powder in the presence of Ni-base alloys, namely crystals, grains and powders. Region 2 is where poly-crystalline Diamond materials are formed: where larger pieces of graphite ceramic can be rapidly transformed into monolithic polycrystalline Diamonds with the help of Ni-base alloys, or Diamond powder can be sintered into large monolithic polycrystalline Diamonds. The line 3 is the boundary where the thermodynamic potentials of graphite and Diamond are equal. This line separates the regions of thermodynamic stability of graphite and Diamond (if the surface energy of the grains is much smaller than the volume energy).

The Ultrahigh Vacuum Pin-on-Disc, UHV-PoD, tribometer [[3]] used in the investigation was mounted in a vacuum chamber capable of reaching ultrahigh vacuum, UHV, conditions (Figure 3). Unidirectional sliding friction experiments were conducted in ambient air (T = 21-23 °C, relative humidity, ~18-20 percent). Vacuum sliding experiments were performed at 3.7×10^{-6} to 2.9×10^{-5} Pa and vacuum cryogenic experiments were performed at 6.0×10^{-3} to 7.7×10^{-3} Pa. Sliding experiments carried out with a liquid nitrogen shroud placed about the disc holder achieved a temperature of -100 °C. Note that the shroud must not contact the specimen holder, hence the specimen often does not achieve the same temperature as the liquid nitrogen in the shroud. Additional experiments were performed under a blanket of CO_2 (27.6 kPa above atmospheric pressure). Most experiments were conducted with ~3-6 mm diameter pins having a tip curvature of radius ~3-10 mm composed of similar material to the discs of ~6-16 mm diameter. All experiments were conducted with a load of 0.98 or 1.96 N using a 6, 8 or 12 mm track diameter at a constant sliding speed of 0.038-0.075 m/s. Sliding speed varies between different experiments and is a function of track diameter and RPM. The friction force was continuously monitored during the sliding friction experiments and expressed as Friction Coefficient in the data output. The sliding wear was determined by measuring the depth and width of the wear track on the discs at 4 locations approximately equidistance from each other and by measuring the flat worn into the curvature of the pins by optical interferometry.

Figure 2. Diagram of HP ranges: I- moderate pressure; II- high pressure; III- very high pressure.

1. Single-crystalline Diamond materials (solid pink): crystals, grains, powders made from graphite powder in the presence of Ni-base alloys;

---

[3] Miyoshi, K., 1998. Lubrication by diamond and diamond like carbon coatings, J. Tribology, 120: 379–384.

2. Polycrystalline Diamond materials (blue hatch): monolithic polycrystalline Diamonds made from polycrystalline graphite ceramic by rapid transformation in the presence of Ni-base alloys, or polycrystalline Diamond compacts sintered from Diamond powders;

3. Line of equal thermodynamic potentials of graphite and Diamond.

Tribometry procedures

The components, tribological couples, fabricated for research on the UHV-PoD tribometer are shown in the drawings of Figure 4 a) and b). These drawings show the guide plate and pin geometries and dimensions required to determine a value for sliding friction in motion and wear for Diamond against Diamond and Diamonite-C against Diamonite-C at the conditions described above.

A second PoD, a dust PoD, was used to evaluate performance of tribological couples (Diamond and Diamonite-C pins vs. steel discs) in dust simulating operation of unlubricated, exposed bearings in the Lunar and Mars environment. This PoD setup was used in the 0.1–10 m/s speed range and was calibrated at $\upsilon \sim 1$ m/s. The range of loads was 0.1-10 N (~10 gf-1 kgf on the sliding samples). It was calibrated for "most heavy loads", and the velocity of ~1 m/s for aluminum-aluminum (mirror polished) couples. This unit was used only in ambient air for unlubricated clean tribocouples and with the addition of lunar dust simulant, JSC-1AF, or sand.

The relationship necessary for the dust pin-on-disc experiments were as follows: Coefficient of friction in motion (K_M) can be expressed as:

$$K_M = (\eta \Delta W)/(\upsilon F), \quad (2)$$

where F is the normal force or load, υ is velocity, η is the coefficient of efficiency for the tribometer and ΔW is the increase of electrical motor power required to rotate the sample while under load. Then $\Delta W = V\Delta J$ if voltage V is constant and ΔJ is the increase of current. The value of ΔW, as well all other values, can be measured with very high precision. This set up is a very reliable and simple tool to measure the sliding friction in any bearing in very a wide range of speed, load and cycling conditions. Most importantly, it can be used to evaluate the effect of dust on friction and wear which is not possible using the UHV-PoD or the tilted plane methods. Once calibrated it can measure the dependence of friction on velocity, load (pressure), temperature, environmental parameters, including dust, and a variety of other conditions.

The sliding friction coefficients were also measured by the tilted plane method in addition to the pin-on-disc method. In the tilted plane method, if α is the angle of a tilted plane when the sample just begins to move, the coefficient of friction at rest, K_R, is equal to $\tan(\alpha)$. When the sample slides smoothly on a flat sheet tilted at angle β, the coefficient of friction in motion K_M can be evaluated from

$$K_M = \tan(\beta) - 2L/(gt^2\cos(\beta)), \quad (3)$$

where L is the length of the sliding sample path, t is the time interval of sliding through L, and $g=9.8$ m/sec^2. The tilted plane method was used in the 0.1 mm/s – 1 m/s sample velocity range.

Results and Discussion

The UHV-PoD tribometer results can be summarized in Tables 1 and 2 and the results of the tilted plane and dust PoD methods are summarized in Table 3.

Abrasion continues to become more pronounced to the point at which the hardness of the abrasive on the Vickers or Knoop scales approaches a value of 1.5 times that of the bearing material where the ability to abrade levels off [4]. Severity of abrasion for metals, ceramics and other regular solid materials can be gauged by comparing the hardness values of the abrasives, the lunar regolith constituents, to the engineering materials [[5]]. Some of the constituents of Lunar regolith have hardness values approaching 8.0 on the Mohs scale with sharp irregular edges and points, consequently, synthetic Diamond and Diamonite-C are attractive bearing materials in that they can be produced by HPHT technology in near net shape, resist wear by abrasive dust, have low density (no porosity, lightweight materials - a weight savings) and can perform unlubricated.

Figure 3. Schematic of an ultrahigh vacuum pin-on-disc (UHV-PoD) tribometer

Figure 4. Specifications for tribological couples used in the UHV-PoD tribometer: a) Disc; b) Pin. Pins and Discs were constructed out of Diamond and Diamonite-C

Table 1. Coefficient of friction for unlubricated sliding bearings in ambient, and simulated Lunar and Mars Environments from UHV-PoD tribometry

Pin Composition	**Disc Composition**	**Condition**	**Average CoF, μ**
Diamonite-C	Diamond	air	0.034
Diamonite-C	Diamonite-C	air	0.01
Diamond	Diamond	air	0.001
Diamonite-C	Diamonite-C	Vac	0.12
Diamond	Diamond	Vac	0.29
Diamonite-C	Diamond	Vac/Cryo	0.106-0.142 *
Diamonite-C	Diamonite-C	Vac/Cryo	0.084-0.148 *

---

[4] Tylczak, J.H., "Abrasive Wear," in ASM Handbook, **18**, Friction, Lubrication, and Wear Technology, edited by S.D. Henry, ASM International, Cleveland, OH, 1992, pp. 187-189.

[5] "Expected Mechanical Characteristics of Lunar Dust: A Geological View." D. Rickman and K.W. Street, Proceedings of the Space Technology and Applications International Forum, Albuquerque, NM 2/12/08. Am. Inst. Phys. Conf. Proc., 969, 949-955 (2008)

Diamond	Diamond	CO_2 (27.6 kPa)	0.002-0.028*

* = unsteady friction coefficient

In the experiments described herein lunar dust simulant JSC-1AF (a Mare type simulant), where F denotes the Fine particle size range. Although the composition of Mare Soil does not exactly coincide with the simulant composition, it is believed at the present time that this simulant provides a good approximation for abrasion and mechanical studies. Figure 5 shows the sharp and pointed edges of the material and the relative large quantity of fine particles. Actual Lunar mare contains the largest amount of particles in the nm range with the mean particle size of approximately 100 nm. It is interesting to notice that the element carbon, so abundant on Earth and being a basic element of all life in air, water and soil, was never found in the samples taken from the surface of Moon. Similarly, hydrogen was not found in any lunar samples as well, although it is postulated that a significant quantity of water exists on the moon, particularly in permanently shadowed craters, like Shackleton crater.

Table 2. Dimensional wear coefficient (wear rate) for unlubricated sliding bearings in ambient, and simulated Lunar and Mars Environments from UHV-PoD tribometry

Pin Composition	**Disc Composition**	**Condition**	**Pin Wear Rate (mm^3/N•m)**	**Disc Wear Rate (mm^3/N•m)**
Diamonite-C	Diamond	air	2E-08	1E-06
Diamonite-C	Diamonite-C	air	1E-08	2E-06
Diamond	Diamond	air	5E-09	1E-03
Diamonite-C	Diamonite-C	Vac	5E-08	7E-05
Diamond	Diamond	Vac	1E-08	9E-05
Diamonite-C	Diamond	Vac/Cryo	1E-07	3E-04
Diamonite-C	Diamonite-C	Vac/Cryo	1E-06	4E-03
Diamonite-C	Diamonite-C	CO_2 (27.6 kPa)	nd	6E-05
Diamond	Diamond	CO_2 (27.6 kPa)	1E-06	1E-04

nd = not determined

Table 3. Wear and Friction of Diamond, Diamonite-C and Other Materials from dust PoD and tilted plane Tribometry

Material*	**Wear** (mm^3/min)**	**Sliding friction coefficients**	
		K_R*-at rest**	**K_M***-in motion**
Diamond and Diamonite-C	≤0.0001	0.12	0.09
WC/Co	0.0002	0.20	0.13
Hardened Steel (440C Stainless Steel)	0.0060	0.21	0.20
Titanium Alloy	0.0542	0.20	0.17

*The material of the pin sliding on a polished steel surface.

**Wear of unlubricated spherical radius 3.175 mm (1/8") pins is measured on the Dust PoD during sliding on a steel surface in sand and Lunar dust simulant with a velocity of 0.47 m/s and load of 10 N/pin for all samples under ambient conditions.

*** Sliding friction coefficients are measured on the tilted plane tribometer at rest (K_R) and in motion (K_M) for spherical pins with radius 3.175 mm (1/8") versus mirror polished steel, unlubricated, dry, with a load of 1 N/pin and average velocity of 0.3-0.5 m/s.

(Note to Table 3: It was observed in the tilted plane method that if the surface is very clean, K_M strongly depends on speed. In this Table we refer to values that correspond to an average sliding velocity of 0.3-0.5 m/sec. The sliding motion on a clean surface can be divided into three zones: 1) very slow and smooth motion at constant velocity, 2) slow motion at constant velocity with possible instability and acceleration-deceleration regions and 3) motion with constant acceleration. If extrapolated from zone 1, very slow motion at constant velocity, to a velocity of 1 μm/s, close to no motion or rest, the sliding friction coefficient at rest of Diamond material is close to $K_R = <0.05$, when the surface is dry and unlubricated. This value is a much lower value than the value of K_M=0.09 in stable motion at high speed. This concept cannot be found in the tribological literature. Its publication would contribute to our understanding of the nature of sliding friction. However, we intend to focus here on practice rather than theory).

Figure 5. SEM Characterization of lunar dust simulant, JSC-1AF. a and b) circles illustrate pointed particles and sharp edges. c) circles show a large number content of fine particles with similar morphology to a and b.

The lunar dust simulant consists largely of basalt and basalt derived glass particles in the 1-10 μm that are not as hard as silica (quartz) particles. Particles of 5-10 μm are especially dangerous and can damage bearings. The silica-quartz powder used for these experiments had this range of grain sizes. The properties of silica and quartz sands are well known. Therefore, both JSC-1AF dust and quartz sand were used for the wear determinations.

Characterization of Diamond as compared to graphite

Diamond is the hardest material (H_{Knoop}=60-100 GPa for single crystal material and depending on position and orientation of the indenter), with a hardness of 10 on Mohs scale. It is also the toughest material (fracture strength in compression is $\sigma_{fc} \geq 10$ GPa for high quality crystals). Graphite is the softest mineral material (hardness of <1 on Mohs scale), and is good for lubrication under ambient conditions but not in vacuum. Nevertheless, the hexagon-based graphite-like structure has unsurpassed tensile strength (up to $\sigma_{ut} \sim 6.2$ GPa reported), which has been exploited in oriented fibers. Single crystalline Diamond has the highest known thermal conductivity at room temperature $\{\lambda_{300\,K}=2300\ W/(m*K)\}$. The thermal conductivity of single crystal graphite depends on the orientation of the hexagonal basal planes with its highest value parallel to the hexagons $\{\lambda_{300\,K}(\perp c) = 2000\ W/(m*K)$, and the lowest value normal to the hexagons $\{\lambda_{300\,K}(\| c) = 5.7\ W/(m*K)\}$.

The high hardness and thermal conductivity of Diamond and Diamond materials make them the best choice for many cutting and finishing operations, especially when high precision machining is required. Machining is enhanced by low friction coefficients for carbon materials. All these properties combine to produce the lowest wearing tools with fastest dissipation of heat from contact surfaces. All of these properties are beneficial for the manufacturing of unlubricated bearings.

Sliding friction coefficients and friction wear also depend on crystal orientation. Friction is directly related to other mechanical properties, as in smoothing and cutting, and can be used for bearing design. Mechanical properties determined by anisotropy can be exploited in cutting and finishing operations.

Figure 6a. Diamond powder and grains of Diamond powders prepared for this investigation. (I -powder; II – bulky crystals; III- irregularly shaped crystals; IV - octahedrons)

Figure 6b. Two types of bulk (monolithic) Diamond prepared for these experiments: polycrystalline Diamond (1-mined; II-sintered; III-sintered and polished on top) and monocrystalline (IV) Diamond.

Diamond crystals grow in cubic, octahedral and rhombo-dodecahedral habit as well as numerous other intermediate forms. The coefficient of sliding friction on crystal Diamond varies with the crystal plane on which the measurement is taken and ranges from μ_1=0.05 on the octahedral edge to μ_4=0.12 on the cubic plane (100) along edge [100]. Wear on the edge of octahedral crystals is minimal and it is maximum on an edge of the cubic crystal that is usually "softer". The Diamond materials prepared for these experiments and the tribological counterparts are depicted in Figure 6 and 7 respectively.

Figure 7. Polycrystalline Diamond a) disc and b) pin mirror polished (flat and spherically) for tribology.

Characterization of Diamonite-C

Diamonite-C is a new C/C composite. It is relatively easy to fabricate using HPHT methods. It can be used in sliding bearings with low or negligible wear over a very wide range of temperature. It is a composite of Diamond particles and pure carbon-based polymer called Diamonite-B that is much harder than all known hydrocarbon based polymers. It is also much more thermally stable (to to 2000 °C in vacuum) than all known hydrocarbon based polymers which decompose and fail at T<300 °C. Monolithic Diamonite-C consists of Diamond particles embedded into a Diamonite-B matrix, and Diamonite-B is a binder that binds Diamond powder grains into a monolithic body. Characterization of Diamonite-B can be found in the literature [[6],[7]].

[6] O.A.Voronov, K.W. Street "Raman scattering in a new carbon material" Diamond & Related Materials 19 (2010) 31-39

[7] K.W. Street, O.A. Voronov, B.H. Kear "Diamond-dispersed Fiber-reinforced Composite for Superior Friction and Wear Properties in Extreme Environments and Method for Fabricating the Same", Patent US 9,574,080 B1, 2017

Of all the material couples tested in the UHV-PoD, Diamonite-C generally performed better than the other materials (See Tables 1 and 3). The friction coefficient for a bearing operating at speed is K_M which needs to be <0.1 for a materials combination to have any value as bearing materials. In practice, K_M <0.01 is a more realistic expectation unless a materials couple with this low of a friction coefficient can't be found for the particular application at hand. For example, operation of an exposed bearing under Lunar conditions requires a materials couple with a hardness of > 8.5 on the Mohs scale to prevent the harder components of regolith (Spinel M_H 8.0-8.5) from abrading the surfaces of the bearing contact region. This requirement eliminates the following from consideration: Aluminum (H_M<3), 304 Type Stainless Steel (M_H <5), Titanium alloy, Ti-6Al-4V (H_M<6), 52100 Bearing Steel and RHC60 440C Stainless Steel (H_M<7). Coatings can be applied to the surfaces to mitigate abrasion such as: Sapphire, Silicon Carbide, Silicon Nitride and Titanium Carbide (H_M 9.0-9.3) and Diamond (H_M = 10). The problem with these materials is that the coatings need to be applied as thin layers - a discussion of reasoning is beyond the scope of this paper. All of these coating materials are brittle, an unattractive feature for operation on the moon where the feasibility of recoating the surface is impractical.

Figure 8. Diamonite-C discs a) and b) mirror polished for tribology: light scatters differently from different areas of composite where Diamond particles are incorporated into the Diamonite-B matrix. c) Mirror polished surface of Diamonite-B reflects light like a perfect mirror. The apparent density of Diamonite-C samples is 2.48 g/cm3 and Diamonite-B is 1.76 g/cm3 as compared to the density of Diamond, 3.51 g/cm3.

Tribometry studies, friction coefficients

The friction coefficient at rest, K_R, and in motion, K_M, (Table 3) are evaluated by the tilted plane method with the slider material sliding on a dry, unlubricated, mirror-polished steel plane using a load of 1 N/pin, and average velocity of 0.3-0.5 m/s on spherical pins having a radius of 3.175 mm (1/8"). In Table 3, the value of K_M is presented for Diamonite-C along with some other potential sliding materials. Low velocity between material couples in relative motion is where the most severe wear often occurs. Even for "mirrored" surfaces, on the microscopic scale, asperities occur and the surface contact between opposing asperities cause extreme abrasion and often galling (adhesive wear between opposing surfaces). Consequently, it is equally important to have very low friction coefficients under this condition. While Diamonite-C performed better than the other materials tested, it was only marginally better than the other couples.

Low friction coefficients for metal on metal tribocouples such as hardened steel and titanium alloy as appearing in Table 3 are not unexpected even though the friction coefficients for nascent metal on nascent metal are expected to be large (approaching 1.0 or greater due to galling and other surface adhesion mechanisms). The data from Table 3 from the tilted plane method has a sliding distance of ~1 meter, barely enough distance to remove surface layers of contamination and oxide coatings which are lubricious in nature. Diamond and tungsten carbide cobalt are less prone to surface oxidation and have correspondingly low friction coefficients although it is not possible to determine the effect of other surface layers.

A complete understanding of tribological behavior depends on many more factors, including chemistry of surfaces, humidity, adsorbed gases, pressure, temperature, etc. In tilted plane experiments with Diamonite-C, ambient parameters included: T = 21 °C, P = 103.2 kPa, humidity 37%, and for the Ti and carbide-based alloys: T = 23 °C, P = 103.9 kPa, humidity 29%. With the tilted plane equipment used for this research, it was not possible to evaluate the effects of surface chemistry, adsorbed gases, etc.

The UHV-PoD, tribometer employed in this research also allows evaluation of the coefficient of friction in sliding, K_M, and in addition provides quantitative information on wear. The PoD is housed in an UHV chamber which allows these tribological properties to be determined in vacuum or in various atmospheres backfilled into the chamber (e.g. dry air, nitrogen, CO_2, etc.). Figures 9 and 10 shows several friction traces obtained on the UHV-PoD used in these studies. Figure 9a is the raw data showing large oscillations due to warpage in samples, samples mounted in a tilted pane, or a large number of other variables, including noise in the actual friction coefficient during each revolution of the disc. Since the computer program that processes the data does not handle negative values there is an initial offset of $\mu = 0.20$ applied to all runs. Figure 9b, the smoothed data from 9a, reveals a break in period in the initial 25 revolutions of the disc followed by steady state friction coefficient. It is this latter region of the friction trace where the average friction coefficient is read from the friction trace. The inset shows the computation of the friction coefficient as follows. The stead steady state friction is 0.54 minus the initial baseline offset. This value is then divided by 10 to account for the signal amplification created by applying an actual load of 200 gf and entering a value of 20 gf for the run. Figure 10a illustrates the reason for signal amplification. In Figure 10a, the first 3800 revolutions represent real data and the rest of the trace represents a malfunction during the experiment. In the steady state friction coefficient, between 500 and 3800 revolutions $\mu = 0.01$ above the offset. Without amplification, the signal would read $\mu = 0.001$ which would be lost in the residual noise after data smoothing. Figure 10b shows a prolonged break-in period followed by steady state friction coefficient, $\mu = 0.02$ (no amplification of data was used during this run). Figure 10c shows unstable friction coefficient data to at least 43,000 revolutions (360 min.). Had this experiment been run longer, data above 360 minutes might have established a region of steady state friction. Occasionally, unusable data is acquired throughout the run (Figure 10c). Figure 10 demonstrates the need to pay careful attention to data acquisition and heavily interpret the friction coefficient data acquired for samples having extremely low friction coefficients. Some experiments never yield useful friction coefficient data as in Figure 10d; however, useful wear data may still be obtained from the experiment.

Results of PoD tribometry are presented in Tables 1 and 2. An advantage of runs performed in vacuum or vacuuming out the UHV chamber down to high vacuum before backfilling with dry, purified gas (for e.g. CO_2) is that all humidity and loosely bound surface molecules are removed from the tribological contact. In addition, during the first few orbits, any residual surface bound layer(s) is removed from the contact zone. All of the runs in air produced low friction coefficients (<0.1) which is in line with literature values for Diamond on Diamond couples (approaching superlubricity, friction coefficients between 0.001 and 0.01), whereas literature values for Diamond on Diamond in vacuum and inert atmospheres are typically much higher (approaching 0.7 or greater). Humidity plays an important role in the friction and wear properties of Diamond. It is also possible to add a considerable amount of hydrogen, H, (H content in the region of 40 Atomic %) to Diamond resulting in hydrogenated amorphous Diamond, a-C:H. It is interesting that a-C:H has almost the exact opposite properties of conventional Diamond in that it has low to super-low friction coefficients in vacuum and high friction coefficients in air (over an order of magnitude higher). The friction coefficients reported herein are similarly elevated in vacuum but not nearly as much as some of the literature citations [[8] and references therein]. Interestingly, the Diamonite-C in vacuum produced an almost acceptable friction coefficient for unlubricated operation. For unlubricated operating of bearings under Lunar conditions temperature is also a consideration. Lunar operation would involve the previously mentioned temperature with the goal of producing bearings capable of operation in Shackleton crater for recovery of water. The Diamonite-C-Diamonite-C combination did not appear to or has a negligible temperature dependency between ambient and the -100 °C, whereas the Diamond-Diamond couple has a large, unusable dependency with an unstable friction trace (the friction coefficient was not determinable). The Diamond-Diamonite-C couple had a slight increase in friction coefficient between ambient and -100 °C which might be attributed to better performance characteristics of the Diamonite-C. Because the Diamonite-C had better performance characteristics in both air and vacuum, it was selected for testing under a high pressure of CO_2 (relative to Martian atmosphere of approximately 933 Pa CO_2). Surprisingly the Diamonite-C couple produced unstable friction traces whereas Diamond performed as well under CO_2 blanket as it did in air. For efficient self-lubricating bearings, the friction coefficient between the moving parts must be <0.1 [[3]].

Figure 9. UHV-PoD friction trace for Diamonite-C pin and disc in air. a) raw data and b) smoothed data

Tribometry studies, wear

Wear rate, also known as the dimensional wear coefficient, is computed from loss of material in the contact zones of both opposing surfaces and carries the units of "volume lost divided by both distance traveled and applied force on the moving surfaces", e.g. $mm^3/N\ m$. For the UHV-PoD experiments, wear on the pin and disc are calculated by determining the wear volume experimentally. For the pin, the volume removed is determined by interferometric profilometry which gives the surface of the flat on the pin and surrounding area (See Figure 11a). A computer program then computes the original curvature over the flat and subsequently computes the volume

Figure 10. Examples of UHV-PoD friction traces. a) Diamond in air, b) Diamonite-C in air, c) Diamond in CO_2, and d) Diamonite-C in CO_2

loss assuming that the pin is spherical (See Figure 11b). The volume loss in the irregular wear track on the disc is also calculated from interferometric profilometry by measuring the cross-sectional depth and width of the track at four arbitrary and evenly spaced locations around the track, averaging the cross sectional loss, and assuming that the track is linear with length equal to the circumference of the original track. Force is derived from the load on the pin and not the actual force per unit area in the contact, the Hertzian contact pressure. In practice, applying any load to such arrangement causes plastic and/or elastic deformation in the contact area resulting in a very small area of contact, Hertzian contact area (typically a few μm in diameter), which can theoretically be calculated from the applied load and physical properties of the materials in contact (Young's modulus and the Poisson ratio). After traveling a large distance, significant material is removed from the spherical area of the pin in contact with the disc. The area on the flat of the pin is then significantly greater than the Hertzian contact area that would be computed

---

[8] Superlubricity, edited by A. Erdemir and J.-M. Martin, Chapter 16, pp 260-61 by A. Erdemir and O.L. Eryilmaz; and Chapter 17, pp 273-75, by J. Fontaine and C. Donnet.

for the start of the experiment, 1.8 mm diameter for the case in Figure 11. The wear track on the disc is then significantly wider than the value expected if only a Hertzian Contact is observed by the end of the experiment which is the case for all pin/disc combinations tested.

Wear data for pins and discs used in these experiments is presented in Table 2. Figure 12 is the optical image of the pin from Figure 11 which clearly shows not only the wear but also the direction of sliding. The leading edge is rounded producing greater light reflection. The data in Table 2 indicates marked differences in both friction (Table 1) and wear resulting from differing

Figure 11. Image of UHV-PoD Diamonite-C pin tip after sliding over 1,256 m of Diamonite-C at a load of 200 gf in CO_2. a) interferometric profilometry of pin tip, and b) height profile across the slice shown in image a.

environmental conditions and combinations of materials. Wear rates for all pins in air and vacuum were in the 10^{-8} range which is considered low wear and in line with other Diamond materials tested on this same PoD rig [[3, 9, 10]]. Under cold conditions, (-100 °C) in vacuum and in CO_2 the wear rates of the pins generally increased by an order of magnitude. While the friction coefficients under these conditions are acceptable, the wear rates are not, especially for prolonged use. In general, the wear rates on the discs were two orders of magnitude or greater than the wear rates for the corresponding pins. Figures 13 and 14 are images of wear tracks on the Diamond discs. Figure 13 shows the Diamond disc with images of two wear tracks, one with lower material removal and the other with more pronounced removal. The irregular profiles of the tracks make the determination of material removed somewhat subjective (see measurement lines given in Figure 13b) which is why dimensional wear coefficients, wear rates, are typically reported only in order of magnitude. The enhanced wear of the discs is possibly due to a plowing mechanism of the pin shearing off asperities on the discs (likely individual Diamond grains) which were quite rough (Ra = 2.8 μm for the Diamonite-C disc and Ra= 4.0 μm for the Diamond pin) as compared to Diamond like surfaces reported in the literature (Ra values in the low 10's of nm). It is also possible that after initial wear on the pin, the enhanced surface contact area increased the shear strength or the surface area of the contact resulting in additional surface removal due to adhesion. Figure 14 shows SEM images of 3 different wear tracks. Images a and b are of the same track at different magnifications. In image a, center, the apparent width of the track is greater than on the left side

Figure 12. Optical image of UHV-PoD Diamonite-C pin tip after sliding over 1,256 m of Diamonite-C at a load of 200 gf.

of the image illustrating the difference in track wear experienced by different areas of the disc which is why a minimum of four profiles were averaged to determine the cross section area of material removed. Images c and d show typical wear debris alongside of and in the track after completion of the experiment. In the upper right corner of image c, particulate matter appears to

---

[9] "Friction and wear properties of a-DLC-based functionally gradient nanocomposite coatings," K. Miyoshi, K.W. Street, J.S. Zabinski, J.H. Sanders and A.A. Voevodin, Pestology, 20 (7), X1-7 (1996).

[10] "Sliding wear and fretting wear of diamondlike carbon-based, functionally graded nanocomposite coatings," K. Miyoshi, B. Pohlchuck, K.W. Street, J.S. Zabinski, J.H. Sanders, A.A. Voevodin and R.L.C. Wu, Wear, 225-229, 65-73 (1999).

have been removed by an adhesion mechanism. These particles range in dimension from a few μm to a few 10s of μm. Removal of any surface contamination would assist the adhesion process. In order to produce an efficient self-lubricating, wear-resistant bearing, the material must have not only a friction coefficient of <0.1, but a dimensional wear coefficient of $<10^{-6}$. With improvements in manufacturing technology and polishing of the parts produced, the wear rate of the discs may be lowered by an order of magnitude or more approaching the wear coefficients for the pins, thus making these materials suitable for a wider variety of bearing applications.

Additional wear data for pins against steel is available from the dust PoD experiments summarized in Table 3. In these experiments, sliding speeds of 0.47 m/s and high loads (10 N/pin) were applied to the sliding surfaces while large quantities of JSC-1AF or sand were introduced into the contacts. Both titanium alloy and even hardened steel are relatively soft with respect to tungsten carbide (WC/Co) and Diamond. Under these conditions, titanium alloy, the softest material tested, had an order of magnitude greater wear than hardened steel. The hardened steel exhibited over an order of magnitude more wear than the WC/Co. Both Diamond and Diamonite-C samples

Figure 13. Image of UHV-PoD Diamond disc. a) interferometric profilometry of two wear tracks, and b) depth profile across the slice shown in image a.

exhibited so little wear that reliable wear numbers could not be obtained and the entry in Table 3 is only a conservative estimate of the maximum wear that these samples might have experienced. The actual wear of Diamond and Diamonite-C could be as little as an order of magnitude less than the value obtained for WC/Co. This data is encouraging in that bearings constructed out of Diamond and Diamonite-C sliding against steel may be operated in extremely dusty environments.

In general, bearings should not be constructed of metal/metal unlubricated couples for fear of severe galling of the nascent metals once the surface oxide (and other) layers are removed by friction wear. Ceramic and hybrid ceramic/metal bearings are commercially available having good friction and wear properties. All of the aforementioned bearing couples work best when properly lubricated. Unfortunately for space application, lubrication is not an attractive option, especially for bearings exposed to the environment. For several decades Diamond materials have been proposed to overcome many of the drawbacks of conventional bearings. As Diamond is the hardest material, machining of complex geometries necessary for anything other than simple sliding bearings is prohibitive. Diamond coatings by chemical vapor deposition or physical vapor deposition applied as thin layers on metal substrates have the drawback of only a thin layer which, once removed uncovers a bare substrate. Further the substrate, often metal, has much higher density than Diamond. Diamond and Diamonite-C are interesting materials since they are virtually pure

Figure 14. SEM images of UHV-PoD wear track on Diamond. a and b) same wear track at 75 times and 500 times magnification respectively. c and d) images of wear debris in wear tracks at 75 times and 500 times magnification respectively.

Diamond (or pure carbon/carbon Diamond-base composite), can conveniently be synthesized into near-net shape, and have all the attractive tribological properties of Diamond, low friction and wear in air at ambient temperature. Diamonite-C has the advantage over Diamond in that the friction coefficient is near the required value for efficient operation in both vacuum and vacuum at temperatures as low as -100 °C. Diamonite-C is yet to be fully investigated both in terms of physical properties and for use as a bearing material, although it is relatively easy to fabricate.

Conclusion

Near net shape Diamond and Diamonite-C components can be readily fabricated by HPHT processing methods. The materials are lighter than most construction metals, an important consideration for space and aeronautic applications. Bearings made from these materials are of particular interest since they can operate without lubrication. The unlubricated tribological properties of the materials are as good as or superior to that of current unlubricated bearing materials. Well-polished Diamond and Diamonite-C surfaces have low friction coefficients and experience negligible wear, even in dusty or sandy environments. With some additional development, Diamond and especially Diamonite-C may be used in extreme Lunar environments for bearing applications as well as for light-weight components in other aerospace applications.

Fig 1a

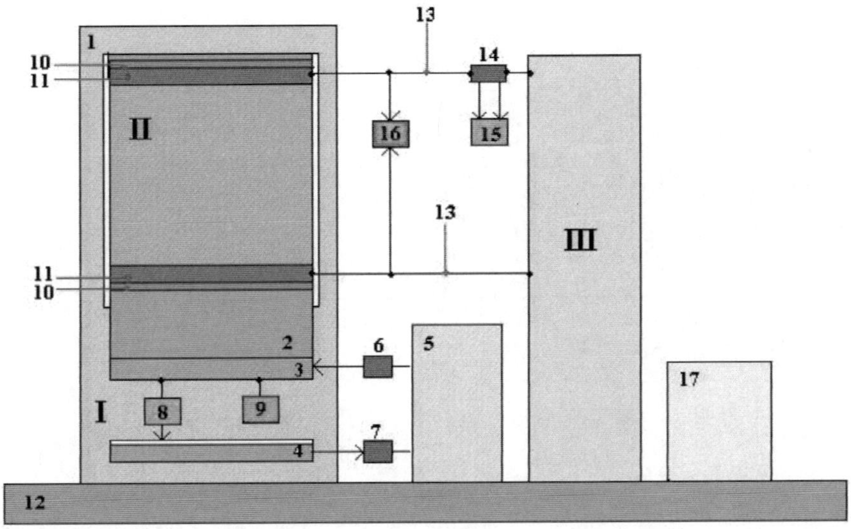

Fig 1b

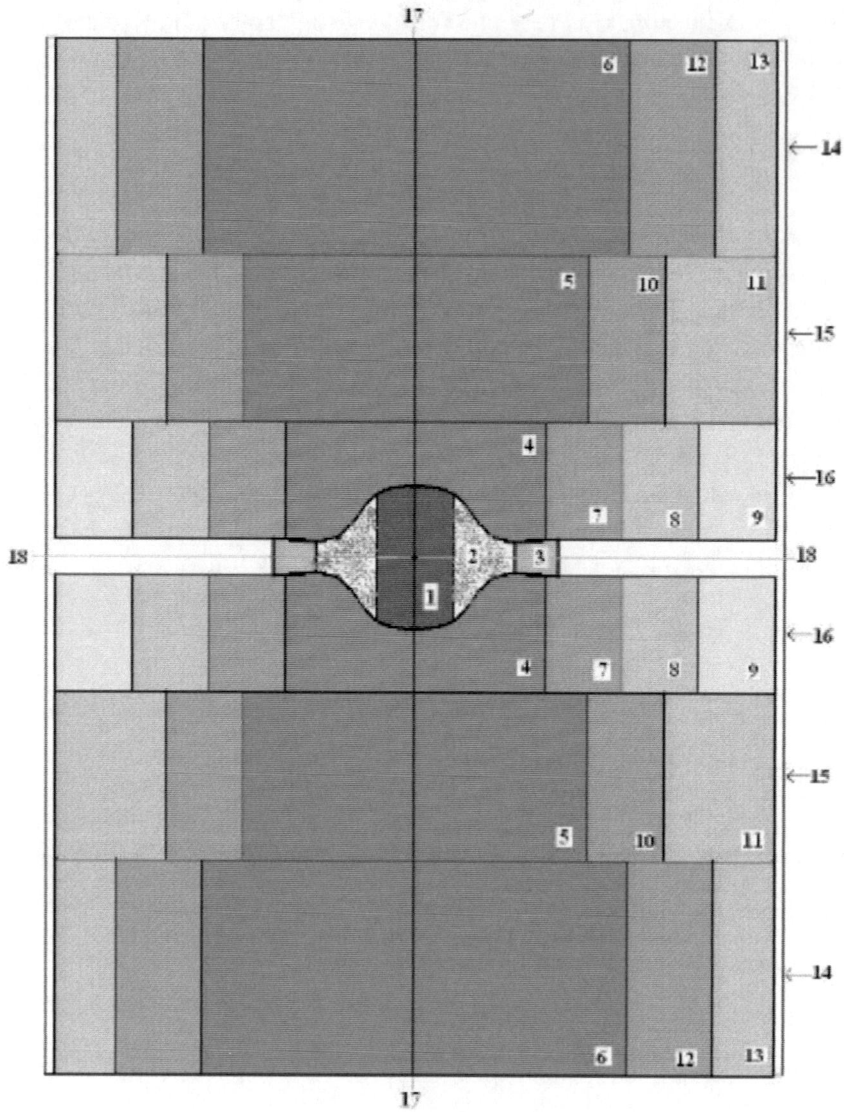
17
6
12
13
14
5
10
11
15
4
16
7
8
9
18
2
3
18
1
16
4
7
8
9
15
5
10
11
14
6
12
13
17

Fig 2

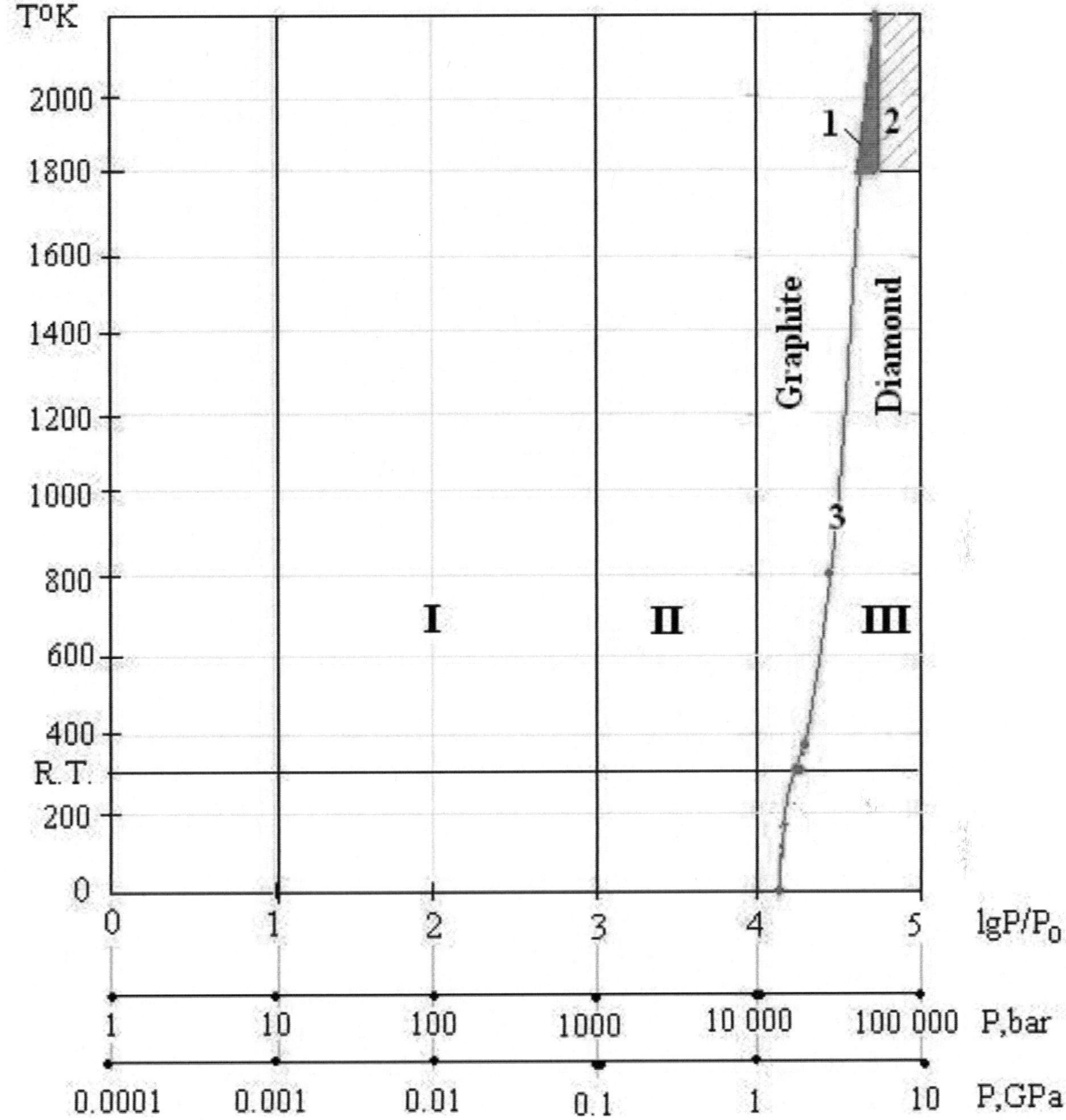

T°K
2000
1800
1600
1400
1200
1000
800
600
400
R.T.
200
0
1
2
3
Graphite
Diamond
I
II
III
0
1
2
3
4
5
lgP/P0
1
10
100
1000
10 000
100 000
P,bar
0.0001
0.001
0.01
0.1
1
10
P,GPa

Fig 3

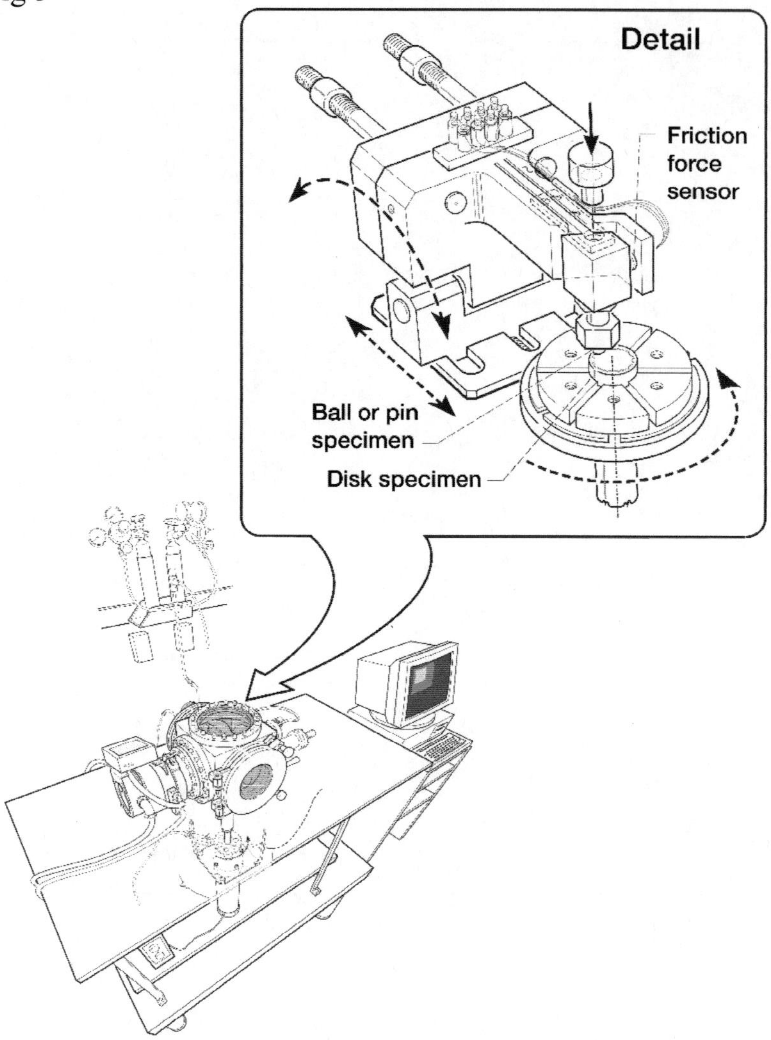

Detail
Friction force sensor
Ball or pin specimen
Disk specimen

Fig 4 a

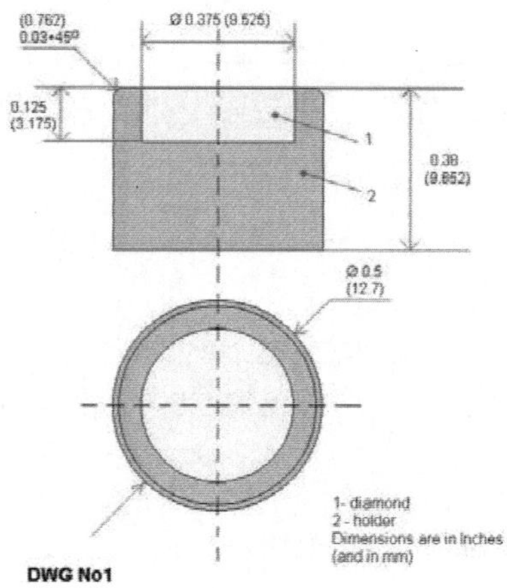

(0.762)
0.03•45º
Ø 0.375 (9.525)
0.125
(3.175)
1
2
0.38
(9.652)
Ø 0.5
(12.7)
1- diamond
2 - holder
Dimensions are in Inches
(and in mm)
DWG No1

Fig 4b

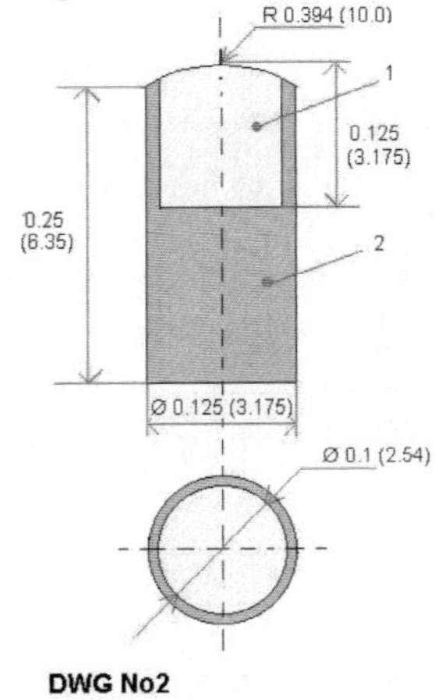

Fig 5

Fig 6a (I-IV)

Fig 6b (I-IV)

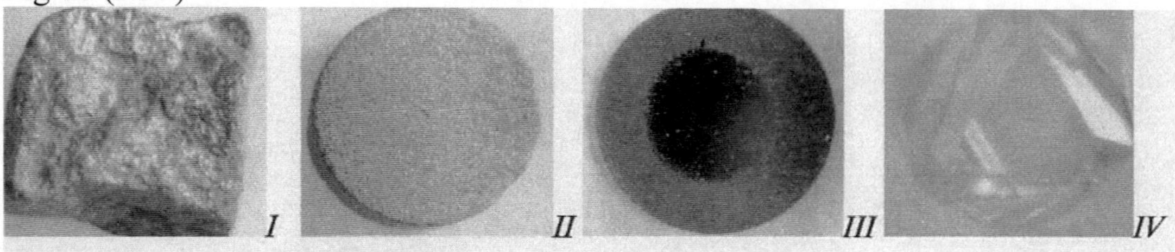

Fig 7 a and b

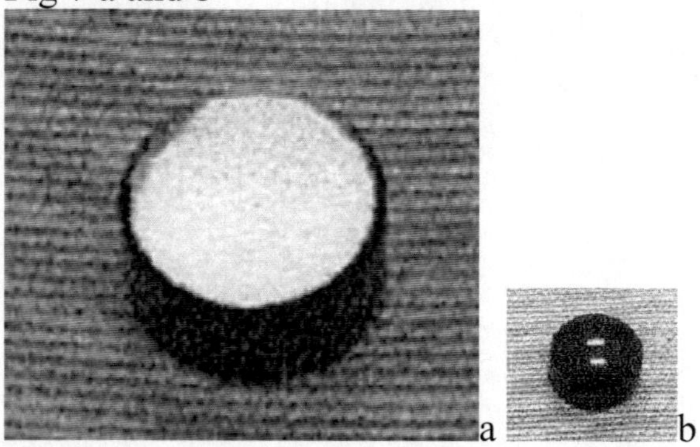

Fig 8 a-c

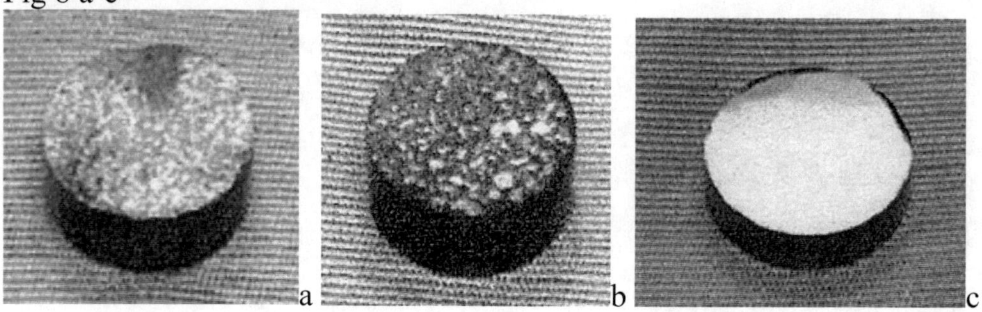

Fig 9

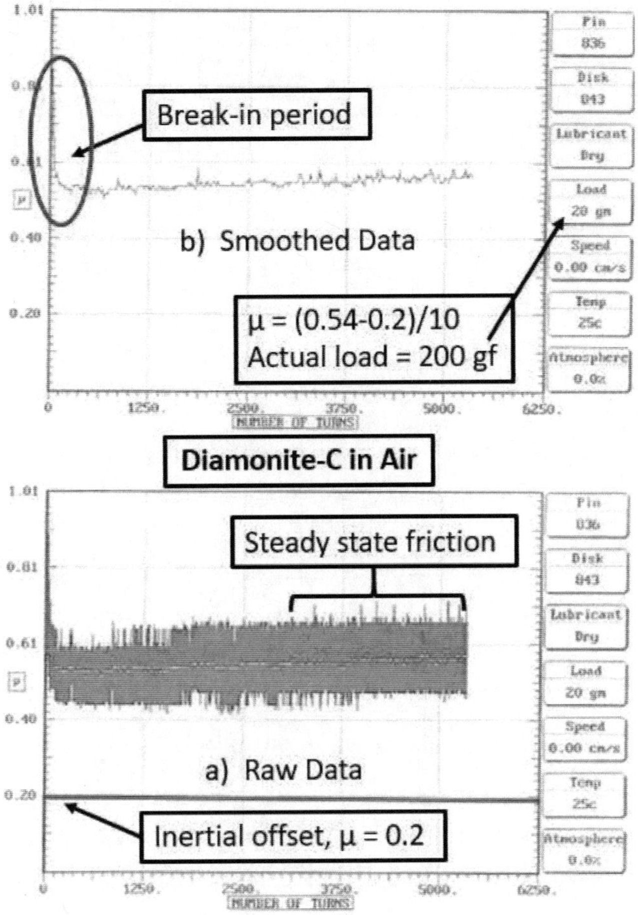

Break-in period
b) Smoothed Data
μ = (0.54-0.2)/10
Actual load = 200 gf
NUMBER OF TURNS
Diamonite-C in Air
Steady state friction
a) Raw Data
Inertial offset, μ = 0.2
NUMBER OF TURNS

Fig 10

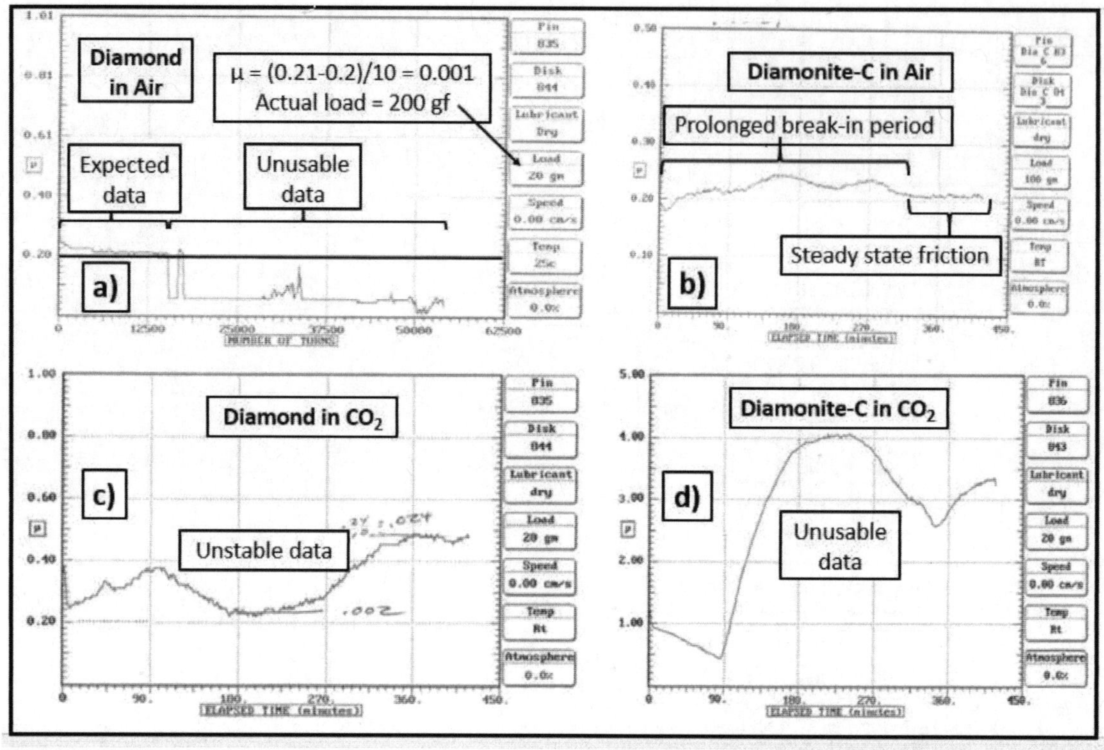

Diamond in Air
μ = (0.21-0.2)/10 = 0.001
Actual load = 200 gf
Expected data
Unusable data
a)
Diamonite-C in Air
Prolonged break-in period
Steady state friction
b)
Diamond in CO2
c)
Unstable data
Diamonite-C in CO2
d)
Unusable data

Fig 11

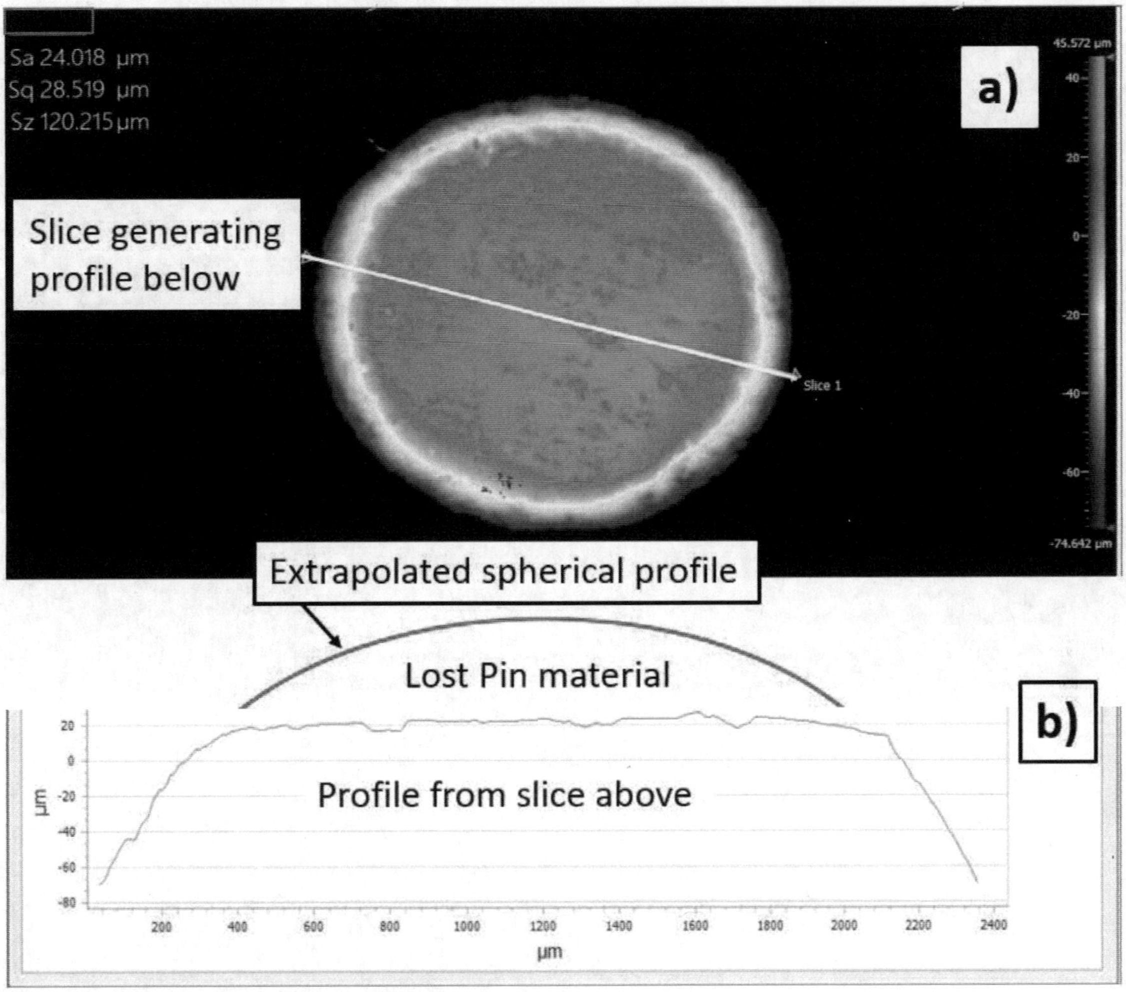
Sa 24.018 μm
Sq 28.519 μm
Sz 120.215μm
a)
Slice generating profile below
Slice 1
45.572 μm
-74.642 μm
Extrapolated spherical profile
Lost Pin material
b)
Profile from slice above
μm
μm

Fig 12

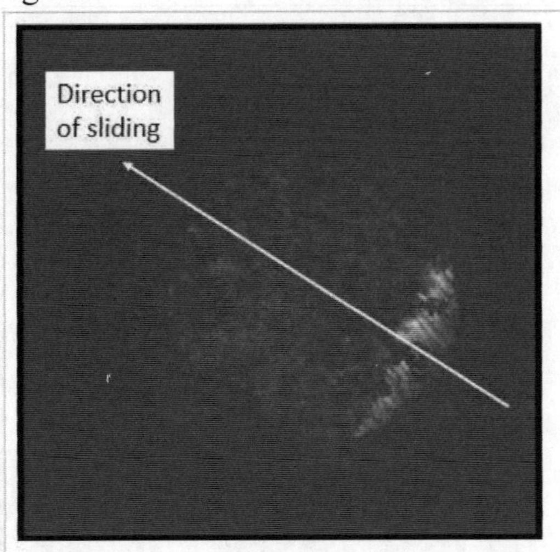
Direction of sliding

Fig 13

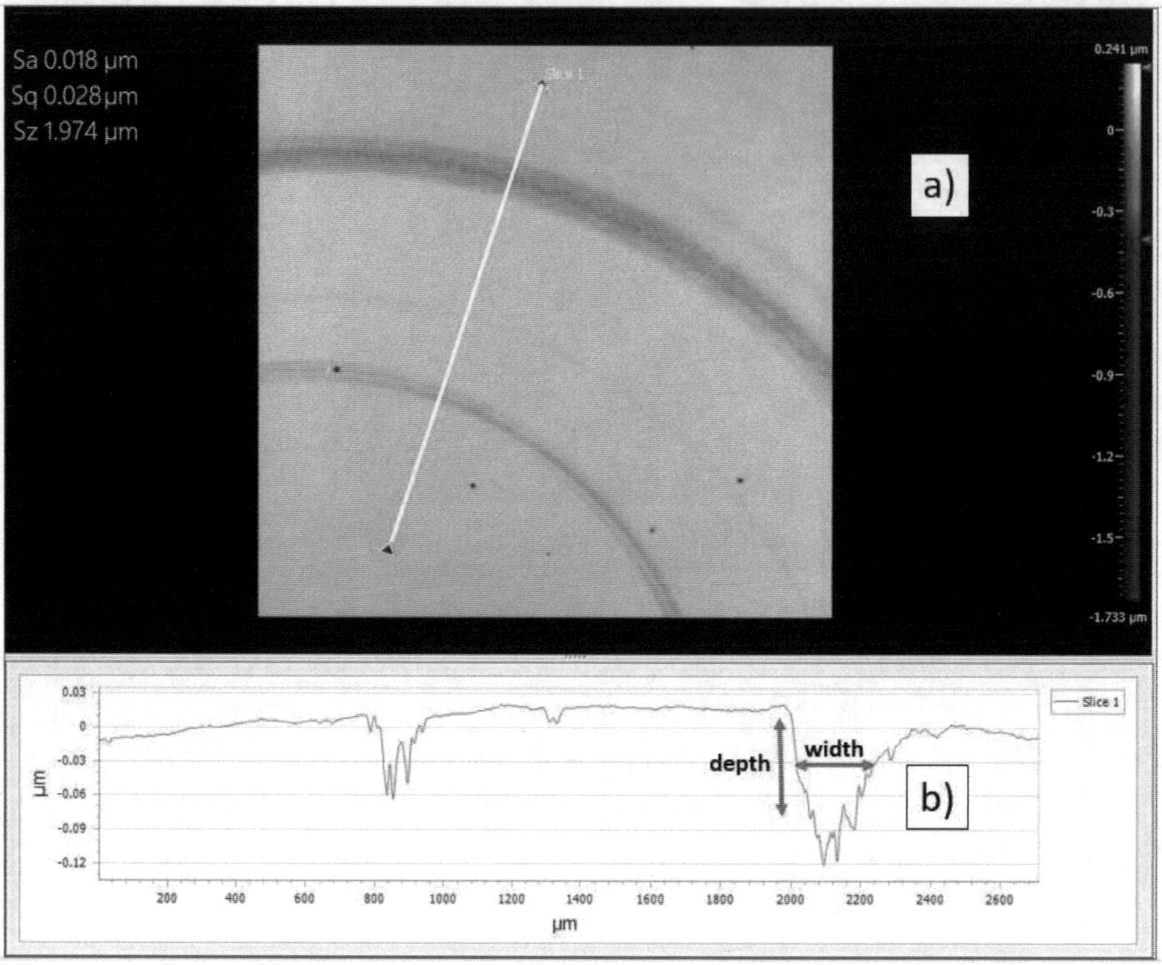

Sa 0.018 µm
Sq 0.028µm
Sz 1.974 µm
a)
0.241 µm
0
-0.3
-0.6
-0.9
-1.2
-1.5
-1.733 µm
Slice 1
depth
width
b)
µm
0.03
0
-0.03
-0.06
-0.09
-0.12
200
400
600
800
1000
1200
1400
1600
1800
2000
2200
2400
2600

Fig 14

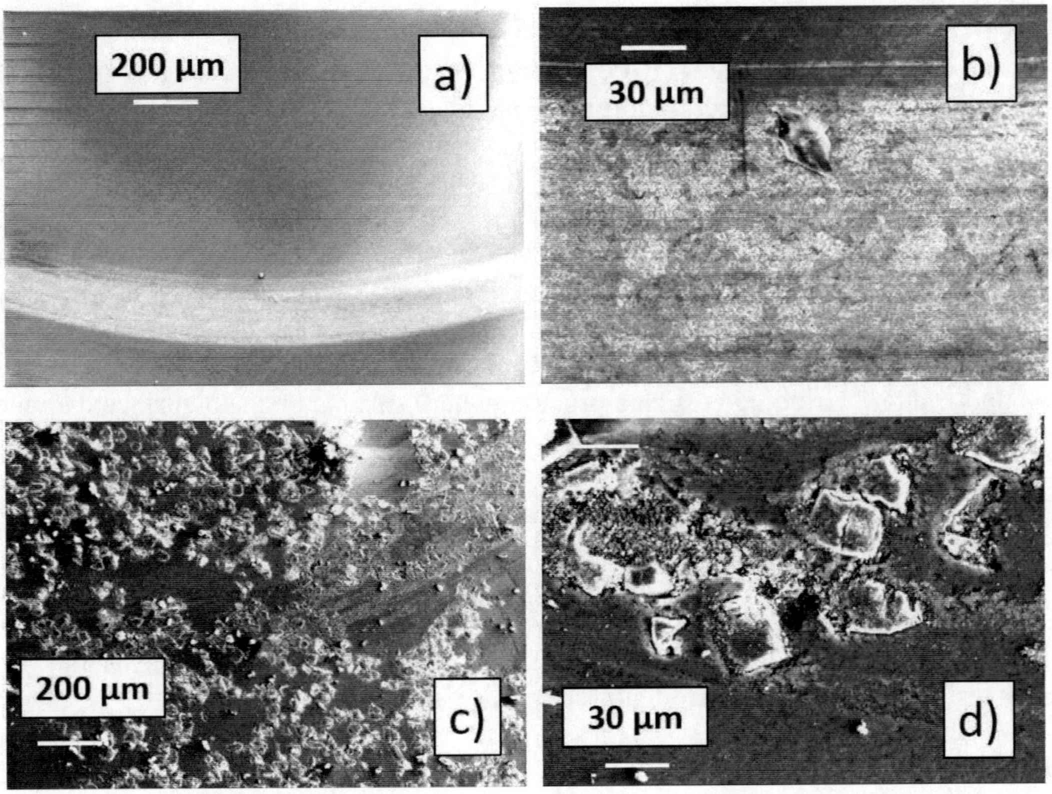
200 µm
a)
30 µm
b)
200 µm
c)
30 µm
d)

Influence of Abrasive Type on Grinding Induced Residual Stresses and Surface Integrity

Dr. John A. Webster, Andreas Planakis, and Chris Winkel
Worldwide Superabrasives, USA

Abstract

Many engineered products are subjected to applied cyclic and static stresses in their designed use. Examples include bearings, transmission shafts, and gas turbine blades/fans. In addition to these operational stresses, the materials often contain residual stresses, cracks, and defects, imparted during the manufacturing process, especially when high temperatures are created by chip formation and friction. Grinding is such a process, where surface and sub-surface damage can result if the process is not adequately controlled.

This paper identifies and compares common techniques that are used to inspect the surface and sub-surface integrity of ground parts, and clearly shows the advantages of X-Ray diffraction (XRD) over Barkhausen Noise. The paper also presents white-layer analysis, hardness change, and crack detection. Comparisons are made between grinding cycles based on superabrasive grain and conventional grain grinding wheels. Results of residual stress measurements and surface integrity tests made at WWSA are featured in this paper

Introduction

The fatigue life of a metallic component that is subjected to cyclic stresses is significantly influenced by the surface and sub-surface condition produced by the manufacturing processes that made the component. As grinding is often the last process to be used to produce the final surface topography and surface integrity, manipulation of the thermal input and mechanical stresses is paramount. Tensile residual surface and sub-surface stresses created by grinding can add to the service stresses in use and lead to surface, or sub-surface cracking, plus tribological issues with bearing surfaces such as Brinelling, spalling, etc. Compressive residual stresses are deemed as desirable at a surface since they help the workpiece surface to resist the service stresses. The type of abrasive, grit size, sharpness of the wheel surface, wheel-speed, material removal rate, coolant delivery, etc, all affect the surface integrity of the final workpiece surface.

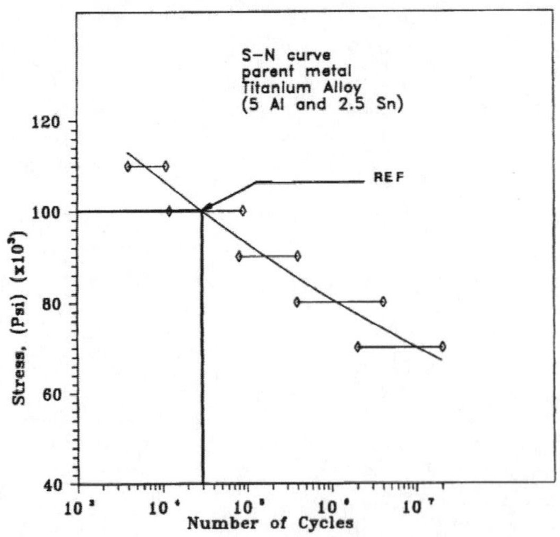

Figure 1. S-N rotating fatigue curve by Rangaswamy [1]

Figure 1 is an example of an S-N rotating fatigue curve by Rangaswamy [1]. The line is downward sloping. We can see from this graph that the smaller the applied stress to the testpiece the more the number of cycles before failure occurs.

A schematic of a rotating stress fatigue tester is shown in Figure 2. The testpiece is continuously subjected to a once-per-rev bending stress as it rotates, and a counter stops the revolution counter at the time of failure.

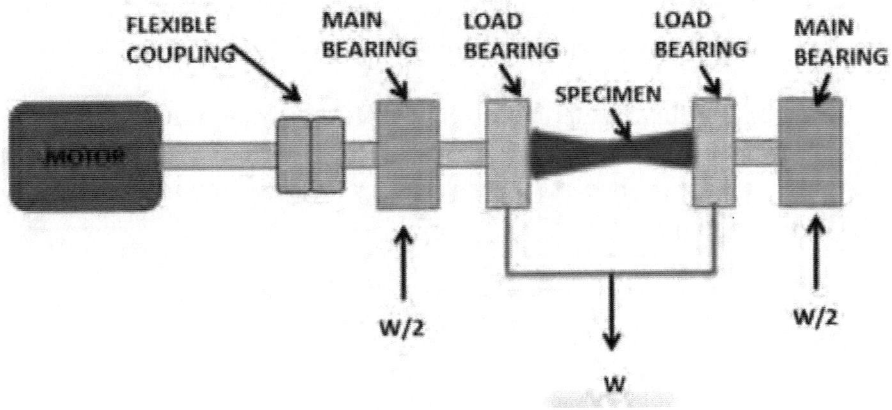

Figure 2. Schematic of rotating fatigue tester

Aero-engine rotating components, wheel bearings, transmission shafts, etc, are categorized as critical components and are therefore subjected to highly stringent dimensional accuracy, surface integrity, and fatigue requirements. It is well documented that the functional performance of parts subjected to machining processes is strongly affected by the resulting material surface/subsurface integrity conditions [2]. Publications detailing investigations relating to the influence of material removal processes on workpiece residual stresses initially appeared in the 1950s, however, Field and Kahles [3] were the first to define the concept of surface integrity and propose associated methodologies for assessment, with particular application to machining processes.

Whilst specific standards of workpiece surface integrity and acceptance criteria for aero-engine components exist, they are commercially sensitive and closely guarded, which precludes full disclosure in the public domain. Nevertheless, surface integrity/metallurgical assessment of the workpiece following grinding is typically undertaken to confirm the absence of various anomalies including cracks, amorphous/recast layers, re-deposited/foreign material, contamination and work hardening, none of which would be acceptable. In addition, the analysis can also encompass residual stress measurement and cyclic life assessment. It is well known that the fatigue life of a cyclically loaded component is a function of its surface and sub-surface residual stress profile after machining.

Residual stress measurements are typically performed using two different approaches, X-Ray diffraction (for all metals) and Barkhausen Noise (just magnetic materials such as steel). X-Ray diffraction systems are considered to be destructive since after the surface stress is measured a layer of material is etched away and the process continued at the new surface. Barkhausen Noise systems use a dedicated sensor that conforms to the profile of the workpiece and after the measurement the workpiece can be ready for its' intended use. In addition, X-Ray diffraction gives the true stresses in the part, whereas Barkhausen Noise gives relative stresses and often uses a sample measured by X-Ray diffraction in order to calibrate the Barkhausen sensor output.

According to Heinzel [4] "residual stresses are rigidly effective, multi-axial stresses in a solid system that remain without external forces and moments and are in a mechanical equilibrium". This means that residual stresses are pure internal stresses of a workpiece or a part component. Furthermore, changes in the residual stress distribution are always combined with form changes of the part because a new mechanical equilibrium has been established.

Residual stresses can be distinguished as macro-stresses (type I) and micro-stresses (Type II + III) which are characterized by the scale at which they exist within a material. Macro-stresses occur over long distances within a material whereby micro-stresses exist only locally between workpiece grains or inside a grain [4].

1. Type I stresses: Macro-stresses are almost homogenous over large distances that involve many workpiece grains.
2. Type II stresses: These micro-stresses are almost homogenous over small distances that involve one grain of the material. They can be caused by differences in the microstructure of a material.
3. Type III stresses: These micro-stresses are in-homogenous and exist inside a grain of the material as a result of crystal imperfections within the grain.

Other surface integrity measurement techniques include: surface white etch layer (WEL), eddy current testing, micro-hardness, and fluorescent die penetrant (FPI).

Thermal Benefits of Superabrasives

According to Kumar [5] a comparison of the thermo-physical properties of CBN and aluminum oxide reveal that CBN has an extraordinary thermal diffusivity. Thermal diffusivity is the ratio of the heat conducted to the heat absorbed, by a body. The thermal diffusivity of CBN is 67 times that of aluminum oxide. This unique property allows CBN to actually absorb over 70% of the heat generated in chip formation by an individual CBN grain when properly used. The heat which remains in the workpiece is not sufficient to soften the steel by exceeding its tempering temperature. CBN effectively cold-works the component surface with no metallurgical changes.

Kumar also states that residual compressive stresses are generated through this process. In order to effectively couple the surface integrity capabilities of CBN with its clearly established productivity capabilities, more information is needed about the: CBN specification; grinding parameters; wheel surface speed; material removal rate; and residual stresses generated.

Toenshoff [6] showed that the surface temperature of the workpiece during grinding with CBN was much lower than other tests using aluminum oxide, confirming Kumar. Toenshoff's test were carried out on 100Cr6 bearing steel, at 30 m/s, with emulsion coolant, using surface grinding. He measured the temperature using a thermocouple, hence the short duration pulse shown in Figure 3.

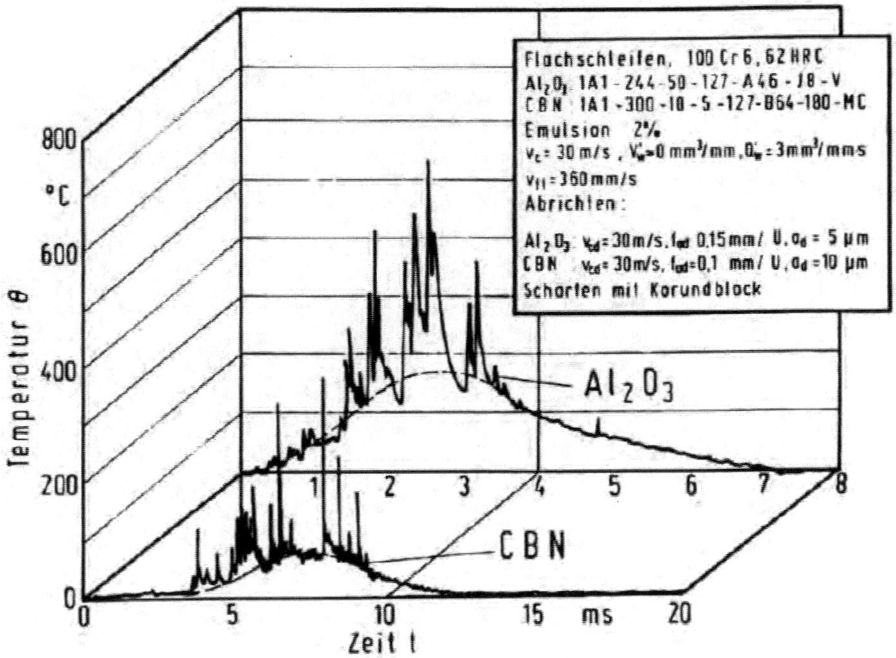

Figure 3. Measured temperature profiles and duration during surface grinding

Measurement of Residual Stresses and Comparison to Other Techniques

X-ray diffraction (XRD) can be used to measure residual stress using the distance between crystallographic planes (d-spacings) as a strain gage. When the material is in tension, the d-spacing increases and, when under compression the d-spacing decreases.

The d-spacings are calculated using Bragg's Law: $\lambda = 2d\sin\theta$.

If a monochromatic (λ) x-ray beam impinges upon a sample with an ordered lattice spacing (d), constructive interference will occur at an angle θ. Changes in strain and thus the d-spacing translate into changes in the diffraction angle θ measured by the x-ray detectors. Stresses can be determined from the measured d-spacings.

Webster, et al, [7] compared several techniques that are typically used to determine thermal damage to a ground component by continuous-dress creep feed grinding, and different coolant application conditions. In order to reduce the number of grinding tests, the workpiece was inclined in the direction of the wheel by 1.5 degrees, creating a depth of cut that increases by 2.5 mm over the length of the 95 mm workpiece. This approach was also used by Toenshoff et al. [8]. In many cases the starting depth of cut was greater than zero to force the point of burn to be close to the centre of the workpiece for the particular flowrate being tested. A model was developed to predict the instantaneous material removal rate at any position along the workpiece, using the initial depth of cut, true wheel diameter, table speed (fixed) and inclination angle (fixed). The model was based on matching the instantaneous contact length to the depth of cut of the equivalent non-inclined workpiece which has the same contact length. The technique gives a range of grinding conditions within one component, dramatically reducing the test matrix.

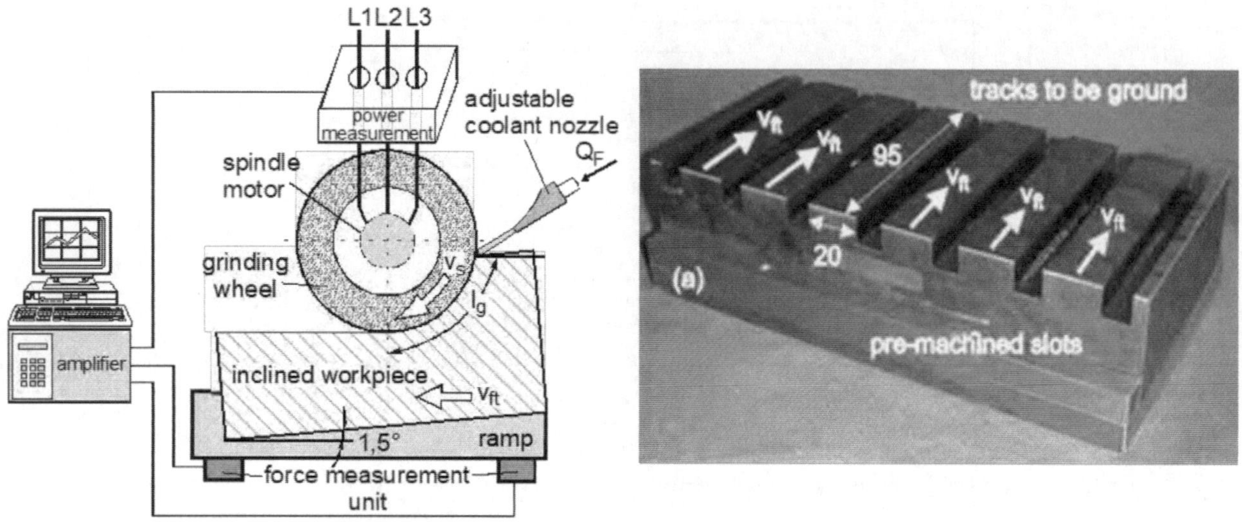

Process creep feed grinding (down grinding)	Dressing CD-dressing q_d = 0.7; 0.8 f_{rd} = 1 µm
Grinding Wheel vitrified Al_2O_3 grit G-grade, 51% porosity d_s = 500 mm b_s = 25 mm	Workpiece steel 4340 570 HV1
Grinding v_s = 20; 30; 35 m/s v_{ft} = 0.2 m/min b_c = 20 mm	Coolant supply coolant: emulsion, 4% Q'_F = 2.4; 4; 6 l/(min mm) free jet nozzle with adjustable aperture

Figure 4: Experimental setup.

A phase-corrected Hall-effect power transducer was fitted to the spindle motor drive to accurately determine the grinding process power (Figure 4). The coolant nozzle was placed 300 mm away from the contact point, oriented approximately horizontal and focused directly into the contact point. The specific power values were obtained by dividing the total power by the active grinding wheel width b_c (20 mm) and relating to the instantaneous depth of cut of the wheel into the workpiece.

The onset of thermal damage of each workpiece was assessed by several well-known methods: visual observation; position of power surge, hardness change, nital etch and residual stress, in order to determine the methods that give the earliest, and the easiest way to detect burn.

A typical example of the top view of a ground track is shown in Figure 5. The visual burn has a curved profile possibly due to the fluid that spills down the relief slots, that are pre-machined between each 20mm section of workpiece in the 6 wide block, quenching the sides of the workpiece and reducing the side temperature. Also, the white layer in the longitudinal micro-section can be clearly seen in relation to the white lines on the photo. The hardness trace shows a re-hardening zone corresponding to the micro section and the visible burn area. The trace of the hardness measurement is not in every case as clear as in this picture.

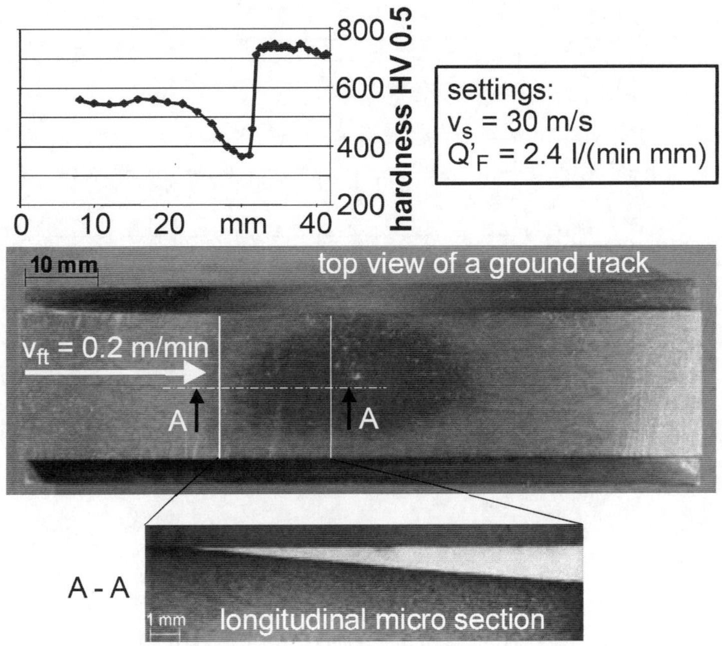

Figure 5: Photo of the ground track and longitudinal micro section.

Several workpieces were measured using the surface integrity assessment techniques mentioned earlier, a typical example being shown in Figure 6. All values are plotted versus the specific material removal rate, to examine the critical value, Q'_{wCRIT} , for the used methods. In this plot the initial depth of cut was greater than zero and the initial wheel contact part of the measured spindle power was not included. The spindle power is increasing linearly for higher specific material removal rates, then at $Q'_w = 15\ mm^3/mm{\cdot}s$ a sudden rapid increase is visible when the critical power flux is reached for the particular grinding fluid and wheel speed used. This point is denoted the 'critical' specific material removal rate at spindle power surge, $Q'_{w.SPI}$. The effect of critical power flux was explained by Malkin [9] to be due to the increase in specific grinding energy when the workpiece surface temperature increases due to film boiling. During this 'dry grinding' phase there will be little lubrication and the chip formed will be more ductile, both recipes for higher specific grinding energy. Malkin also indicated that some wheel loading may also occur, however, this may be masked by the continuous dressing. Andrew [10] also observed this power surge in CDCF grinding at 'burnout'.

At around $Q'_w = 15.5\ mm^3/s{\cdot}mm$, a re-hardening of the workpiece can be seen from the hardness data. The trace of the surface hardness serves to find the initial point of change, i.e., - the critical $Q'_{w.H}$. The area of visible burn, as well as the area detected by nital etching, starts at the critical material removal rate $Q'_{w.VB}$ and $Q'_{w.NEB}$ respectively of $15.6\ mm^3/mm{\cdot}s$ in both cases and is in the same range as the re-hardening point. Both methods are typical methods used on the shop floor to determine burn. However, once these visible signs of burn are just seen by the operator, thermal damage of the subsurface has already happened, which has been proved by the x-ray residual stress measurements.

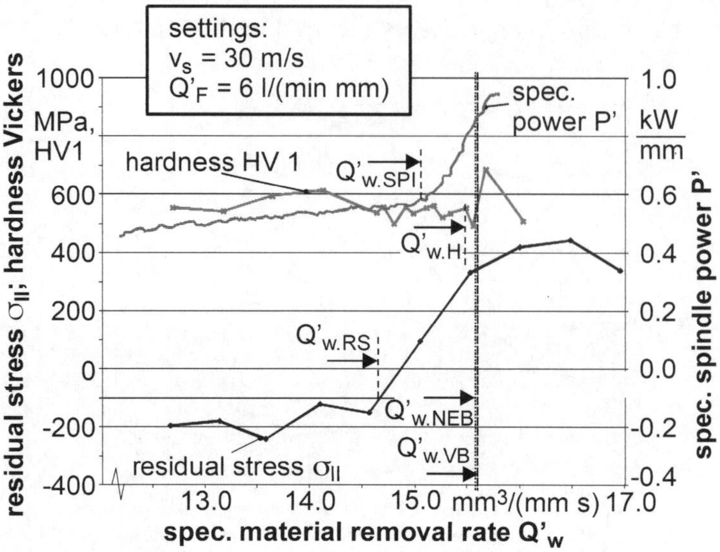

Figure 6: Comparison of methods to detect thermal impacts.

Residual stress measurement is a suitable method to detect alterations of the surface layer [11]. The generated residual stresses reflect the thermal and mechanical influences of the grinding process on the surface layer. The thermal impact of the grinding process can raise the residual stresses up to the non-desirable tensile residual stress domain (residual stress > 0 MPa). This increase must be prevented by the fluid supply, because tensile residual stresses promote crack formation and spreading – increasing over the lifetime of the workpiece. Often the thermal influences on the surface layer are detected earlier by residual stresses than by hardness measurement or metallographic analyses of the micro structure [12].

The residual stress XRD trace in Figure 6 ($Q'_{w\cdot RS}$) shows a transition from compressive to tensile stress far ahead of the other forms of detection. If the residual stress XRD slope increase was used instead of the +/- transition, then an even earlier indication of damage would result since the fatigue life of a ground workpiece would be improved if the stresses remained compressive as opposed to neutral. Therefore, to determine the critical specific material removal rate, $Q'_{w.RS}$, using the residual stresses, the increase in the trace slope is used with the crossing of the threshold of σ_{II} = -100 MPa used for this event represented on the plot.

A further method for detecting changes in the surface layer of ferromagnetic workpieces is the micro-magnetic Barkhausen Noise technique. This method determines the changes in the ferromagnetic properties, caused by the grinding process, in the surface layer. A variation of the ferromagnetic properties results in an increasing or decreasing of the micro magnetic quantities. This quantity often corresponds very well to the residual stress XRD measurement [13]. Figure 7 shows the comparison of the Barkhausen noise amplitude with the X-Ray residual stress and the spindle power trace. The slope of the Barkhausen noise amplitude shows a similar behaviour as the XRD residual stress trace but lags behind. There is a change in the slope at the zero residual stress condition. The other studied specimens shows the same results. So the workpiece analysis by residual stress measurement is by far the best way to determine the start of thermal influence on the workpiece surface layer.

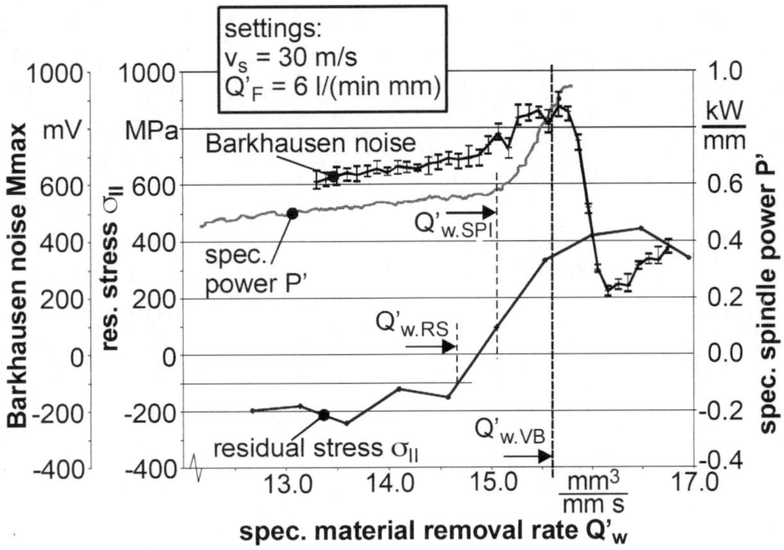

Figure 7: Comparison of Barkhausen Noise, X-Ray Diffraction and spindle power.

Figure 8 shows a comparison between spindle power, P, the spindle power related to the contact width, P', and the spindle power related to the contact area, P'' (power flux). The y-axis has a logarithmic scale to show all three traces in one graph. The spindle power is increasing linearly with increasing specific material removal rate up to the critical Q'_{wSPI} as mentioned before a rapid rise. The same increase is seen for the specific spindle power P'. In comparison, the trace of the specific spindle power related to the contact area is on an almost constant level up to the critical $Q'_{w.SPI}$. This phenomena indicates the influence of the contact length on the power increase.

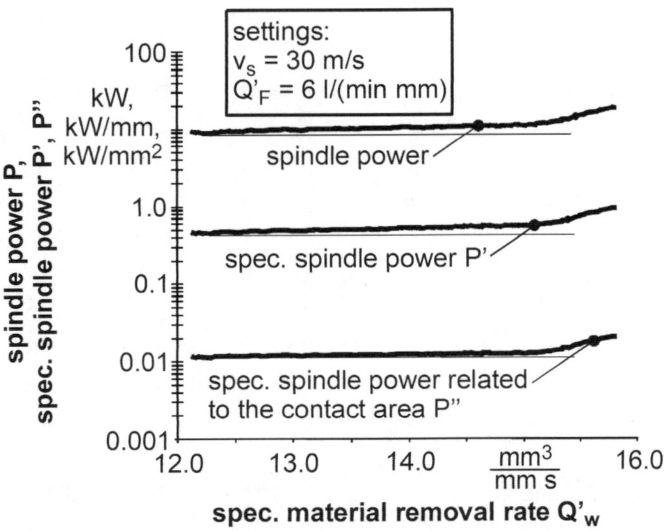

Figure 8: Influence of the contact length on spindle power.

Table 1 is a summary of the results obtained for the specified wheel. The specific power values are obtained by subtracting the power just prior to contact, to eliminate hydrodynamic effects and idle running power. The critical specific material removal rates detected by all five methods are shown in the table. In most cases the residual stress increase was the earliest method of detecting changes in the workpiece material due to thermal impact (up to 10% earlier in one case).

vc [m/s]	Q'F [l/(min mm)]	specific power/ energy values at start of increase			critical specific material removal rate Q'wCRIT [mm3/(mm s)]				
		P'SPI [W/mm]	P''SPI [W/mm2]	U = P'SPI / Q'w.SPI [J/mm3]	Q'w.SPI	Q'w.VB	Q'w.NEB	Q'w.H	Q'w.RS
20	2.4	461	10.0	41.9	11.0	11.0	11.1	-	10.6
20	4	454	11.2	36.8	12.4	12.5	12.5	12.3	11.1
20	6	448	11.0	40.2	11.1	11.1	11.0	-	10.7
30	2.4	463	10.9	40.9	11.3	11.8	-	11.4	10.8
30	4	584	12.9	38.8	15.1	15.6	15.5	14.8	15.0
30	6	566	12.7	37.6	15.1	15.6	15.6	15.5	14.7
35	2.4	421	11.0	45.1	9.3	-	-	9.9	9.4
35	4	543	13.2	45.9	11.8	12.6	12.5	12.5	12.1
35	6	445	12.6	43.8	10.2	11.6	11.3	11.1	10.3

SPI: start of spindle power increase H: hardness change
NEB: start of burn after nital etching VB: start of visual burn
RS: residual stress increase (>100 MPa)

Table 1: Critical material removal rates and power/energy values

The specific power values, P'_{SPI}, at each grinding condition, based on the power surge, are included in the table. Also, the specific grinding energy is shown for each grinding condition. The average values of specific grinding energy for each wheel speed showed no change from 20 m/s to 30 m/s (at 39 J/mm^3) and a large increase from 30 m/s to 35 m/s (at 45 J/mm^3). This small speed increase should not have such a pronounced effect on specific grinding energy, the reason for it will be explained later. The critical power flux at burnout is shown as P''_{SPI}, obtained by dividing P'_{SPI} by the contact length at burnout. The range of power flux values at burnout achieved during these tests was 10-13 W/mm^2. By comparison, Malkin [9] recorded between 13-26 W/mm^2, and Andrew [10] recorded a range of 26-35 W/mm^2, both with similar process conditions.

The burn study shows that residual stress assessment of burn is more sensitive than visual burn, nital etch, power surge and hardness change. However, power surge is much more convenient to use for process monitoring, providing the user considers that burn occurs prior to visual burn and the removal rate used must be up to 10% lower than the critical value.

Titanium Alloys

A recent engine failure on a Southwest Airlines jet was caused by metal fatigue of one of the 24 fan blades of the Turbofan engine leading to detachment at the root, near the central hub [Figure 9]. The engine in question is one of the most popular models in the world, with over 6,700 aircraft being fitted with them. The fan material in question is made from Gamma Titanium Aluminide (γ-TiAl). This alloy has high specific strength, high-temperature stability and oxidation resistance relative to conventional titanium and nickel alloys. Grinding is a key process for manufacturing turbine blade roots and typically fewer integrity problems are encountered during abrasive machining of γ-TiAl alloys [14, 15] than for any other chip-forming processes.

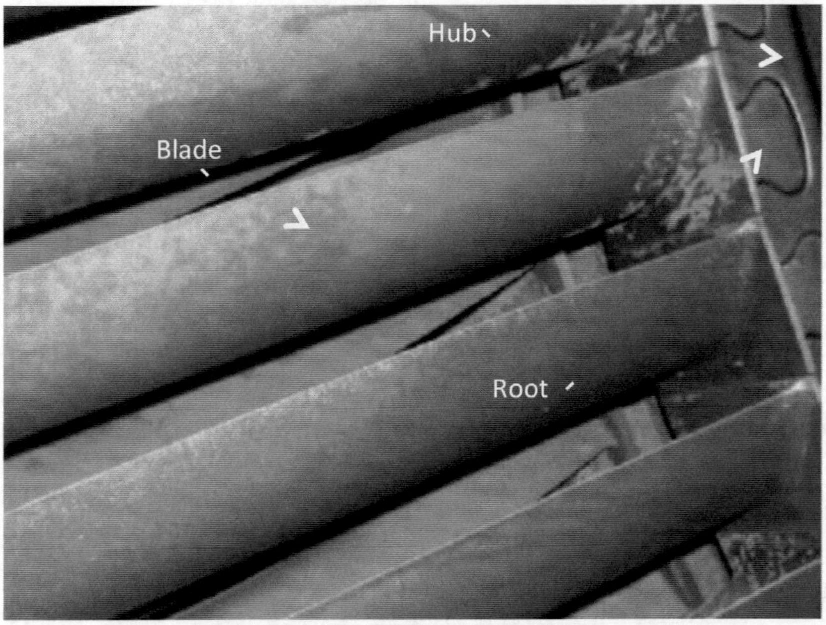

Figure 9. Turbofan blades assembled to the hub

The traditional grinding wheel abrasive used for this material is silicon carbide (SiC), in a vitrified bond, using creep feed grinding, with additional finishing and sparkout passes. When employing such a regime, workpiece surface damage was reported [16] to include minimal surface burn and discoloration, with sub-surface deformation and cracking maintained within acceptable limits. Furthermore a hardened layer ~100 μm deep was observed, with a maximum hardness of 575 HK (compared to the bulk value of 350-400 HK), and compressive residual stresses of up to 300 MPa at the workpiece surface. Stone [17] showed that diamond grinding wheels are preferred over conventional grit wheels due to lower operating temperatures and levels of workpiece damage less than 80 μm deep. When creep-feed grinding with diamond abrasive wheels [29], workpiece residual stress measurements by Hood [28] showed compressive values of up to 600 MPA in the transverse direction, with G-Ratios typically ~1000% higher than with SiC at similar operating parameters.

According to Biro [18], when comparing the grinding of γ-TiAl using electroplated diamond, electroplated CBN, and silicon carbide abrasive wheels, with careful selection of grit sizes, the diamond wheel drew consistently the lowest grinding power for up to 1800 mm^3 of material removed, without burn. With the CBN wheel the power increased a little more than the diamond wheel, again without burn. The silicon carbide wheel grinding power increased dramatically, resulting in burn after 1000 mm^3 of material removed. The results are shown in Figure 10.

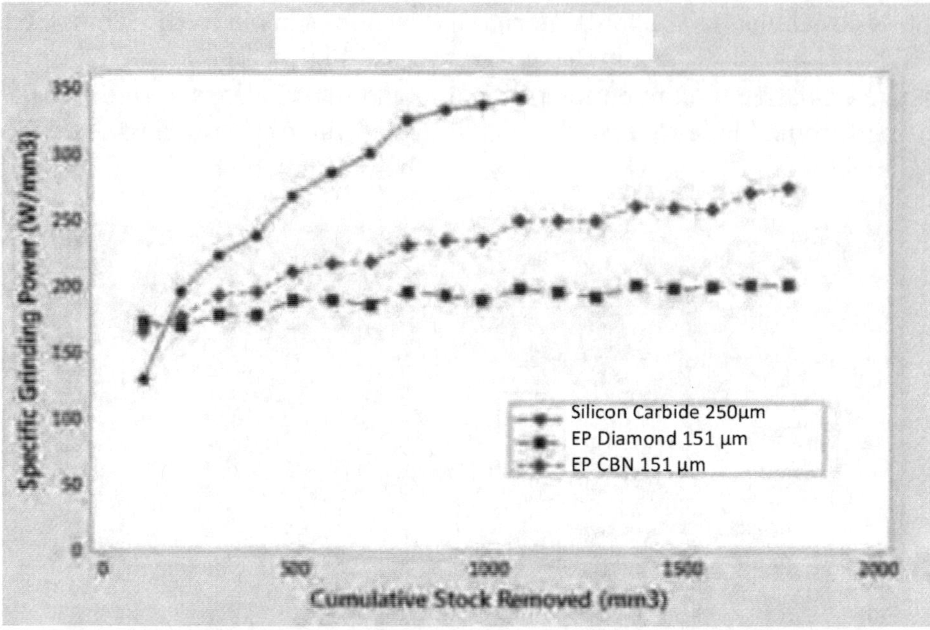

Figure 10. Comparison of grinding power for different abrasive grain types (Biro)

According to Hood [19] the recommended operating conditions for surface grinding of Ti-45AI-2Mn-2Nb + 0.8 vol.% TiB2 using vitrified SiC wheels includes a wheel speed of 30 m/s, a feed rate of 0.4 m/s, with 80 mesh/ grade H products employed for roughing and 120 mesh/grade K for finishing [Bentley 20]. Additionally, roughing depths should be a maximum of 25 µm, reducing to between 5 to 10 µm for finishing [Aspinwall 21] with sparkout used in the final passes [Bentley 22]. When employing such a regime, workpiece surface damage was reported to include minimal surface burn and discoloration, with subsurface deformation and cracking maintained within acceptable limits. Furthermore, a hardened layer ~100 µm deep was observed, with a maximum hardness of 575 HK (compared to the bulk value of 350-400 HK) and compressive residual stresses of up to 300 MPa at the workpiece surface. Superabrasive diamond wheels are preferred over conventional grit wheels due to lower operating temperatures and levels of workpiece damage (up to 80 µm deep) [Stone 17, Razavi 23]. The advantages of using surface grinding include lower cost of equipment and wheels that are often available in most machine shops, however the low depth of cut and feed rate necessary to produce surfaces of acceptable integrity, result in a relatively low material removal rate, typically ~2-4 mm^3/mm.s [Bentley 20, Bentley 22].

Creep feed grinding (CFG) of Ti-45AI-2Mn-2Nb + 0.8 vol.% TiB_2 using conventional abrasives has been shown to be successful under a wheel speed of 30 m/s with continuous dressing, provided that both depth of cut and table speed are kept relatively low (1 mm and 100 mm/min, respectively) [Bentley 24]. The use of continuous dressing reduced the incidence of wheel wear-flats and hence the area of contact, causing both the sliding component of the grinding force and the energy imparted into the workpiece to decrease, thus improving surface integrity by minimizing workpiece burn. When CFG with superabrasive wheels [Bentley 25], workpiece surface residual stress measurements showed compressive values of up to 600 MPa in the transverse direction, whilst tensile residual stresses of up to 700 MPa were recorded along the feed direction. Both these stresses were reduced by the application of a sparkout pass, which left a crack-free surface [Bentley 26]. More recent published research relating to CFG has focused on the alloy Ti-45AI-8Nb-0.2C [Hood 27], where a G-ratio of ~30 was achieved using high grit content/low porosity SiC abrasive wheels at a low wheel speed of 15 m/s, depth of cut of 1.25 mm and feed rate of 150 mm/min. The subsequent assessment of workpiece surfaces revealed the presence of a hardened layer, albeit with only a moderate increase in hardness of up to 520 HK (bulk value 350-400 HK) and damage consisting of intermittent deformation/bending of the lamellae to a depth of ~4μm, with no visible cracking [Hood 29]. Increasing the wheel speed and/or depth of cut and feed rate, caused significantly greater workpiece surface burn and cracking, along with increased surface hardening of up to ~ 150 HK above the bulk value and bending of lamellae up to ~20μm [Hood 28]. Tests using more open SiC wheels showed lower forces and power levels along with a reduction in surface burn and cracking, however, G-ratios were reduced to <5 [Hood 29]. With superabrasive diamond wheels, G-ratios were typically ~ 1000% higher than when utilizing SiC wheels at similar operating parameters as a consequence of the increased grit hardness and thermal conductivity. Additionally, increasing grinding speeds up to 50 m/s resulted in workpiece surfaces that were comparable in terms of microhardness and surface/subsurface damage to those produced using SiC wheels at low wheel speeds [Hood 30].

Nickel-Based Alloys

Nickel-based super-alloys constitute 40-50% of engine weight in current commercial aircraft and are deployed particularly within the combustor and turbine sections where operating temperatures can exceed 1250C. Grinding is a key process for the manufacture of blades and vanes as economic machining of cast nickel-based super-alloys using defined cutting edges (milling) is difficult [15]. Depending on the part design, up to 12 distinct abrasive cutting operations are required to machine a blade, including face, plunge, profile, and arc grinding. Arguably, the most important feature on a blade is the 'fir tree' shaped profile at the root that locates it into the rotor, and has to withstand immense centrifugal forces.

The most common approach in industry used to grind the turbine blades is continuous dress creep-feed grinding, using low table speed, low wheel speed (15-30 m/s) and large depth of cut, with an aluminum oxide grinding wheel. Osterle et al [31] evaluated the influence of operating conditions on workpiece residual stress when creep-feed grinding IN738LC. The residual stress profile shown in Figure 11a for continuous dress creep-feed grinding shows a tensile stress of 400 MPa at the surface, rapidly changing to a compressive stress of -800 MPa 10μm below the surface, and becoming tensile again at 70 μm below the surface. By comparison Figure 11b shows the surface ground using a CBN wheel, with intermittent dressing, which had lower tensile stress at the surface and remained as a compressive stress until 90 μm below the surface. The relatively low tensile stresses using the CBN wheel were attributed to the high thermal conductivity of CBN versus aluminum oxide.

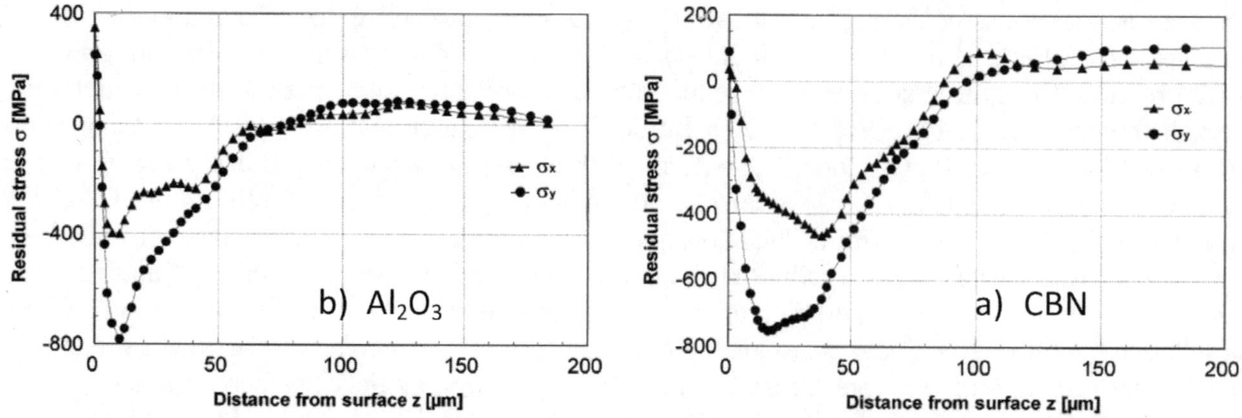

Figure 11. Residual stress depth profiles for creep feed grinding using aluminum oxide and CBN abrasive (σx means measured in the grinding direction, and σy the transverse direction)

Guo [32] showed that thermal damage in the form of a WEL is of particular concern and needs to be avoided for aerospace components. For evaluating the dependence of WEL on wear flat area and temperature rise, he ground 17 specimens under various grinding conditions when the CBN electroplated wheel was at various stages of wear. In Figure 12, only one specimen (c) exhibits WEL. Micrographs of two specimens with no WEL present are also shown in Figure 13

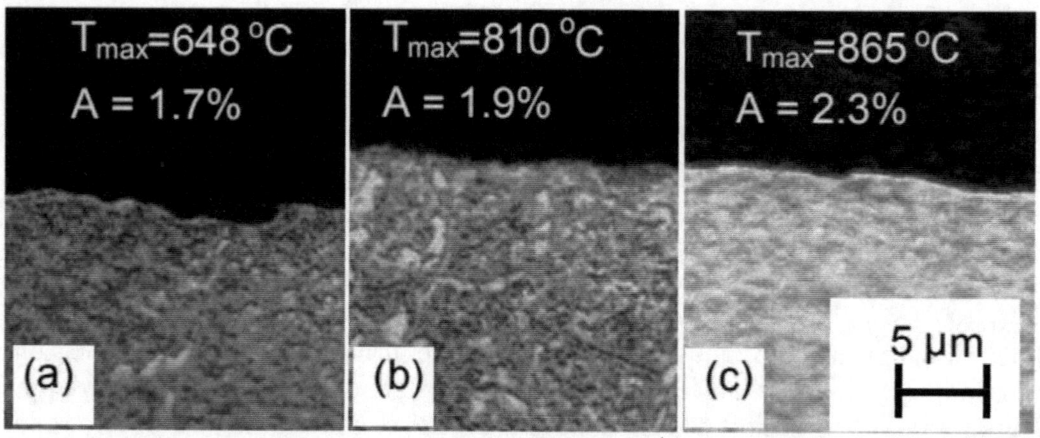

Figure 12. Micrographs of Ground Specimen Section Showing WEL in (c)

Tests done at WWSA on an ANCA TX7, using an electroplated wheel with AMB Type CBN grain, at 40 m/s, and using water-based coolant, confirmed the presence of high compressive residual stresses (-800 MPa) at the surface of INCO718 testpieces, as shown in Figure 13. The maximum compressive stress is at 20 µm below the surface, and no tensile stress at all is in the bulk of the material. The result also shows that for the four different specific removal rates: 1, 2, 3, and 4 mm^3/s.mm. The residual stress profile does not change dramatically as productivity is increased.

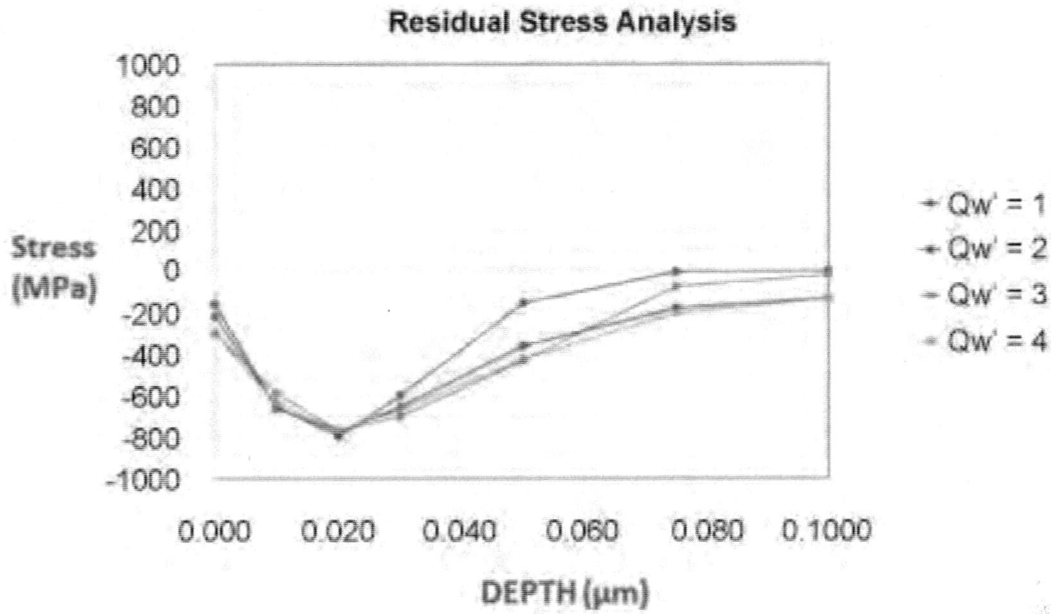

Figure 13. Residual stress profile using CBN at 4 different Q_w' on inconel

Steel Alloys

Sawicki [33] presented his results of a study aimed at determining the residual stresses which result from grinding carburized 17CrNi6-6 gear and bearing steel, using both CBN and aluminum oxide grinding wheels. Figures 14 and 15 show a comparison of the residual stress distribution for samples carburized by traditional methods and then ground using a 38A60K8V aluminum oxide wheel, and RNB80/63B75V CBN wheel. Three different cooling strategies were used: A. dry grinding; B. flood oil emulsion; and C. MQL oil. In all cases with the aluminum oxide wheel the surface residual stresses were tensile. In the case of the CBN wheel all surface residual stresses were compressive with a transition to tensile approximately 0.4 – 0.5mm below the surface

Brinksmeier [11] compared the surface residual stresses of aluminum oxide and CBN grinding wheels when surface grinding 100Cr6 bearing steel, hardened to 62 Rc. In both cases the grinding wheel speed was 30 m/s, depth of cut 7 µm, and table speed 400 mm/s. In the case of the aluminum oxide grinding wheel the surface residual stress was approximately 480 MPa, increasing to 800 MPa at 20 µm depth, and was always tensile. In the case of the CBN grinding wheel the surface residual stress was approximately -400 MPa (compressive) and decreasing to zero at 20 µm depth. Figure 16 shows the residual stress curves relative to depth below the surface. Both transverse and longitudinal stresses are shown.

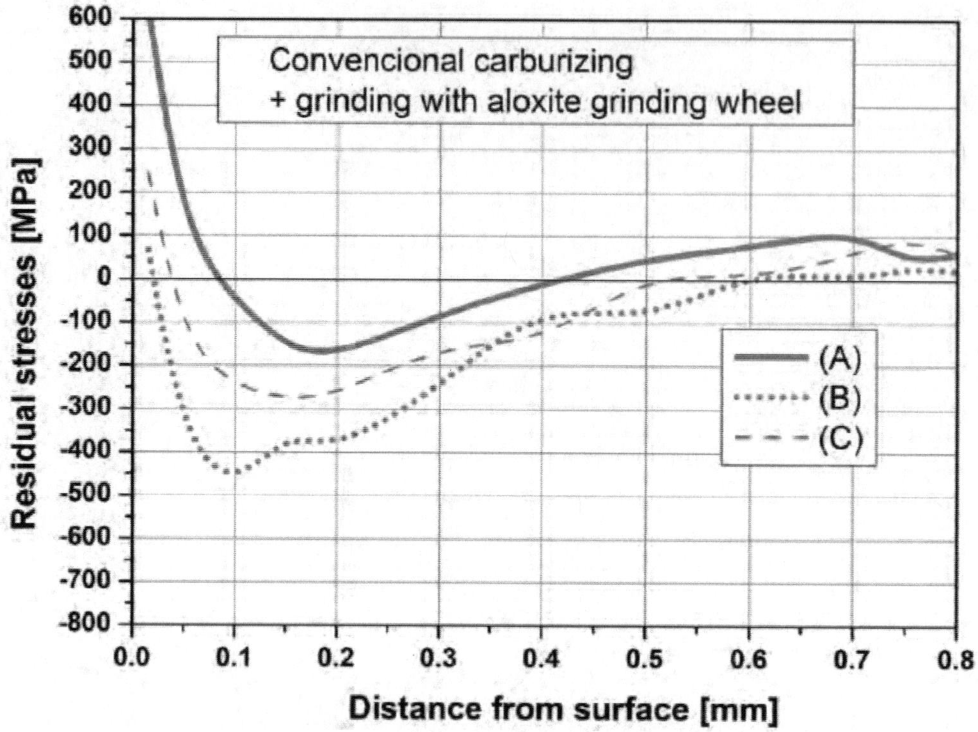

Figure 14. Distribution of residual stress using aluminum oxide wheel

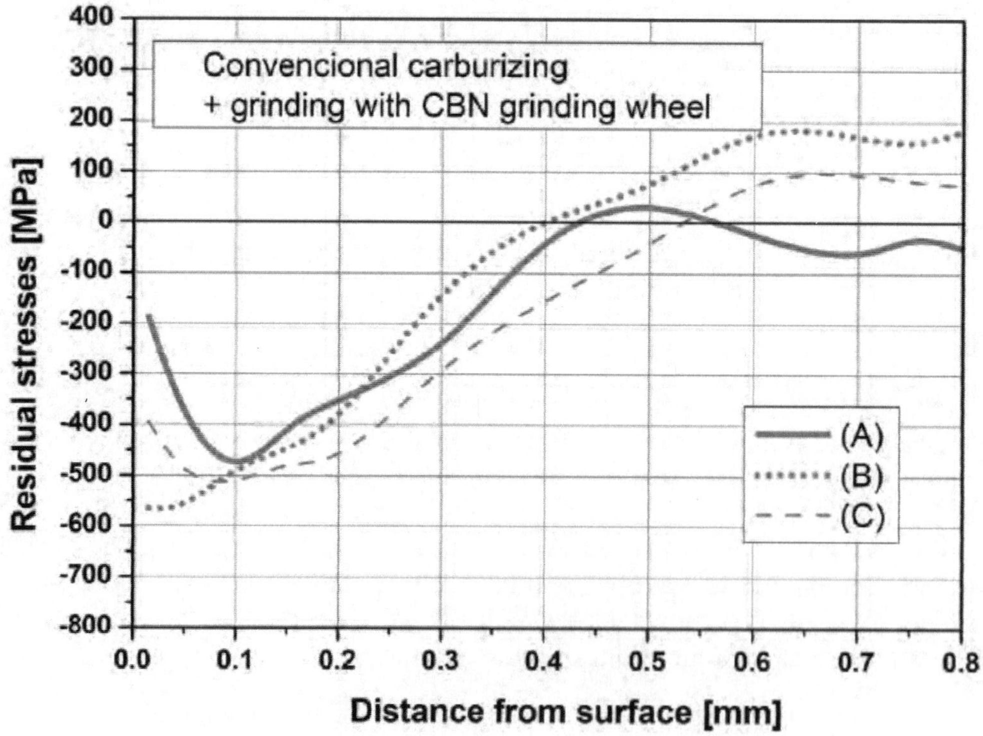

Figure 15. Distribution of residual stress using CBN wheel

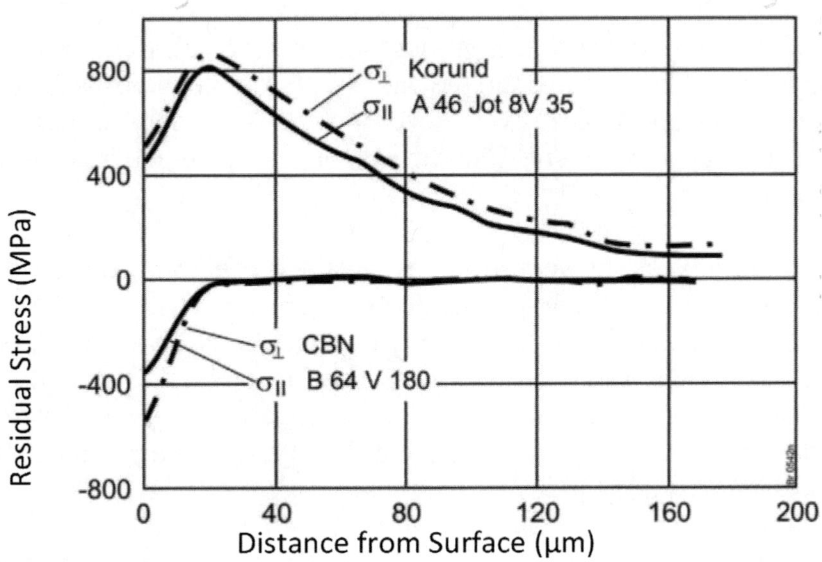

Figure 16. Residual stress comparison between aluminum oxide and CBN grinding wheels

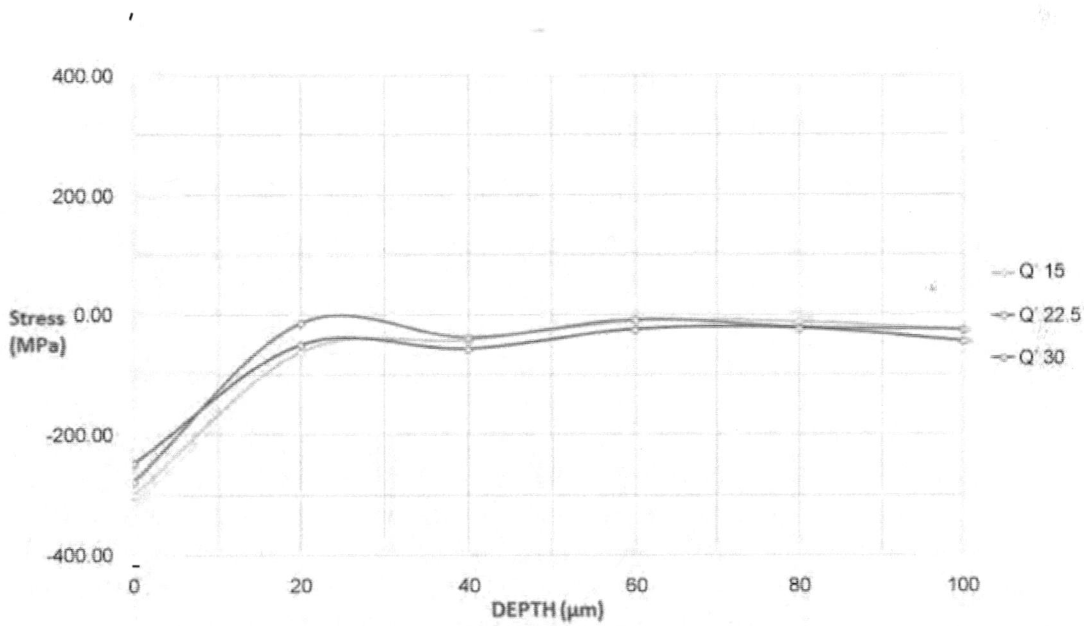

Figure 17. Residual stress profile using CBN AMA-1 grain at 3 different Q_w' on steel

Tests done at WWSA on an ANCA TX7, using a vitrified bond wheel with AMA Type CBN grain, using water-based coolant, confirmed the presence of medium compressive residual stresses (-300 MPa) at the surface of the 52100 hardened steel testpieces, as shown in Figure 17. The maximum compressive stress is at the surface, and no tensile stress at all is in the bulk of the material. The result also shows that for the three different specific removal rates of: 15, 22.5, and 30 mm^3/s.mm, the residual stress profile does not change dramatically as productivity is increased.

Conclusions

The favorable residual stresses, surface finish, productivity, and wheel life associated with superabrasive grinding wheels shows considerable value with rotating aerospace components. Other critical components that also endure cyclic loading such as bearings, transmission shafts, crankshafts and camshafts, can also benefit from the excellent surface integrity produced by superabrasive wheels.

In this paper X-Ray diffraction has shown to be an extremely useful tool to assess the onset grinding damage in advance of other traditional techniques, such as visible burn, nital etch, micr-hardness, fluorescent dye penetrant, eddy current, etc. Barkhausen Noise is also a useful technique, however it gives a relative assessment of surface stresses, takes skill to use correctly with a custom sensor geometry, and only works with magnetic materials, such as steels. For most aerospace materials that are subjected to cyclic loading, X-Ray diffraction evaluation of residual stresses should be the standard for the industry.

The use of ramped grinding tests to screen the burn limit for a process based on: wheel type; bond; grit spec; wheelspeed; coolant application, etc, is a cost and time effective method. Once the ramp test has screened the test conditions specific verification tests should be done.

References

1. Rangaswamy, P., Terutung, H., and Jeelani, S., Effect of grinding on the fatigue life of titanium alloy under dry and wet conditions, NASA Technical Library, Document N91-28121, 1991
2. Jawahir, I., Brinksmeier, E., M'Saoubi, R., Aspinwall, D., Outerio, J., Meyer, D., Umbrello, D., and Jayal, A., 2011. "Surface Integrity in Material Removal Processes: Recent Advances". CIRP Annals-Manufacturing Technology 60(2):603-626.
3. Field, M., Kahles, J., 1964. "The Surface Integrity of machined and Ground High Strength Steels". DMIC Report 210:54-77
4. Heinzel, C., Residual Stress (Abrasive Processes), CIRP Encyclopedia of Production Engineering
5. Control of Residual Stress in Production Grinding With CBN, K.V. Kumar, R.R. Matarrese and E. Ratterman, SAE Transactions, Vol. 98, Section 5: JOURNAL OF MATERIALS AND MANUFACTURING (1989), pp. 819-830
6. Toenshoff, H. K., Brinksmeier, E., and Choi, H. Z., Abrasives and their influence on forces, temperatures and surface integrity, Proc of the 15th NAMRC, 1987, 70-89
7. Webster, J., Brinksmeier, E., Heinzel, C., Wittmann, M., and Thoens, K., Assessment of grinding fluid effectiveness in continuous-dress creep feed grinding, CIRP Annals - Manufacturing Technology, 51/1/2002, 235 – 240
8. Toenshoff, H. K.; Brinksmeier, E.; Hetz, F., 1987, Detection of Microcracks, Annals of the CIRP, 36/2: 545-552
9. Malkin, S., 1989, Grinding Technology - Theory and Applications of Machining with Abrasives, John Wiley and Sons, ISBN 0-85312-756-5
10. Andrew, C.; Howes, T.; Pearce, T., 1985, Creep Feed Grinding, Holt, Rinehart and Winston

11. Brinksmeier, E., 1982, Randzonenanalyse geschlif¬fener Werkstuecke, Dr.-Ing. Diss., University of Hannover
12. Heinzel, C., 1999, Methoden zur Untersuchung und Optimierung der Kuehlschmierung beim Schleifen, Dr.-Ing. Diss., Univ. of Bremen
13. Friemuth, T.; Mandrysch, T., 1999, Determination of Hardness and Residual Stresses Depth Profiles Using Micromagnetic Techniques. 2nd International Conference on Barkhausennoise and Micromagnetic Testing, October 25-26, Newcastle, UK
14. Bentley, S., Aspinwall, D.,1997. "A Comparison of the Grinding of IMI318 Titanium Alloy and a Gamma Titanium Aluminide Intermetallic". Proceedings of the 32nd International MATADOR Conference, UMIST (Manchester), pp. 331-336. ISBN: 1-333-7295-8.
15. Klocke, F., Soo, S. L., Karpuschewski, B., Webster, J., Novovic, D., Elfizy, A., Axinte, D., and Tonissen, S., 2015, "Abrasive Machining of Advanced Aerospace Alloys and Composites", CIRP Annals Keynote-Manufacturing Technology 64(2015):581-604.
16. Bentley, S., Goh, N., Aspinwall, D., 2001, "Reciprocating Surface Grinding of Gamma Titanium Aluminide Intermetallic Alloy". J. Material Processing Technology, 53: 22-28. (ISSN: 0924-0136).
17. Stone, W., Kurfess, T., 2007, "Grinding Titanium Aluminide: Subsurface Damage", Int. Journal of Manufacturing Technology Management, 12/1-3: 60-71. (ISSN: 1368-212).
18. Biro, A., Varghese, P., Grinding Gamma Titanium Aluminide, Aerospace and Defense Manufacturing, Society of Manufacturing Engineering, September 2015
19. Hood, R., Cooper, P., Aspinwall, D.K., Sool, S.L., and Lee, D.S., Creep-feed grinding of γ-TiAl using single-layer electroplated diamond superabrasive wheels, Annals of CIRP-Manufacturing Technology, CIRP Journal of Manufacturing Science and Technology, 11 (2015) 36-44
20. Bentley, S.A., Aspinwall, D.K., 1997, Identification of significant operating factors when conventional abrasive creep feed grinding of a gamma titanium aluminide intermetallic, in: Proceedings of the 4th International Charles Parsons Turbine Conference, Advances in Turbine Materials, Design and Manufacturing (Newcastle), pp.474-488. ISBN 511250533.
21. Aspinwall, D.K., Dewes, R.C.. Mantle, AL., 2005, The machining of 'Y-TiAi intermetallic alloys, Ann CIRP, 54/1: 99-104. (ISSN: 007 8506).
22. Bentley, S.A.. Lim, S.Z., Aspinwall, D.K., 2000, The effect of wheel dressing parameters on the surface integrity of surface ground gamma titanium aluminide intermetallic alloy, Intermetallics Superalloys, 10:216-221. ISBN 3527 30575. (Proceedings of the EUROMAT 99).
23. Ali Razavi, H., Kurfess, T.R., Danyluk, S., 2003, Force control grinding of gamma titanium aluminide, Int J Mach Tools Manuf, 44:50-56. (lSSN: 0890-6955)
24. Bentley, S.A., Aspinwall, D.K., 1997, Identification of significant operating factors when conventional abrasive creep feed grinding of a gamma titanium aluminide intermetallic, in: Proceedings of the 4th International Charles Parsons Turbine Conference, Advances in Turbine Materials, Design and Manufacturing (Newcastle), pp.474-488. ISBN 511250533.
25. Bentley, SA, 2000, Surface and Creep Feed Grinding of Gamma Titanium Aluminide, School of Manufacturing and Mechanical Engineering, The University of Birmingham, Birmingham. (Doctoral Thesis).

26. Bentley, SA, Aspinwall, D.K., 2000, Creep feed grinding of γ- TiAl, Proceedings of the 33rd MATADOR Conference Manchester, 449-446. ISBN 50233 3235.
27. Hood, R., 2010, Machinability of a Gamma Titanium Aluminide Intermetallic, School of Manufacturing and Mechanical Engineering, The University of Birmingham, Birmingham. (Doctoral Thesis).
28. Hood, R, Aspinwall, D.K., Voice, W., 2007, Creep feed grinding of gamma titanium aluminide and burn resistant titanium alloys using SIC abrasive, Int J Mach Tools Manuf, 47/July (9): 1486-1492.
29. Hood, R., Aspinwall, D.K., Voice, W., 2007, Creep feed grinding of a gamma titanium aluminide intermetallic alloy using SiC abrasives, J Mater Process Technol, 191/1-3: 210-214. (2007).
30. Hood, R., Aspinwall, D.K., Voice, W., Dando, P., Tuffy, K., 2007, Creep feed grinding of a gamma titanium aluminide intermetallic using superabrasives, Advances in Abrasive Technology X Proceedings of the ISAAT 2007 Conference 25-29 September 2007 Detroit USA, 397-403.
31. Osterle, W., Li, P., Nolze, G., 1999, "Influence of Surface Finishing on Residual Stress Depth Profiles on a Course-Grained Nickel-Based Superalloy", Materials Science and Engineering A238: 308-311
32. Guo, C., Shi, Z., Attia, H., McIntosh, D., Power and wheel wear for grinding nickel alloy with plated CBN wheels, Annals of CIRP, 56/1/2007, 343-346
33. Sawicki, J., Kruszynski, B., and Wojcik, R., The influence of grinding conditions on the distribution of residual stress in the surface layer of 17CrNi6-6 steel after carburizing. Advances in Science and Technology Research Journal, VOL 11, ISSUE 2, June 2017, pp. 17-22

GRINDING OF ADDITIVELY MANUFACTURED AEROSPACE ALLOYS

Intertech 2019 – Paper #1940

K. Philip Varghese, PhD, Group Leader, Saint-Gobain Abrasives

John Hagan, Sr. Application Engineer, Saint-Gobain Abrasives

Address: 9 Goddard Road, Northboro, MA, 01532

Keywords

Aerospace, Additive Manufacturing, Superalloys, Residual stresses, grinding

Abstract

Additive manufacturing of components in the aerospace industry are being pursued by a number of different organizations. As the part families expand both in type and volume, there is a growing need to modify the surfaces of these parts to meet the final application requirements in a cost-effective manner. Some of these manufactured components have a simple geometry and some are very complex. Since the additive process produces components in near-net shape, there is often very little stock that remains on these components for providing final finish. Therefore, any post-processing finishing methods that are employed need to be extremely consistent in achieving part tolerances and surface quality. This paper demonstrates the results from a finish grinding study conducted on additively manufactured nickel-based superalloy (IN718) and titanium alloy (Ti64) specimens that are used for various applications in the aerospace industry. Results include grinding process performance measures such as power, material removal rates, grinding forces as well as 3-D surface finish maps, and metallurgical analysis data of the ground specimens.

Background

Growth of Additive manufacturing (AM) in the aerospace and medical markets is creating an increasing need to finish the surfaces of additive parts to meet final application requirements. Since the AM process is capable of producing components close to near-net shape, there is often little stock that remains for providing final finish. This causes a reduction of waste in manufacturing, but it also means that post-processing has limited room for errors and inconsistencies. In such a scenario, post-processing needs to be consistent in achieving part tolerances and surface quality. At the state –of-the-art Higgins Grinding Technology Center in

Northboro, Massachusetts, engineers from Norton/Saint-Gobain conducted a finish grinding study on Inconel 718 and Ti-6Al-4V specimens made additively. The goal of the study was to find answers to the following pertinent questions:

1. What is the minimum volume that needs to be ground off an AM component?
2. What is the range of surface finishes obtained from grinding an AM component with latest generation of abrasive grinding wheel?
3. What, if any, is the impact on surface residual stresses on an AM component after finish grinding?

Finishing of AM IN-718 Alloy

AM IN718 specimens manufactured using the direct metal laser sintering (DMLS) process were obtained from Stratasys Direct Manufacturing. Figure 1 shows the build direction of the AM test specimen. After AM, these specimens were media blasted and went through stress relief, hot isostatic pressing, solution treatment/anneal and precipitation hardening. The heat treatment parameters used for the AM specimens as well as the standard cast specimens is shown in Figure 1. The hardness of the specimens measured to be approximately 40 HRC. The specimens were then ground using a Norton NQX60E24VTX2 grinding wheel on a Magerle MFP-125.50.65 creepfeed grinder located at the Norton | Saint-Gobain Higgins Grinding Technology Center. Figures 2 shows a picture of the test set-up.

At the beginning grinding was done at progressively increasing depths of cut to determine the minimum stock that needs to be removed to clean up an additively manufactured surface. After each trial at a specific depth of cut, the surfaces were looked at for the presence of visual imperfections. Grinding was continued until the surfaces looked free of visual imperfections. The trial results determined that in order to eliminate any geometrical inconsistencies and surface imperfections related to the AM process, approximately 0.30 - 0.45 mm of stock needs to be removed from the component.

For this test, the grind length was 50.8 mm, and the wheel diameter was 462 mm, operating at 23 m/sec. Tables 1, 2 and 3 document test details and grinding conditions used in the study. Multiple material removal rates (MRR') were tested ranging from 0.1 mm^3/sec./mm to 2.0 mm^3/sec./mm to determine the effect of low to medium MRR's on surface finish and grinding power. The MRR' was adjusted by altering the feedrate (mm/min.) or the depth of cut (mm). The MRR's selected in this study were lower, based on the typical finishing rates used in industrial processes now to finish standard Inconel workpieces.

Federal Contact profilometer system 5000 was used to measure the surface finishes of the workpieces before and after grinding. Measurements were taken at several locations and in two directions on the workpiece. Non-contact profilometer surface measurements using Nanovea 3D surface profilometer, white light chromatic aberration technique was also taken in multiple locations. Figure 3 and 4 show the variation of surface finish Ra and Rz respectively before and after grinding as measured in parallel and across the direction of grinding. Figure 5 shows the surface maps of the specimen before and after grinding captured using non-contact profilometry. An evaluation of the surface finish of the incoming AM material before grinding found it to be considerably coarser along the direction of z-build. This large difference would not typically be observed when measuring a standard, as-cast Inconel 718 material. Evaluation of the surface finish on the AM material before grinding versus after grinding found that grinding improved the surface finish significantly. This was true for measurements in both the parallel and perpendicular directions. The AM IN718 had slightly coarser finish as compared to the standard as cast material at each of the MRR' tested.

To understand the impact of grinding on the overall surface integrity of the AM IN718 specimen, surface residual stresses of the AM specimens were measured using Bruker-AXS D8 Discover microdiffraction equipment in the "as received" condition as well as after grinding. Figure 6 shows a comparison of the measurements taken in both conditions at six different locations on the same specimen. Results show that near-surface residual stresses are negative/compressive in both the "as-received" and "after-grind" condition. This is typical of components after finish grinding, where near-surface residual stresses are found to be compressive in nature, an effect that tends to help fatigue life by retarding surface crack propagation.

Finishing of AM Ti6Al4V Alloy

AM Ti6Al4V specimens manufactured using the direct metal laser sintering (DMLS) process were obtained from Stratasys Direct Manufacturing. Figure 7 shows the build direction of the AM test specimen. After AM, these specimens were media blasted and went through stress relief, hot isostatic pressing, solution treatment/anneal and precipitation hardening. The heat treatment parameters used for the AM specimens as well as the standard cast specimens is shown in Figure 7. The hardness of the specimens measured to be approximately 35 HRC. The specimens were then ground using a Norton Paradigm Diamond SD220C168P100X grinding wheel on a Magerle MFP-125.50.65 creepfeed grinder located at the Norton | Saint-Gobain Higgins Grinding Technology Center. Figures 8 shows a picture of the test set-up.

As in the case of AM IN718 specimens, grinding of AM Ti6Al4V specimens was done at progressively increasing depths of cut to determine the minimum stock that needs to be removed to clean up an additively manufactured surface. The trial results determined that in order to eliminate any geometrical inconsistencies and surface imperfections related to the AM process, approximately 0.30 - 0.45 mm of stock needs to be removed from the component.

Tables 4, 5 and 6 document test details and grinding conditions used in the study. Multiple material removal rates (MRR') were tested ranging from 0.1 mm^3/sec./mm to 2.0 mm^3/sec./mm to determine the effect of low to medium MRR's on surface finish and grinding power. The MRR' was adjusted by altering the feedrate (mm/min.) or the depth of cut (mm).

Figure 9 and 10 show the variation of surface finish Ra and Rz respectively before and after grinding as measured in parallel and across the direction of grinding. Figure 11 shows the surface maps of the specimen before and after grinding captured using non-contact profilometry. An evaluation of the surface finish of the incoming AM material before grinding found it to be considerably coarser along the direction of z-build. This large difference would not typically be observed when measuring a standard, as-cast Ti6Al4V material. Evaluation of the surface finish on the AM material before grinding versus after grinding found that grinding improved the surface finish significantly. This was true for measurements in both the parallel and perpendicular directions. The surface finish measurements of the AM Ti6Al4V was found to be very similar to that of the standard as cast material at each of the MRR' tested.

To understand the impact of grinding on the overall surface integrity of the AM Ti6Al4V specimen, surface residual stresses of the AM specimens were measured using Bruker-AXS D8 Discover microdiffraction equipment in the "as received" condition as well as after grinding. Figure 12 shows a comparison of the measurements taken in both conditions at three different locations on the same specimen. Results show that near-surface residual stresses are negative/compressive in both the "as-received" and "after-grind" condition. This is typical of components after finish grinding, where near-surface residual stresses are found to be compressive in nature, an effect that tends to help fatigue life by retarding surface crack propagation.

Conclusion

The results of this study proved that finish grinding of additively manufactured, nickel-based superalloys such as Inconel 718, and Titanium alloys such as Ti6Al4V can be successfully done to achieve a significant reduction in surface roughness. Additionally, grinding wheels with finer abrasive grains or other abrasive products, such as engineered abrasive belts, can be used to generate a super-finished surface if required. The study also showed that approximately 0.30 mm - 0.45 mm of stock needs to be removed from the component to generate improved surface finishes and clean up any geometrical inconsistencies related to the 3D manufacturing process. The minimal stock requirement is critical information for additive manufacturers as they strive to make AM components as close to near-net shape as possible to be cost-effective. Moreover, the residual stress data showed that the surface stresses both before and after grinding were compressive in nature in case of both AM IN718 and Ti6AL4V specimens.

Figures

Figure 1: Build direction of the AM IN718 specimens and heat treatment parameters of both AM and standard cast IN718 specimens

Standard Inconel 718

Anneal
Annealed 1,800°F for 3 hours
Age Hardened (Vacuum Furnace aged - Precipitation hardening)
Vacuum age at 1,325°F for 8 hours
Drop Temperature to 1,150°F in 2 hours
Hold at 1,150°F for 8 hours
Total Time 18 hours
Cool at any rate

Hardness: 39.9 HRC

Additively Manufactured Inconel 718

Stress Relieve
Vacuum Stress Relieved at 1,925°F. Cool at rate equivalent to air cool or faster
Hot Isostatic Press
1950°F for 3-5 hours @ 14.5K PSI minimum, cool in autoclave to 1200°F or below. Cool from 1200°F at any rate.
Anneal
Solution Anneal in vacuum at 1900°F for 1 hour and air cool to 1200°F or below within 30 minutes, cool at a rate equal to air cool or faster when at or below 1200°F.
Age Hardened (Precipitation hardening)
1450°F for 6 to 8 hours, followed by rapid or air cooling.

Hardness: 40.4 HRC

Figure 2: Test Set-up for finish grinding of AM IN718 test specimens

Figure 3: Variation of Surface Finish (Ra) before and after grinding (as measured in parallel and across grind direction)

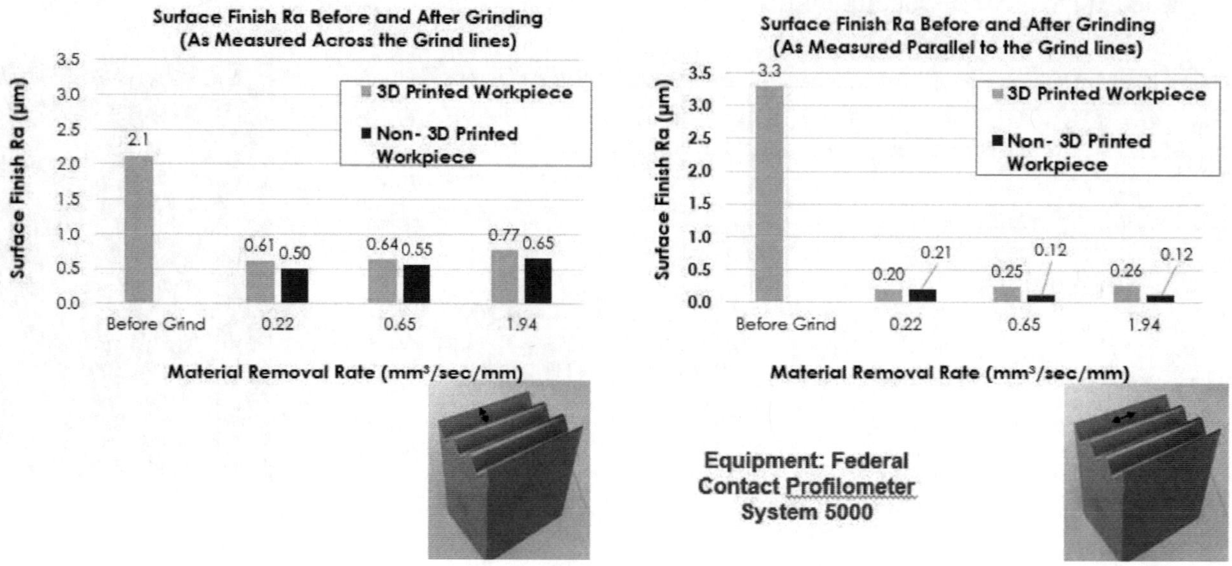

Figure 4: Variation of Surface Finish (Rz) before and after grinding (as measured in parallel and across grind direction)

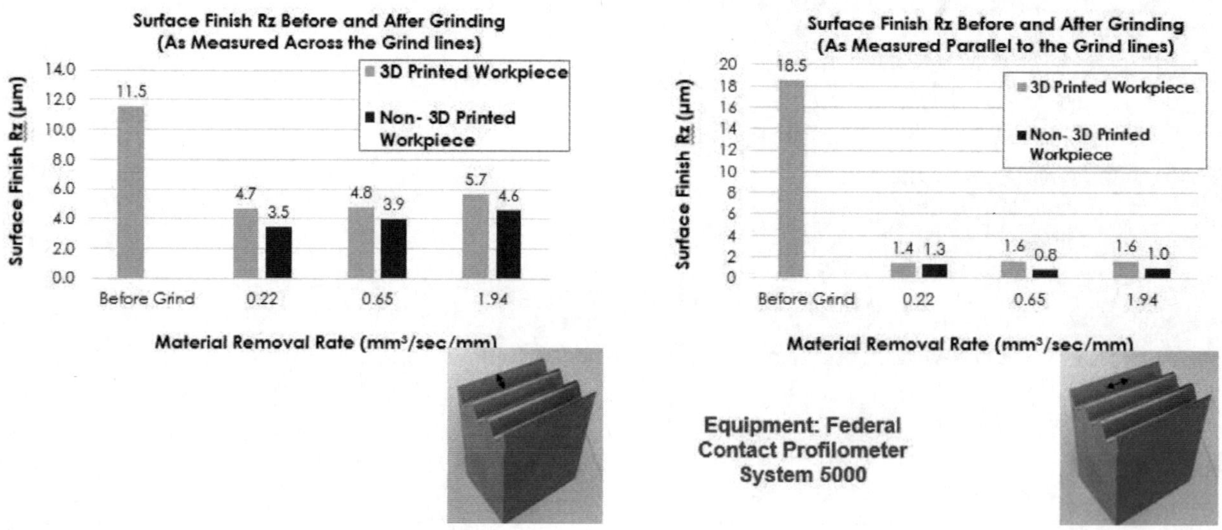

Figure 5: Surface maps of the specimens before and after grinding

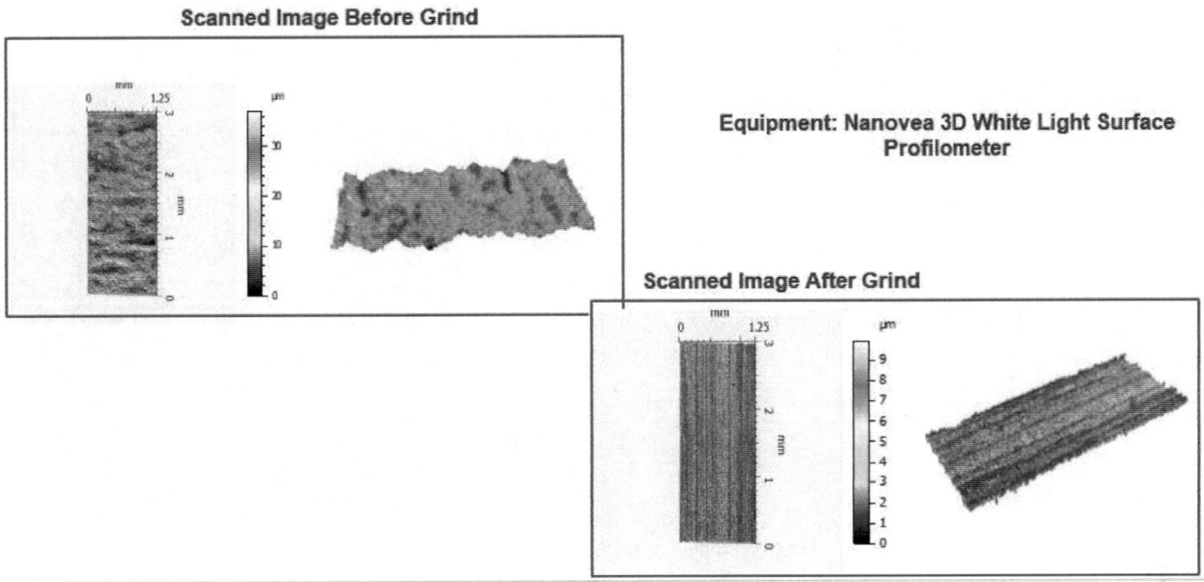

Figure 6: Comparison of surface residual stresses on AM IN718 specimens

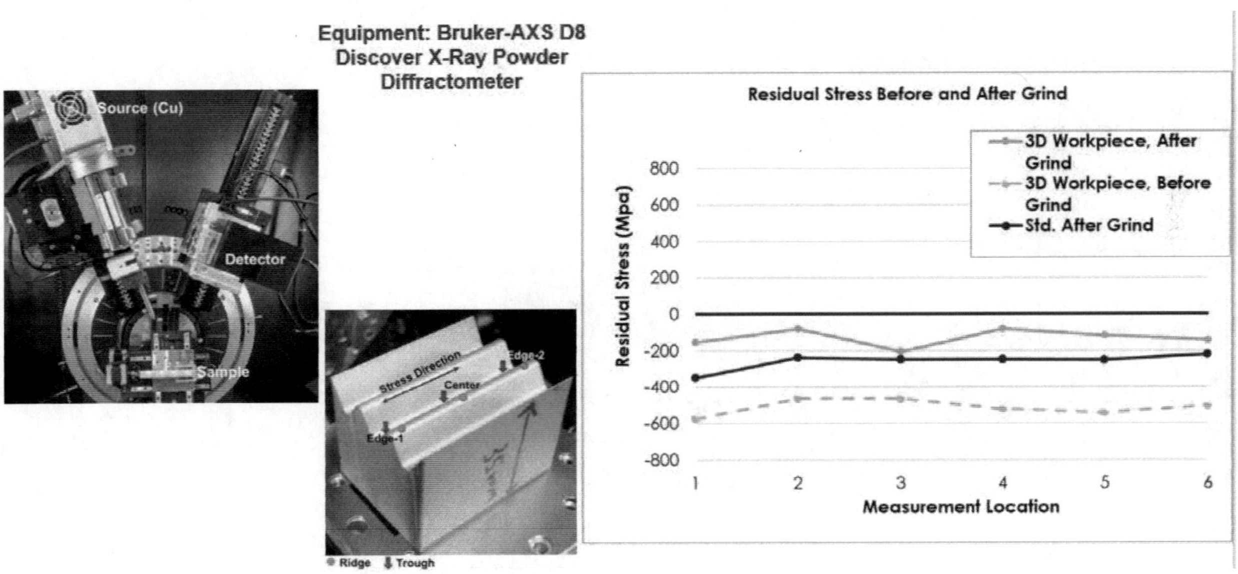

Figure 7: Build direction of the AM Ti6Al4V specimens and heat treatment parameters of both AM and standard cast Ti6Al4V specimens

Standard Ti64 **Hardness after Alpha Case Removal: 36 HRC**

Anneal
Solution heat treated at 1,675 -1,750°F; (913 - 954°C), 1 hour, water quench
Age Hardened (Vacuum Furnace aged - Precipitation hardening)
At 975 -1,025°F; (524 - 552°C), 4 to 8 hours - air cool

3D Printed Ti64 **Hardness after Alpha Case Removal: 35 HRC**

Stress Relieve
1100°F for 2 hours +/- 15 minutes in a vacuum or inert atmosphere
Cool at a rate equivalent to air cool or faster.
Hot Isostatic Press
1650°F for 2 hours +15/-0 minutes at 14,500psi minimum.
Cool in autoclave at 1200°F or below. Cool from 1200°F at any rate
Anneal
Solution heat treat in a vacuum or inert environment at 1750°F for 1 hour +/- 10 minutes and water, polymer or oil quench.
Age Hardened (vacuum aged - Precipitation hardening)
In a vacuum or inert environment at 1300°F for 4 hours followed by air or fan cooled.

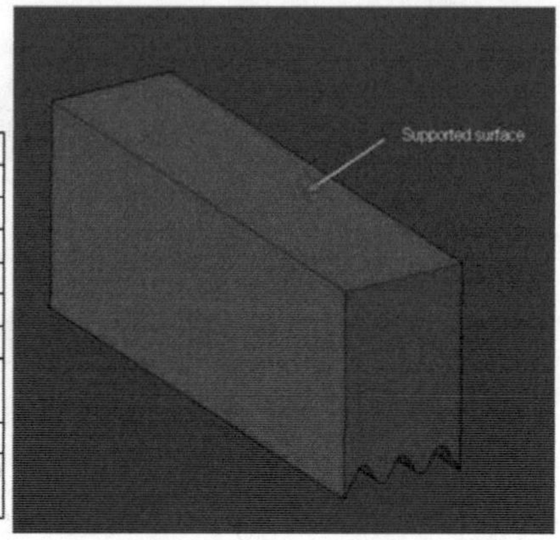

Figure 8: Test Set-up for finish grinding of AM Ti6Al4V test specimens

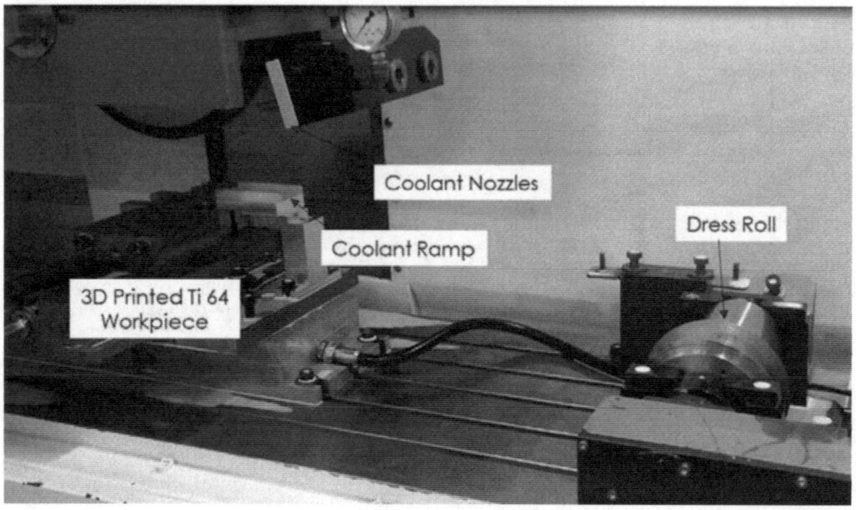

Figure 9: Variation of Surface Finish (Ra) before and after grinding (as measured in parallel and across grind direction)

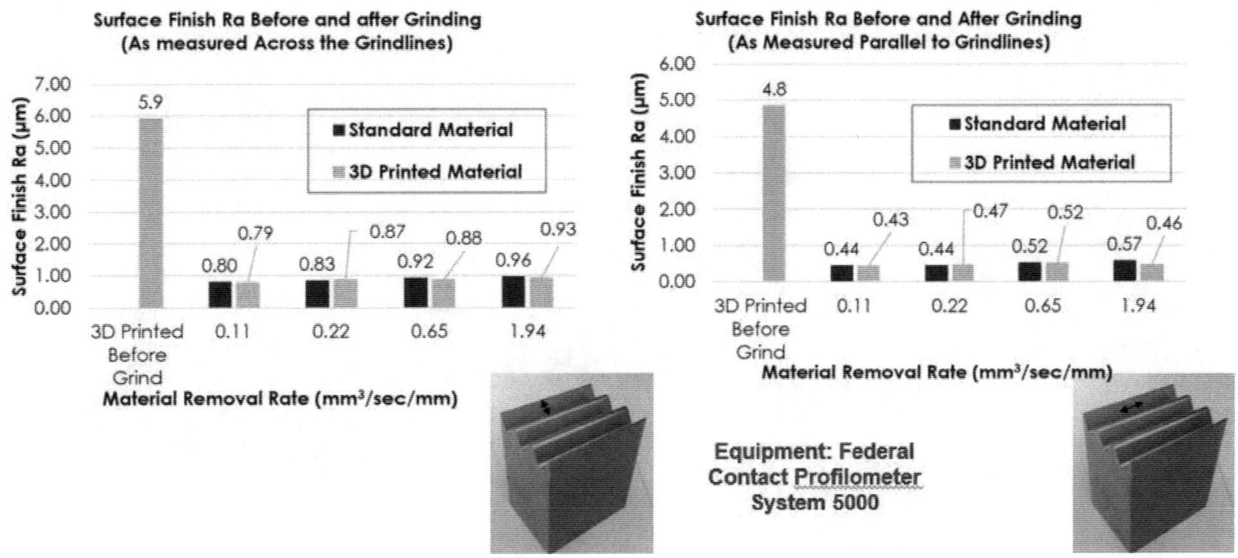

Figure 10: Variation of Surface Finish (Rz) before and after grinding (as measured in parallel and across grind direction)

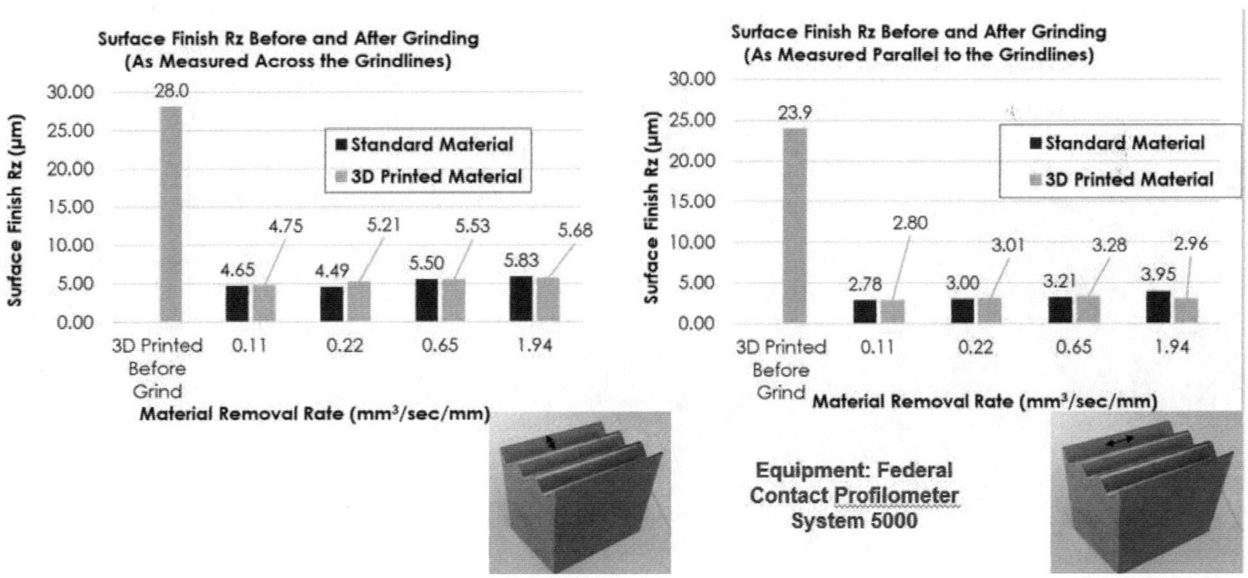

Figure 11: Surface maps of the specimens before and after grinding

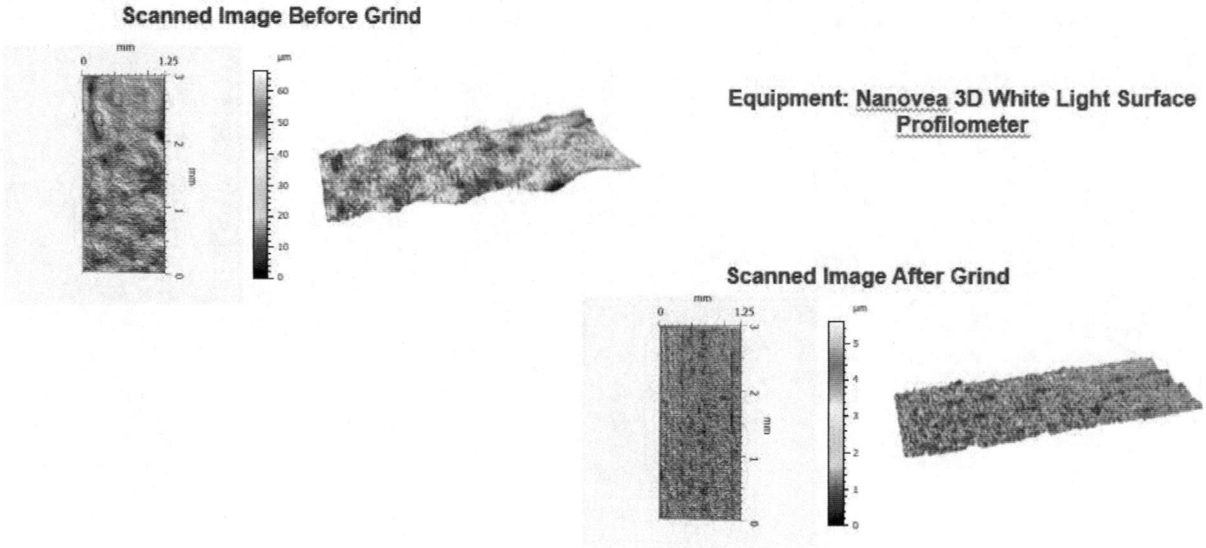

Figure 12: Comparison of surface residual stresses on AM Ti6Al4V specimens

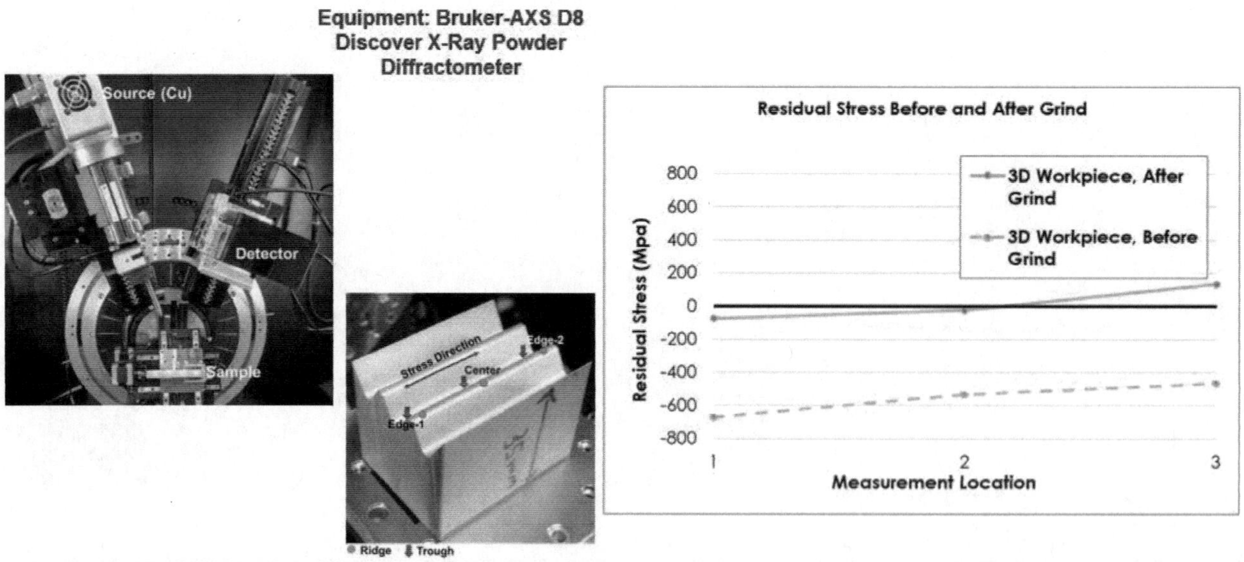

Tables

Table1: Abrasive and workpiece details for finishing AM IN718

Wheel Spec	3NQX60E24VTX2
Grind Length (mm)	50.8
Workpiece width (mm)	38.1
Wheel Diameter (mm)	502
Wheel Speed (m/s)	23
Wheel (rpm)	870

Table 2: Coolant details for finishing AM IN718

Coolant Type	Water Soluble Oil at 5-10%
Flow Rate Volume	52 gpm –197 L/m
Flow Rate Speed	4,500 fpm – 23 m/s
Nozzles Type	13 Nozzles positioned to match the contour of the workpiece
Coolant Ramp	used to prevent coolant deflect off the end of the workpiece

Table 3: Dressing details for finishing AM IN718

Speed Ratio	0.9	
Wheel Diameter	502	mm
Roll Diameter	152.4	mm
Wheel Speed	23	m/s
Wheel Speed	870	rpm
Roll Speed	21	m/s
Roll Speed	2578	rpm
Dress depth per rev (Radial)	0.254	μm/rev

Table 4: Abrasive and workpiece details for finishing AM Ti6Al4V

Wheel Specification	SD 220 C168 P100X
Grind Length (mm)	50.8
Workpiece Width (mm)	38.1
Wheel Diameter (mm)	353
Wheel Speed (m/s)	30
Whee Speed (rpm)	1650

Table 5: Coolant details for finishing AM Ti6Al4V

Coolant Type	Water Soluble Oil at 5-10%
Flow Rate Volume	60 gpm – 227 L/m
Flow Rate Speed	6,000 fpm – 30 m/s
Nozzles Type	13 Nozzles positioned to match the contour of the workpiece
Coolant Ramp	used to prevent coolant deflect off the end of the workpiece

Table 6: Dressing details for finishing AM Ti6Al4V

Speed Ratio	-2	
Wheel Diameter	355.6	mm
Roll Diameter	152.4	mm
Wheel Speed	14.3	m/s
Wheel Speed	766	rpm
Roll Speed	-28.5	m/s
Roll Speed	3574	rpm
Dress depth per rev (Radial)	0.076	μm/rev

Fixed abrasive three-dimensional plate for lapping and polishing of advanced materials

Benjamin Rosczyk, Clement Onyenemezu, Ph.D., Hyeseung Rachel Park, and Ion C. Benea, Ph.D.

Engis Corporation, Wheeling, IL 60090, USA

1. Abstract:

A fixed abrasive three-dimensional plate is provided, which comprises micron size diamond beads, ranging in size from a few microns to a few tens of microns, incorporated into a matrix consisting of one or more inorganic binders and fillers. Said composition is cast in a mold, followed by heat assisted hardening, or room temperature hardening, or hot pressed into a rigid plate blank. The fixed abrasive three-dimensional plate is then mounted on a base substrate to form a lapping/polishing plate. Said fixed abrasive three-dimensional plate is capable of delivering high material removal rates coupled with reduced surface roughness when lapping and polishing of advanced materials, such as sapphire, titanium carbide reinforced alumina (AlTiC), silicon carbide (SiC), gallium nitride (GaN), aluminum nitride (AlN), zinc selenide (ZnSe), and other compound semiconductor materials.

The micron size diamond beads incorporated in the fixed abrasive three-dimensional plate, are composed of diamond particles ranging in size from a few nanometers to a few tens of microns, bonded with the aid of one or more inorganic binders and additives.

2. Introduction:

The most common industry practice used to polish materials such as AlTiC and other semi-conductor materials is chemical-mechanical planarization (CMP) of the work piece on a pad or rigid plate. In the hard disk drive industry, the surface finishing process may also include rough lap and/or fine lap polishing. Rough lapping may use a free/fixed abrasive lapping method where some of the diamond grains are embedded into the plate to form a 2-body system while others roll between the plate and the work piece in a 3-body system. The use of both free and fixed abrasive diamond allows a higher material removal rate but at the cost of a higher surface roughness. After rough lapping, the surface roughness may be improved by adding fixed abrasive nano-grinding, using only pre-embedded fixed abrasives and lapping with a lubricant vehicle. Fixed abrasive nano-grinding gives lower lap rate but better surface finish. While the use of a free/fixed abrasive combination has resulted in improvements in surface finish of many hard materials, there are some disadvantages of handling free diamond slurries such as charging efficiency and uniformity. It has been estimated that only about 10% of diamond particles embeds into the lap plate; the rest run off of the plate and are discarded. Most embedded (charged) diamond particles are simply pushed further into the lap plate as lapping continues, resulting in a decrease in the lap rate and necessitating plate reconditioning. Further, during lapping a small fraction of the charged diamonds may dislodge from the lap plate and scratch the work piece.

In order to improve the efficiency of lapping, fixed abrasives plates have been developed, but they have generally utilized thin films with limited life. In this study, development of a thick 3-dimensional fixed abrasive lap plate was designed to slowly wear and expose new abrasive. Thereby avoiding frequent plate reconditioning while delivering the surface finish of fixed abrasive lapping.

3. Experimental:

1. Preparation of Spherical Diamond Composites (SDC)

Diamond powder was dispersed into metal oxide binders such as colloidal silica, ceria, zirconia, alumina, titania and mixtures thereof using a propeller mixer to produce a diamond/metal oxide slurry. The slurry was then spray dried using Yamato ADL311 spray dryer (Yamato Scientific America Inc., Santa Clara, California) equipped with a 1530 µ nozzle assembly. During spray drying, the slurry was under constant agitation and pumped into the spray chamber with a peristaltic pump with the inlet temperature of the spray dryer maintained at 170°C. The slurry was atomized as it entered the drying chamber and water striped from each slurry droplet to form discrete particles. The discrete powder particles were separated by a cyclone into a collection jar while the steam was exhausted, while oversized particles and aggregates collected at the bottom of the drying chamber were discarded as waste.

The collected particles were composed of diamond particles held together by a continuous metal oxide binder matrix. Analysis by scanning electron microscopy (SEM) showed the particles had a nearly spherical shape with distinct features of porosity and diamond grains finely dispersed within each composite particulate as shown in Fig. 1. As spray dried spherical diamond composite particles exhibited a wide particle size distribution (Fig. 2) and a median particle size of ~17µm when measured with a Multisizer 3 Coulter Counter (Beckman Coulter Inc.). Spherical diamond composite powders were subsequently graded to a median particle size of >20µm and used to make 3-dimensional fixed abrasive SDC lapping/polishing plates.

2. Hot-pressing of 3-D Fixed Abrasive SDC Plate

The 3-dimensional fixed abrasive plates were produced by dry mixing spherical diamond composites (SDC), phenolic binder, hollow glass spheres, and fillers. The mixture was then poured into a metallic mold and hot pressed for a specified time and temperature to fully cure the phenolic binder. After curing, the SDC abrasive plate was mounted atop an aluminum base plate and ground to flatness with a carbide grinding wheel. The new SDC plate as shown in Fig. 3 was characterized by plurality of non-interconnected and irregular surface pores.

During development testing, surface porosity was found to be essential for high lapping and polishing efficiency. Surface porosity was analyzed with a color thresholding of an image using a high resolution 3-D optical microscope. As represented in Fig. 4, the percent surface area with pores were mapped and calculated to give the percent porosity of the plate surface at a given magnification (i.e. a measure of the pore area compared to total surface area of the plate). Surface porosity of the plates tested for this work ranged from 3% to 60% at 500X magnification.

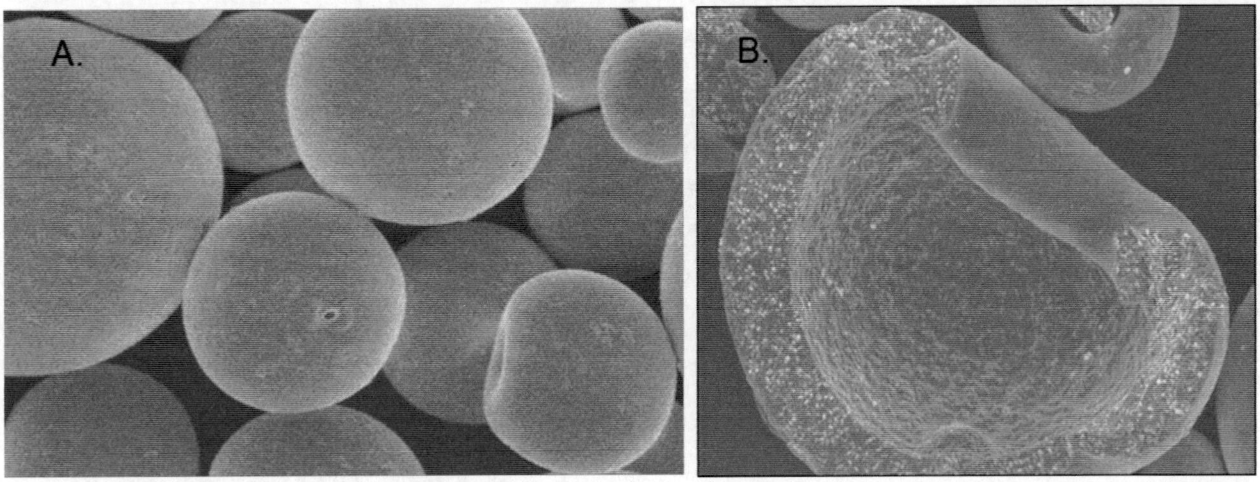

Fig. 1: SEM micrographs of SDC particles with a (A) rounded particle shape and (B) hollow shell structure with well dispersed diamond particles.

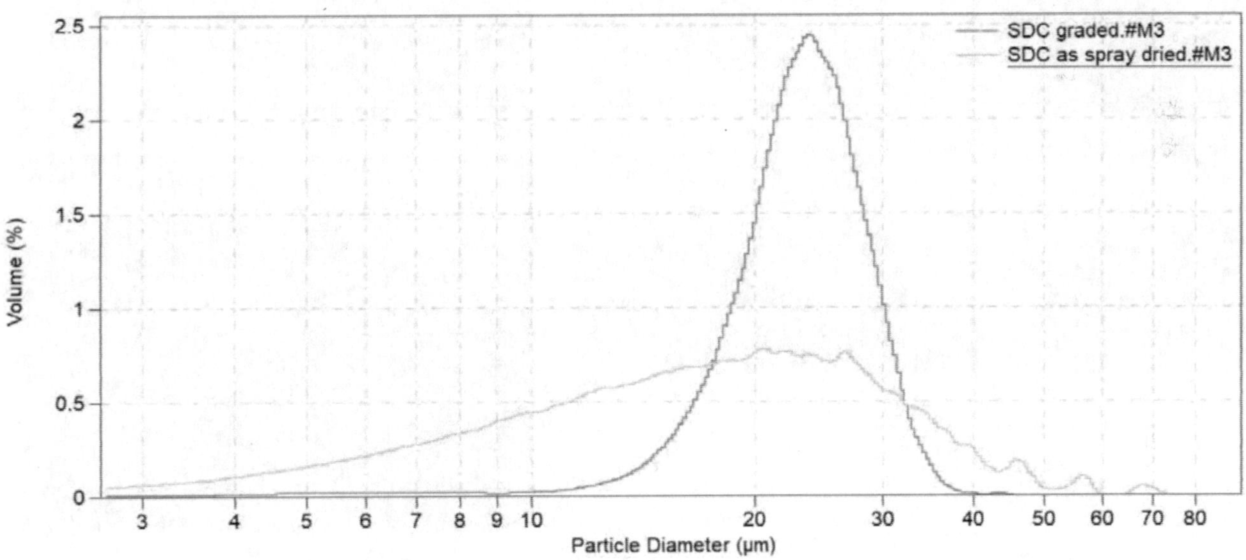

Fig. 2: Particle size distribution of as spray dried SDC particles (D50~17μm) and after sizing (D50>20μm).

3. Lapping/Polishing Tests

The planarized SDC plate was tested on a modified 15-inch Engis Corporation Fast Lap machine, FL-15. An 8-inch diameter SDC plate was centered atop a 15-inch diameter platen and held in place with three studs. The workpiece consisted of samples of AlTiC or SiC bonded to a 4-inch diameter ceramic block in a triangular pattern (Fig. 5). The workpiece was positioned 2 inches from the center of the SDC plate and both SDC plate and workpiece rotated counterclockwise at a set RPM. Lapping of the workpiece was performed in six successive runs of 10 minutes each with a constant application of lubricant (Glycol lubricant from Engis Corporation, Wheeling, IL). After each run, the difference in thickness of each workpiece was measured with a Fowler Sylvac Ultra-Digit Mark IV drop indicator with accuracy $\pm 2 \mu m$ in order to determine the average material removal rate (MRR) over 6 runs. Surface roughness was measured using Zygo NewView 6K Optical Profilometer.

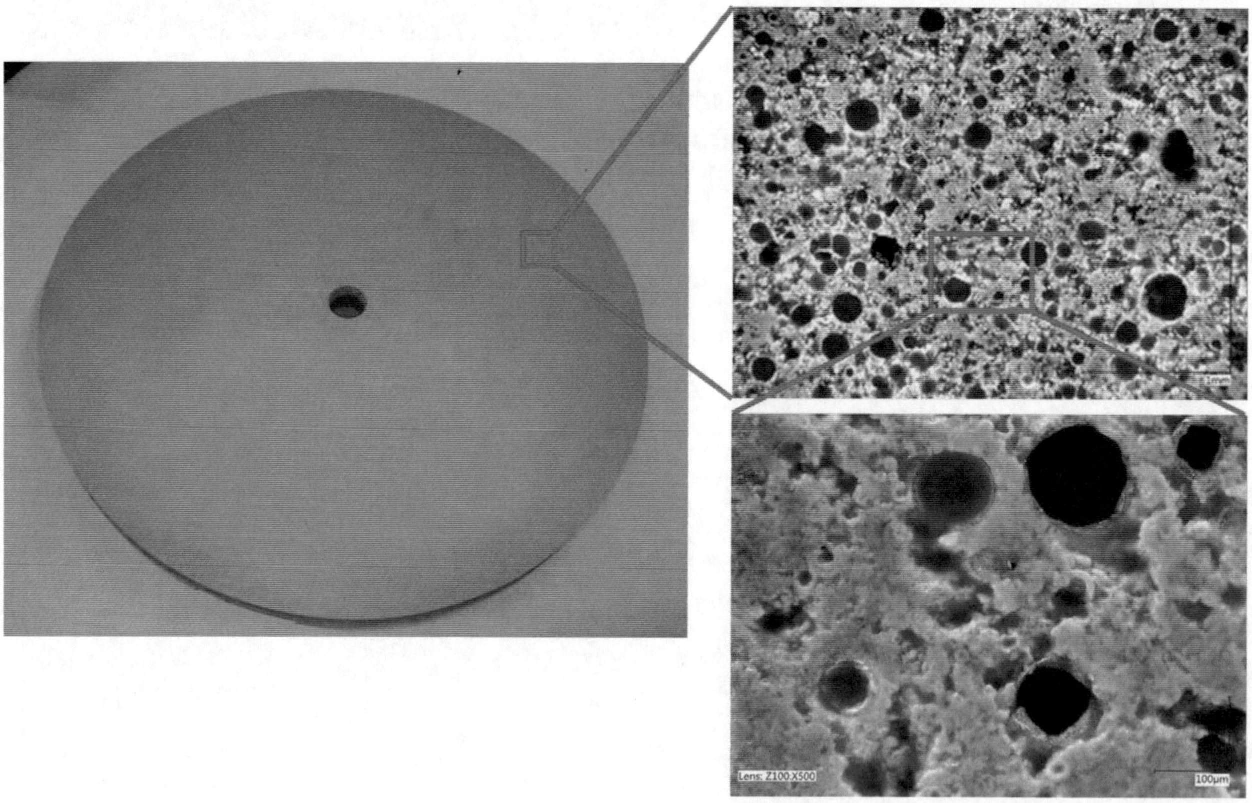

Fig. 3: Optical microscope picture of SDC plate at different magnifications.

4. TEM subsurface damage analysis

Samples of lapped SiC were prepared and analyzed by TEM by the staff of the Nuance Center at Northwestern University. TEM samples were prepared using SEM/FIB according to Fig. 6. After sample preparation, TEM images near the lapped surface of the SiC were collected to characterize subsurface damage.

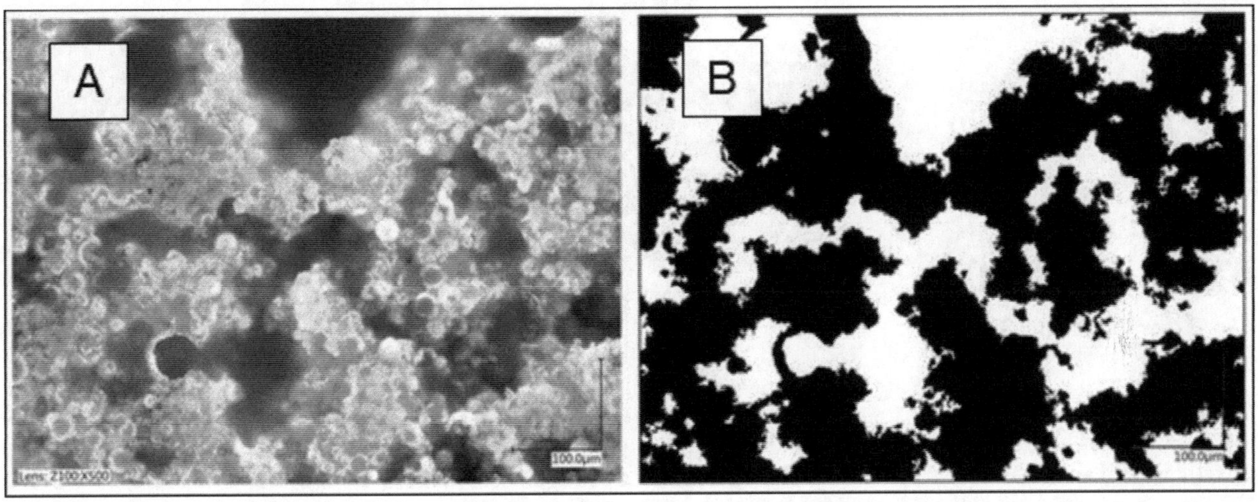

Fig. 4: Optical microscope pictures of (A) SDC plate at 500X magnification and (B) after thresholding image analysis processing to distinguish the pore area (white).

Fig. 5: AlTiC samples bonded to ceramic block for testing.

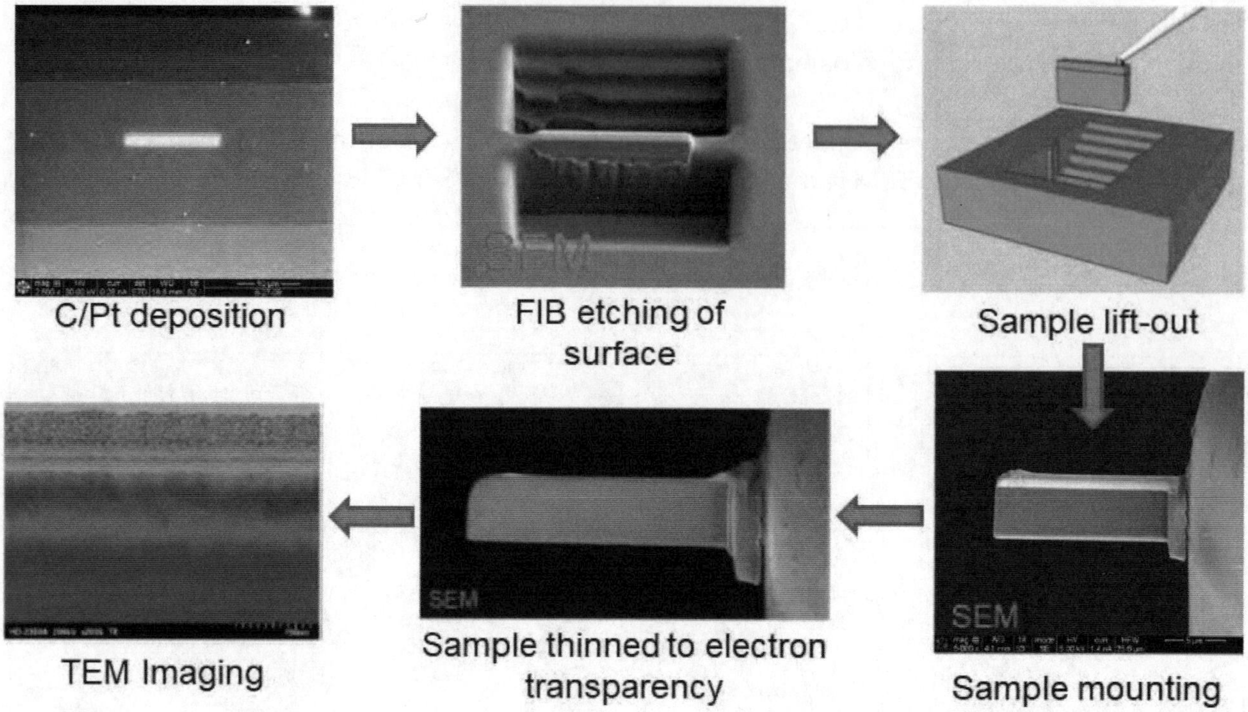

Fig. 6: Processing steps used to prepare SiC TEM samples after SDC lapping to evaluate subsurface damage. *Images courtesy of B. Myers at Northwestern University*

4. Results and Discussion

1. Lapping of Titanium Carbide reinforced Alumina (AlTiC) bars

Three (3) AlTiC bars were bonded to a 4-inch diameter ceramic block as shown in Fig. 5. Four lap plates made with SDC of diamond sizes 1.25μm, 3μm, 4.5μm and 6μm were tested under the process conditions shown in Table 1. The work piece was lapped under constant pressure of 14psi and constant plate speed of 60rpm. Results of lap rate and surface roughness of the work piece are shown in Fig. 7. As expected, the material removal rate (MRR) of the AlTiC and surface roughness as measured by Ra decrease with smaller diamond size in the SDC lap plates.

SDC fixed abrasive process may increase the efficiency of the AlTiC HDD head lapping by eliminating the need for facing, grooving, shaving, and charging of diamond during plate preparation frequently encountered in the current free abrasive lapping process. With SDC plates, plate preparation only involves planarization and plate cleaning which may result in improvement of production, without the loss of surface quality.

Fig. 8 compares the lap rate and surface finish of free abrasive lapping to fixed abrasive lapping. SDC plate with 0.25μm diamond grains gave lap rates similar to 0.25μm free abrasive lapping with comparable surface finish. For faster removal, 1.25μm diamond SDC plate was used to give lap rate of 0.5μm/min about double that from free abrasive without compromising the surface finish as indicated by the Ra of 2.9nm.

A common issue with diamond charged lapping plates used in hard disc industries was the reduction in lap rate over time. Embedded diamonds become dull or further imbed into the plate and become ineffective for lapping. With SDC fixed abrasive lapping plate, the lap rate remains consistent over longer times. As shown in Fig. 9, lapping AlTiC bars with 4.5μm SDC plate provided consistent removal rate over time. Because SDC plates do not need to be reconditioned at the end of each run, plates last much longer than Sn-Bi plate and lowers processing cost. The breaks in the data from Fig. 9 indicated when the plate was stopped and cleaned of swarf. The removal tended to increase due to plate wear and exposure of new diamond cutting points. Compared to a competitor fixed abrasive lap plates and a diamond lapping film (Fig. 10), SDC plates gave superior surface finish and comparative lap rates. Competitor A is a 1 μm diamond in a resin bonded fixed abrasive film used for rough lapping, while Competitor B is a 4.5micron diamond in a vitrified bonded fixed abrasive plate. Lower lap rate of Competitor A may be due to lower diamond size, but Ra as a measure of surface finish is high. The superior surface finish of the SDC plate may be due to a better distribution of diamond grains within each SDC particle; thereby avoiding diamond aggregates that cause deep scratches.

For efficient removal of the swarf and maintenance of constant cutting rate, the new 3-dimensional fixed abrasive plate was designed with plurality of pores in the plate. Pores on the surface of the plate, as shown in Fig. 3 and 4, form discontinuous lapping lands and provide channels for lapping lubricant to flow and prevent hydroplaning of the workpiece. The resultant debris/swarf is washed into the pores to avoid scratching of the workpiece. Control of the surface porosity of the plate is critical to SDC plate performance, and hollow glass spheres were added into the formulation of the 3-dimensional fixed abrasive plate to increase and control the porosity of the plate. SDC particles also had a hollow sphere structure with a non-sintered, porous silica/diamond shell. The SDC particulates are friable and easy to abrade to expose diamond. There were also pores due to entrapped air during mixing and/or curing that was controlled by the molding pressure. Excessive molding pressure squeezed out entrapped air and decreased the interstices and overall porosity of the SDC plate. Stock removal rate efficiency of the workpiece depended on total porosity of the plate as shown in Fig. 11. 3-dimensional fixed abrasive SDC plates with little to no porosity did not provide a good stock removal. Fig. 11 summarizes the effect of surface porosity on the AlTiC lapping removal rate. It was found that with a surface porosity of about 30% or greater, high lap rates were achieved.

Table 1. Processing Parameters for SDC Plate Lapping of AlTiC

	Process conditions
Plate conditioning:	Engis Diaflex diamond pad
Workpiece:	3 pieces (58.6mm x 2.5mm) AlTiC bars attached to 4" ceramic plate
Diamond size:	6µm; 4.5µm; 3µm; 1.25µm
Plate speed (rpm):	60
Pressure (psi):	14
Total Process time (min):	60
Lube:	Glycol lube
Lube flow rate (ml/min):	2.4

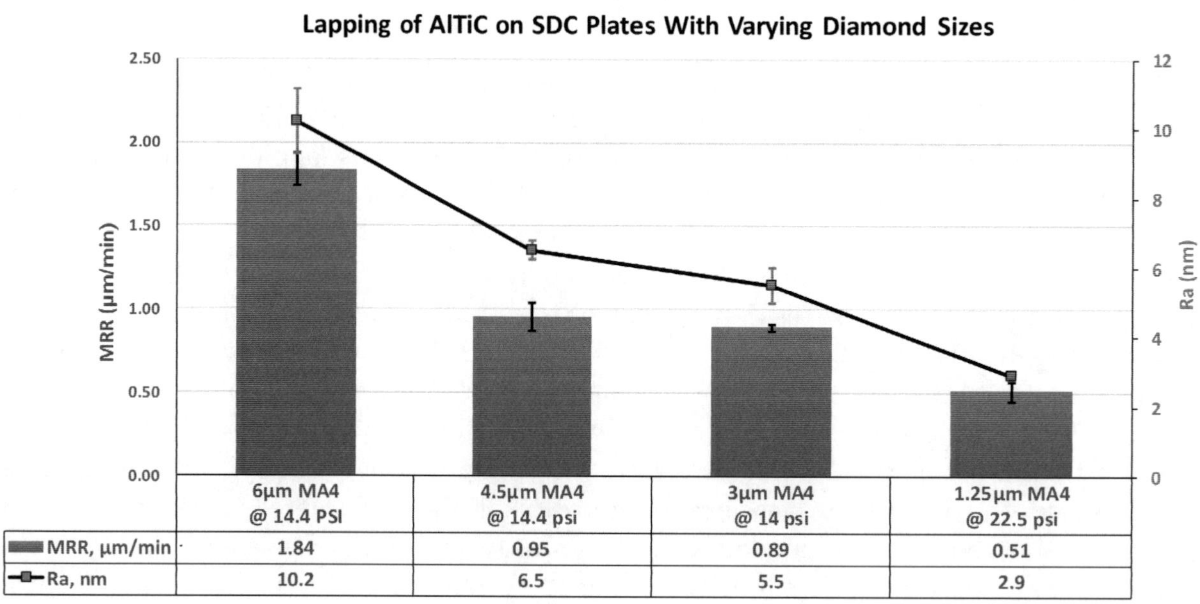

Fig. 7: Results of AlTiC lapping on SDC plates.

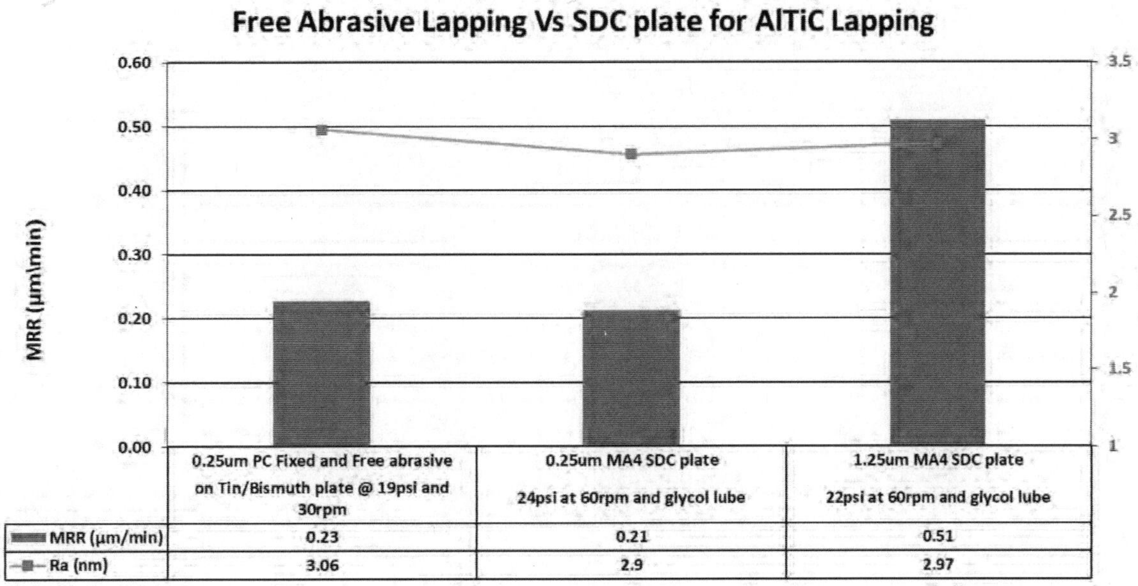

Fig. 8: Comparison of conventional free abrasive lapping and SDC plate lapping of AlTiC.

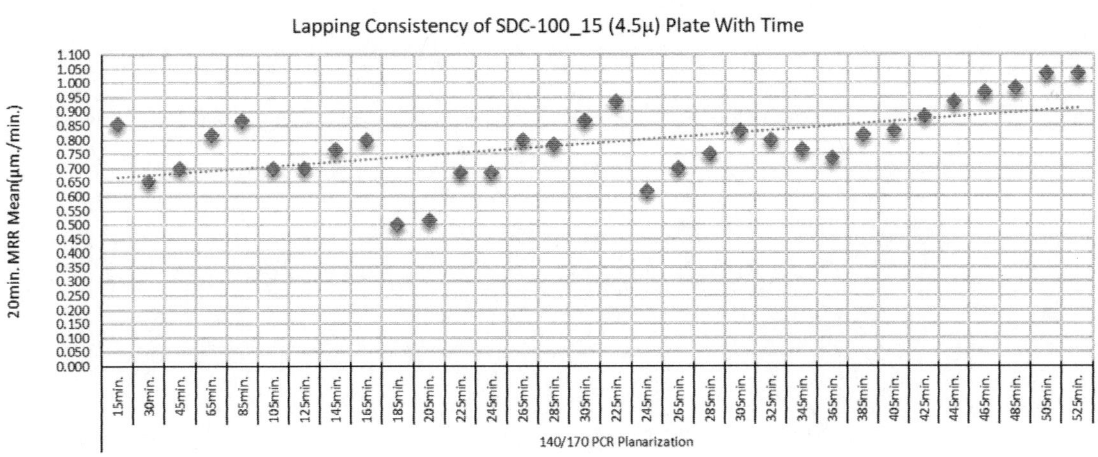

Fig. 9: AlTiC lapping with 4.5µm SDC plate.

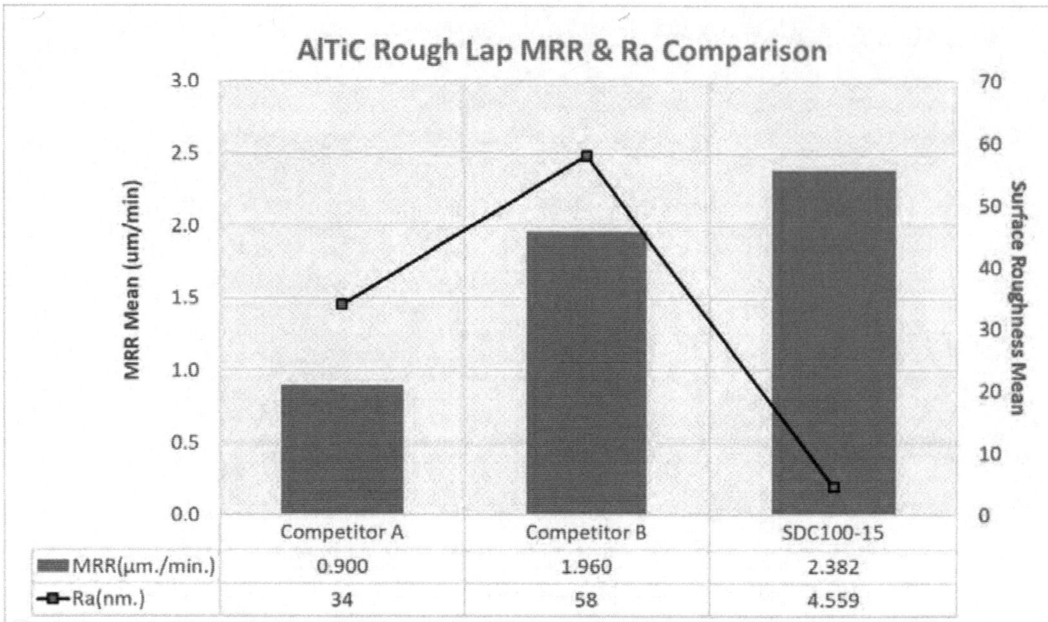

Fig. 10: Comparison of MRR and surface finish (Ra) of AlTiC samples lapped with a competitor fixed abrasive film (Competitor A), a fixed abrasive vitrified plate (Competitor B) and an SDC plate.

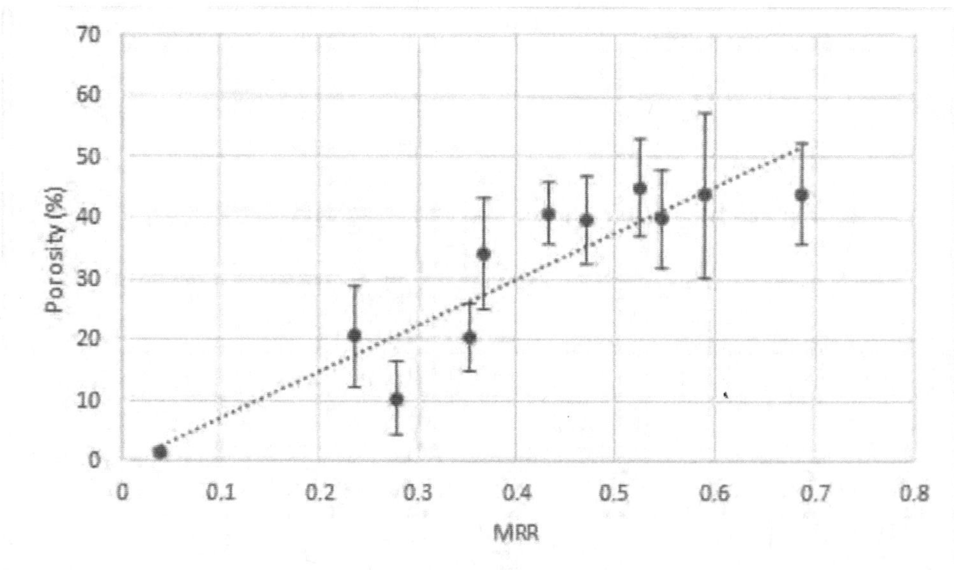

Fig. 11: SDC plate porosity with MRR during AlTiC lapping. MRR seems to maximize at surface porosity >30%.

2. Lapping of Silicon Carbide Wafer

Silicon carbide (SiC) is one of the fastest growing materials for semiconductor and optoelectronic applications due to its hardness, resistance to abrasion, relatively low thermal expansion, and high thermal conductivity. These good physical characteristics also make it very difficult to process. Processing single crystal SiC often involves, grinding, lapping with free abrasive slurries and chemical mechanical planarization (CMP) on a pad as illustrated in Fig. 12. The lapping step often involves two or three diamond lapping steps before CMP. Often the CMP step is notoriously long and uses harsh chemical to achieve the desired surface finish. Also, due to the long CMP step, edge rounding is often a problem in processing SiC. The present work presents an alternative lapping process with fixed abrasive spherical diamond composite plates. The wafers are lapped with 3μm and 0.25μm fixed abrasive SDC plates to Ra of about 0.4nm without edge rounding problem and then cleaned up with CMP step for a short time if necessary, as illustrated in Fig. 12.

In this work, planarized SDC plate was used to lap a silicon carbide workpiece shown in Fig. 13 under the lapping conditions of Table 2. The workpiece consists of 4-pieces of 1 square inch 4H- N type SiC wafers bonded to a 4-inch diameter glass block. The fixed abrasive plate was mounted atop a 15-inch platen in the center and held in place in three places with studs. The workpiece was brought into contact with the fixed abrasive plate at a pressure of 5.2 psi. The SDC plate rotates counterclockwise at 90 rpm, while the workpiece which is off center from the plate by 2 inches rotates counterclockwise at about 30 rpm. Lapping of the workpiece was done in three successive runs of 10 minutes each, at a constant flow rate of 2.4 ml/min of 20 wt.% L6037 lubricant (glycol lubricant from Engis Corporation, Wheeling, IL) in deionized water. After each run, the difference in thickness of each silicon carbide workpiece was measured with a Fowler Sylvac Ultra-Digit Mark IV drop indicator with an accuracy of ± 2μm in to determine the average material removal rate (MRR). Surface roughness was measured using Zygo NewView 6K Optical Profilometer.

The results from lapping with different sizes of diamond grains in SDC plates are shown in Fig. 14. Removal rate and surface finish varied proportionally with diamond size with submicron diamond SDC plates producing Ra less than 0.5nm. A surface Ra of <0.5nm may shorten the CMP step to achieve an Epi-ready surface. One major concern of diamond abrasive lapping of SiC wafers subsurface damage which may deteriorate the epitaxial layer. The CMP process is supposed to remove any subsurface damage remaining from the lapping steps, but due to very low material removal rates, the damaged layer may require a very long process time. It is preferred to minimize subsurface damage before the CMP step.

SiC wafer lapped with a 0.25μm SDC plate was analyzed by TEM to characterize any subsurface damage after lapping. During the sample preparation (Fig. 6), the surface was coated with carbon/Pt to protect the surface from any damage from subsequent sample preparation steps. Fast Ion Bombardment (FIB) was used etch out a small sample from the surface. The sample was removed and attached to the TEM sample holder where it was thinned to electron transparency. TEM micrographs of the SiC wafer surface lapped with 0.25μm SDC plate showed only about 4-5nm thick damaged layer on the wafer surface (Fig. 15). This thin subsurface damage layer could significantly reduce or eliminate the CMP step.

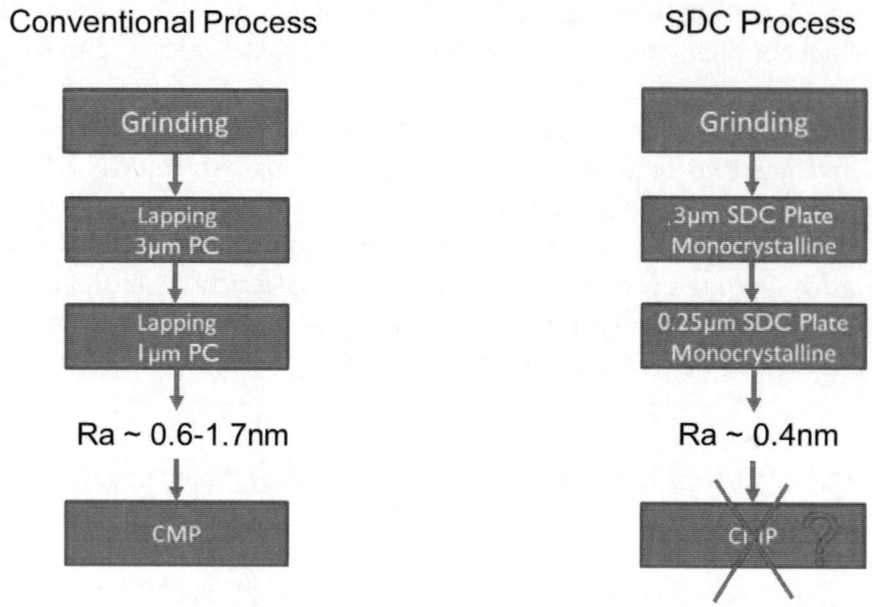

Fig. 12: Comparison of a conventional SiC wafer processing and SDC SiC wafer processing. Surface finish after 0.25μm SDC plate may allow reduction or elimination of CMP step.

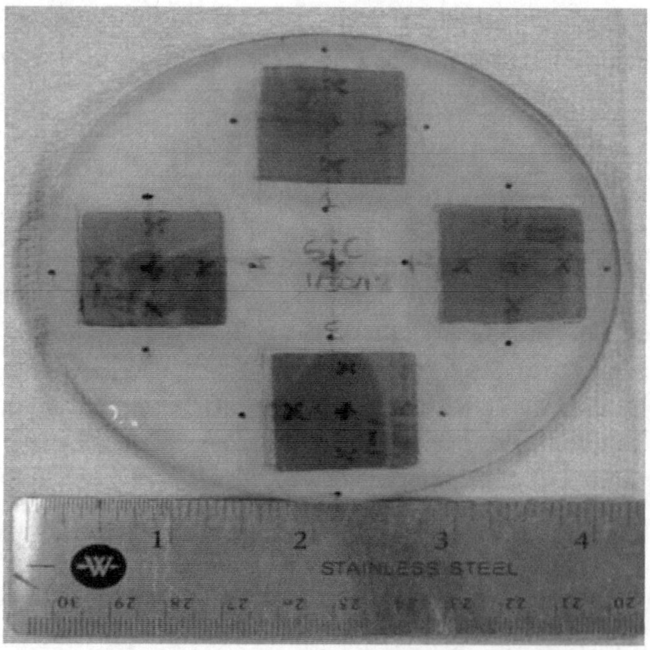

Fig. 13: SiC samples bonded to a glass block used for SDC plate lapping.

Table 2. Processing Parameters for SDC Plate Lapping of SiC

	Process conditions			
Plate Conditioning:	Engis Diaflex diamond Pad			
Workpiece:	4 pieces 1" square SiC wafer attached to 4" glass plate			
Diamond Size:	3µm	1.25µm	0.4µm	0.25µm
Total Process time (min):	30	60	90	60 or 90
Plate Speed (rpm):	90			
Pressure (psi):	5.2			
Lube:	Hydrogen peroxide lube			
Lube flow rate (ml/min):	1.2			

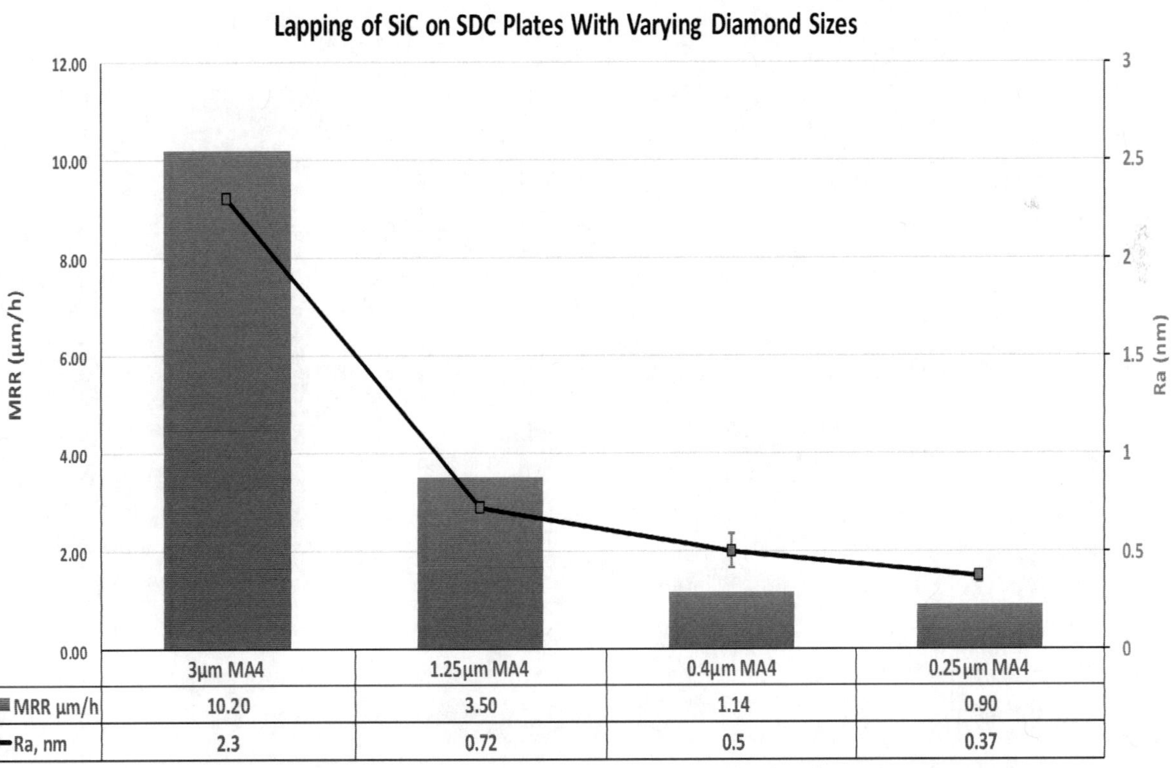

	3µm MA4	1.25µm MA4	0.4µm MA4	0.25µm MA4
MRR µm/h	10.20	3.50	1.14	0.90
Ra, nm	2.3	0.72	0.5	0.37

Fig. 14: Material removal rate (MRR) and surface finish (Ra) of SiC after SDC plate lapping.

5. Conclusions

This study found that the newly developed fixed abrasive SDC lap plate can deliver higher lap rates and comparable surface finishes (Ra~3nm) when lapping AlTiC as compared to a conventional free/fixed abrasive lapping process. SDC lap plates can also maintain high lap rates for longer than a conventional lap plate charged with a single layer of diamond abrasives. SDC lap plates wear slowly, expose fresh cutting points, and eliminate the need for frequent plate facing and charging.

The CMP processing step in the convention lapping and polishing of monocrystalline SiC wafers can be the most time consuming and inefficient in terms of slurry usage. A process using SDC fixed abrasive plates made with fine 0.25μm monocrystalline diamond was found to produce SiC wafers with surface roughness of 0.4nm with subsurface damage of 4-5nm. When compared to a conventional process with an after-lapping surface roughness of 0.6-1.7nm, the processing time needed for the CMP step in an SDC process may be significantly reduced or eliminated altogether. Further improvements in surface roughness and subsurface damage results may be possible in monocrystalline SiC lapping with SDC plates manufactured with diamond sized <0.25μm.

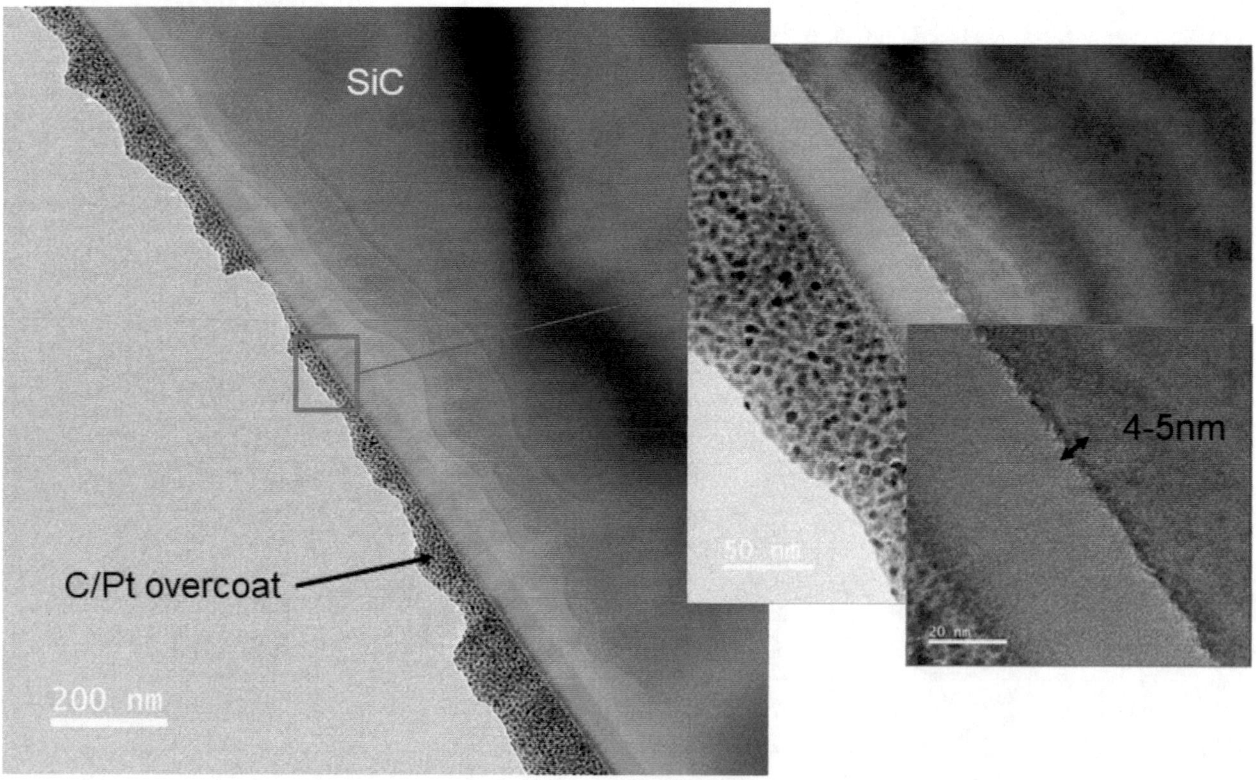

Fig. 15: TEM images of SiC lapped with 0.25μm SDC plate showing 4-5nm of subsurface damage.

6. Acknowledgements

This work made use of the EPIC facility of Northwestern University's NU*ANCE* Center, which has received support from the Soft and Hybrid Nanotechnology Experimental (SHyNE) Resource (NSF ECCS-1542205); the MRSEC program (NSF DMR-1720139) at the Materials Research Center; the International Institute for Nanotechnology (IIN); the Keck Foundation; and the State of Illinois, through the IIN.

Authors:

1. Mohammed Y. Tharwan: research associate at University of Toledo and Lecturer at Jazan University, Mechanical department, Saudi Arabia .

   Email: kli10therwan@hotmail.com

2. Ioan D. Marinescu: Professor at University of Toledo USA

   Email: Ioan.Marinescu@utoledo.edu

Theoretical–Experimental Study of Silicon Carbide Grinding

Abstract

The application of hard and brittle materials, commonly exhibited by materials that are produced by burning of minerals at high pressure, has attracted attention with the goal of heightened performance. Traditional grinding generates large cracks and damages surfaces. However, we anticipated that when using a high-precision controlled micro-grinding machine, the improved process of material removal will reduce the surface roughness enhance the surface of brittle materials via increased plastic deformation (ductility). In this experiment, a unique sloped bracket was created to obtain a set of comparison results in one grinding process. Analyses of surface morphology, including scanning electron microscope images of subsurface side cross-sections are shown in this paper. The results show the importance of keeping the depth of cut as small as possible in order to prevent subsurface damage.

Keywords

Advanced ceramics, brittle materials, grinding process, grinding force, plastic deformation, fracture damage, SEM, subsurface fracture, critical depth of cut, Silicon carbide and cross-section

INTRODUCTION

Brittle material characteristics and properties

Materials are generally classified by two characteristics: brittleness and ductility (Bifano,1988). Brittle materials may be further classified into three major groups: amorphous glasses (formless non-crystalline), hard crystals (resistant grains) and advanced ceramics (materials produced by burning under high pressure). Of these three types of brittle material, the advanced ceramics produced under high pressure are some of the hardest and most brittle (Bifano, 1988). Advanced ceramic products are different from those produced using older methods because of their particular mechanical features and correspondingly complex manufacturing processes (Chen et al., 2010).

Examples of advanced ceramic materials produced by burning under high pressure include silicon carbide (SiC), silicon nitride (Si_3N_4), aluminium oxide (Al_2O_3), zirconia (ZrO_2), zirconia-toughened alumina (ZTA), boron carbide (B_4C), and polycrystalline diamond. Machining these materials is challenging due to their hardness and brittleness (Chen et al., 2010).

Ceramic atoms are bonded with high-energy bonds (covalent bonding, ionic bonding, and association of the two classes), while metal atoms are typically connected by low-energy metallic bonding. Considering metal oxide ceramic (Al_2O_3), the relative magnitude of covalent and ionic bonding is about 4:6. However, the relative magnitude is 9:1 in SiC, an average non-oxide ceramic (Horvath el al., 2002).

The potency of materials with a wide range of ratios of ionic bonding are significantly altered using a measure of heat or cold. Forms of matter that contain covalent bonds, however, are not altered by abnormally high temperatures. Moreover, low thermal and depressed temperature coefficients of expansion and proportionally high thermal conductivity have also been proven to be distinct properties of ceramics with covalent bonding (Horvath el al., 2002).

In engineering, a material is termed brittle and resistant if it never demonstrates plastic transformation before cracking. Where there is a need to transform an inflexible material by rubbing it with an abrasive surface such as industrial diamonds, the plastic transformation will not be demonstrated as the diamond grit plows through it to remove material, nor will it exhibit the residual stress profile characteristic of a ductile material (Chandrasekar and Sathyanarayanan, 1987). Instead, the material will break apart, resulting in smaller cracks (microcracks), which remain beneath the surface or subsurface. The most important aspect of grinding brittle materials is to reduce the destruction below the surface by following a set of procedures and using the most efficient equipment which will give desired results.

Ductile regime grinding principle

Ductile-regime grinding is the most recently developed material removal technique (Bifano T., 1992). The working principle is straightforward as it requires a highly stiff apparatus that has an accurate grinding such that exceptionally small-scale material removal is espunged, then materials that are brittle can stand a chance to be ground in a manner that is ductile. Following this, workpieces that are brittle can be machined in a way that can be determined and at the same time producing surface finishes that conform to those that are attained in non-deterministic, ductile procedures like polishing and lapping. This form of grinding can be attained by ensuring that the apparatus used in grinding a structurally in-place loop is stiff, has realist grinding infeed time control, can be isolated from interferences from the environment and in place state of the art wheel trueing mechanisms (Bifano T., 1992).

If a brittle material is subjected to pressure, it can be transformed. If the critical shear stress at any point within the material exceeds the elastic yield stress, the mechanism of deformation will change from one of reversible energy storage via elastic stretching to one of irreversible energy dissipation. In one example of irreversible deformation, micro-crack creation, phase deformation, displacement of motion (in crystals), and continuous motion between molecules (in amorphous substances) (Bifano, 1988; Chen el al., 2010). But then, removal methods for brittle and ductile materials differ. For brittle materials, material is removed by means of the propagation and intersection of cracks. On the other hand, the ductile mechanism involves inducing a flow of plastic material in severely sheared machining chips. The grinding ductility of chosen materials has been compared to their intrinsic properties to establish a model that correctly informs users about the critical depth of cut (Bifano et. al., 1988). This model basically related the measured the rate of critical infeed for about ten percent fracture to a computed critical of depth based on the properties of the materials in place. It is essentially derived from an established formula that illustrates the critical depth for fracture when hard materials are identified (Bifano, 1988):

$$d_c \propto \left(\frac{K_c^2}{H^2}\right)\left(\frac{E}{H}\right)$$

$$d_c = 0.15\left(\frac{K_c^2}{H^2}\right)\left(\frac{E}{H}\right)$$

where d_c is the critical depth, E the elastic modulus, K_c the fracture toughness, H the hardness, and the term $\left(\frac{K_c^2}{H}\right)$ is considered an effective measurement of brittleness(Marshall and Lawn, 1986).

Since advanced, precise engineering can facilitate monitored grinding in feed rates as low as several nm per wheel revolution, brittle materials can be ground at very low rates (Blake & Scattergood,1990). Thus, the most dominant method of extraction of materials is plastic-flow rather than by breaking down materials through the ductile regime of grinding. Grinding brittle materials using plastic transformation method, the results obtained are comparable to those of materials that have been polished (Bifano, 1988; Chen el al., 2005). However, unlike polishing, grinding is a deterministic process which permits fine control of contour and creation of complex shapes. A ductile regime of grinding is another choice that can be used to make materials finer (Bifano, 1992).

Several investigations are conducted in ductile regime grinding using the Scanning Electron Microscopy. Zhipeng Li (2007) observed that damage free machining of RBSiC using ultra-precision grinding is possible under small depth of cut.Also, Zhichao (2017) in his study concluded with beautiful statement ‘it requires the grain depth of cut of the wheel to be less than the critical cut depth og RBSiC’.

This paper demonstrates experimental application grinding of Silicon Carbide using ductile regime grinding principles.

EXPERIMENTAL DESIGN

A very important question that this study intends to answer is the applicability of ductile regime grinding to silicon carbide (SiC). To answer this question, this study attempts to find out the critical deformation point according to relevant surface morphology. In this study, a critical depth of cut model developed by Blake in 1988 and 1990 was used.

In this work, a total of 6 SiC workpieces were processed. The workpieces were divided into two groups, each processed at a different table speed (feed speed). Within each group, the workpieces were assigned to three different wheel speeds. The minimum step size of depth for each grinding was 2.54 μm. After the grinding wheel first touched the peak of the workpieces, grinding was repeated twice to equilibrate the height of all workpieces and make them perfectly flat before data collection began. Then, all workpieces were cut to three equal lengths: 18 mm, 4 mm, and 5 mm After polishing all the cutting surfaces, these three short workpieces were tightly piled together, as shown in Figure 1, and were inserted into a specially designed tilt holder with a workpiece slot length of 60 mm. On the longitudinal direction of this slot, a height of 10 μm was designed from the left end (zero points) to the right end. Additionally, three further points were defined in the middle of each divided piece and designated A, B, and C for purposes of further calculation and data analysis.

A schematic drawing of this design is displayed in Figure 2. This setup was designed to provide an opportunity to observe the cutting sections. During the next grinding process, with the parameters shown in Table 1, the workpieces on this holder experience a small grinding edge angle (Equation 2). Such small grinding edge angles can lead to various depths of cut on the workpiece and various normal force. The left end is set as 0 and the first cut section point to left end is 18 mm (from Equation 2). The depth of cut at this point is 3 μm. The second cut section point to left end length is 36 mm, and the depth of cut is 6 μm, and so on.

Grinding edge angle $\alpha = \sin^{-1}(10\ \mu m/60\ mm) = 0.00017° \approx (2x10^{-4})°$

Table 1: Grinding conditions

Work material	Silicon carbide (SiC)
Grinding wheel	Winter D1500 75 K+730B RB 4637
Wheel speed	47.87 m/s, 39.89 m/s, 31.92 m/s
Feed rate	0.0105 m/s, 0.0211 m/s
Depth of cut	3 μm, 6 μm, 10 μm.

EXPERIMENTAL SETUP

The SiC used in the workpieces was initially created via a high-temperature electrochemical reaction of carbon and sand.

Silicon carbide is a perfect example of an abrasive material. Thus, it was utilized in producing grinding chips together with other abrasive materials. SiC has been developed into a high-quality technical grade ceramic with outstanding mechanical characteristics. SiC ceramics have not only splendid mechanical characteristics at room temperature (i.e. high flexural strength, high resistance to oxidation and oxidative erosion, elevated resistance to wear and abnormal temperatures) (Huerta & Malkin, 1976) but they also maintain their strength when subjected to temperatures of up to 1600 °C.

Coolant is used for lubrication, removing chips, and reducing temperature to keep workpiece away from burning. The TRIM C270 with 9 pH, Viscosity of 1.51 MPa, and concentration 5% produced by Master Chemical Cooperation.

Grinding wheel is dressed every time after each workpiece grinding, with diamond dresser and dressing wheel speed is 1000 RPM. The TRIM C270 coolant was used for both grinding and dressing. The dressing takes off 0.005 Inches wheel materials each time.

Procedure and Techniques

To conduct the experiment, the tasks conducted sequentially are listed and described below:

1. Pre-polishing each surface of workpiece
2. Cut workpiece
3. Fine polishing each cutting section

4. Measurement
5. Grinding
6. Measurement
7. Coating
8. SEM analysis

1. Polishing:

Pre-polishing is very necessary before everything start. The workpiece has slight powder and other mechanical impurities attached. Pre-polishing before cutting and measurement will help to get accuracy weight. A traditional 600 grit sand paper was used for pre-polishing

2. Cutting workpieces:

The material workpiece samples were cut to three equal lengths by a low speed-cutting saw. This low speed-cutting saw equipped a diamond-wafering blade. Because of SiC property, diamond-cutting saw is the only way to cut it. Moreover, low speed cutting will help us to observe the cutting depth and process. Cut of specification in table 2

Table 2: Cutting condition for workpiece surface preparation

Wheel type	Buehler IsoMet Blade, 15HC	Diamond, 240 grit
Wheel RPM	500 r/min	
Downforce	400 g	
Coolant Type	water	

3. Cross Section finishing:

Fine polishing of cutting section will be performed after cutting. There are three steps for fine polishing. First, rough polishing; take off the salient point from cutting section edge. Second, fine polishing cutting section surface by different grit sand paper. Last, super fine polishing by 0.3 μm diamond polishing powder paste. The smooth cutting section surface is not only for all workpiece integration tightly, but also good for later SEM observation.

4. Mounting the sample

Workpiece holder is fixed on top of Kistler 9257B dynamometer shown in figure 3. Dynamometer will collect all of forces during the grinding process. The actual grinding force is an important measurement in evaluating the grind potential of materials, grinding force can help achieve high precision and efficiency of a work piece.

5. Grinding machine

The grinding machine adopted in this study is Thompson Surface Grinder with a horizontal-spindle, Fanuc 3M–Model C controller was used control operation of the grinding machine during experiment.

The primary feed motion is the reciprocity of the worktable on which the work is mounted (X); this moves the worktable forward and backward as feed motion, by moving it up and down to the column (Y) or by moving it perpendicular on the direction of main motion (Z).The Thompson surface grinding machine specifications are

Model: FSH CNC 30/90

Work Capacity: 11" (width) x 35" (length) x16" (height)

Wheel size (max.): 12" (OD) x5"(ID) x1.5"(width)

Wheel rotation: 0-3000 RPM

Wheel head: max rapid traverse speed: 500 in./min.

Least input increment: 0. 0001 in.

Wheel in feed: max rapid speed: 300 in./min.

Least input increment: 0.0001 in.

Wheel motor: 5.5 kW.

Feed rate: 0 - 6 m/min

To provide real-time dynamic balancing, a dynamic balance system was equipped with the wheel spindle. The system excludes disequilibrium of the grinding wheel and reduces grinding wheel vibration while maintaining an optimum grinding process.

The computer located on the right side is connected to the dynamometer set on the grinder table. Between computer and grinder is the control area and grinding wheel auto balance monitoring system.

RESULTS AND DISCUSSIONS

From Figure 4, we can determine that the normal force continued to increase with longer grinding times. This trend is inconsistent with our workpiece holder design: as the barrier (workpiece) height increases across the longitudinal direction from left to right, the grinding wheel experienced an increasing counterforce, proportional to the normal force. Moreover, the force data acquisition system keeps the grinding process known; when then wheel is in contact with workpiece.

A comprehensive overview of grinding factors' effect on surface roughness such as feed speed, wheel speed, and depth of cut can be observed in Figures 5–7. Several issues could be inferred from those results: First, the surface roughness became greater with faster feeding speed. Moreover, Higher-feeding speed leads to insufficient contact between the workpiece and the grinding wheel. In figure 6, as the depth of cut decreases the better roughness will have, as greater depth will result in higher material removal in brittleness mode, which the essential concept of this study. Similarly, the increased wheel speed could enlarge the contact efficiency between the grinding wheel and the workpiece. Therefore, as shown in Figure 7, the coarseness decreased as the speed of the wheel was increased. Moreover, as the deepness of cut increases, coarseness also increases. In the meanwhile, the outcome of using these three grinding wheel speeds suggest that better surface roughness will be obtained when using a grinding wheel with higher speed. Lower material removal rate is linked with the higher grinding speed. Figure 5 is an example of how different wheel speed and feed speed affect the surface roughness. These effects are independent of each other.

From Figure 8, it is realized that as the depth of cut increases the normal force will increase. Since the depth of cut is proportional to the material removal rate (MRR), according to a force model established by (Fan & Miller, 2006), the normal force can be divided into vertical and horizontal components:

$F_z = k_z \,.MRR^{m_z} + F_{z0}$

$F_x = k_x \,.MRR^{m_x} + F_{x0}$

Here, k_z, k_x, m_z, and m_x are constants, while F_{z0} and F_{x0} are the corresponding rubbing components. Based on this model, it is clear that a higher material removal rate leads to a higher normal force.

Figure 9 shows that the normal force is reduced with higher wheel speed. This phenomenon can be explained according to our previous discussion: a higher wheel speed reduces coarseness. Therefore, the horizontal counterforce, which is one component of the total normal force, was reduced due to reduced friction.

A Veeco Dektak 150 was used to measure the surface roughness of the samples, with the scan directions taken normal to the long axis of the samples (perpendicular to the grinding direction). Five scans were taken and the average and standard deviations of the RMS roughness (R_q) and average roughness (R_a)

Model

The aim of this section is to derive and describe a model for future prediction for the surface roughness (R_a), and normal force (F). It can be noted that the natural logarithm transformation was sufficient to obtain the desired linearity of the process and thus first order models for both roughness and force were proposed in order to obtain the relation between input and output variables. Minitab software (v.17.3.1) was used to create both models. The empirical model to be used in this study for the surface roughness is the following:

$$Ra = C * (V_S)^{\alpha} * (V_W)^{\beta} * DC^{\gamma}$$

Where Ra is the roughness, C is a constant, V_S (m/s) is wheel speed, V_W (m/s) is table speed, and DC (μm) is depth of cut. The powers α, β, and γ are to be found experimentally

In order to solve for the model proposed, the natural log was taken for all terms and so the model will look like:

$$\text{Log (Ra)} = \text{Log (C)} + \alpha \ \text{Log} (V_S) + \beta \ \text{Log} (V_W) + \gamma \ \text{Log (DC)}$$

Substituting the converted data into Minitab, we get the first regression model as

$$Ra = 3.673\ V_S^{0.03548}\ V_W^{5.7}\ DC^{0.0275}$$

The empirical model to be used in this study for the surface roughness is the following:

$$F = C * (V_S)^{\alpha} * (V_W)^{\beta} * DC^{\Upsilon}$$

Where F is the force, C is a constant, V_S (m/s) is wheel speed, V_W (m/s) is table speed, and DC (μm) is depth of cut. The powers α, β, and γ are to be found experimentally

In order to solve for the model proposed, the natural log was taken for all terms and so the model will look like

$$\textbf{Log (F)} = \textbf{Log (C)} + \alpha\ \textbf{Log} (V_S) + \beta\ \textbf{Log} (V_W) + \gamma\ \textbf{Log (DC)}$$

Substituting the converted data into Minitab, we get the first regression model as

$$F = 2.841\ V_S^{0.01386}\ V_W^{29.26}\ DC^{0.0622}$$

The significant factors for both models roughness and force are Wheel speed (V_s), Table speed (V_w) and the depth of cut (*DC*).

Also, the natural logarithm transformation will help reduce multi-collinearity between the variables considered in the model as well as making it easy to compute and examine the effect (in terms of magnitude and direction)of each independent variables on the dependent variable which in this case is roughness.

Validation

To validate the model, a random point between two parameters of the wheel factor was chosen. The result from the Ra model was 1.6097 compare to experimental that is 1.406 with error of 14.48%.

As for F model result was 1.93314 compare the experimental that is 1.63 with error of 18.59%

Scanning Electron Microscopy analysis

The Hitachi S-4800 High Resolution Scanning Electron Microscope, a cold-field emission high-resolution scanning electron microscope with many advanced features, was used to view the surface (SD) and subsurface (SSD) damage to the processed workpieces. The goal of the Scanning Electron Microscopy (SEM) analysis was to estimate the grinding ductility by observing the microstructure of a ground workpiece. Figure 10 (left and right) shows the surface morphology of a workpiece at both point A and B. The localized micro-plastic deformation has been observed, where is the depth of cut is smaller than 4.5 um (Figure 10, left), and Figure 11 (left) gave a better estimation that such deformation only happened within several micrometres from the surface of the workpiece (subsurface). In this region, few structural defects could be observed, indicating that lower depth of cut has less impact on the internal tissues. Moreover, this change on the shoulder edge of a grinding surface could be identified as a plastic flow on lower depths of the cut workpiece. Therefore, a shallow ductile regime has been demonstrated on the shoulder edge of the grinding surface under lower depths of cut; in essence, this results in lower material removal rates.

Figure 10 (right) displays the surface morphology between points B and C. Here it was found fracture cracks due to a higher depth of cutting. The appearance of cracking around the cutting section surface at (point B) shows changes induced by the material removal from ductile change to hard brittle material grinding. The ductile–brittle transition phase is termed a critical parameter where the surface and subsurface damages begins. Relative to the grinding depth, brittle material extraction and removal is categorized into three types: ductile grinding, ductile-brittle grinding, and brittle grinding. The critical depth of cut exists between our experimental ranges. As point B stood for a depth of cut of 4.5 μm, it can be inferred that the 4.5 μm depth of cut is likely the critical point.

CONCLUSION

Recall, this study was aimed to show the experimental application of the theoretical grinding process for brittle materials such as SiC. In this study, the experiment focused on the effects of several vital factors (feed speed, depth of cut and rotational speed of the wheel) when exposed to a normal force, as well as processed surface roughness.

It was observed that the surface roughness after grinding was increased by higher feed speed, lower wheel speed, and higher depth of cut. Additionally, normal force increased with increasing the depth of cut or reduced wheel speeds.

Surface morphology was assessed to estimate a range for proper grinding. Through SEM analysis, this study concluded that within a shallow depth of cut (less than 4.5 μm), the workpiece displays a ductile regime with a plastic flow only until several micrometres from the top surface, leaving the internal tissue unchanged. However, as the depth of cut continues increasing, fracture cracking is triggered at both shallow surface and bulk areas of the workpiece, which indicates the occurrence of an unwanted brittle regime. It can be further inferred that as the cut depth increases, the brittle regime is enhanced and might replace the ductile regime. Based on the experimental design, it is suggested that a 4.5μm depth of cut is likely the critical point for grinding on a SiC workpiece.

REFERENCES

Blake, P. N. (1988) 'Ductile-Regime Diamond Turning of Germanium and Silicon', Ph.D. Thesis, North Carolina State University, United States.

Blake, P. N., & Scattergood, R. O. (1990) 'Ductile-Regime Machining of Germanium and Silicon', *Journal of the American Ceramic Society,73*(4), 949-957. doi:10.1111/j.1151-2916.1990.tb05142.x

Bifano, T.G. (1988) *Ductile-regime grinding of brittle materials*. PhD Thesis, NC State University, Raleigh, NC. United States.

Bifano, T. G. (1992) 'Ductile-regime grinding. A new technology for machining brittle materials', *Precision Engineering,14*(1), 54. doi:10.1016/0141-6359(92)90161-o

Chandrasekar.S., and Sathyanarayanan.G. (1987) 'An Investigation into the Mechanics of Diamond Grinding of Brittle Materials'. *15th North American Manufacturing Research Conference Proceedings*, Vol. 2, Manufacturing Technology Review, pp. 499-505

Chen, J. et al., (2010) 'Grinding characteristics in high speed grinding of engineering ceramics with brazed diamond wheels', *Journal of Materials Processing Technology*, 210(6-7), pp.899–906.

Chen, M., Zhao, Q., Dong, S., & Li, D. (2005) 'The critical conditions of brittle–ductile transition and the factors influencing the surface quality of brittle materials in ultra-precision grinding', *Journal of Materials Processing Technology,*168(1), 75-82. doi:10.1016/j.jmatprotec.2004.11.002

Cheng, J., & Gong, Y. D. (2013) 'Experimental study on ductile-regime micro-grinding character of soda-lime glass with diamond tool', *The International Journal of Advanced Manufacturing Technology,69*(1-4), 147-160. doi:10.1007/s00170-013-5000-3

Dai, J., Su, H., Yu, T., Hu, H., Zhou, W., & Ding, W. (2018) 'Experimental investigation on materials removal mechanism during grinding silicon carbide ceramics with single diamond grain', *Precision Engineering,51*, 271-279. doi:10.1016/j.precisioneng.2017.08.019

Fan, X., & Miller, M. H. (2006) 'Force Analysis For Grinding With Segmental Wheels', *Machining Science and Technology,10*(4), 435-455. doi:10.1080/10910340600996142

Horvath, M., Kundrak, J., Mamalis, A., & Gyani, K. (2002) 'On the Precision Grinding of Advanced Ceramics', *The International Journal of Advanced Manufacturing Technology,20*(4), 255-258. doi:10.1007/s001700200150

Huerta, M., & Malkin, S. (1976) 'Grinding of Glass: The Mechanics of the Process', *Journal of Engineering for Industry,98*(2), 459. doi:10.1115/1.3438907

Krauskopf, B. (1984) 'Diamond Turning: Reflecting Demands for Precision', *Manufacturing Engineering*, 92 (5), 1984, 90-100

Marshall, D. B., and Lawn, B. R., (1986), 'Indentation of Brittle Materials,' *Microindentation Techniques in Materials Science and Engineering,* ASTM STP 889, P . J. Blau and B. R. Lawn, eds., ASTM, Philadelphia, pp. 26-46.

Shearer & R., T. (2006) 'Diamond wheel grinding 101: understanding and controlling the grinding parameters are essential prerequisites to machining advanced ceramics', *Ceramic Industry*.[online]

Zhipeng, Li., & Feihu Zhang, & Yong Zhang, & Xicchun Luo (2017)' Expreimental Iverstigation on the surface and subsurface damages characteristics and formation machanisms in ultra-precision grinding of SiC' *The International Journal of Advanced Manufacturing Technology* DOI 10.1007/s00170-017-0267-4

Zhichao Dong, & Haobo Cheng (2017) ' Ductile mode grinding of reaction bonded Silicon Carbide mirrors' *Applied Optics*. Vol 56. No 26

https://www.ceramicindustry.com/articles/88009-machining-finishing-diamond-wheel-grinding-101 (Accessed 20 November 18)

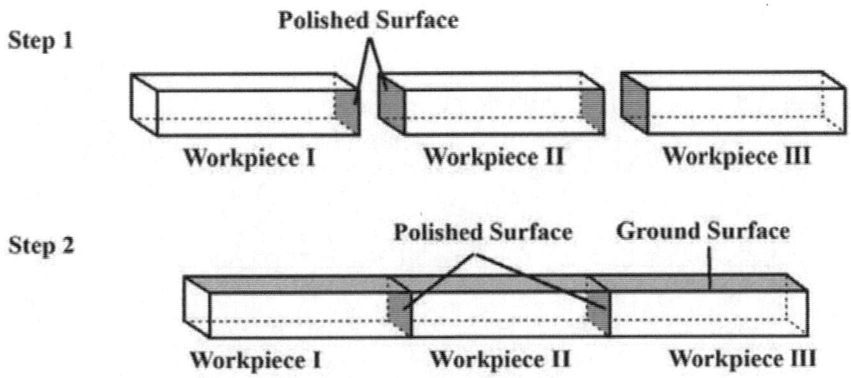

Figure1: ***Schematic of the workpieces and their various surfaces.***

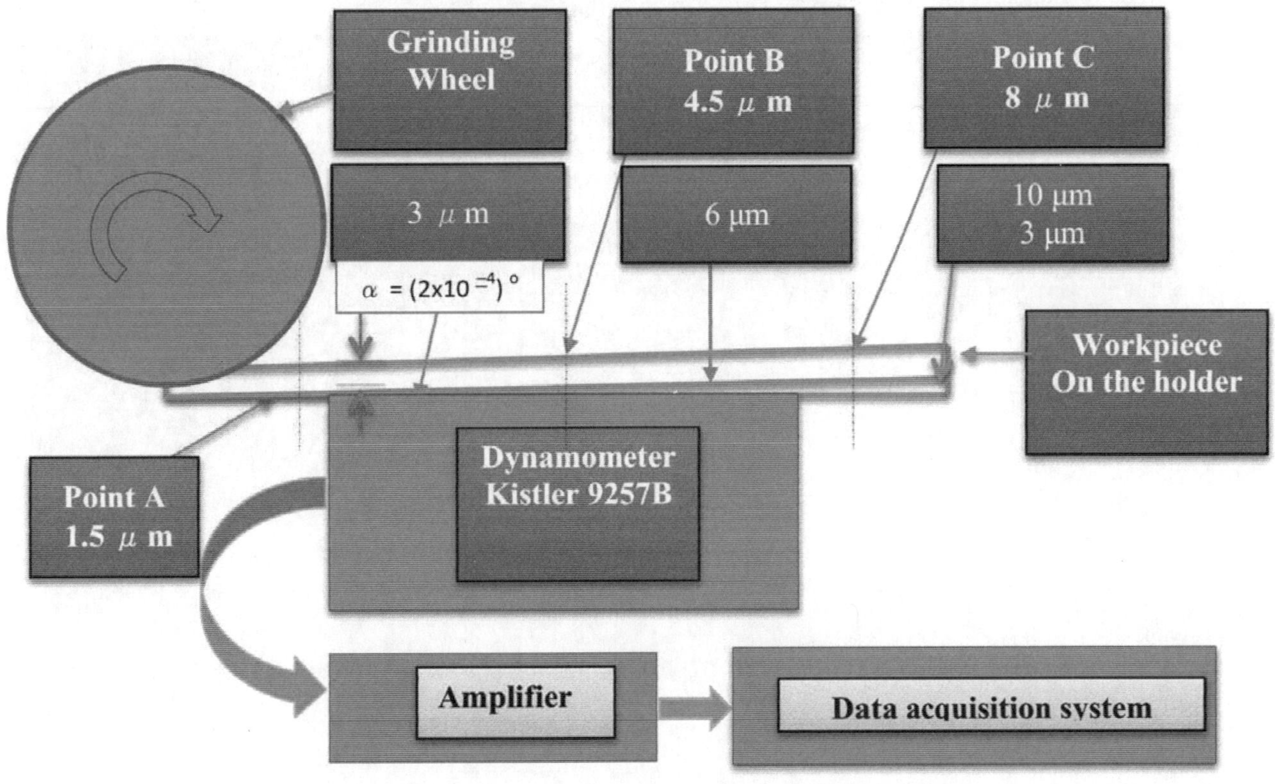

Figure 2: ***A schematic drawing of the equipment used in this experiment***

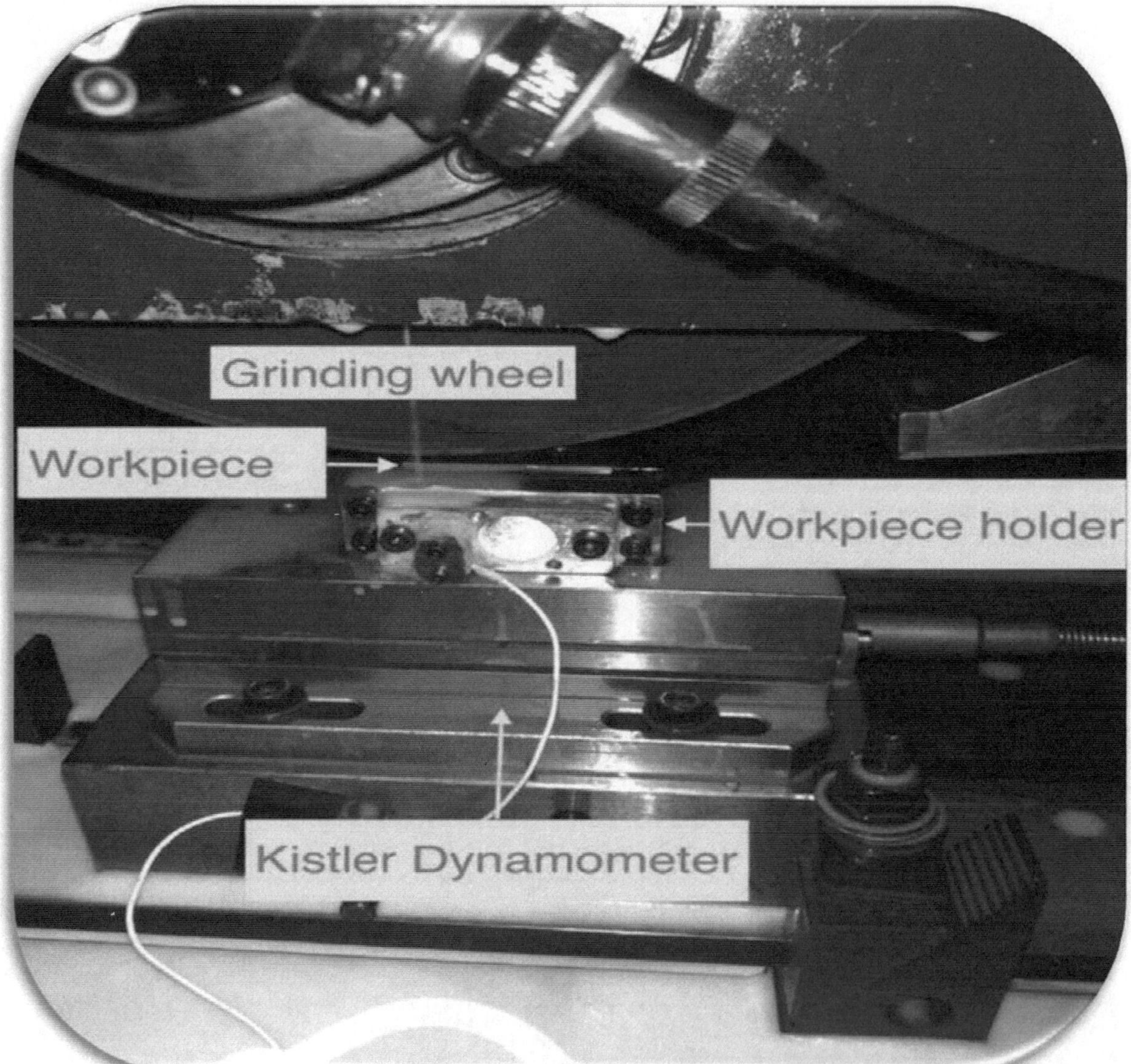

Figure 3: ***The Kistler dynamometer type 9257B used in this work.***

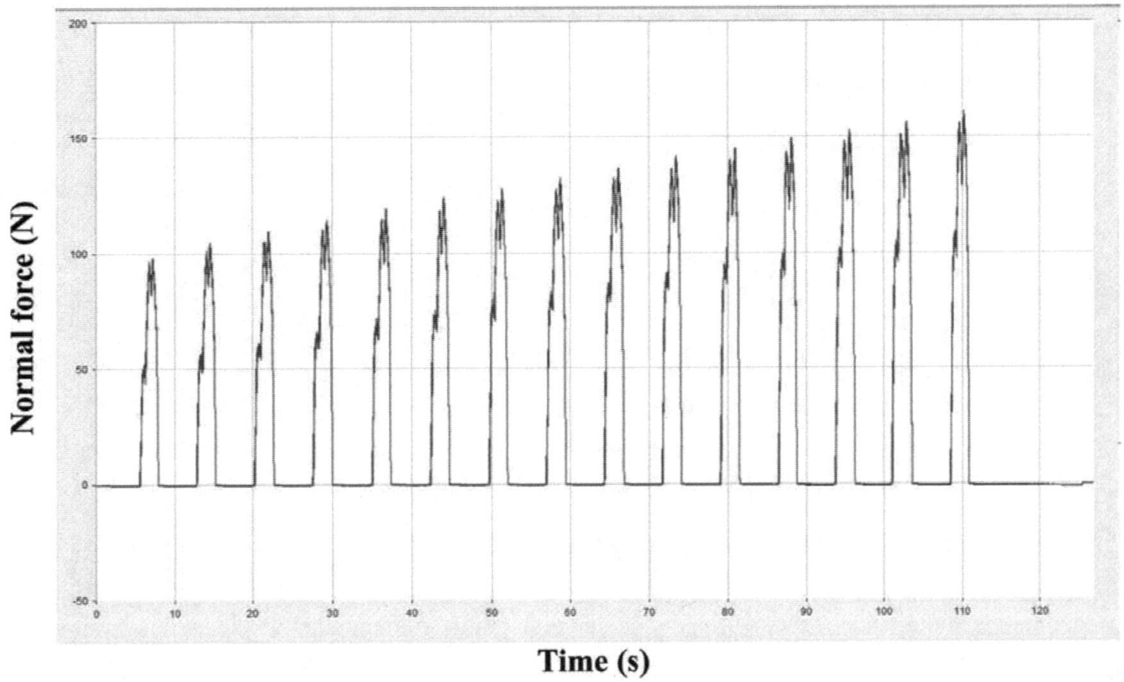

Figure 4: ***Grinding normal force plotted via data analysis spread sheet (DADisp 6.7)***

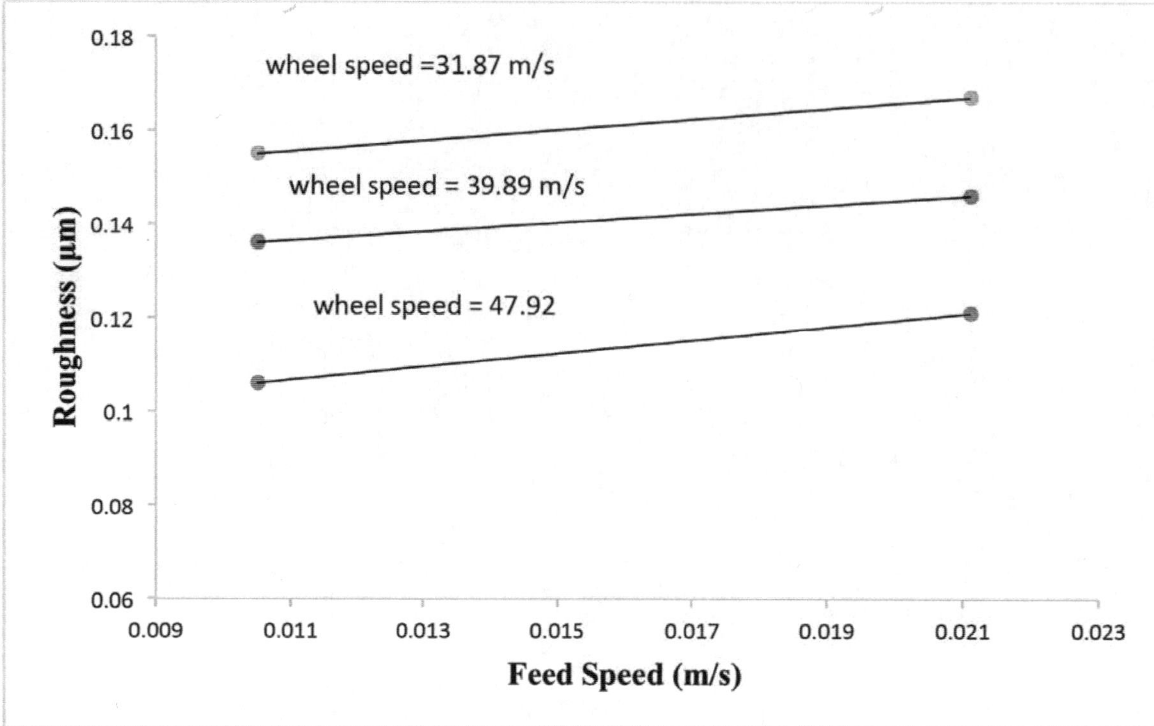

Figure 5: ***Influence of feed speed on surface roughness***

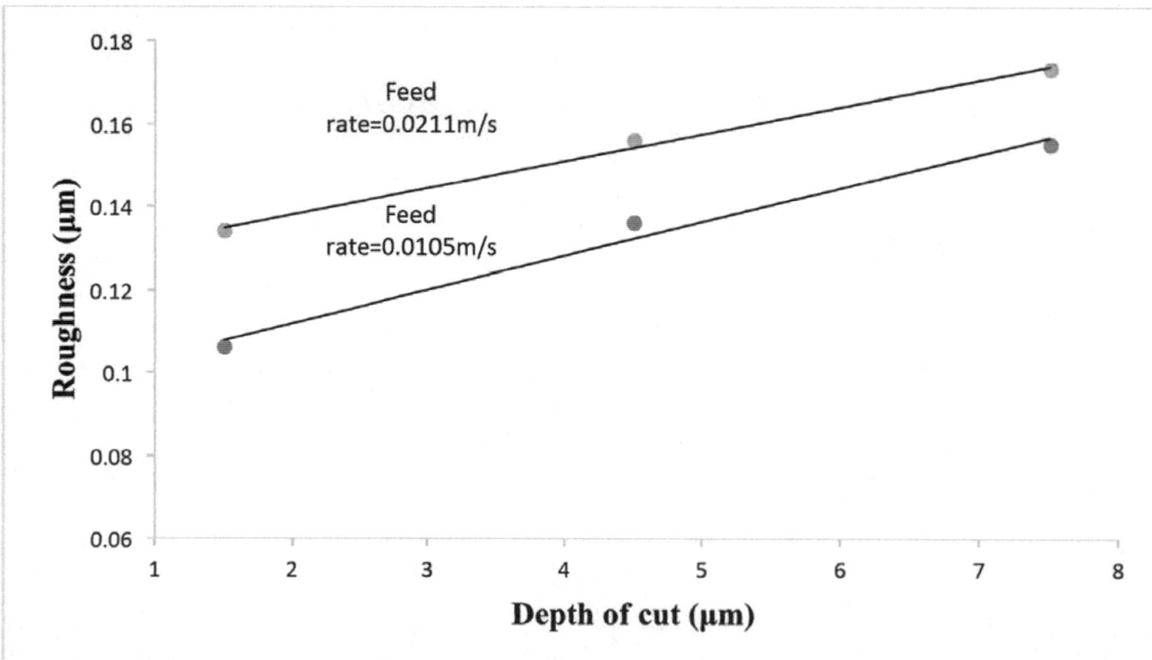

Figure 6: ***Influence of depth of cut on roughness***

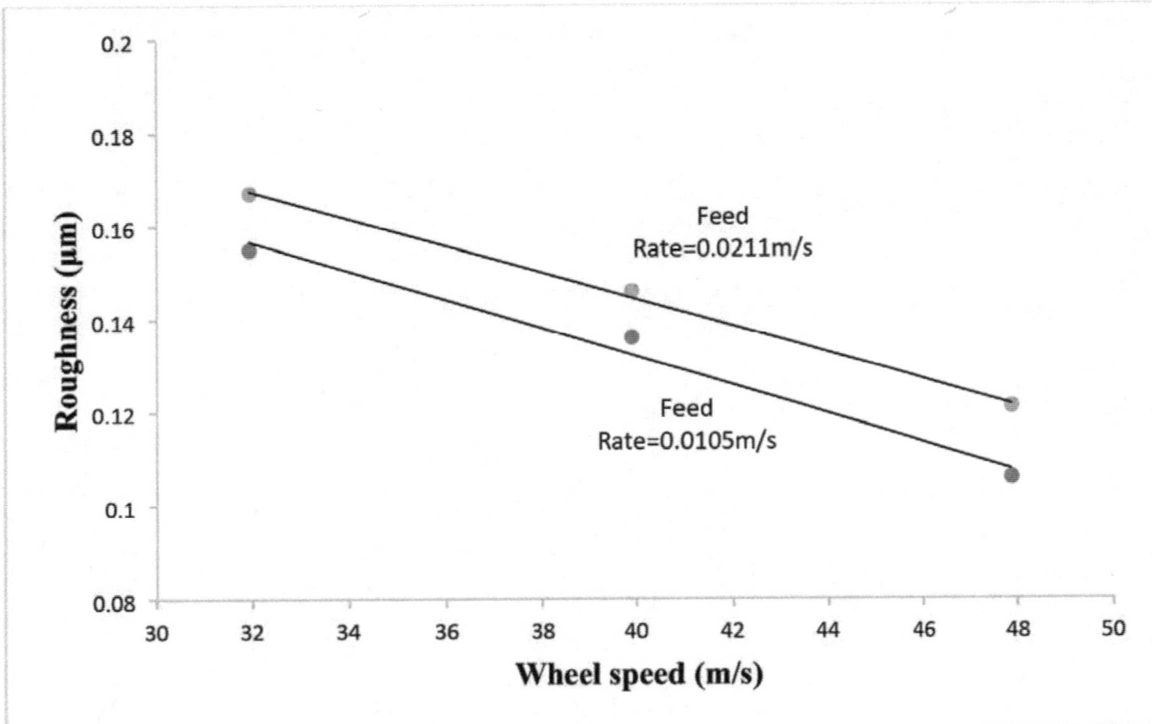

Figure 7: ***Influence of wheel speed on roughness***

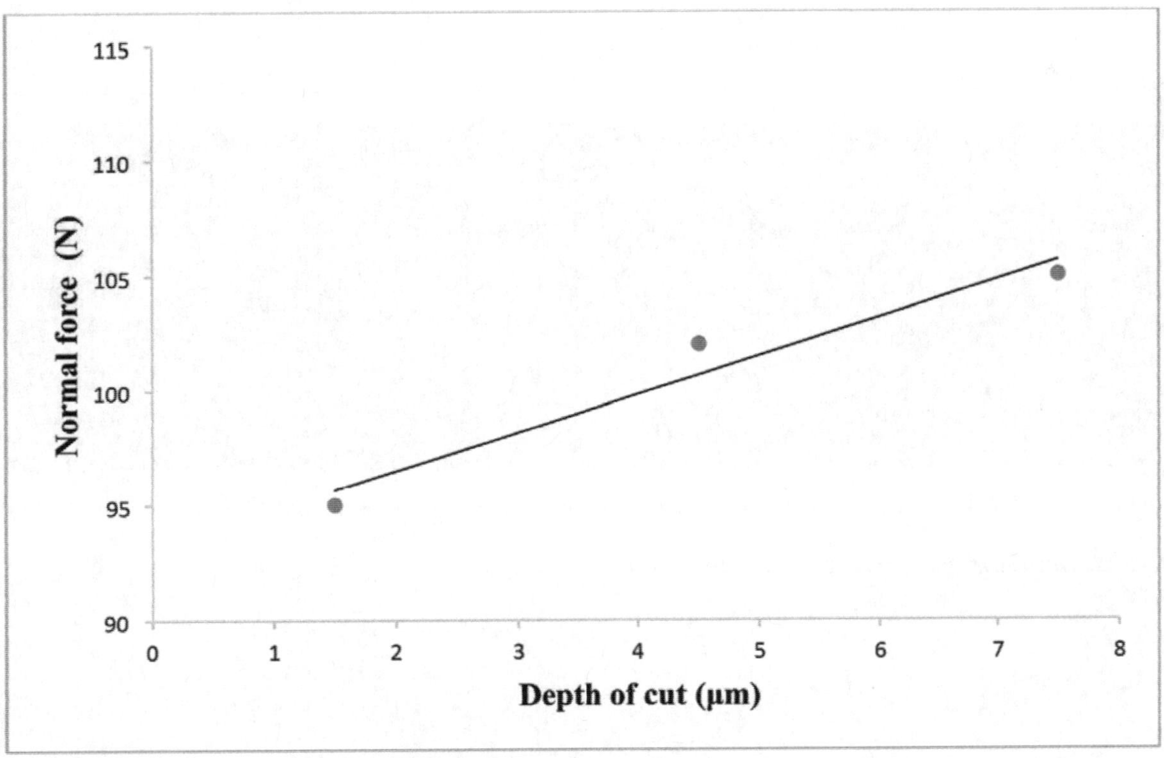

Figure 8: ***Influence of depth of cut on normal force***

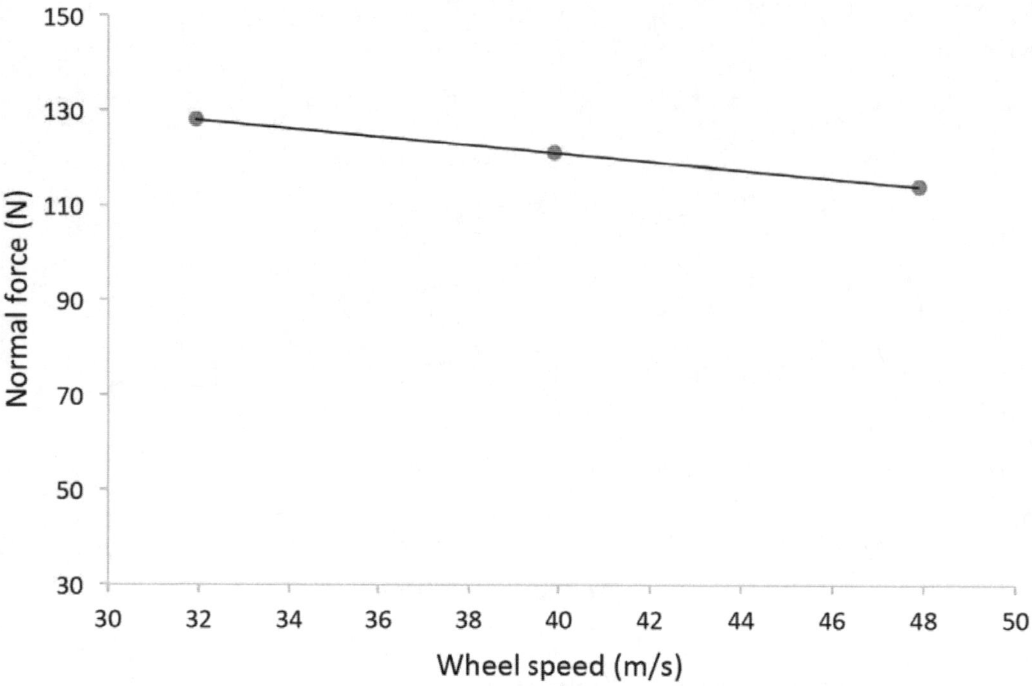

Figure 9: ***Influence of wheel speed on normal force***

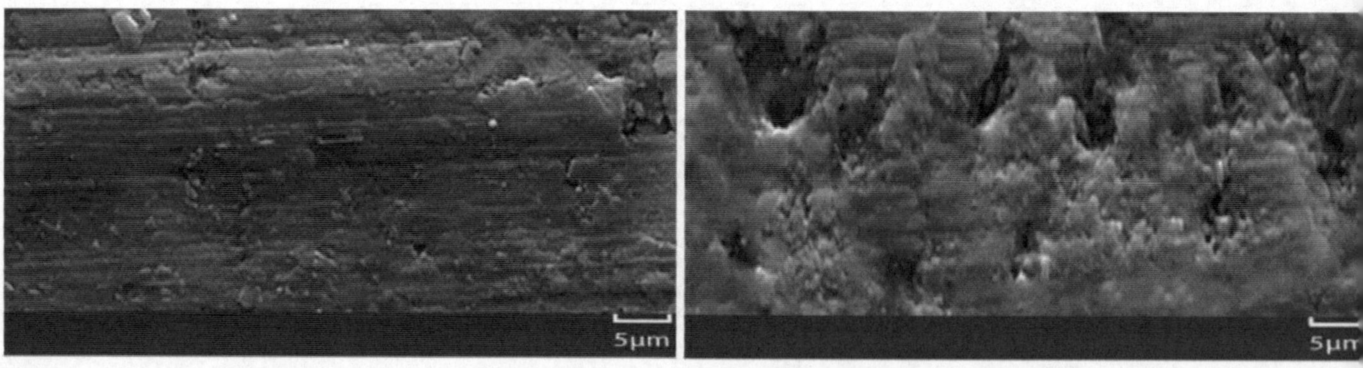

Figure 10: ***SEM surface between points A and B images for the (left) ductile and (right) brittle modes between points B and C***

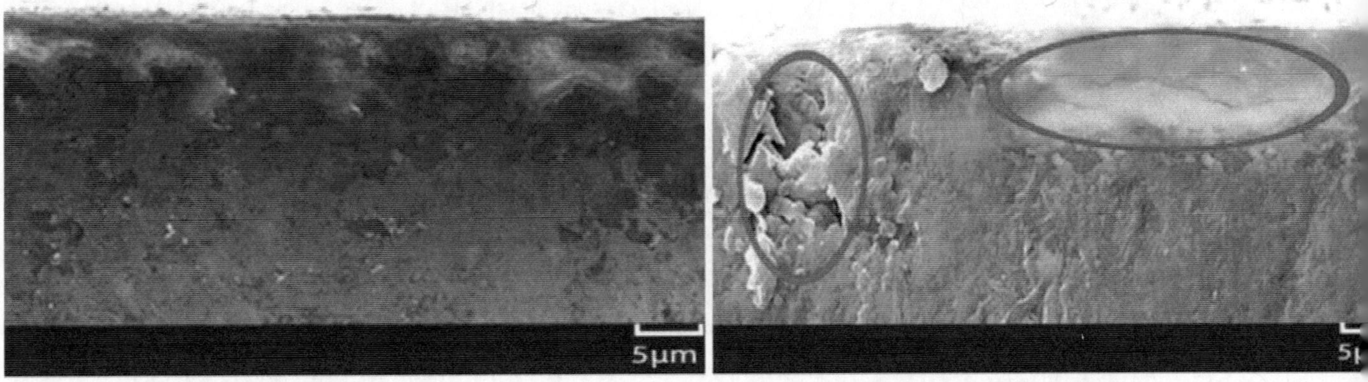

Figure 11: ***SEM subsurface between points A and B images for the (left) ductile and (right) brittle modes between points B and C***

GENIS 2 - The next generation of Vitrified CBN for cylindrical grinding in the Automotive Industry

Grinding as one of the oldest manufacturing methods is very often the last step in the production of components. The geometry and surface quality are precisely manufactured in very tight tolerances. The main field of application of these newly developed products in the automotive industry is the grinding of shafts (crankshafts, camshafts, transmission shafts). The properties of these grinding tools required for this purpose are explained both from a development and application point of view.

An important aspect of the quality of these components is the surface damage caused by heat, which is inevitable during grinding.

The example (Fig. 1) shows the energy balance during grinding.

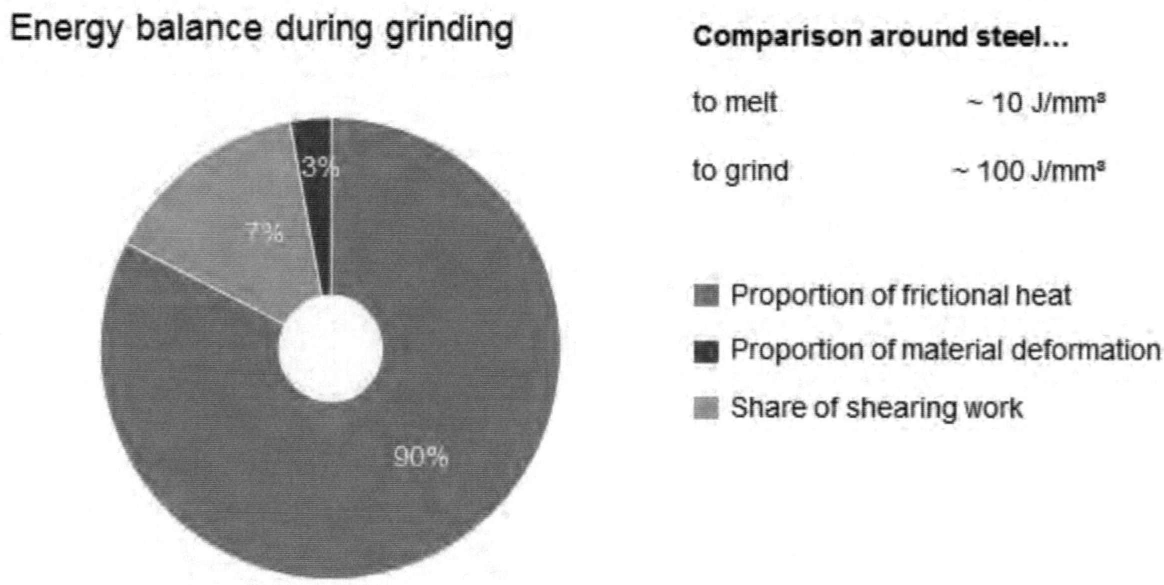

Figure 2 shows the effect of grinding heat on the material to be ground.

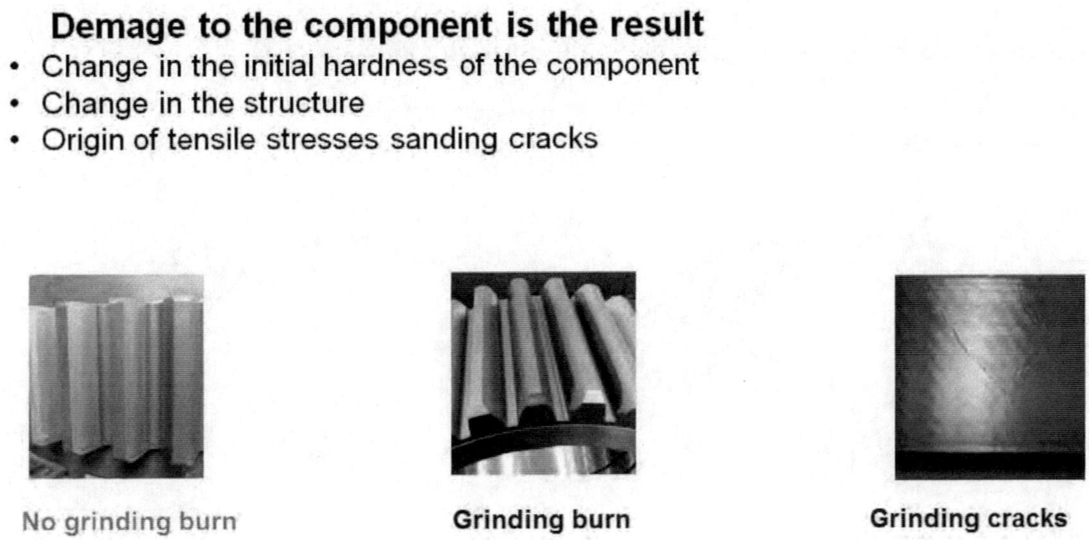

Fig. 2: Influence of the heat input in the grinding process

For developing new grinding products one has only limited variables (grain, bond, carrier body) which can be influenced. To create a grinding system which offers low cutting forces and heat generation during the use also the manufacturing technology has to be taken into account.

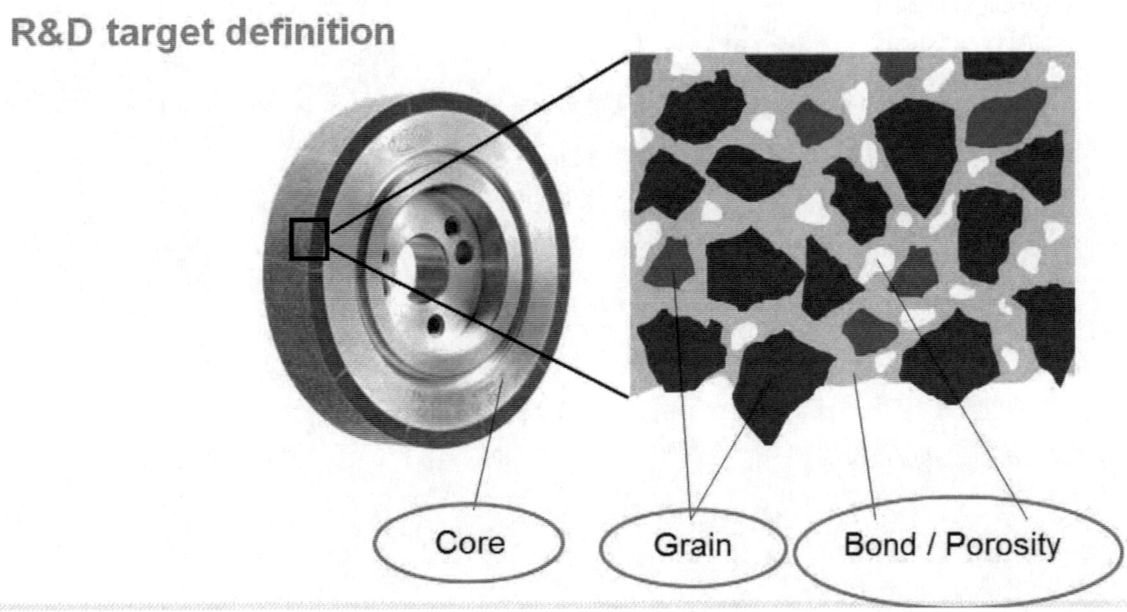

Fig. 3: Components of grinding wheels

Also very important is the accuracy and the run out of the whole system (grinding wheel + core).

Fig. 4: The two product line generations Genis and Genis 2

Both product lines in particular but Genis 2 products meet the requirements to a high degree.

What is the difference between both products?

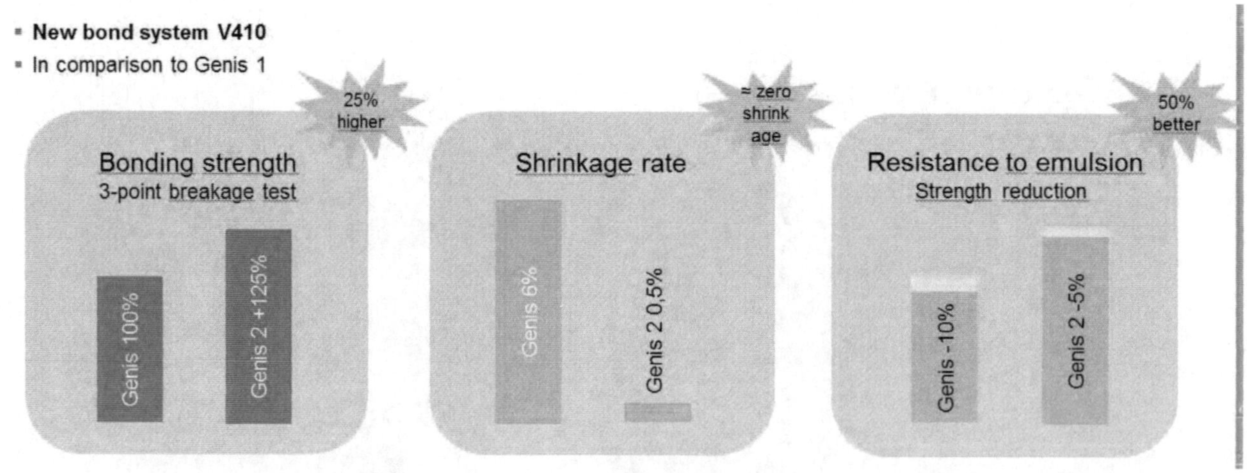

Fig. 5: Product comparison

The physical properties, especially the enhanced bonding strength, are particularly evident on the fracture surfaces of bending samples. This leads to visible breakage of grains as the adhesion of the bond to CBN is very good - also under the influence of coolants.

The excellent wetting of the grain surface by the ceramic bond can be seen in the pictures below.

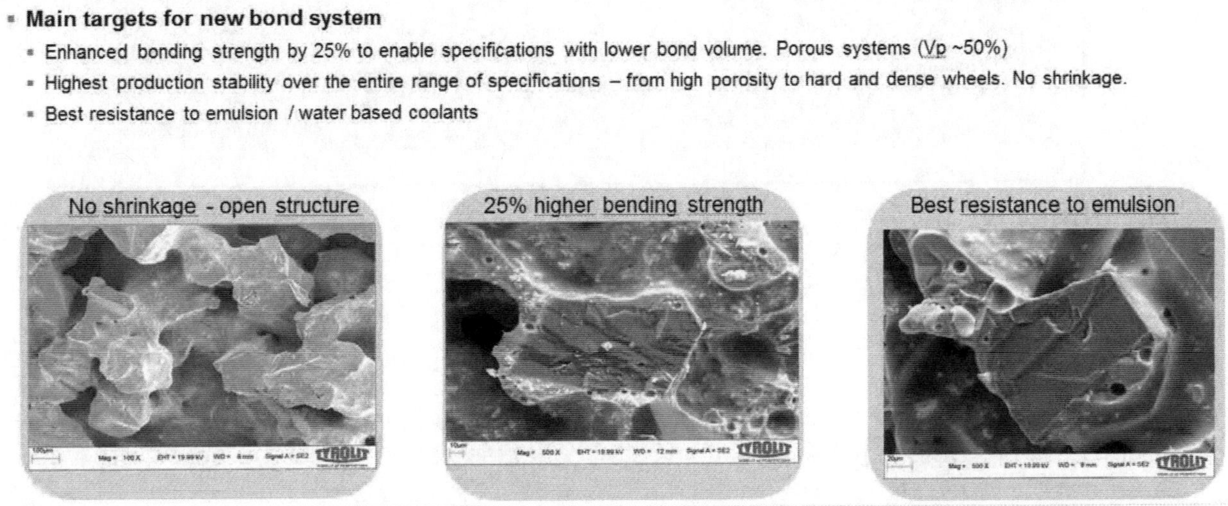

Fig. 6: wetting behavior of Genis 2 bond

An extraordinarily important development goal was to assure a stable and repeatable production process with a specification range as wide as possible.

The variance of the grain size, grain volume, binding volume, concentration of the CBN fraction in the volume leads to a very great complexity of abrasive material possibilities, which must be safely producible.

Due to the very low degree of shrinkage we can assure that the desired specification can be produced very consistent and repeatable. This allows to utilize also very fine adjustments to the grinding wheel specification which enables us to create the best possible grinding solution for our customers.

Furthermore this allows a much larger spectrum of producible specifications than any previous developed bond system.

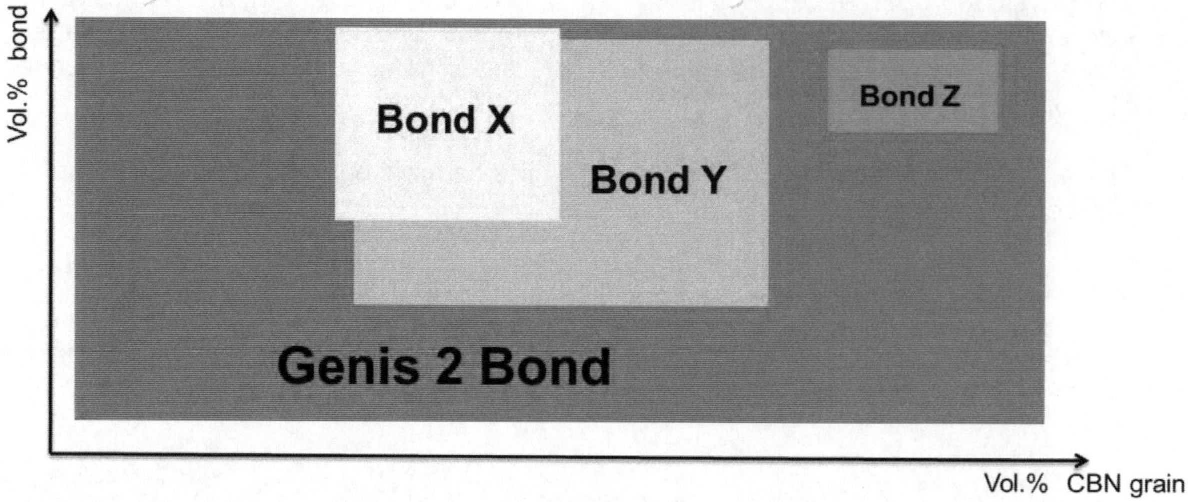

Fig.7: Production range Genis 2 vs. older bond systems

To fully characterize a CBN grain you need to have a comprehensive test and comparison system in place. This in return enables you to select the best CBN type for the customer application.

Selecting the „suitable" grain:

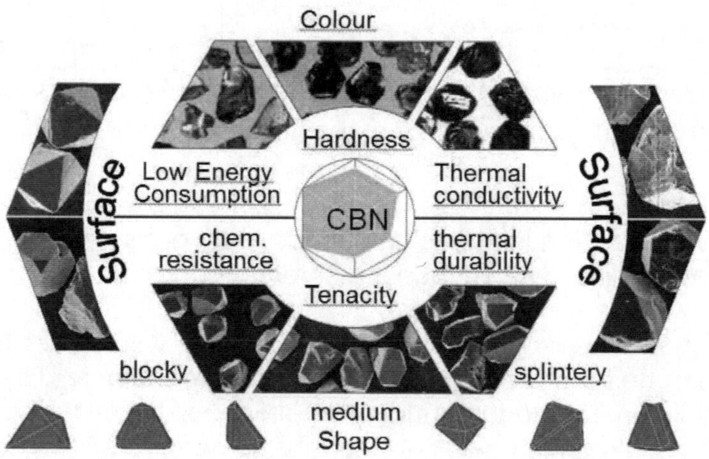

Fig. 8: CBN grain properties

In order to accurately characterize grain properties, comparative loop investigations are required. For this purpose we have 6 grinding machines in our internal Test Center at Tyrolit.

Fig. 9: Cylindrical grinding machine from Tyrolit Test Center

In addition to these internal grinding tests, extensive field trials at selected customers are conducted to verify that our internal results meet the requirements.

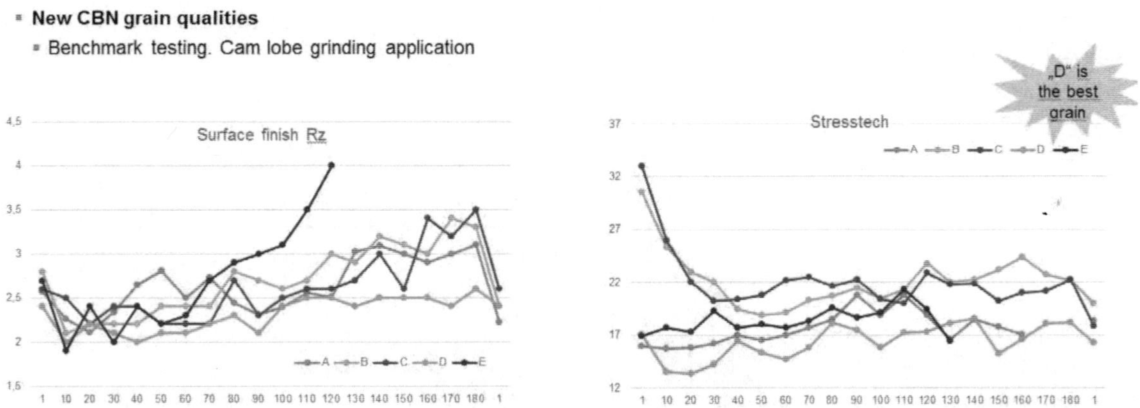

Fig. 10: Test results grain comparison cam lobe grinding

These experimental data determined under real conditions are particularly important for the selection of optimal CBN grain qualities.

For a practice-oriented use of CBN grades, a system has been developed.

This combines measured values of physical properties so that an application-oriented choice of the most suitable CBN grain is possible.

Figure 11 shows the characteristics of different CBN qualities.

- **New CBN grain qualities**
  - Updated CBN assortment

Grain	Toughness	Shape	Performance	Price	Main Application
I	69	Splintery	48	++++	Highest performance steel
II	67	Splintery	47	+++	Peel grinding steel
III	65	Medium blocky	44	+++	Cam lobe steel
IV	63	Splintery	43	++	Standard OD steel
V	58	Blocky	39	++	Cast iron
VI	45	Blocky	29	+	Easy to dress

Fig. 11: Application – related properties of CBN

Another extremely important component of a precision grinding wheel is the carrier body (Fig.11)

In general: The carrier influences the grinding behavior. In addition to the illustrated carrier materials, light weight carriers, in particular made of steel and/or aluminum, are part of the product portfolio (Fig.12)

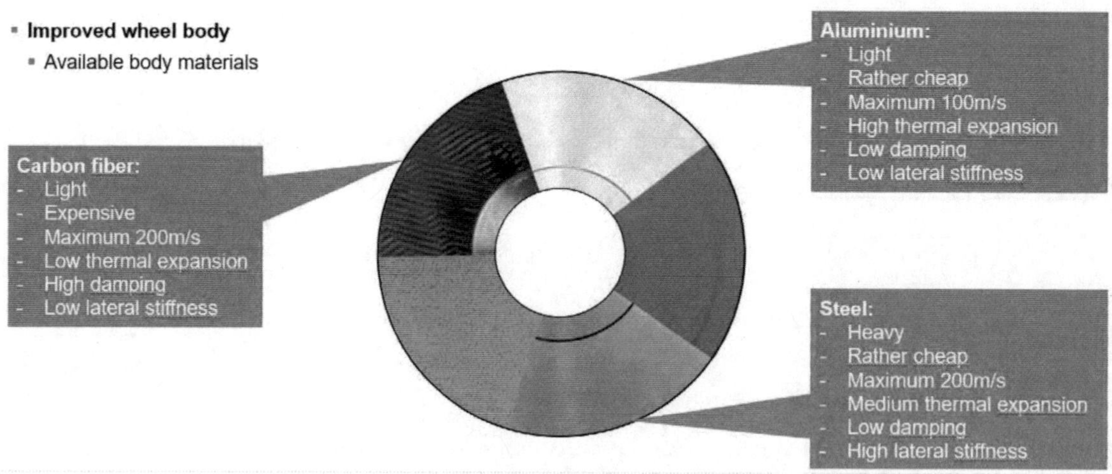

Fig. 12: Frequently used wheel hub materials and their advantages and disadvantages

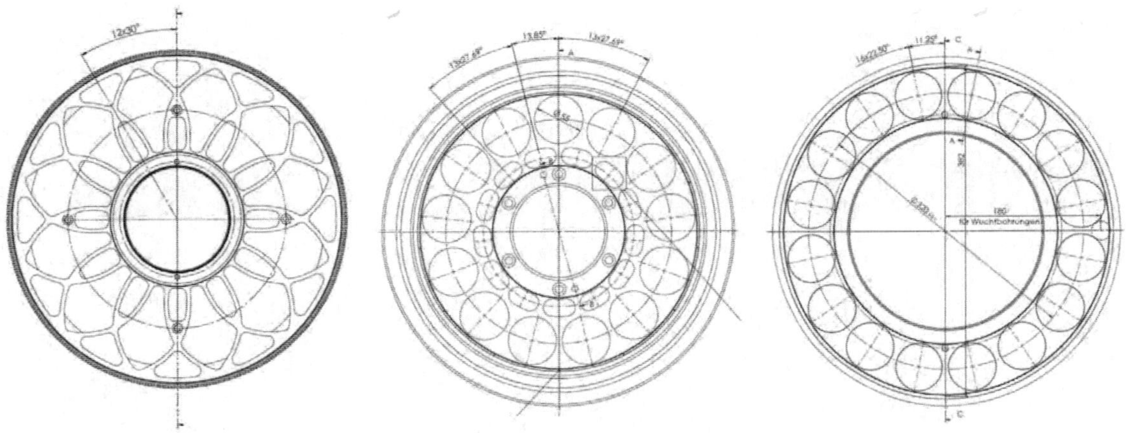

Fig. 13: Lightweight versions are shown in this illustration

CF as a carrier material is also used in many applications (Fig.14).

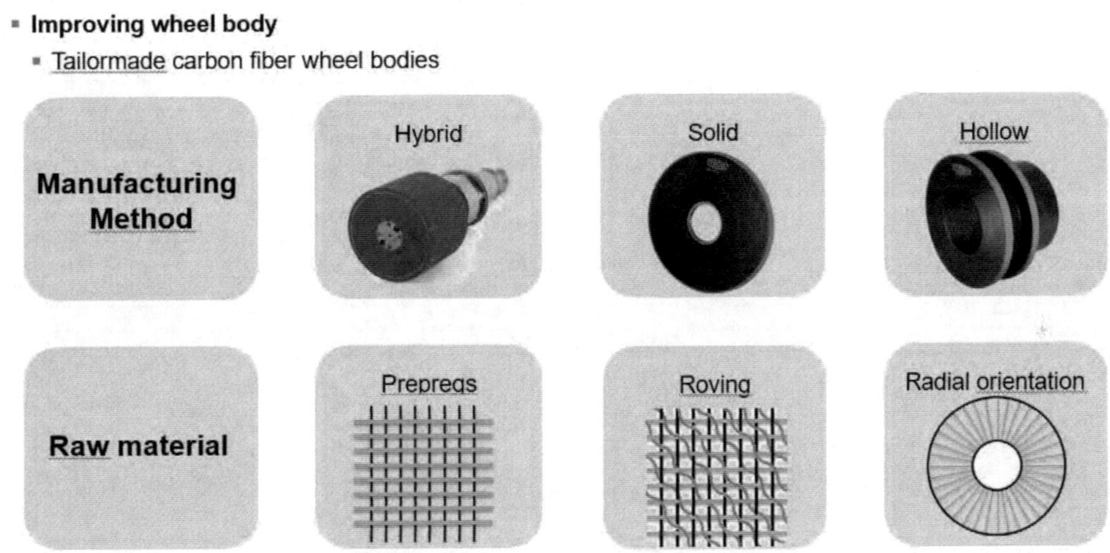

Fig. 14: Examples for CF hub versions

Without a mechanical processing of the carrier body to very narrow tolerances and an excellent balancing quality, the performance of these high-precision tools will be compromized (Fig.14 and 15).

- **Improving wheel body**
  - Higher accuracy of the body if necessary
  - Hardened and ground bore
  - Especially for Japanese machines like Toyoda and NTC

Accurate body	
Flange surface flatness	5µm
Bore roundness	2µm
Bore cylindricity	2µm
Layer concentric run-out	5µm
Layer axial run-out	5µm

Fig. 15: Wheel tolerances

- **Improving wheel body**
  - Improved balancing quality
  - New balancing machine

Diameter [mm]	Rest unbalance [gmm]
75	10
100	10
125	13
150	15
175	19
200	24
250	30
300	36
350	42
400	48
450	54
500	60
600	72
700	84
800	96
> 800	108

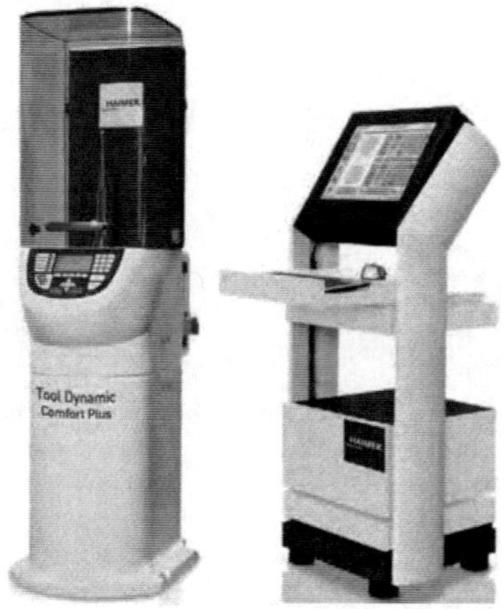

Fig.16: Balancing quality and balancing equipment

To successfully master grinding applications the perfect interaction between raw materials, bond technology and production needs to be assured.

Performance gains from our internal test could be seen in the field tests at customer applications with real parts which is the final prove that our new Genis 2 system has a higher performance level.

Figures 16 and 17 give an overview of the most common grinding applications on automotive shafts.

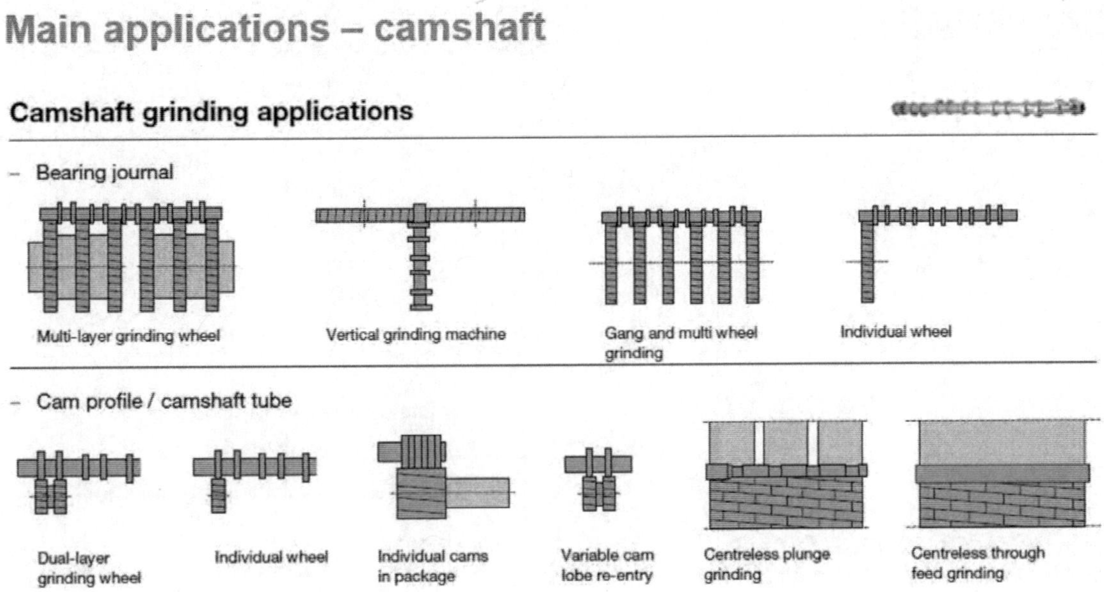

Fig. 17: Examples cam grinding

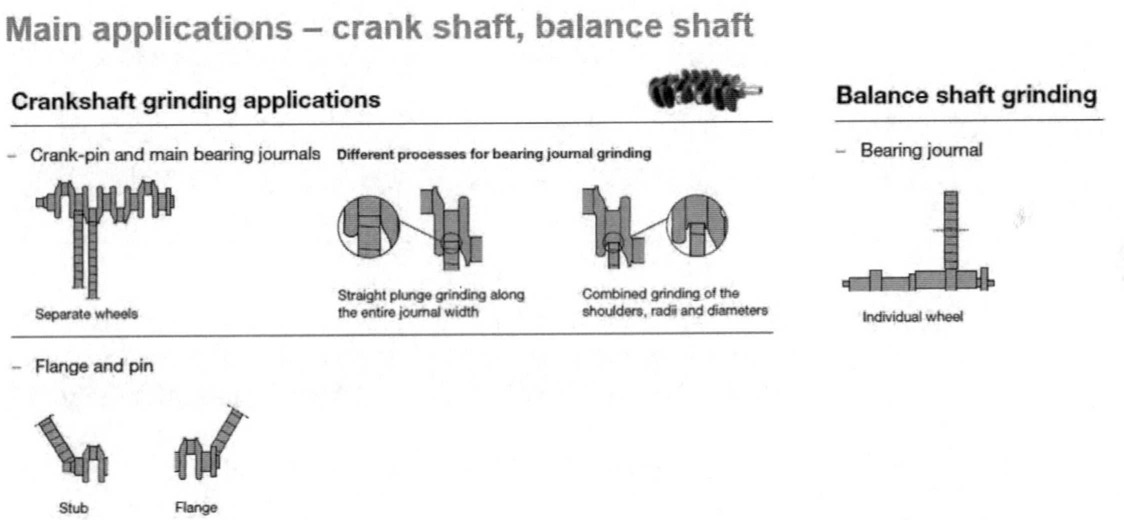

Fig. 18: Examples crank shaft grinding

The benefits of GENIS 2 grinding tools are presented on the basis of these actual grinding applications at various customers.

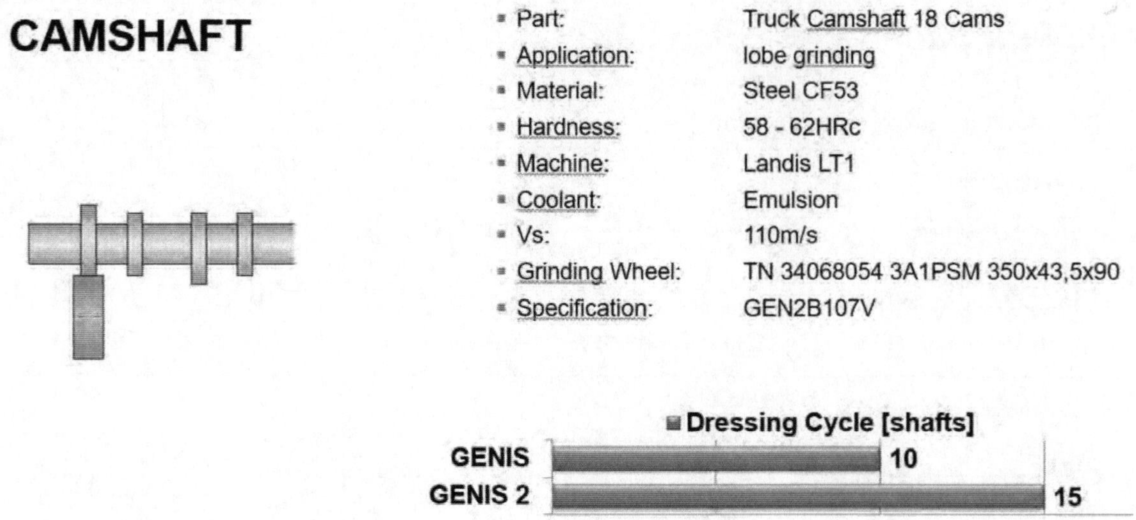

Fig. 19: Camshaft grinding results

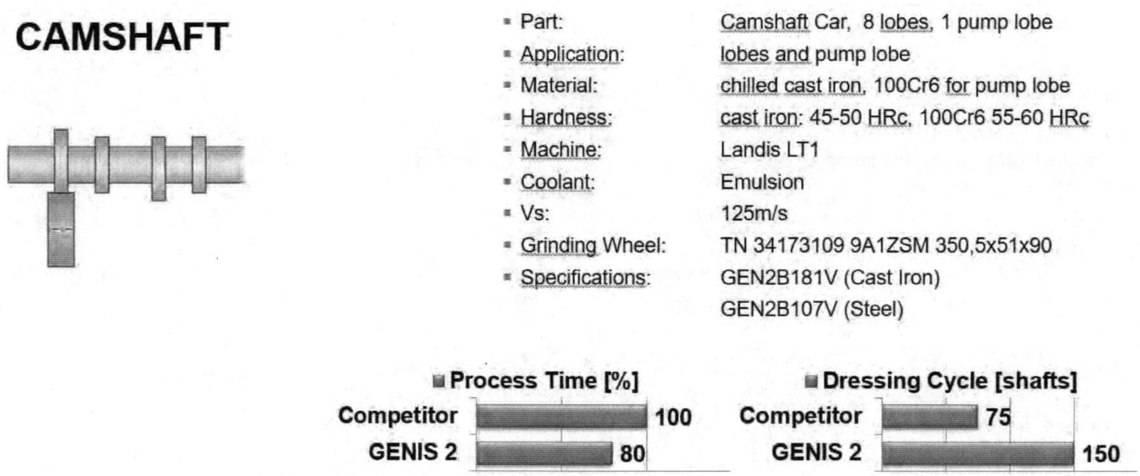

Fig. 20: Camshaft grinding comparing competitor over Genis 2 product

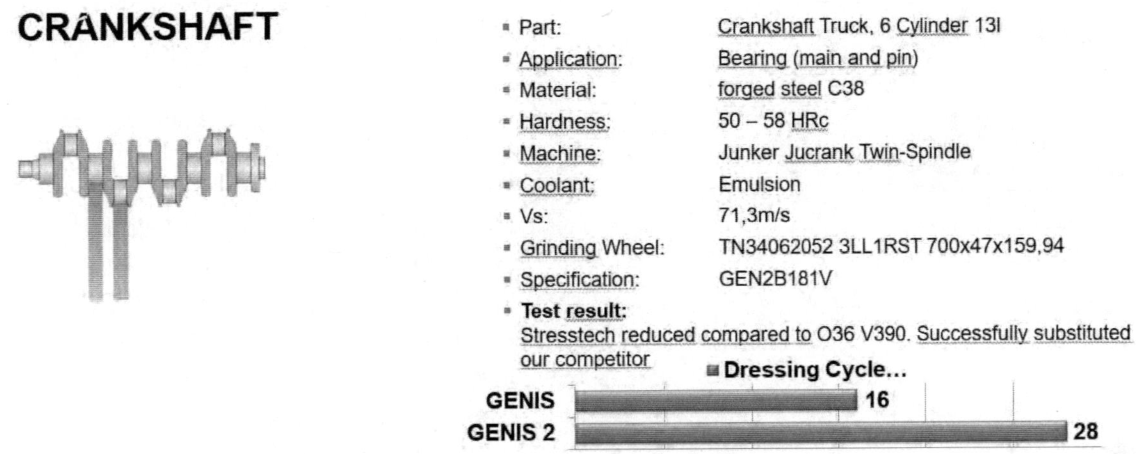

Fig. 21: Improvement by using Genis 2 for crankshaft grinding

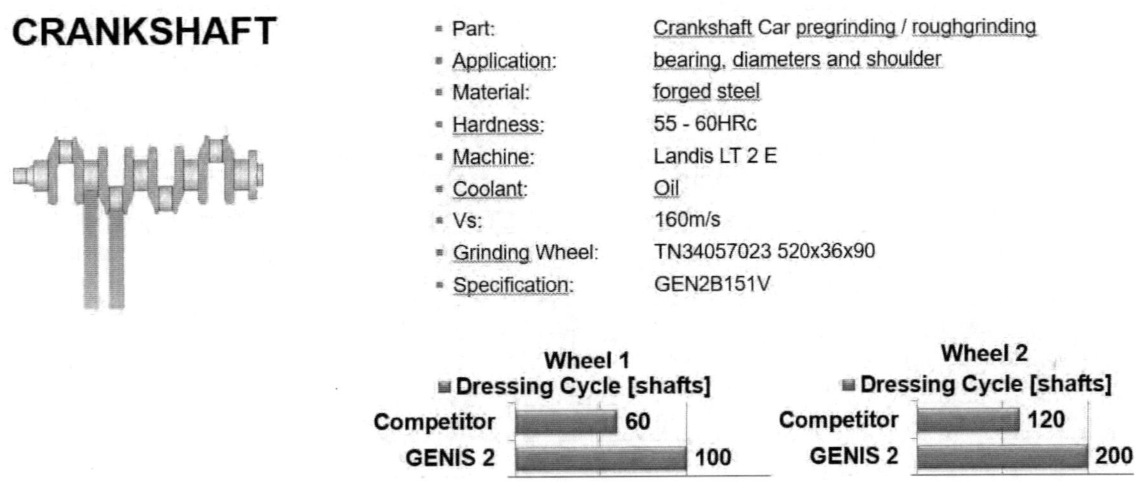

Fig. 22: Advantage of Genis 2 product over a competitive product

GEARSHAFT

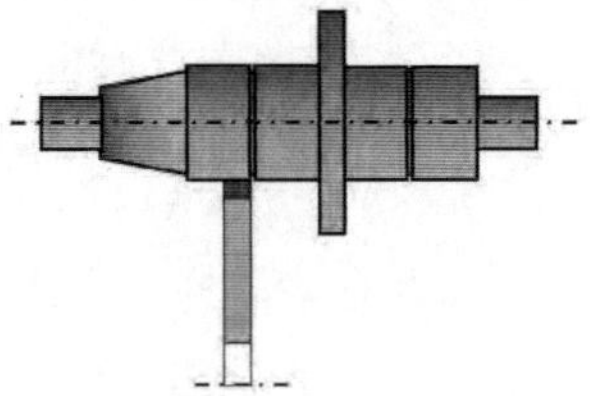

- Part: Gearshaft (output shaft)
- Application: Peelgrinding Bearing and Diameter
- Material: forged steel
- Hardness: 50 - 58HRc
- Machine: Junker Quickpoint
- Coolant: Oil
- Vs: 100m/s
- Grinding Wheel: TN 34060864 3M1ST 350x25x126,94
- Spezi: GEN2B126V

- **Test result:**
  Up to now we had no feasible Specification (tool wear / chatter marks)

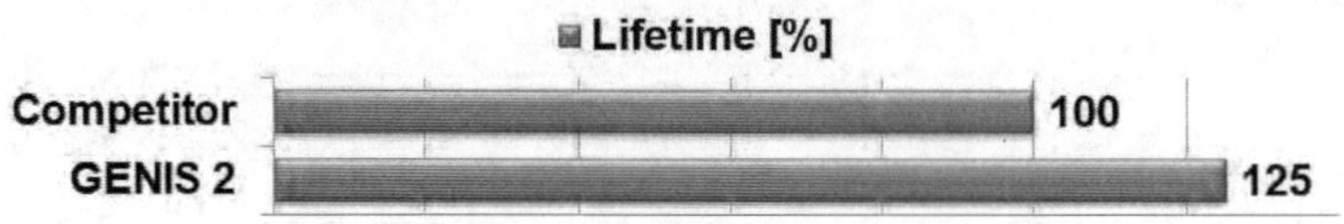

Fig. 23: Advantage of Genis 2 product over a competitive product

Conclusion:

The products of the GENIS 2 product line are very powerful grinding tools for OD grinding within the automotive industry. The application results in the different applications show this very clearly. The high performance and excellent reproducibility of this new generation of grinding tools have more than convinced our customers to use these tools.

Investigation and modeling of ball-burnishing factors affecting the residual stress of AISI 8620 steel

Abstract:

Ball-burnishing is a superfinishing operation performed to improve surface and subsurface integrity. This work created a model to confirm that ball-burnishing of AISI 8620 steel affected its microstructures, roughness, and residual stress. As part of the process, the influence of different parameters on the residual stress surface in the axial direction was investigated. Samples were prepared by heat treatment and turning. Residual stress after ball-burnishing was found to be affected by pressure and feed rate to a statistically significant extent. Stress is found to be in the form of compression. Steel displayed roughly a 60% improvement in surface roughness following burnishing. The linear regression models derived from the data for the relationship between burnishing factors and residual stress offer R^2 values of 80.88%.

Keywords: Residual stress, ball-burnishing, AISI 8620 steel, modeling

1. Introduction:

Surface modification and treatment is crucial to improving the service life of various components used in engineering and structural functions. Surface engineering methodology includes the use of thermal, mechanical and chemical treatments and coatings (hard and soft) (Loh and Tam, 1988). Hardened materials machining is an efficient and economical method used in numerous processes in high-tech manufacturing companies (Grzesik and Zak, 2012), e.g when manufacturing hydraulics, bearings (Davim, 2009), and automotive parts (Grzesik, 2008).

The ball-burnishing process is frequently used for post-machining and metal finishing (Revankar et al., 2014). This process is distinguished by its advantageous and high compressive residual stress and surface hardening, a distinctive combination of elements that reduces surface roughness (Okada et al., 2017). Hassan (1997) has described it as the final process performed on a mechanical component, performed by rolling a hard ball against the metal surface while applying pressure. It is a non-chip process performed at low temperatures, which creates a flat, hardened surface via plastic deformation (Hassan et al., 1998). The resulting smooth, finished surface possesses high hardness and good corrosion resistance (Konefal et al., 2013). Fig. 1 shows the plastic deformation and the features that differentiate ball-burnishing from other processes used for the same purpose, e.g. grinding, polishing, honing and lapping of the surface.

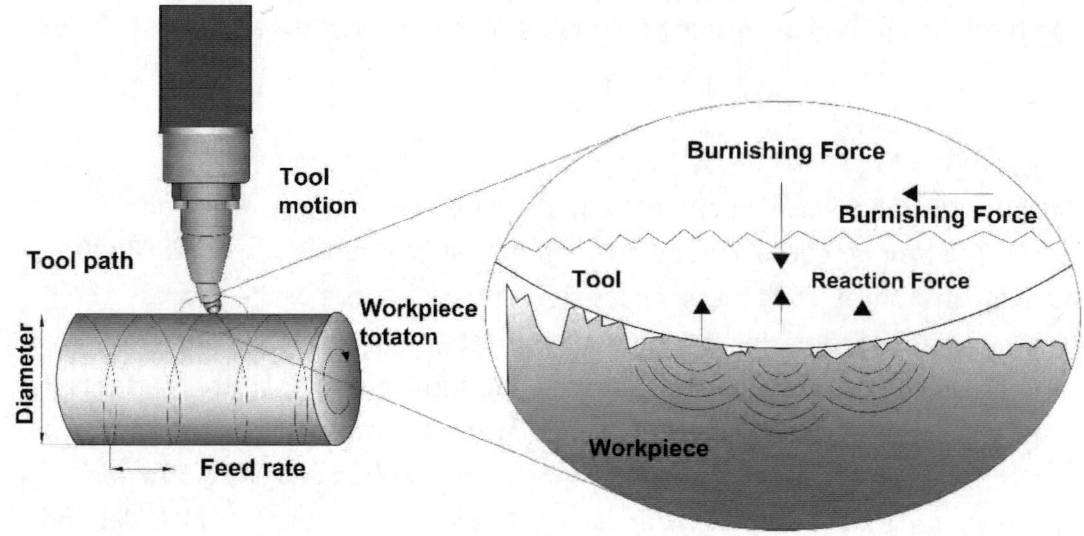

Fig. 1. Plastic deformation during burnishing

Ball-burnishing has several advantages: it is a relatively low-cost process (El-Axir and Ibrahim, 2005), requires less-skilled labor than other surface processing options, and it can be performed in conventional machine shops (Revankar et al., 2014). Ball-burnishing also reduces surface defects, modifies surface roughness, and produces compressive stress (El-Axir et al., 2008).

Many researchers have examined the mechanical effects of ball burnishing. For instance, a recent study reported that surface treatment by burnishing decreases the surface roughness and induced high compressed residual stress, which influenced the corrosion and fatigue life resistance (Jawahir et al., 2011). Korzynski et al. (2011) performed a slide diamond burnishing experiment on 42CrMo4 alloy steel shafts and reported that the roughness improved by 0.05–0.18 μm, with an increase in compressive stress of up to 400 MPa of the absolute value on the surface layer. They also tested a Ti3SiC2 material using ball-burnishing and reported that the effects were similar to those obtained by slide burnishing experiment. Another experiment shows that the difference in average roughness after centerless burnishing and turning processes was three to six times the original value (Korzynski and Pacana, 2010). Chomienne et al. (2016) tested the effects of ball burnishing on the microstructure of a 15-15PH stainless steel material. They found that the modified ball burnishing process had a limited effect on the thickness (10 to 15 μm). Zhang et al. (2015) described the transformation of the tensile stress into compressive stress with higher pressure and examined the effects of ball burnishing on roughness and residual stress.

Most of the published research to date has discussed the results of influence of burnishing on surface roughness. For example, Zhang et al. (2015) performed literature surveys related to the surface roughness of cylindrical parts, and López De Lacalle et al. (2005) discussed the burnishing process for the complex parts on the milling machines. In contrast, the research to date on the evolution of microstructure is limited. Most of those studies focused on ductile materials such as aluminum (Mhaede, 2012), magnesium (Pu et al., 2012) or ferritic-perlitic steels (Abrão et al., 2014). In contrast, only a limited number of papers have discussed the burnishing of hard martensitic steels (e.g. Kermouche et al., 2010) commonly used in making shafts for mechanical systems that require extreme fatigue resistance.

Empirical models are the most common choice to determine the characteristics of workpiece surfaces (Daniel and Wood, 1980). Sagbas (2011) presented a second-order regression technique to model the roughness of the aluminum alloys surface by employing the technique of surface analysis incorporation with the fundamental design of composite. The model they presented was compatible with results obtained from experimentation; it relies on parameters of applied force, operation speed, feed rate, and the number of passes.

Generally, residual stress is significant in the simulation of finite element models, as residual stress affects the fatigue life of mechanical components. The 2D finite element model established by Murthy et al. (2005) simulated the residual stress distribution acting in the axial and tangential directions and have qualitative relation with the results obtained from the experimentation. Yen et al. (2005) studied surface geometry following ball-burnishing performed on hardened AISI 52100 steel; in their work, 2D and 3D finite element models were used. They found while the 2D model offers better residual stress prediction, the 3D model is more realistic.

This study focused on developing empirical models for residual stresses on AISI 8620 steel. To that end, three major parameters of burnishing—speed, feed rate, and pressure—were examined.

2. Experimental

2.1 Experiment overview

A standard 3^3 full-factorial design was used to study the effects of burnishing on different parameters such as pressure, speed, and feed rate on the residual stresses. AISI 8620 steel samples were prepared, after which images were acquired using a digital microscope VHX-5000 and Proto iXRD along the longitudinal direction of the affected surface after turning and after turning + burnishing. These images were used to assess the surface microstructure and to determine the residual stresses.

2.2. Parameter selection

Burnishing parameters were selected based on the recommendations of manufacturers and our industrial partners. The process parameters used, including burnishing pressure, speed, and feed rate, are shown in Table 1.

Table 1 Experimental levels of burnishing parameters.

Level	Pressure (MPa)	Speed (m/min)	Feed rate (mm/rev)
Low	30	100	0.05
Middle	35	125	0.1
High	40	150	0.15

Residual stresses were measured using a sin^2 ψ method. Table 2 shows the parameters for acquiring the x-ray diffraction (XRD) measurements used to estimate the residual stresses.

Table 2. X-ray diffraction parameters used in this study

Material: STEEL - FERRITIC (Cr_Ka) (BCC, hkl-211)	Units: MPa
Tube: Cr_K-Alpha	Bragg Angle:156.41
Wavelength: 2.291	Peak Location: Gaussian 80%
KV:20.00; mA:4.00	Gain Correction: P/G(s)
Beta Osc:2.00; Phi Osc:0.00; X Osc:0.00; Y Osc:0.00;	Peak Shift: Absolute Peak
Exp. time: 1.00 1.00	(1/2) S2:5.92 E-6 [1/MPa] -S1:1.28 E-6 [1/MPa]
Nr of exp: 20 20	D Spacing (Angstroms): 1.1702087
Aperture: 1	LPA Correction: YES
Filters: 2	Psi Zero Assignment (Pixel): 216.42 224.33
Background static fit:	
Left Start Pixel: 2 1	Left Stop Pixel: 32 31
Right Start Pixel: 448 456	Right Stop Pixel: 478 486
ROI Left: 40 18	ROI Right: 434 480

2.3 Sample preparation and post-experiment processing

Twenty-seven samples of AISI 8620 steel bar (Fig. 2) were heat-treated. Each sample was hardened, then quenched in oil and tempered to a surface hardness of 50/52 HRC. After being carburized to a total case depth of 1.625 mm, each bar was turned by facing and center-drilling to fix it on the lathe. The turning process was first carried out by the ceramic insert with parameters cutting speed (V_c), feed rate (f_C), and depth of cut (a_c), as described in Fig. 3. These turning conditions were selected based on the recommendations of our industrial partner.

Fig. 2 Steel sample after heat treatment

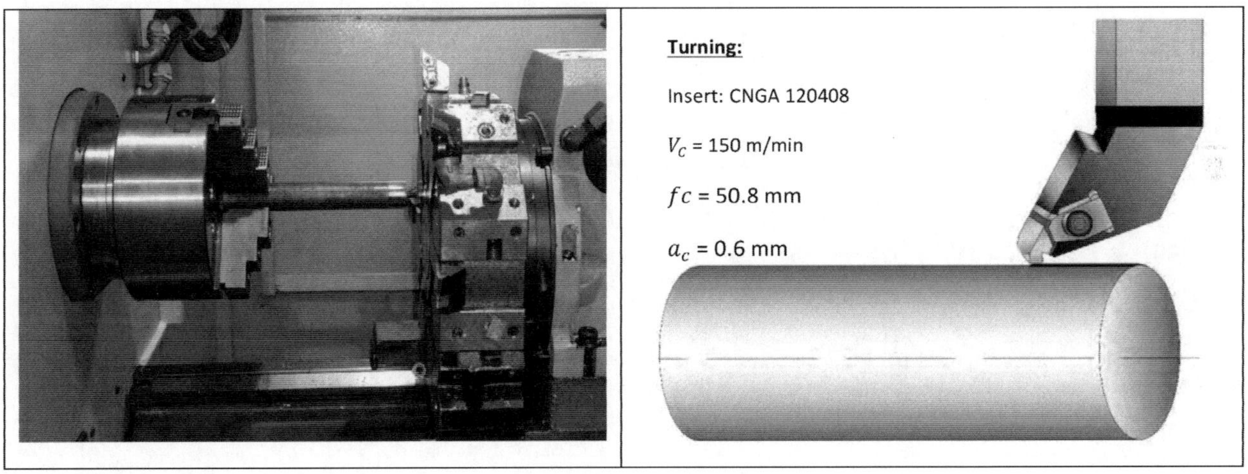

Fig. 3 Turning process

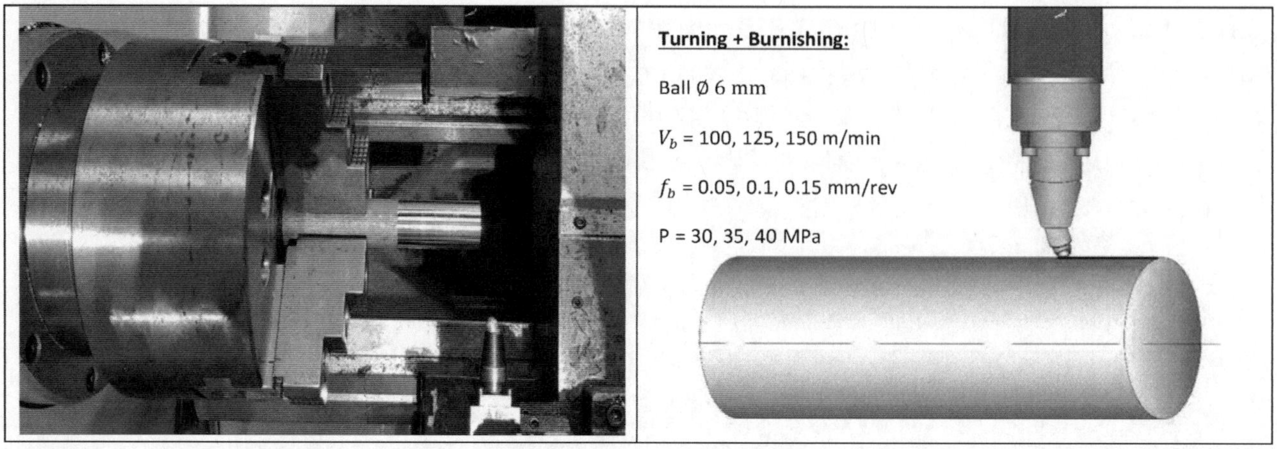

Fig. 4 Burnishing process

2.4. Experimental setup

A CNC lathe machine was equipped with a burnishing tool (Fig 4) and hydraulic pump units were plugged into the burnishing tool. With this set-up, the pressure (P) can be modified by adjusting the pressure gauge on the hydraulic pump units. The burnishing speed (V_b) and the feed rate per revolution (f) can be modified at the CNC interface.

Detailed specifications for the experimental setup

	Specification
Sample	AISI 8620 steels, 101.6 mm (L) x 25.4 mm (D)
Chemical composition (wt%)	0.20 C, Max. 0.40 Si, 0.85 Mn, 0.45 Cr, 0.45 Mo
CNC lathe machine	Romi Centur 35E
Turning	Single-sided tipped ceramic insert (CITCO CNGA120408)
Burnishing tool	ECOROLL (HG6-9L15-SLK25) equipped with a Ø 6 mm silicon nitride ceramic ball with a 15° angle
Pump	ECOROLL hydraulic pump unit (HGP 4.3), 40 MPa maximum pressure

Coolant	Master Chemical Trim ® C270
Roughness tester	HOMMEL ETAMIC W5
Roundness measurement	Surtronic R- 125
Residual stress measurement	Proto iXRD
Digital microscope	VHX-5000

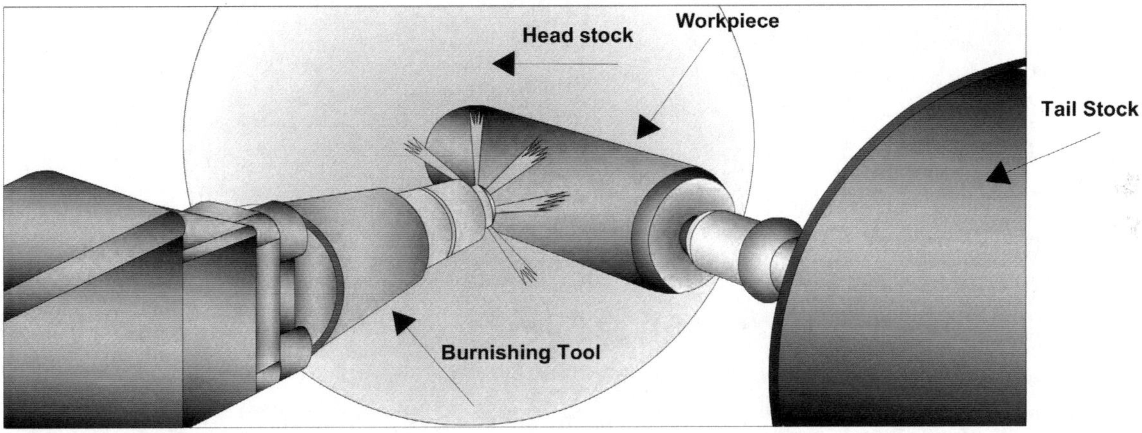

Fig. 5 Sketch of the ball-burnishing process.

2.5. Experimental Procedure

Samples were heat-treated, then installed in the lathe. The parameters were adjusted, and the turning process started. After turning, the burnishing process was started using the ball-burnishing tool (Fig. 5). Then, the sample was removed from the lathe machine (Fig. 6) and the surface microstructure imaged. The image data was used to measure the surface roughness and to determine the residual stress using XRD.

The surface average roughness (R_a) before and after burnishing was measured using a HOMMEL ETAMIC W5 (see table 3) roughness tester. For each workpiece, the average R_a was obtained by taking five readings at different positions along the longitudinal axis of the workpiece.

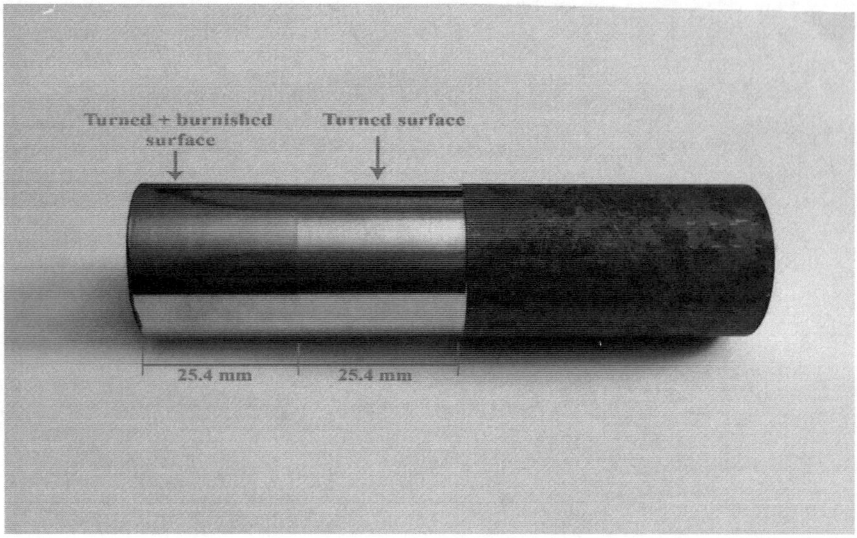

Fig. 6 Sample after turning and turning + burnishing

3. Results and discussions

Surface microstructure analysis

Images were acquired using a VHX-5000 digital microscope along the longitudinal direction. Fig. 7. Illustrated the affected layer after turning (a) and after turning + burnishing (c); panels (b) and (d) show 3D images of the affected layers. The changes on the thickness microstructure were approximately 12 to 18 µm. The burnishing operation did not affect the thickness of the nano-sized grains layer at the surface due to the cold nature of the operation; temperature did not play a major role. There was insufficient thermo-mechanical energy to induce additional dynamic recrystallization in the material (Chomienne et al., 2016).

(a) 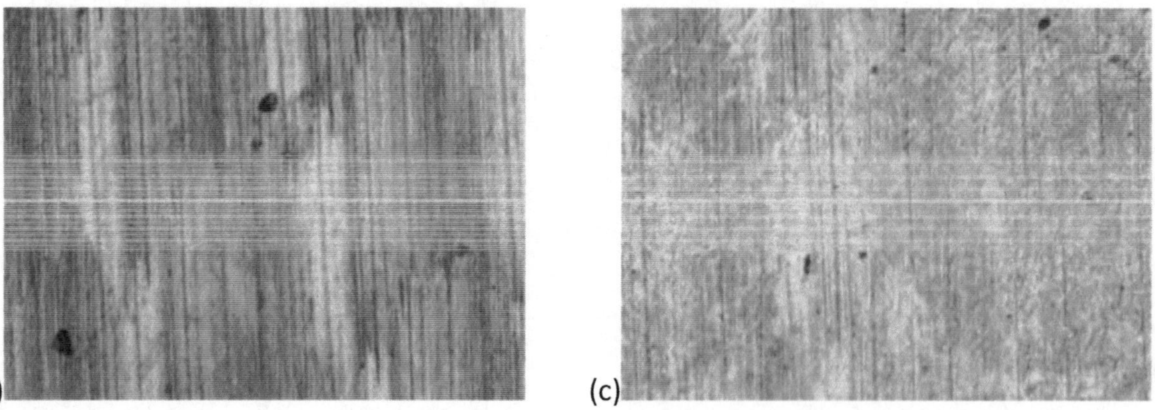 (c)

(b)

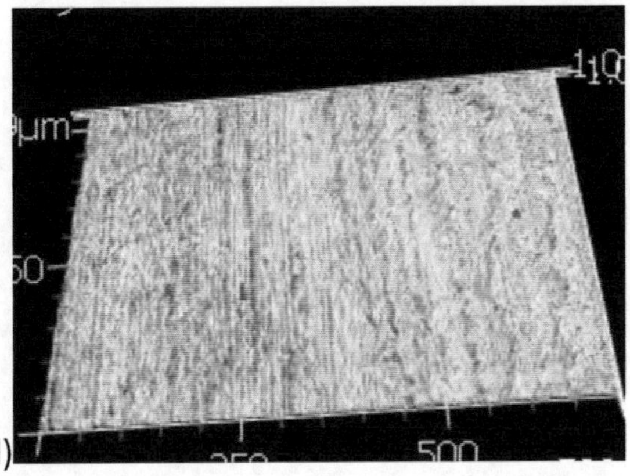

(d)

Fig. 7. Analysis of the surface microstructure obtained after turning (a, b) and turning + burnishing (c, d) with the reference conditions.

Surface roughness

The roughness on the affected layer after turning and after turning + burnishing was improved by more than 60%. Fig. 8 illustrates the roughness profile for run number nine, with P = 35 MPa, F = 0.05 and S = 125 m/min. The improvement after the turning + burnishing process was notable.

(a)	Avg. R_a= 0.240 µm Avg. R_z= 2.167 µm Avg. RSm= 64.753 µm	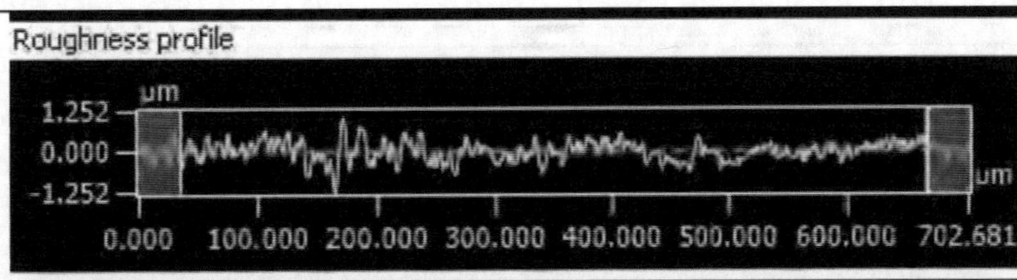
(b)	Avg. R_a = 0.125 µm Avg. R_z = 1.082 µm Avg. RSm= 57.762 µm	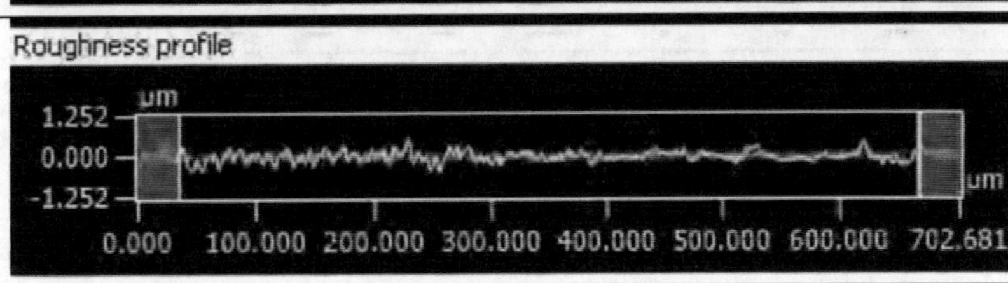

Fig. 8. Surface roughness profiles obtained (a) after turning (b) after turning + burnishing.

Residual stress analysis

The data depicted in Fig. 9 confirm that burnishing can affect the value of residual stress: The stress state was in compression in the external layer after the turning + burnishing operation and then the maximum compressive state occurred below the subsurface, in line with the observations of (Klocke and Liermann, 1998).

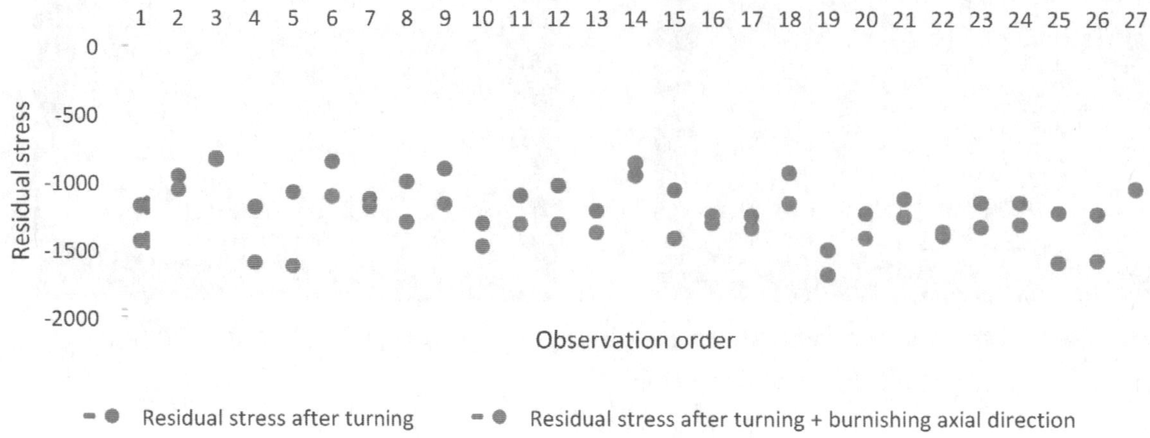

Fig. 9. Residual stress for each sample after turning RS_T and after turning + burnishing RS_{TB} in the axial direction

Modeling

A general ANOVA analysis was performed using Minitab 18 to generate a model for final residual stresses. Two factors, pressure (*P*), and feed rate per revolution (ƒ) had the greatest influence in the model; the burnishing speed (V_b) has less of an effect. An analysis of the main effects is shown in Fig. 10 and plots of residuals to validate the model are provided in Fig. 11. A gross average was taken from all residual stresses measurements in each run to simplify the molding process.

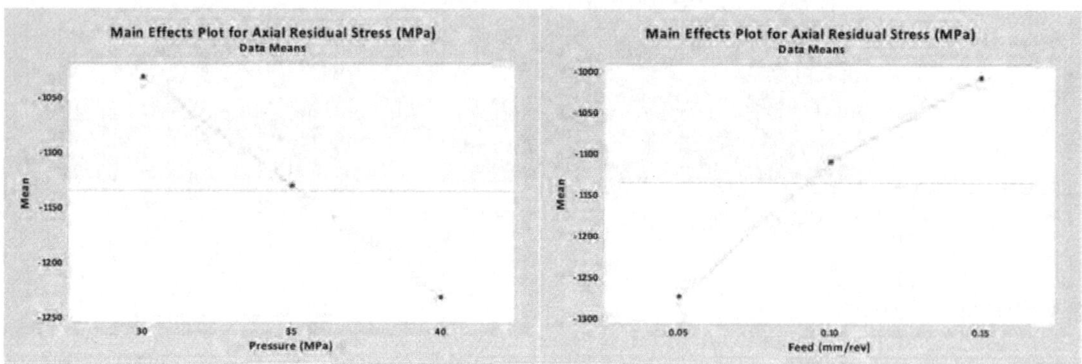

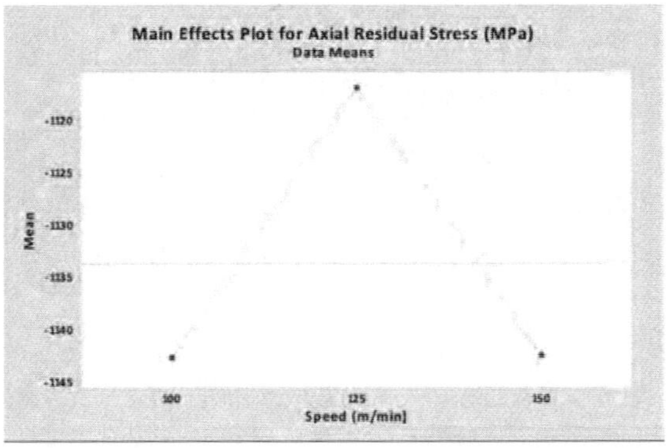

Fig. 10 Main effect of (a) Pressure (b) Feed rate and (c) Speed on Residual Stress

The profile of the residual stress is plotted with three different pressure values. Pressure is shown to have a significant influence on the residual stress values as can be seen in figure 10(a). As pressure value increases, the induced plastic deformation increases which causes the residual stress to decrease. This finding fits in with the work of (Loh and Tam, 1988). The effect of feed rate levels on the residual stress is plotted in figure 10(b). As shown, the correlation appeared to be linear as with the increase of feed rate there is an increase in the residual stress. This can be attributed to the quality of surface roughness produced after burnishing. According to (Loh and Tam, 1988), higher (rougher) surface roughness is associated with increasing the feed rate in burnishing. This can also be linked to the correlation found by (Mezlini et al. 2013) in their working conditions where it was found that as surface roughness improves (decrease) the subsurface properties such as residual stress improves (decrease). Therefore, combining the two findings only strengthens the correlation found in this study. The range of speed selected for this study is within the recommended values found in literature as shown in (Loh and Tam, 1988) work. The effect of this speed range on the values of residual stress is shown in figure 10(c). The trend shows an increase of the residual stress to a maximum value followed by a steady decrease as the speed gets faster. This trend follows the correlation between speed and surface roughness in burnishing as shown in (Mezlini et al. 2013) and (Loh and Tam, 1988) work. Since the mechanical properties, namely residual stress, are linearly correlated with surface roughness as shown by (Mezlini et al. 2013), it can be said that the effect of speed on residual stress is expected to follow the same path. However, (Loh and Tam, 1988) have shown that surface roughness increases as burnishing speed increases until a reversed effect is noticed after which the speed causes a decrease in Ra. Therefore, the greater the burnishing does not necessarily constitute improved mechanical properties of materials.

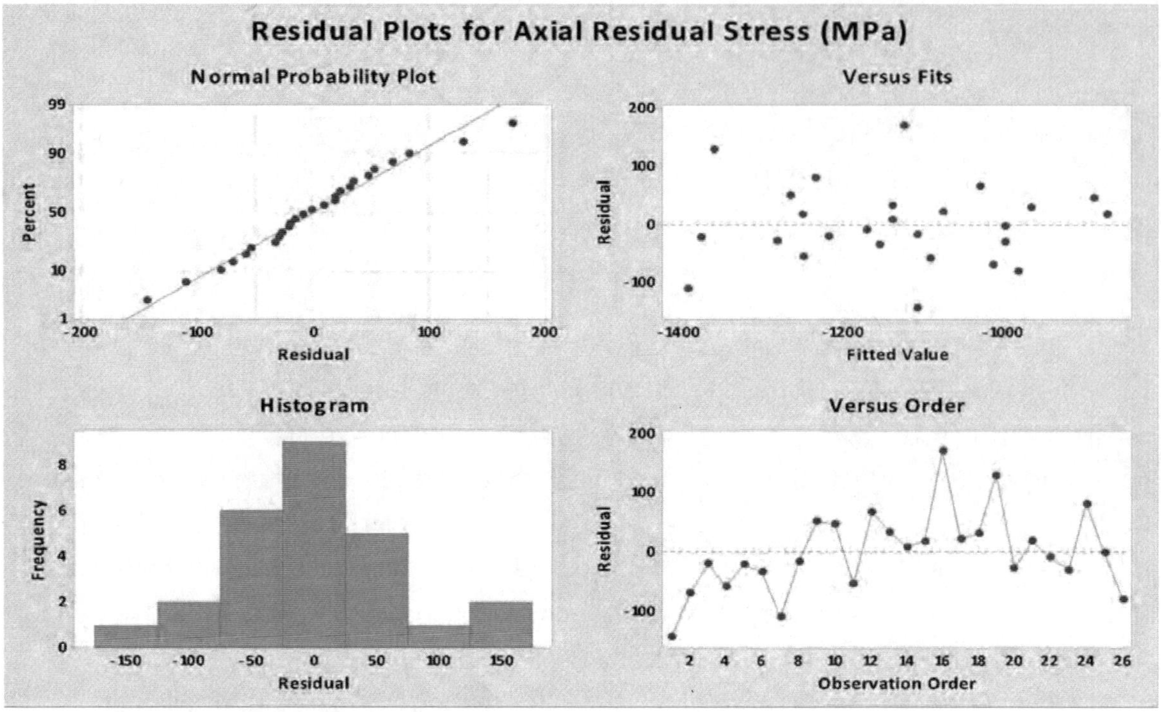

Fig. 11 Plots of the residuals for model validation

A graph was generated to compare the original data and the values predicted from this model (Fig. 12). The obtained R^2 value is 80.88%,

$$RS_{TB} = -718 - 21.94P + 0.640V_b + 2825f$$

Other models were used to fit the model with polynomial and logistic regressions; however, these models yielded smaller R^2 values. Linear regression was found to be a better fit than higher-order models.

Zhang et al. (2015) and Chomienne et al. (2016) agree that the application of higher pressure increases the residual stress on burnished surfaces. This study confirmed that the pressure had a positive effect on the residual stress; however, the feed rate had a negative effect on the residual stress and the burnishing speed had no significant effect. Considering the parameters used here were chosen from industrial practices, this study produced results that can be meaningfully applied in industry.

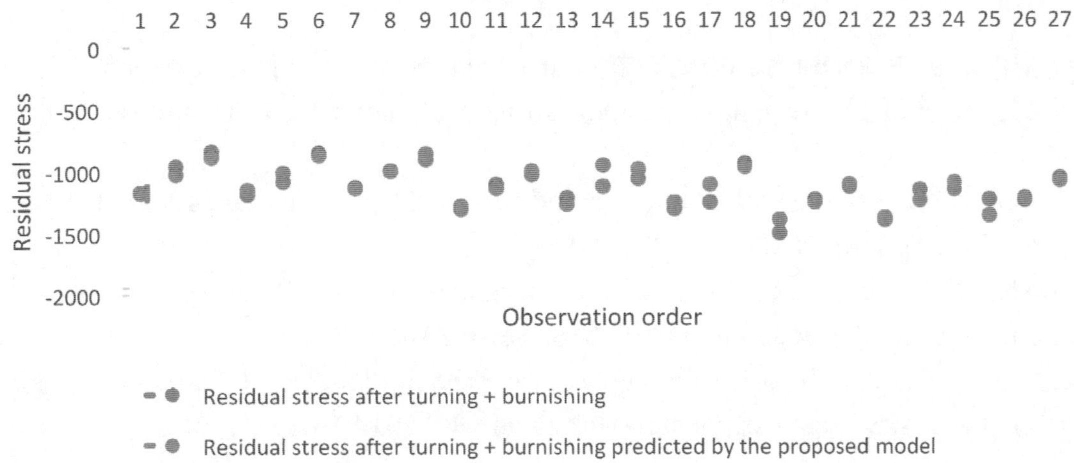

Fig. 12. Comparison between original residual stress data and the residual stresses predicted for the same parameters by the model.

4. Conclusions

Ball-burnishing improves the surface roughness of AISI 8620 steel by more than 60% when applied after turning. The linear regression model derived in this work indicates a strong relationship between the burnishing factors and residual stress with an 80.88% of R^2 values. This model is generally applicable to the residual stress profile of all AISI 8620 steel, because the parameters were all derived from industrial requirements and standards.

Of the three factors examined here, the effects of pressure and feed rate were statistically significant, while, speed had no influence on the residual stress. Pressure positively affected the residual stress, but the feed rate had a negative effect on residual stress.

These experiments show that burnishing speed significantly affects the residual stress within the workpiece. Thus, it is recommended that AISI 8620 steel be burnished using the maximum speed supported by the machine, reducing processing time without increasing residual stress in the workpiece.

References

Chomienne V, Valiorgue F, Rech J, Verdu C. CIRP Journal of Manufacturing Science and Technology Influence of ball burnishing on residual stress profile of a 15-5PH stainless steel 2016;13:90–6.

Klocke F, Liermann J. Roller burnishing of hard turned surfaces. Int J Mach Tools Manuf 1998;38:419–23. doi:10.1016/S0890-6955(97)00085-0.

Zhang T, Bugtai N, Marinescu ID. Burnishing of aerospace alloy: A theoretical-experimental approach. J Manuf Syst 2015;37:472–8. doi:10.1016/j.jmsy.2014.11.004.

Abrão AM, Denkena B, Köhler J, Breidenstein B, Mörke T. The influence of deep rolling on the surface integrity of AISI 1060 high carbon steel. Procedia CIRP 2014;13:31–6. doi:10.1016/j.procir.2014.04.006.

Avilés R, Albizuri J, Rodríguez A, López De Lacalle LN. Influence of low-plasticity ball burnishing on the high-cycle fatigue strength of medium carbon AISI 1045 steel. Int J Fatigue 2013;55:230–44. doi:10.1016/j.ijfatigue.2013.06.024.

Chomienne V, Valiorgue F, Rech J, Verdu C. CIRP Journal of Manufacturing Science and Technology Influence of ball burnishing on residual stress profile of a 15-5PH stainless steel 2016;13:90–6.

Daniel C, Wood FS. Fitting equations to data: computer analysis of multifactor data. John Wiley & Sons, Inc.; 1980.

El-Axir MH, Ibrahim AA. Some surface characteristics due to center rest ball burnishing. J Mater Process Technol 2005;167:47–53. doi:10.1016/j.jmatprotec.2004.09.078.

El-Axir MH, Othman OM, Abodiena AM. Study on the inner surface finishing of aluminum alloy 2014 by ball burnishing process. J Mater Process Technol 2008;202:435–42. doi:10.1016/j.jmatprotec.2007.10.040.

Griffiths B. Manufacturing surface technology: surface integrity and functional performance. Elsevier; 2001.

Grzesik W. Advanced Machining Processes of Metallic Materials: Theory, Modelling and Applications. Elsevier Science; 2008.

Grzesik W, Zak K. Modification of surface finish produced by hard turning using superfinishing and burnishing operations. J Mater Process Technol 2012;212:315–22. doi:10.1016/j.jmatprotec.2011.09.017.

Hassan AM. The effects of ball- and roller-burnishing on the surface roughness and hardness of some non-ferrous metals. J Mater Process Technol 1997;72:385–91. doi:10.1016/S0924-0136(97)00199-4.

Hassan AM, Al-Jalu HF, Ebied AA. Burnishing force and number of ball passes for the optimum surface finish of brass components. J Mater Process Technol 1998;83:176–9. doi:10.1016/S0924-0136(98)00058-2.
Jawahir IS, Brinksmeier E, M'Saoubi R, Aspinwall DK, Outeiro JC, Meyer D, et al. Surface integrity in material removal processes: Recent advances. CIRP Ann - Manuf Technol 2011;60:603–26. doi:10.1016/j.cirp.2011.05.002.
Kermouche G, Rech J, Hamdi H, Bergheau JM. On the residual stress field induced by a scratching round abrasive grain. Wear 2010;269:86–92. doi:10.1016/j.wear.2010.03.012.
Konefal K, Korzynski M, Byczkowska Z, Korzynska K. Improved corrosion resistance of stainless steel X6CrNiMoTi17-12-2 by slide diamond burnishing. J Mater Process Technol 2013;213:1997–2004. doi:10.1016/j.jmatprotec.2013.05.021.
Korzynski M, Lubas J, Swirad S, Dudek K. Surface layer characteristics due to slide diamond burnishing with a cylindrical-ended tool. J Mater Process Technol 2011;211:84–94. doi:10.1016/j.jmatprotec.2010.08.029.
Korzynski M, Pacana A. Centreless burnishing and influence of its parameters on machining effects. J Mater Process Technol 2010;210:1217–23. doi:10.1016/j.jmatprotec.2010.03.008.
Loh NH, Tam SC. Effects of ball burnishing parameters on surface finish-A literature survey and discussion. Precis Eng 1988;10:215–20. doi:10.1016/0141-6359(88)90056-6.
López De Lacalle LN, Lamikiz A, Muñoa J, Sánchez JA. Quality improvement of ball-end milled sculptured surfaces by ball burnishing. Int J Mach Tools Manuf 2005;45:1659–68. doi:10.1016/j.ijmachtools.2005.03.007.
Mhaede M. Influence of surface treatments on surface layer properties, fatigue and corrosion fatigue performance of AA7075 T73. Mater Des 2012;41:61–6. doi:10.1016/j.matdes.2012.04.056.
Murthy RL, Kotiveerachari B, Loh NH, Tam SC, Boyer RR, Pu Z, et al. Surface improvement of shafts by the deep ball-burnishing technique. Int J Mach Tools Manuf 2005;206:144–55. doi:10.1016/j.surfcoat.2011.11.045.
Novovic D, Dewes RC, Aspinwall DK, Voice W, Bowen P. The effect of machined topography and integrity on fatigue life. Int J Mach Tools Manuf 2004;44:125–34. doi:10.1016/j.ijmachtools.2003.10.018.
Okada M, Shinya M, Matsubara H, Kozuka H, Tachiya H, Asakawa N, et al. Development and characterization of diamond tip burnishing with a rotary tool. J Mater Process Technol 2017;244:106–15. doi:10.1016/j.jmatprotec.2017.01.020.
Pu Z, Song GL, Yang S, Outeiro JC, Dillon OW, Puleo DA, et al. Grain refined and basal textured surface produced by burnishing for improved corrosion performance of AZ31B Mg alloy. Corros Sci 2012;57:192–201. doi:10.1016/j.corsci.2011.12.018.

Revankar GD, Shetty R, Rao SS, Gaitonde VN. Analysis of surface roughness and hardness in ball burnishing of titanium alloy. Meas J Int Meas Confed 2014;58:256–68. doi:10.1016/j.measurement.2014.08.043.
Sagbas A. Analysis and optimization of surface roughness in the ball burnishing process using response surface methodology and desirabilty function. Adv Eng Softw 2011;42:992–8. doi:10.1016/j.advengsoft.2011.05.021.
Sasahara H. The effect on fatigue life of residual stress and surface hardness resulting from different cutting conditions of 0.45%C steel. Int J Mach Tools Manuf 2005;45:131–6. doi:10.1016/j.ijmachtools.2004.08.002.
Yen YC, Sartkulvanich P, Altan T. Finite element modeling of roller burnishing process. CIRP Ann - Manuf Technol 2005;54:237–40. doi:10.1016/S0007-8506(07)60092-4.
Zhang T, Bugtai N, Marinescu ID. Burnishing of aerospace alloy: A theoretical-experimental approach. J Manuf Syst 2015;37:472–8. doi:10.1016/j.jmsy.2014.11.004.

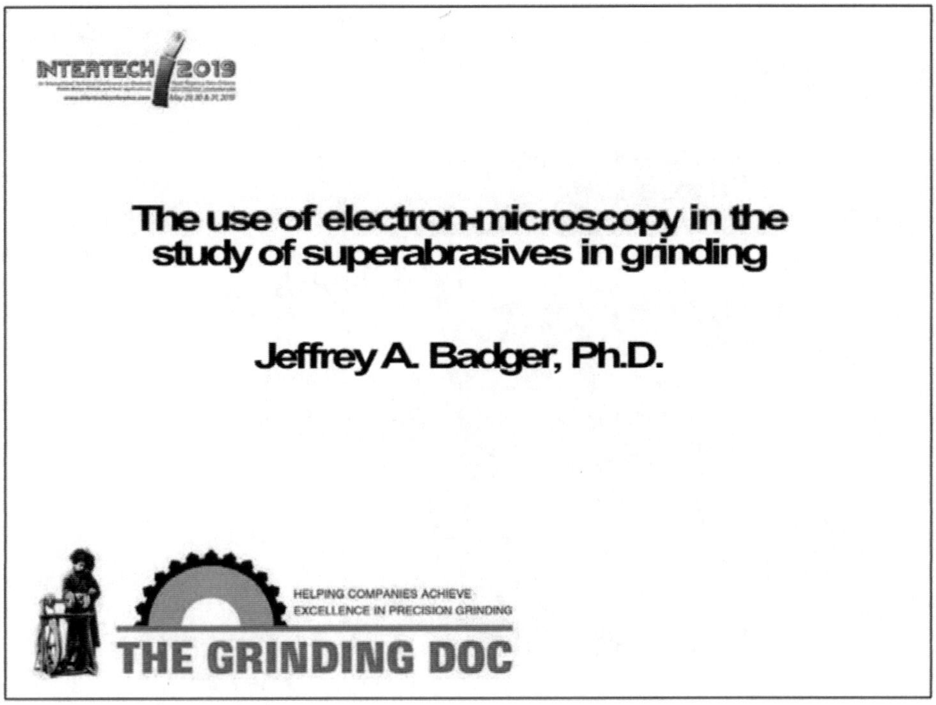

Background of Dr. Jeffrey Badger, ***The Grinding Doc***.

I have spent 20 years collecting grinding swarf, worn grinding wheels, worn dressing tools and ground workpieces in order to take electron-microscope photos. Why? During my grinding courses, I frequently show these electron-microscope photos. When I do, course attendees' eyes light up. They've worked in grinding for years, but have never

been able to visualize what's actually happening.

These photos are very popular. Let's take a look at some of them.

These photos are all in ***THE BOOK OF GRINDING***.

An SEM photo of a worn, resin-bonded CBN wheel after grinding high-speed steel. We can see several grits in the background where the bond material has been stripped away, exposing the CBN grits. We also see some grits in the foreground that are still covered in resin bond. Notice that there is no porosity beneath the surface. In fact, after

dressing, this wheel had almost no porosity. It was only after grinding for a while that we were able to create some of the porosity that we see.

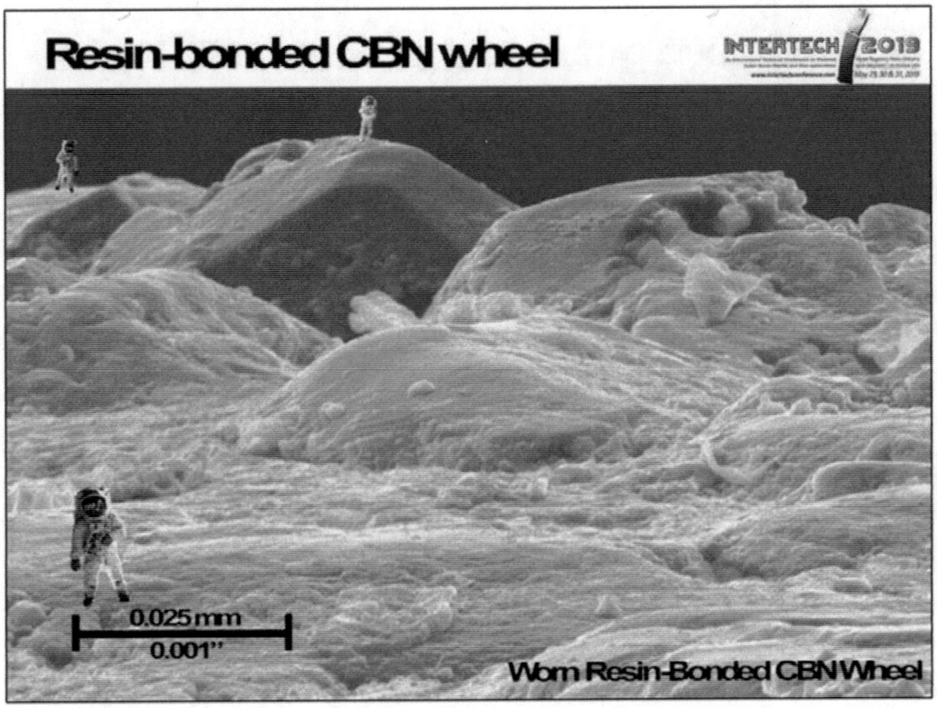

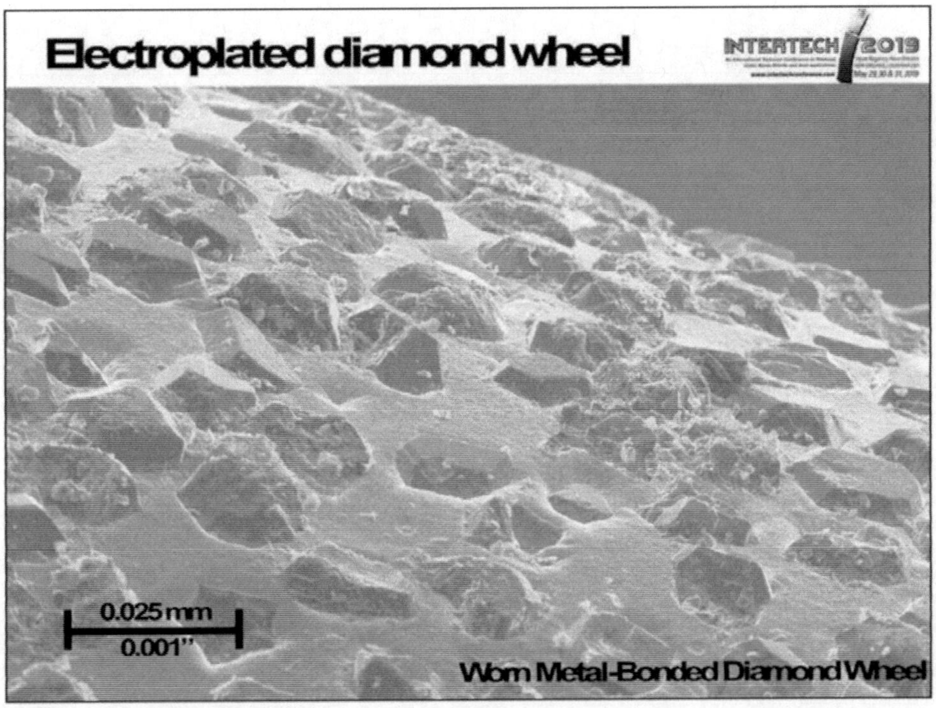

Scanning Electron Microscope (SEM) photo of an electroplated diamond wheel. Electroplated wheels are a single layer of diamond held in a metal bond. Here we can see numerous individual diamonds standing above the bond material. Electroplated wheels have a lot of porosity.

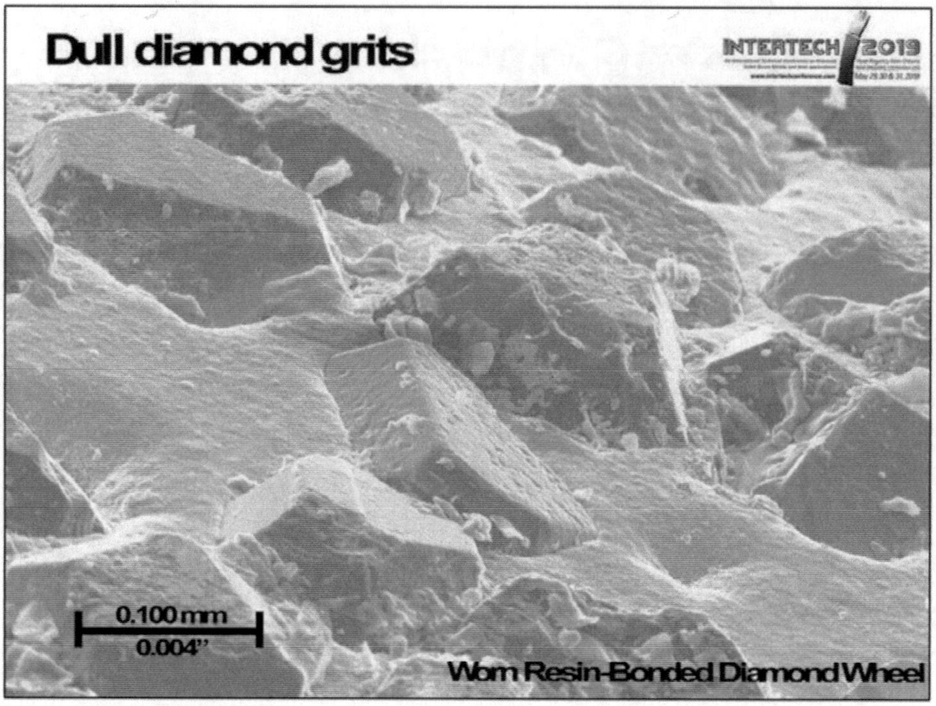

The same electroplated wheel. Here we can see that grit-dulling is occurring, although it is more difficult to see than dulling in Al_2O_3 wheels.

Let's zoom in to a single, dull diamond. The dull diamond area isn't smooth like it is in dull Al_2O_3 grits. It's sort of "crumbly". But it's definitely getting dull.

Here's a photo from the same wheel. It looks like there's a large CBN grit sticking his head above the resin.

Or is it? I was once at a large diamond manufactured in the UK talking to their smart Ph.D. diamond experts. They looked at it and said, "I don't think that's a diamond. I think that's a *cavity*, a hole, left by a diamond." A year later I showed it to some smart Ph.D. diamond experts at another company and they said the same thing.

And I think they're right, as we're getting good, high-resolution surface at that point indicating that the material has a high atomic number, indicating that it's metal. I think it's actually the metal coating we're looking at, and the diamond has been ripping out of the coating, and the coating has remained in the resin.

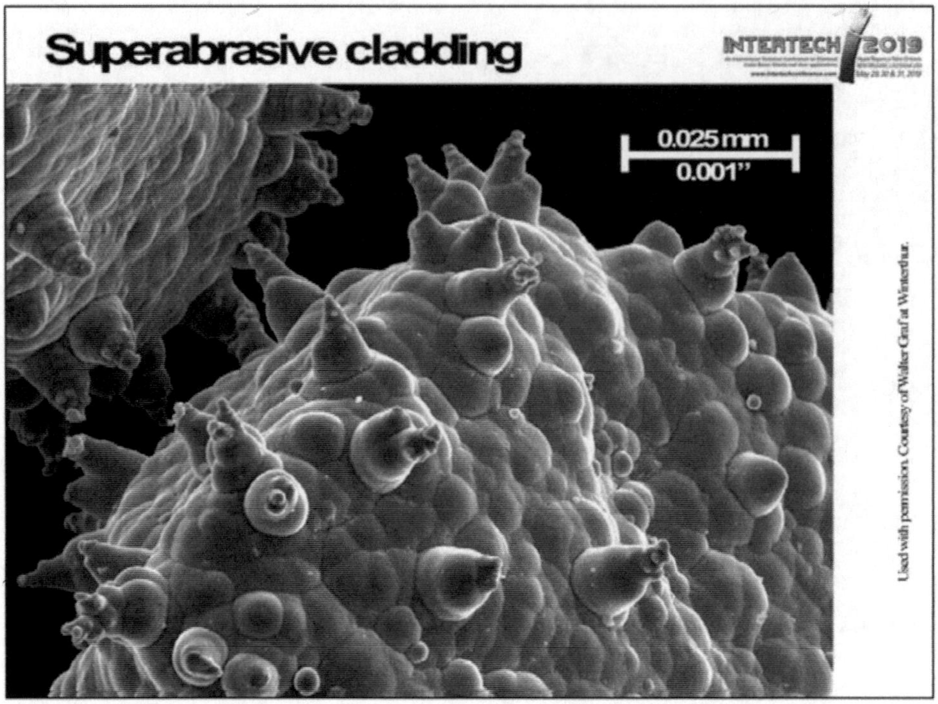

Diamond manufacturers come up with ways to make the coating more effective. One method is to produce "spikey" coatings. The surface area is larger and the coating holds better in the bond.

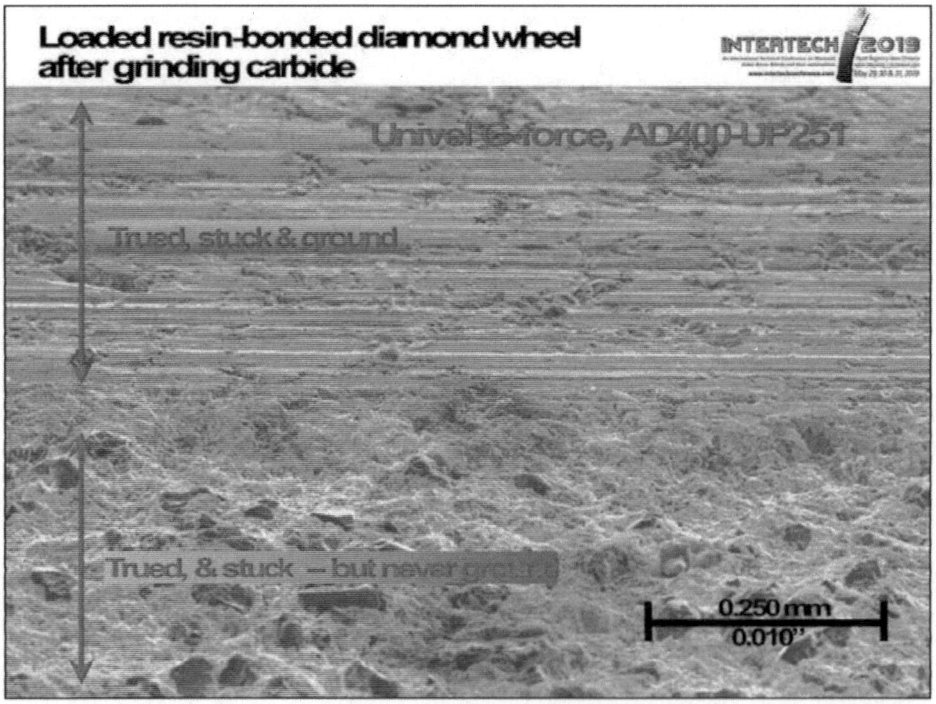

Loading is a major challenge when grinding carbide. Or nuisance. Or problem. Or, whatever you want to call it, it's annoying. Let's take a look at what loading looks like. Above is an SEM photo of a flute-grinding wheel. The wheel was trued and stuck. Then it was used to grind flutes in carbide endmills.

However, only half of the wheel was in contact, and half of the wheel never did any grinding The top portion shown here was trued and stuck and then ground a lot of carbide. The bottom portion was trued and stuck and didn't do any grinding. →

On the bottom we see diamond grits poking their heads about the resin+metal bond. But what do we see on the top portion of the SEM photo? A sea of carbide. Gobs and gobs of carbide. In fact, it's amazing to think that this wheel was able to grind at all. But it was. How is that possible? I don't see any diamond grits.

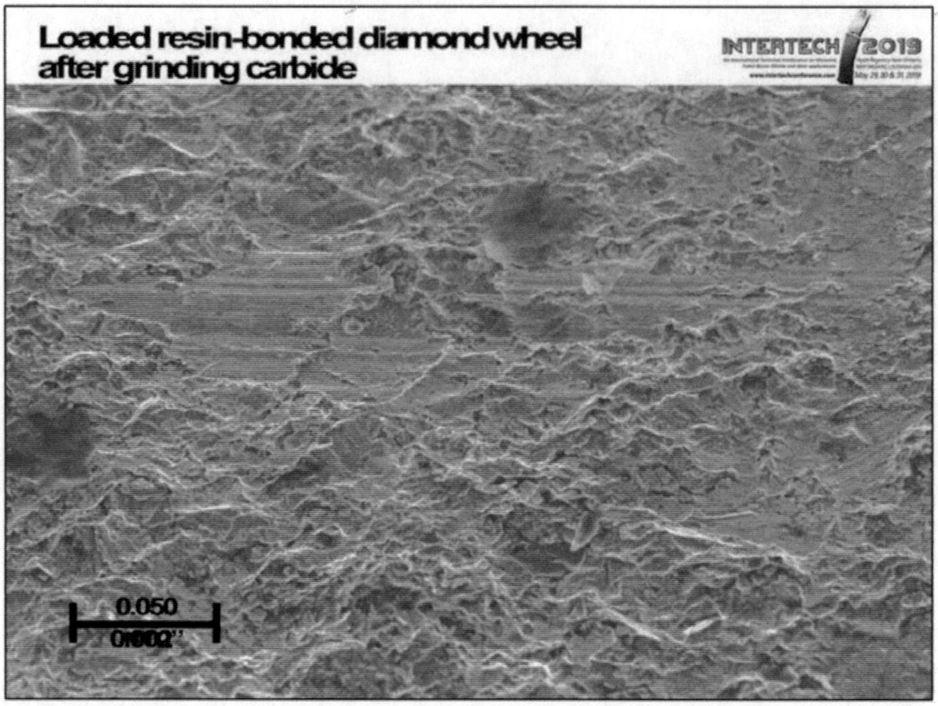

But now let's look at that same exact wheel, but in the portion that's been grinding. We see gobs and gobs of loading. In fact, we hardly see any diamonds at all.
And we need to clear out that loading. Otherwise we're going to get a lot of rubbing and a lot of heat generation.

What does the workpiece look like when we bust out material? It may have grinding scratches, or it may not. But if you look closer, what you'll see is a rough surface that's all cracked up. In this case, part of the glass had been ground (the rough patches here), and part of the material was unground (the smooth patches).

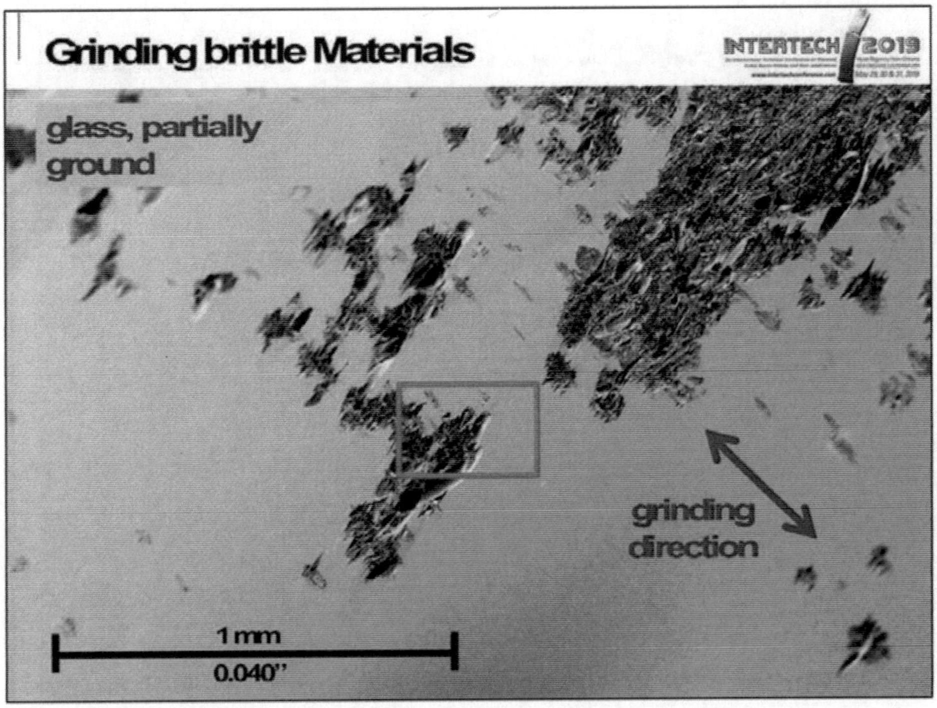

Let's zoom in closer. Here we see the virgin glass surface, which is nice and smooth after it'd been blown at high temperatures. But we also see the ground surface, which is all cracked up. Let's look closer at a cracked-up region. →

An SEM photo of swarf from grinding of 65 HRC (Rockwell C hardness) high-speed steel with an Al_2O_3 wheel. Hardened high-speed steel is very hard. And it has lots of super-hard tungsten-molybdenum-carbides and vanadium-carbides. In spite of that, it still produces long, stringy chips.

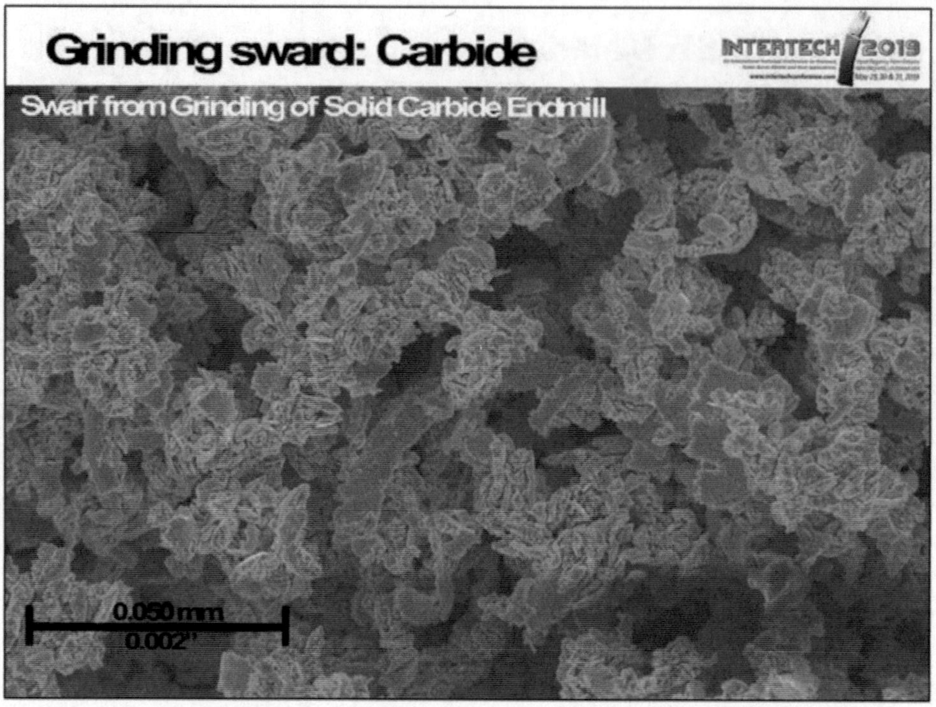

Swarf from grinding of tungsten-carbide/cobalt with a diamond wheel. Notice that the "solid carbide" chips are much sorter and blockier. This is because carbide is much harder than steel, or, more specifically, it has less *ductility*. Nevertheless, it still produces chips.

Grinding swarf: glass
INTERTECH 2019
swarf, rough grinding of glass
2 mm
0.080"

Grinding swarf: wood
INTERTECH 2019
0.050 mm
0.002"
Aluminum-oxide flap disc loaded with wood swarf after "grinding" wood dowels.

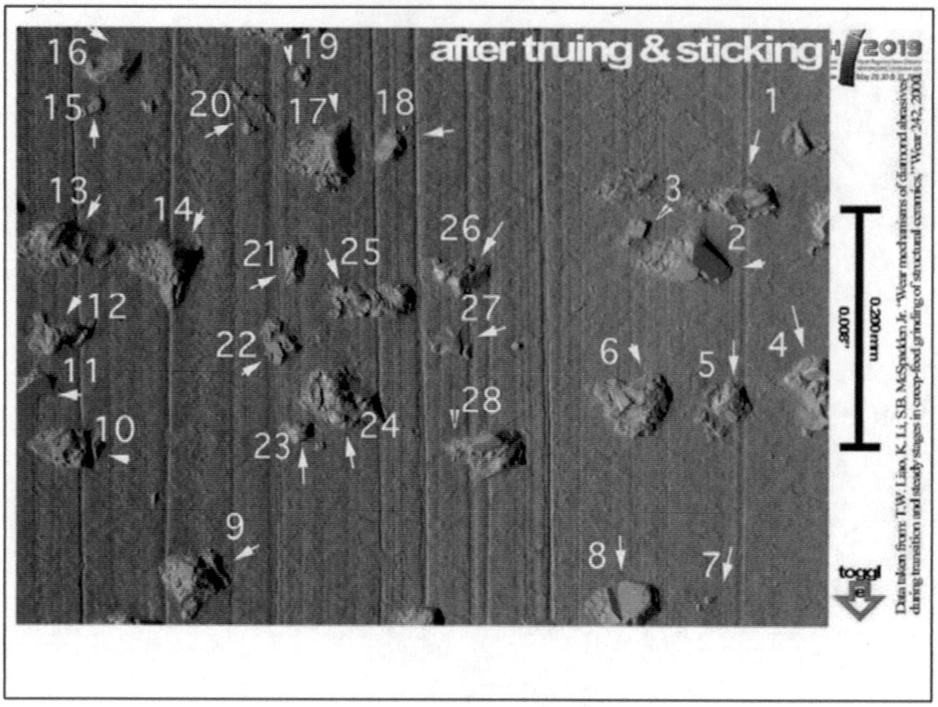

after truing & sticking
2019
0.200mm
0.008"
toggle
Data taken from: T.W. Liao, K. Li, S.B. McSpadden Jr. "Wear mechanisms of diamond abrasives during transition and steady stages in creep-feed grinding of structural ceramics," Wear 242, 2000

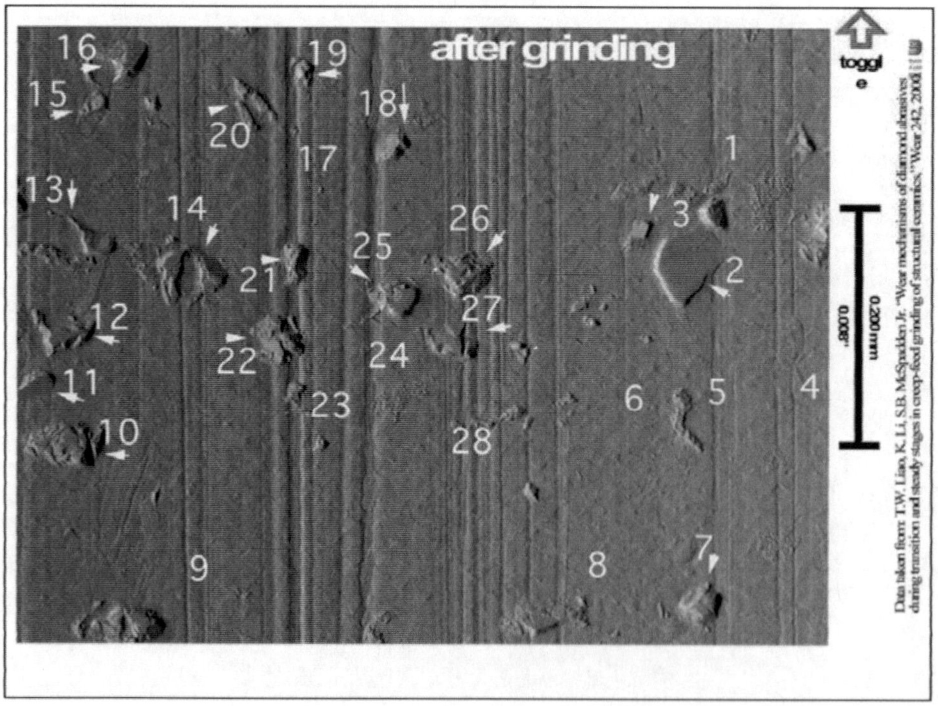

after grinding
toggle
0.200mm
0.008"
Data taken from: T.W. Liao, K. Li, S.B. McSpadden Jr. "Wear mechanisms of diamond abrasives during transition and steady stages in creep-feed grinding of structural ceramics," Wear 242, 2000

Loaded CBN wheel: Steel

Burr

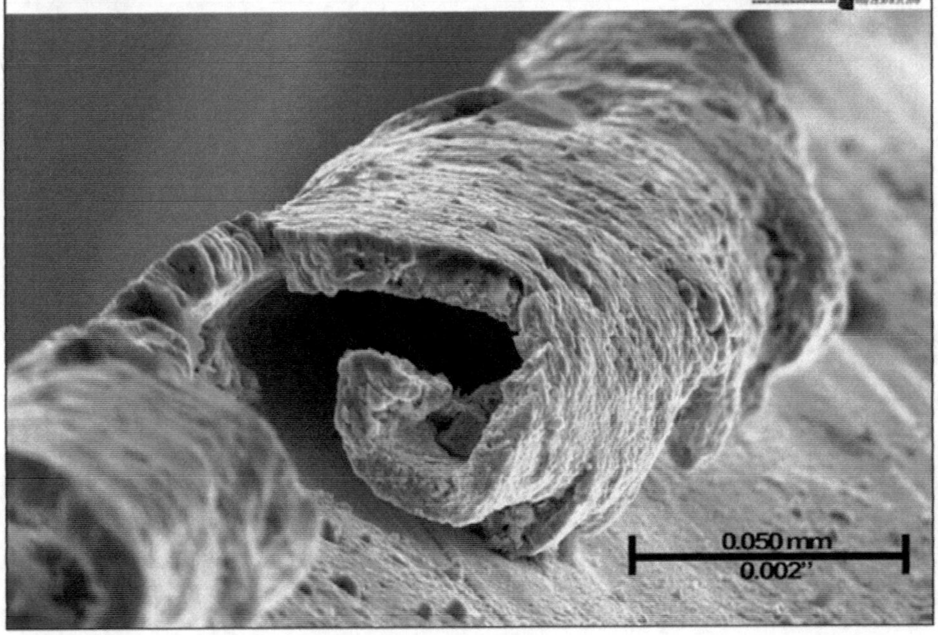

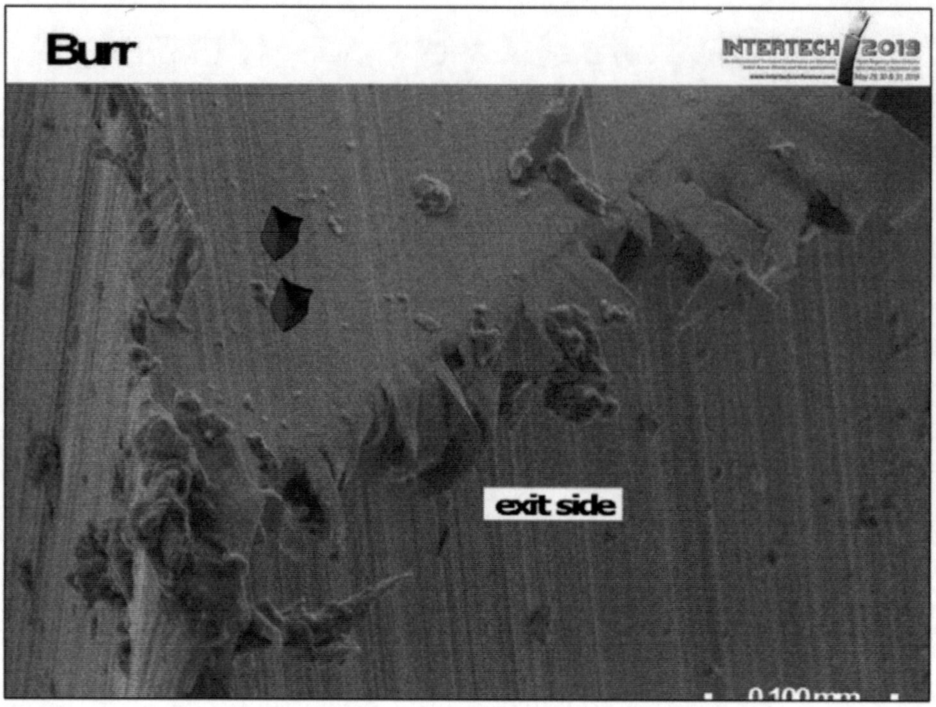
Burr
INTERTECH 2019
exit side
0.100 mm

Burr, backscatter
INTERTECH 2019
exit side
0.100 mm

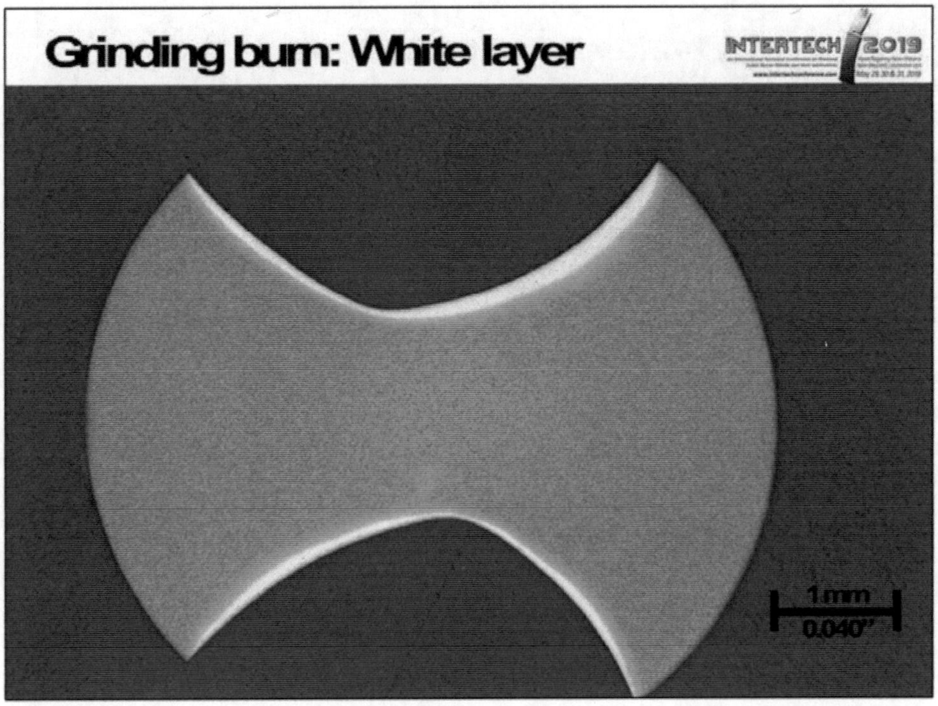
Grinding burn: White layer
INTERTECH 2019
1 mm
0.040"

Dull diamonds after grinding cermets
INTERTECH 2019

Dull diamonds after grinding cermets

Cubitron II Al_2O_3 grits

Taken from 3M.com

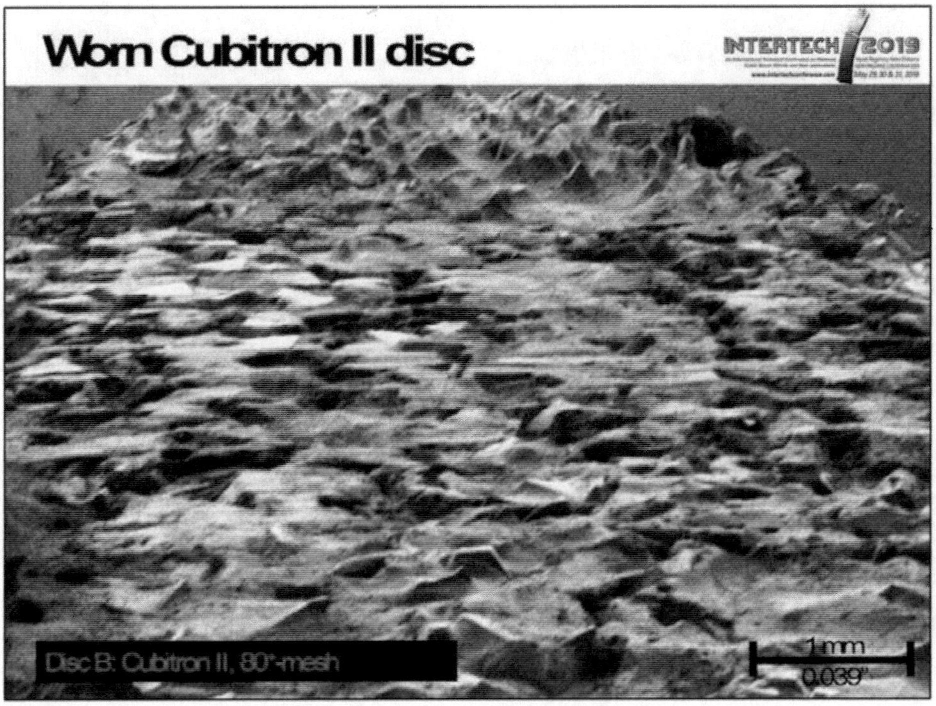
Worn Cubitron II disc
INTERTECH 2019
Disc B: Cubitron II, 80+-mesh
1 mm
0.039"

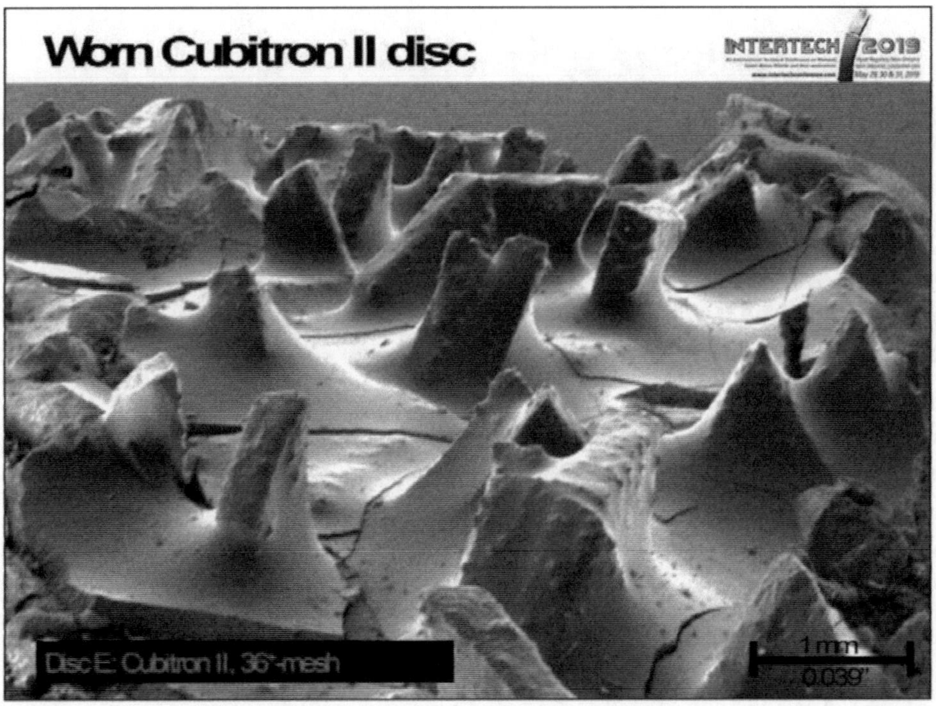
Worn Cubitron II disc
INTERTECH 2019
Disc E: Cubitron II, 36+-mesh
1 mm
0.039"

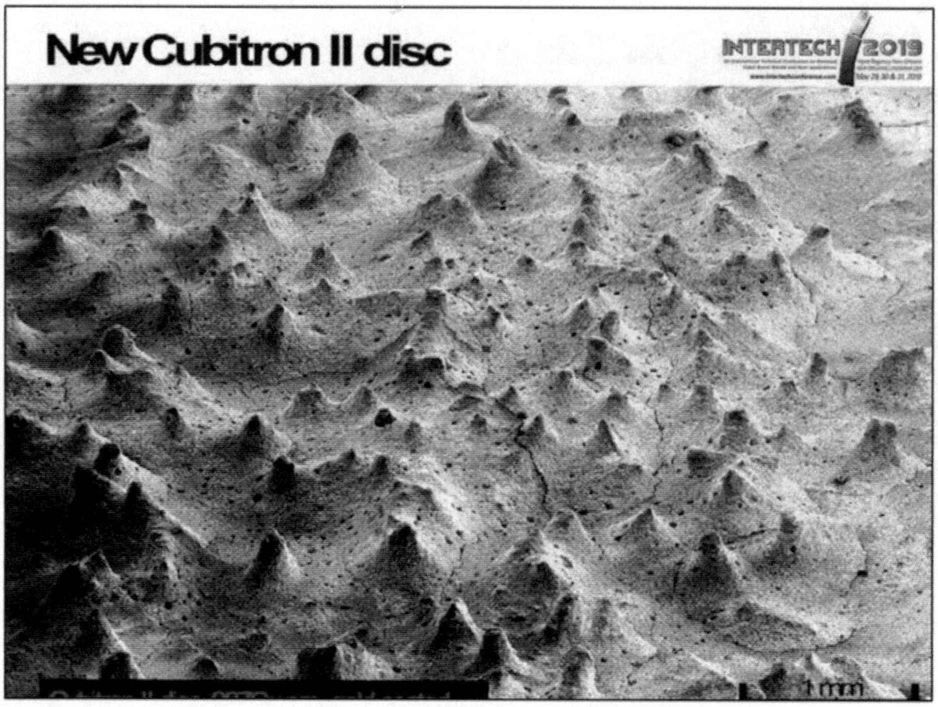
New Cubitron II disc
INTERTECH 2019
1 mm

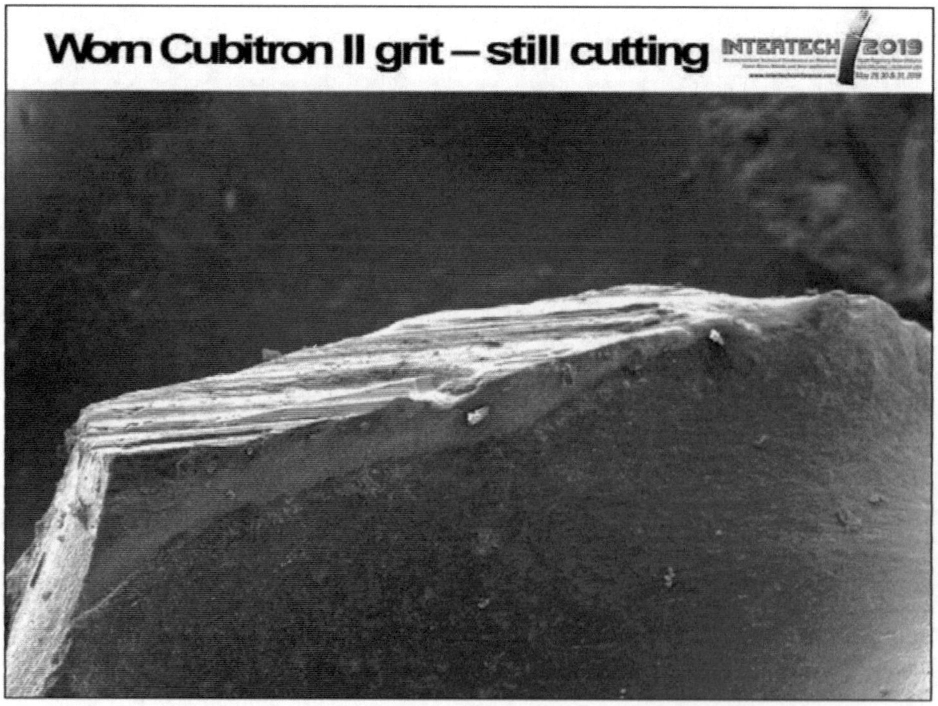
Worn Cubitron II grit – still cutting
INTERTECH 2019

Dull regular grit – not cutting

Trizact

Bubble Alumina: porosity for cooling
0.1
0.0
Photo No. = 1771
Mag = 75 X
100µm
EHT = 20.00

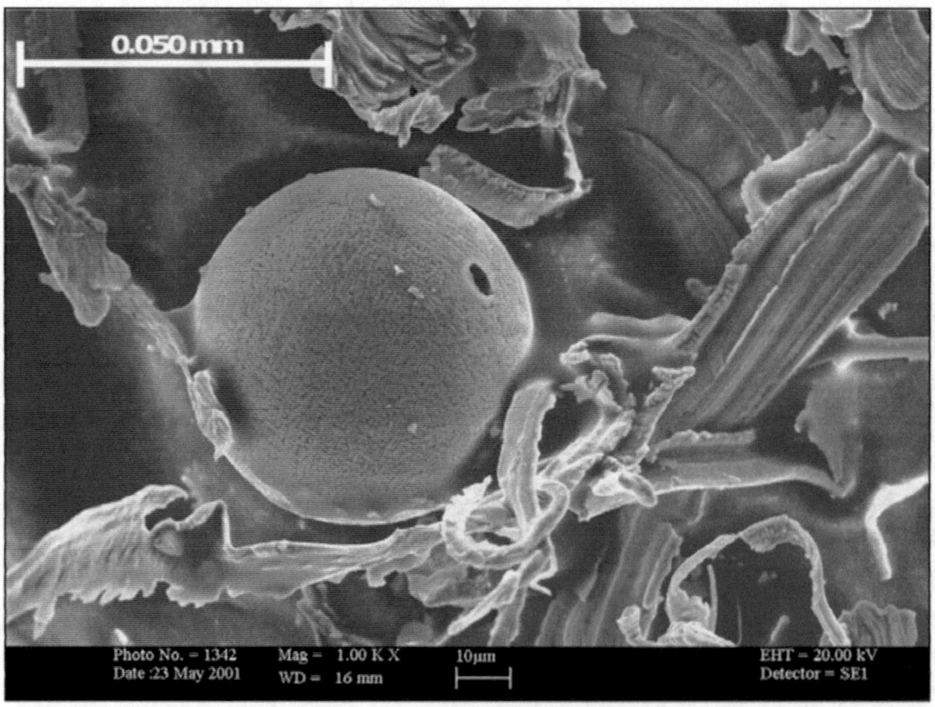
0.050 mm
Photo No. = 1342
Date :23 May 2001
Mag = 1.00 K X
WD = 16 mm
10µm
EHT = 20.00 kV
Detector = SE1

Loading, nickel alloy
INTERTECH 2019
1 mm
0.040"

Dressing
A
B
C
D
G
F
H

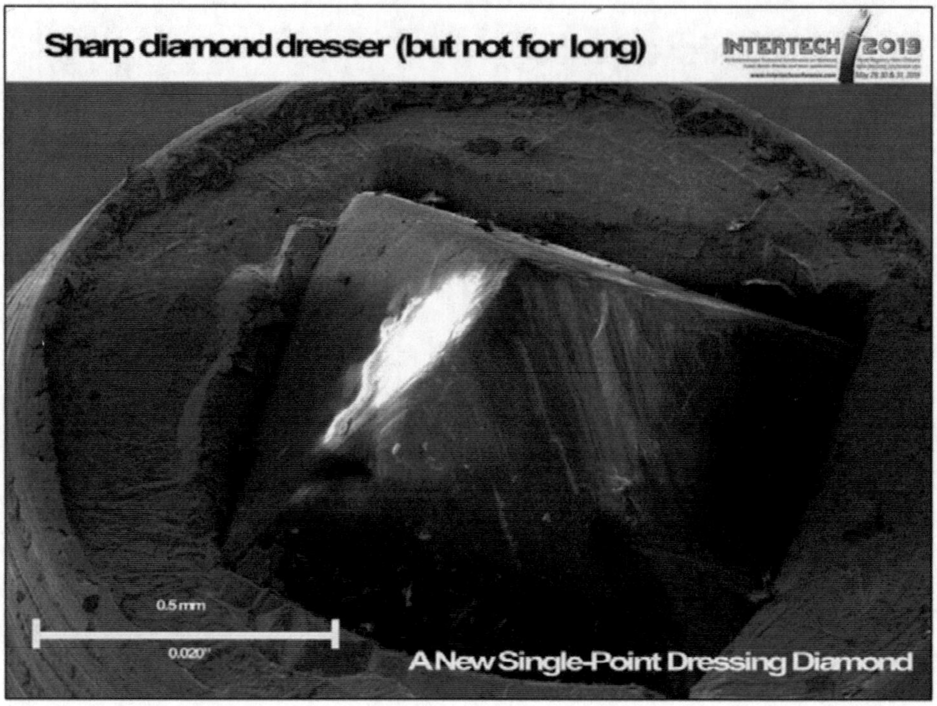
Sharp diamond dresser (but not for long)
INTERTECH 2019
0.5 mm
0.020"
A New Single-Point Dressing Diamond

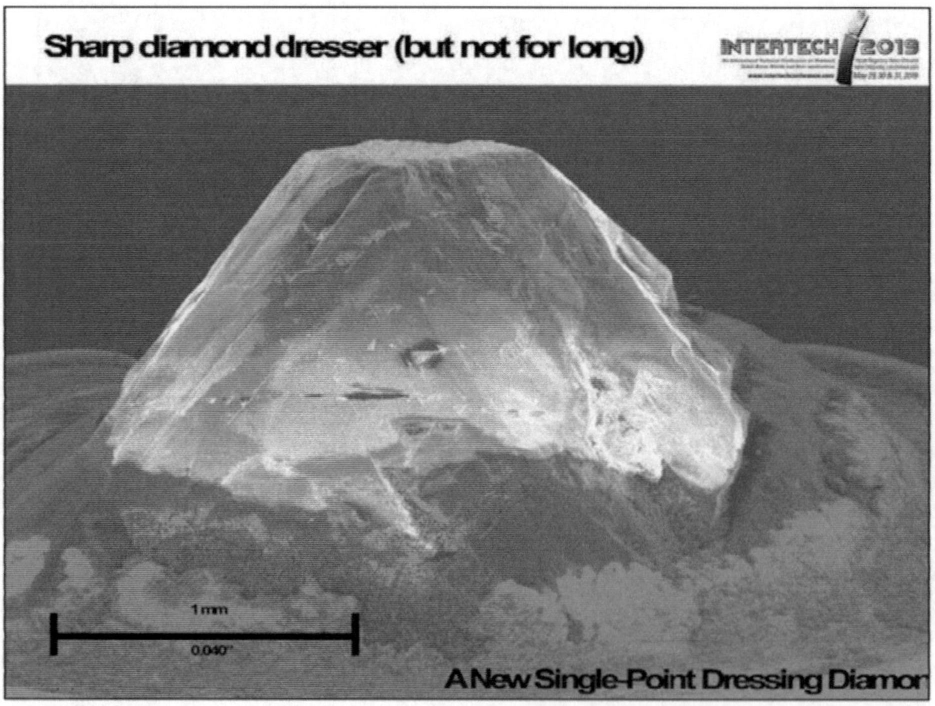
Sharp diamond dresser (but not for long)
INTERTECH 2019
1 mm
0.040"
A New Single-Point Dressing Diamor

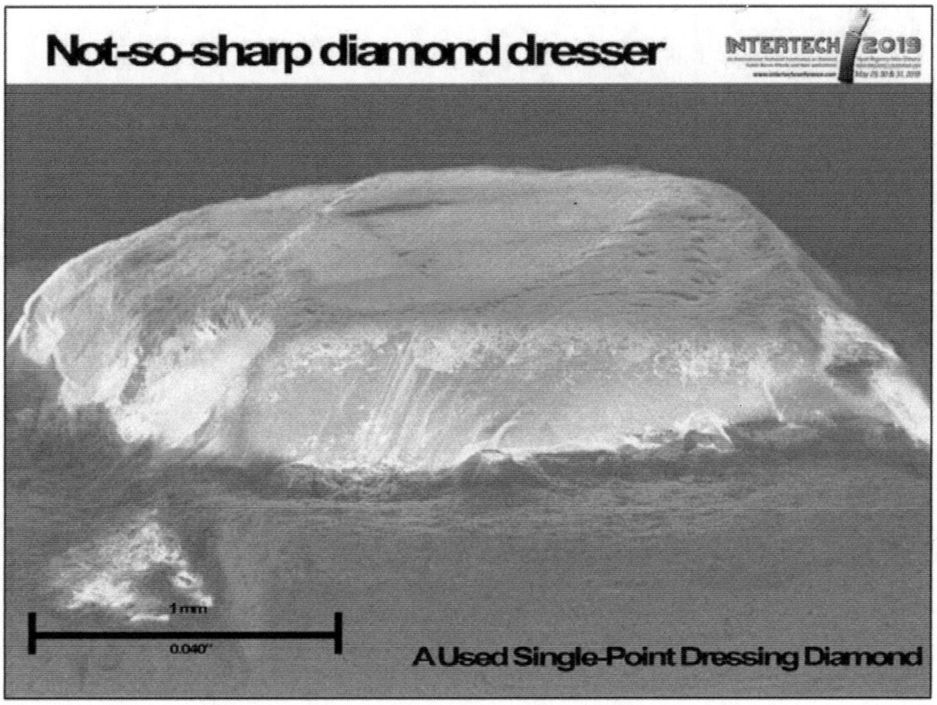
Not-so-sharp diamond dresser
INTERTECH 2019
1 mm
0.040"
A Used Single-Point Dressing Diamond

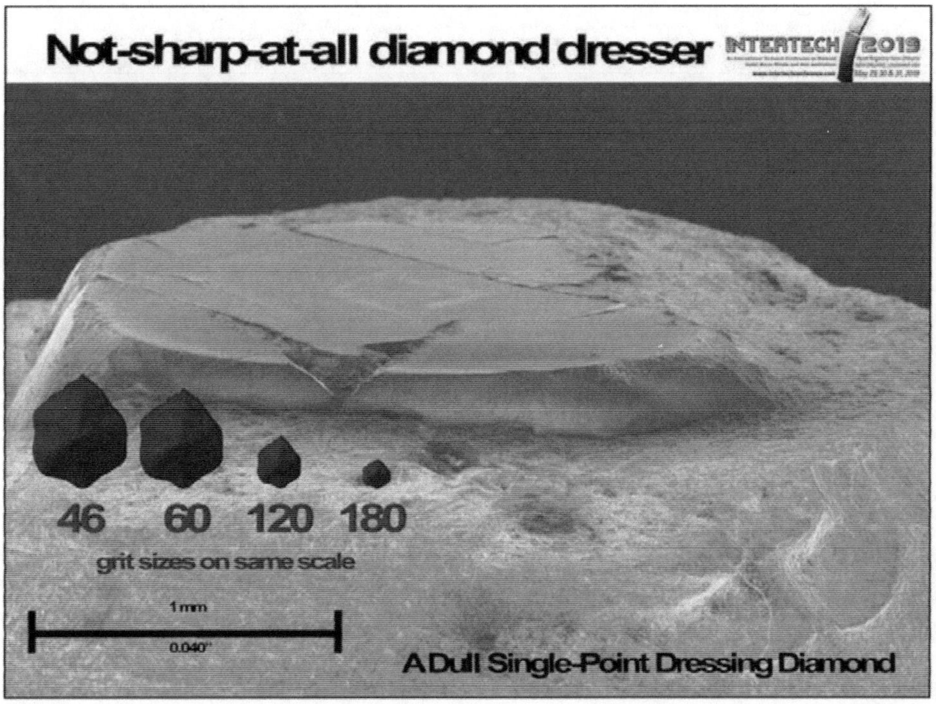
Not-sharp-at-all diamond dresser
INTERTECH 2019
46 60 120 180
grit sizes on same scale
1 mm
0.040"
A Dull Single-Point Dressing Diamond

Chiseldiamond dresser, wear flat
INTERTECH 2019
1.0 mm
0.040"

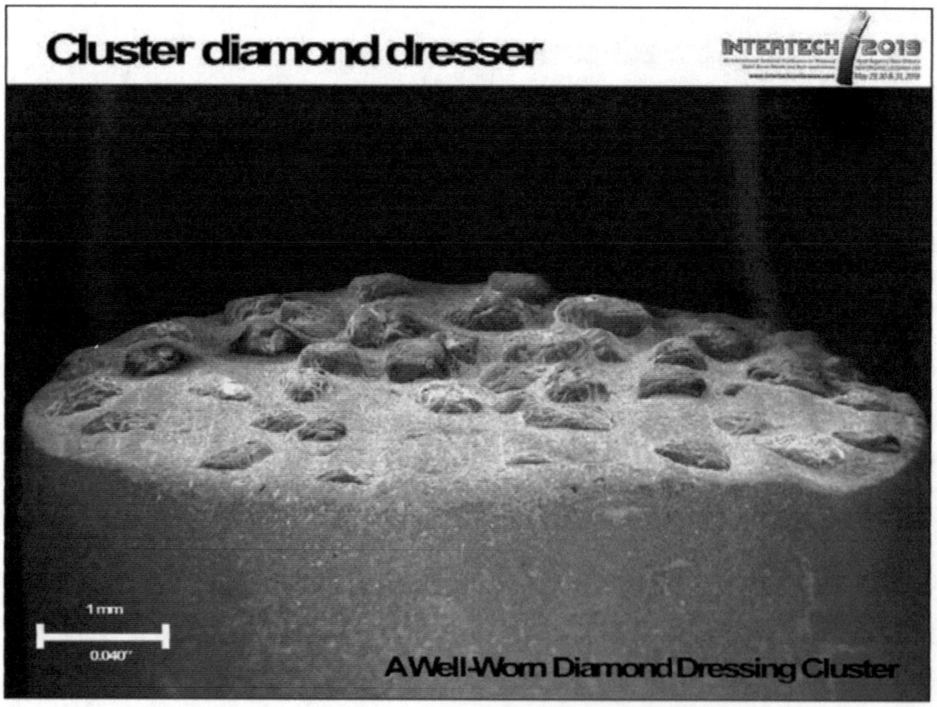
Cluster diamond dresser
INTERTECH 2019
1 mm
0.040"
A Well-Worn Diamond Dressing Cluster

Single-point diamond, rotation

Rotation: Finding a previous flat?

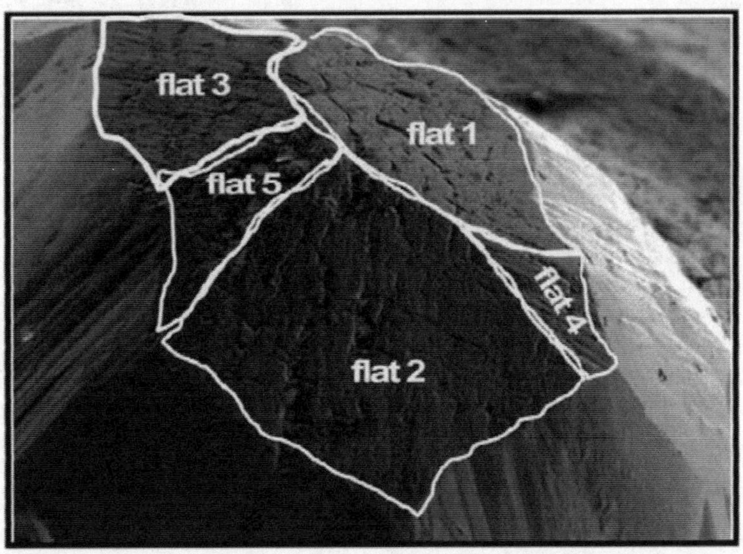

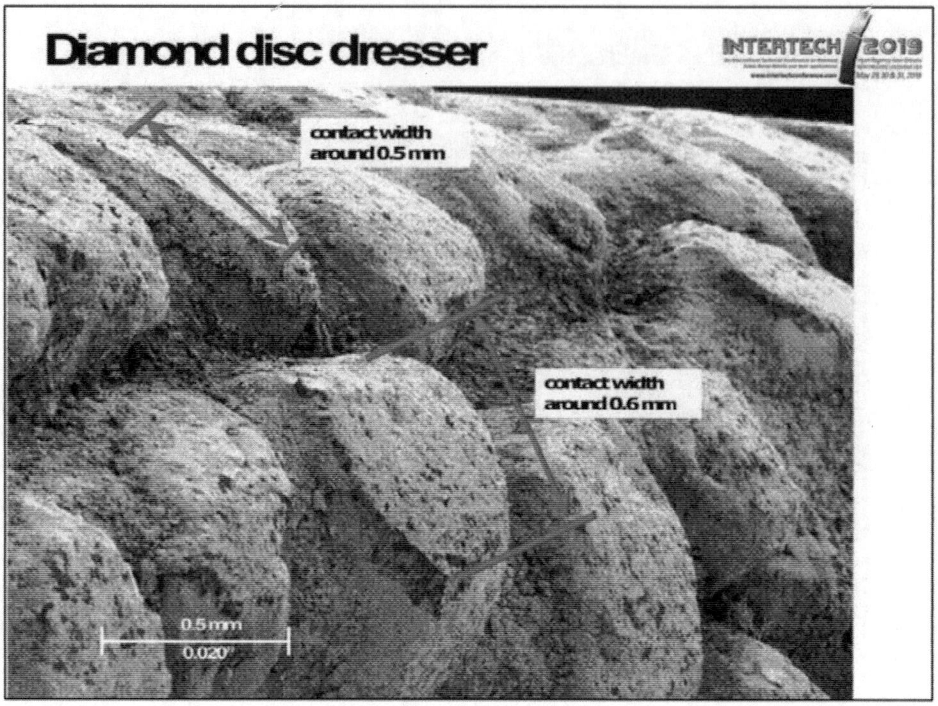
Diamond disc dresser
INTERTECH 2019
contact width
around 0.5 mm
contact width
around 0.6 mm
0.5 mm
0.020"

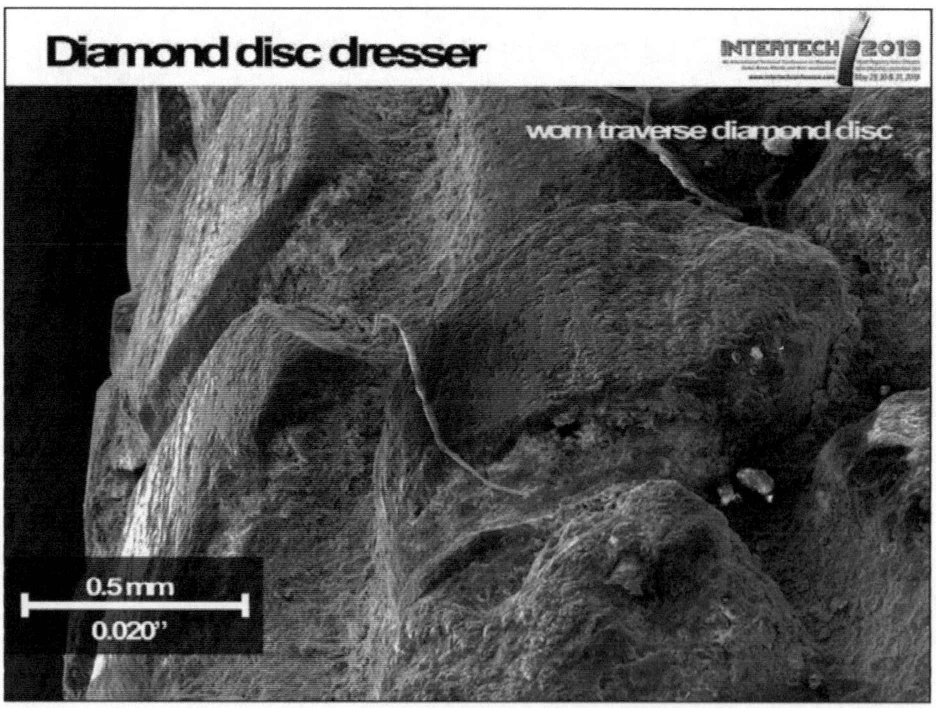
Diamond disc dresser
INTERTECH 2019
worn traverse diamond disc
0.5 mm
0.020"

Rotary plunge diamond roll

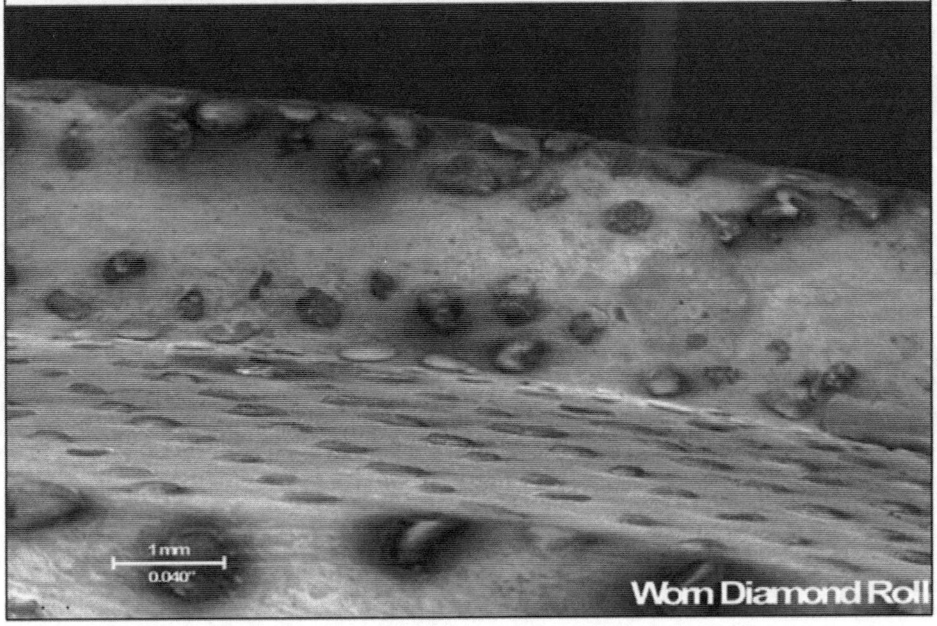

Title/Description of SEM

-

Case study: Worn diamond traverse disc in aerospace industry, dressing Al_2O_3 wheels for nickel alloys, examined under an SEM.

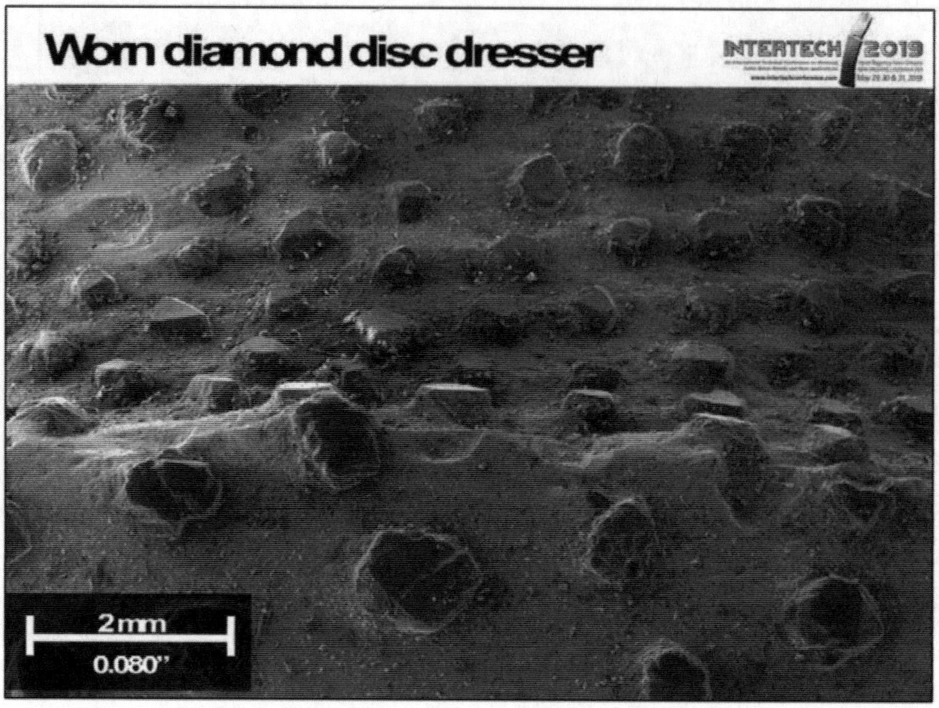
Worn diamond disc dresser
INTERTECH 2019
2mm
0.080"

Worn diamond disc dresser
INTERTECH 2019
0.2 mm
0.008"

INTERTECH 2019

The Book of Grinding

- 4000 electronic pages
- Description with each page
- 20+ years, 20,000 hours
- A serious Investment for Serious Grinders

  $US 4350

  €3790

  £3

INTERTECH 2019

High Intensity Grinding Course

Upcoming Courses

October 8-10, 2019 **Grand Rapids, Michigan**
Hosted by SL Munson & Co. and IGA Abrasives

September 11-13, 2019 **Biberach, Germany**
Hosted by Vollmer **Focus on grinding of tungsten-carbide**

High Intensity Grinding Course, 2009, Hasle-Rüegsau, Switzerland. Hosted by Blaser Swisslube.

INTERTECH 2019

The use of electron-microscopy in the study of superabrasives in grinding

Jeffrey A. Badger, Ph.D.

Thank you for your time!
Questions?

HELPING COMPANIES ACHIEVE EXCELLENCE IN PRECISION GRINDING

THE GRINDING DOC

SINGLE-CRYSTAL DIAMOND GRIT IN CONSTRUCTION: APPLICATIONS AND CHARACTERISATION

Dr. S. Nailer, Dr. S. Moseley, Dr. S. Huth and Dipl.-Ing. T. Klein
Hilti Corporation, Feldkircherstrasse 100, 9494 Schaan, Liechtenstein

ABSTRACT

Synthetic diamond in particulate 'grit' form remains the material of choice for the sawing and core drilling of construction materials such as reinforced concrete. Some sixty years since the first successful synthesis of diamond, diamond systems for construction applications continue to achieve ever-higher performance levels, deriving not only from increased capabilities of the power tools, but also from advances in the design and manufacturing of segments for 'inserts' (core bits and saw blades). One of the most significant developments in construction inserts in recent years has been the ability to position diamonds at specific locations in the segment.

This paper begins with an overview of the applications of diamond power tools and their inserts in the construction industry, with a focus on the operating conditions of the applications and diamond particle wear and fracture behaviours. The paper then discusses the principal characteristics of diamond grit products for construction applications, provides insights into appropriate laboratory characterisation techniques, and describes how their results may be best interpreted to enable optimal diamond selection for today's advanced segment designs.

INTRODUCTION

Particulate ('grit') diamond is the most commonly-used abrasive material in demanding construction applications such as the coring and sawing of steel-reinforced concrete. In addition to reinforced concrete in its various compositions, diamond is also used to drill and cut masonry materials and epoxy coatings. Diamond is found in such diverse inserts as tile bits of diameter 6mm and wall saw blades over 1m in diameter.

The principal method of retaining the diamond particles in the coring/sawing process is the metal-matrix segment. In this multilayer system, diamonds are incorporated at various heights within the segment, and as the diamonds on the exposed 'working' surface of the segment are consumed and removed, fresh diamonds are revealed for use. Each diamond exposed on the segment surface may be considered to have a particle 'lifecycle', from newly emerging through to removed.

In conventional segment manufacture for construction applications, diamond particles are mixed with metal powders and pressed to form segments where the diamonds are considered to be 'randomly distributed'.

Recent developments in segment assembly technology allow diamond particles to be positioned at specific locations in the segment. As well as eliminating the possibility of diamond clustering, such 'arranged-diamond segment' technologies enable the diamond positioning to be designed for the application conditions, such as the expected drilling/sawing performance, drive parameters of the tool motor, and 'base material' (workpiece) to be drilled/sawn.

The performance of the diamond 'insert' (such as core bit or saw blade) can be optimised when the diamond particles are themselves selected to best exploit the design of the arranged-diamond segment and the application conditions. Particulate diamond products used in stone and construction sawing and drilling applications (commonly known as 'saw grit') each comprise particles of a wide range of sizes, shapes, strengths and purities.

Consequently, the effective selection of diamond types for use in arranged-diamond segments is dependent on the quality of the laboratory measurements used to quantify the particle characteristics. Whilst some of these laboratory techniques provide simple single-value 'figures-of-merit', other techniques yield results from every individual particle, and so effective interpretation is dependent on the correct statistical analysis of these 'distributions' of results.

1. AN OVERVIEW OF THE USE OF DIAMOND INSERTS IN CONSTRUCTION APPLICATIONS

'Inserts' featuring diamond grit-impregnated segments may be used to saw, cut, drill or grind a variety of construction 'base materials' (workpieces). A principal construction base material that places especial demands on the diamond insert is steel-reinforced concrete. Concrete is comprised of cement and water, sand, and aggregates, and these may be mixed in varying proportions. As the aggregate materials themselves (such as gravel) also vary around the world due to different geological conditions, so the properties of concrete and their consequent demands on the diamond insert also differ around the world.

Concrete is commonly reinforced with steel bars of diameters 6mm to 59mm, with 26mm diameter 'rebars' often used in buildings. Concrete may be considered highly-reinforced when the total rebar content exceeds 1.5%. The cutting actions of diamond grit-impregnated inserts in the brittle concrete and the ductile rebars are very different, with the wear rate of the diamond segments many times higher in rebar.

Other construction base materials include masonry blocks of sand-limestone, bricks, tiles and epoxy floor coatings. These are easier to cut than reinforced concrete, but they present their own challenges such as the fast removal of drilling dust (for example, in soft brick), whilst epoxy floor coatings are most effectively removed by 'defined-edge' cutters such as polycrystalline diamond.

As the base materials to be cut are diverse, so are the material removal processes. One method of classifying construction applications for diamond is to divide them into 'heavy' and 'light' uses.

'Heavy' diamond construction applications involve large tools which are often rig-based (that is, the power tools are held in position using fixation methods such as anchors). These tools, with their associated inserts, require operation by specialists and therefore such 'heavy' applications are mostly performed by specialist diamond contractors. Examples of heavy diamond applications include rig-based operations such as wall sawing, wire sawing and large-diameter core drilling, in addition to the use of heavy mobile machines such as in floor sawing and floor polishing. Heavy diamond applications increasingly benefit from the convenience and optimal performance offered by tool functionalities such as remote control and automated in-feed.

'Light' diamond construction applications involve smaller power tools that are predominantly handheld. Their portability and ease-of-use enables them to be operated by general tradesmen as well as specialist contractors. Examples of light diamond applications include the handheld coring of masonry materials and concrete, concrete grinding (surface preparation), chain sawing and the cutting of concrete with electric or gas-powered tools with small diameter blades.

Across these diverse material removal processes, the operating conditions of the diamond can vary considerably. Core drilling is a process where the diamond insert is in 'constant contact' with the base material in a confined space. When coring concrete, the use of water is required to cool the drilling zone and flush the drilling debris from the drilling zone. Without water, the drilling zone becomes very hot and the heated concrete debris can adhere to the surface of the diamond segments, reducing the protrusion of the diamonds above the segment surface, and thus impeding drilling progress. Masonry materials such as brick and sand-limestone do not pose the same problem, and are thus able to be drilled 'dry' (without water).

Those applications in which the individual diamond segments experience 'interrupted contact' with the base material (namely, circular sawing and cutting) tend to allow more effective ejection of debris from the cutting zone, and thus the cutting of concrete can be performed without water.

In addition to the conditions of constant or interrupted contact and wet or dry operation, the diamonds may experience surface speeds from 2m/s up to 100m/s, and be driven by tools with power ranging from 1kW to 75kW.

2. THE BEHAVIOUR OF DIAMOND SEGMENTS IN CORING REINFORCED CONCRETE

Whilst 'single-layer' inserts (comprising single layers of diamond attached to supporting bodies by brazing or electroplating) may be used in wire sawing, gas-powered sawing and handheld coring, the most widely-used method for retaining diamond particles in construction inserts is the diamond grit-impregnated metal-matrix segment.

Sintered diamond grit-impregnated segments may be considered 'multilayer' systems in that there are diamond particles located at various depths in each segment – as the uppermost particles on the segment surface are removed in use, the segment wears down to reveal further particles. The segment matrix itself, typically composed of metals such as cobalt, iron and bronze, performs the function of retaining working diamonds whilst releasing worn or broken diamonds. To do this optimally, the segment matrix must wear at a rate that matches the wear rate of the diamond in a given application.

2.1. DIAMOND WEAR STATES IN THE CORING OF REINFORCED CONCREETE

A diamond particle may experience a number of different 'wear states' during its progression from being first exposed on the working surface to being lost from the segment. In one well-established methodology [1] there are six defined wear states. These are shown pictorially in Fig. 1.

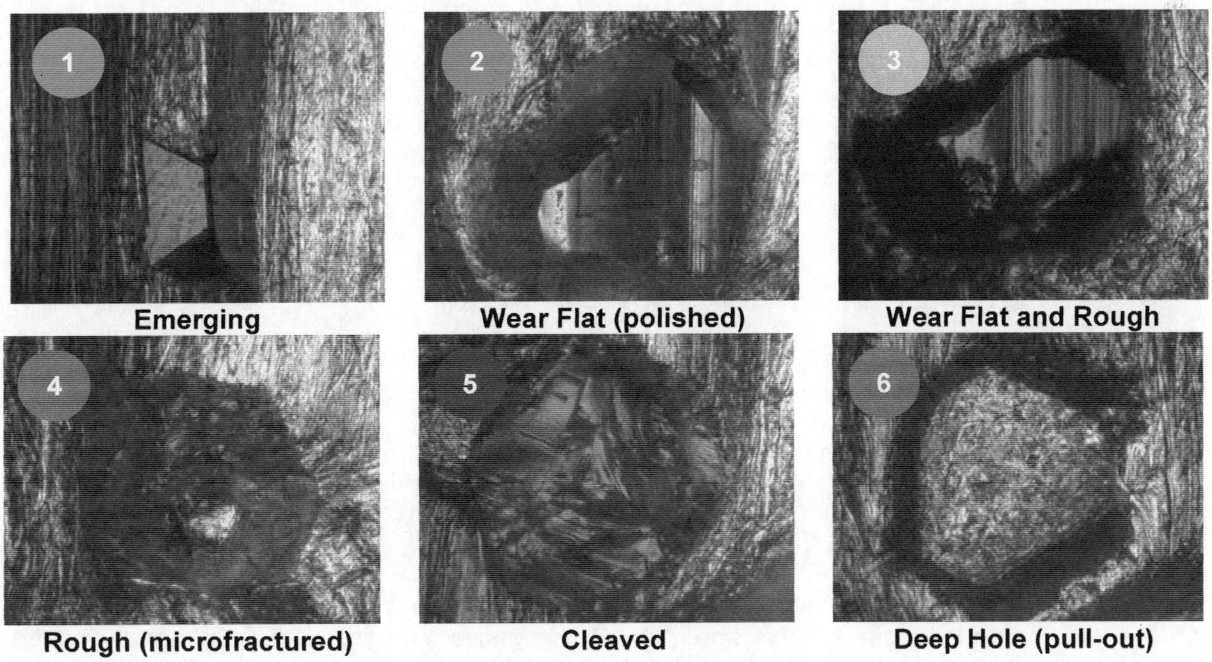

Fig. 1. Six wear states in the 'lifecycle' of a diamond particle in the coring of reinforced concrete.

A diamond is considered 'emerging' when it is exposed on the segment surface by segment erosion but is not yet actively in contact with the base material. The 'wear flat' or 'polished' state is when wear is induced on the diamond from active contact with the base material. The diamond may then move into the 'wear flat and rough' state resulting from fatigue-induced localised microfracture, the progression of which will lead to the 'rough' or 'microfractured' stated. A diamond which has experienced macrofracture, often below the segment surface, is considered 'cleaved', and the final state is when the diamond is removed from the segment, leaving a crater called a 'deep hole' or 'pull-out'.

The 'lifecycle' of a diamond, in terms of how quickly it proceeds through each wear state (or indeed which wear states it will experience), is influenced by the characteristics of the diamond (its size and strength, and how many diamonds are in the segment), the operating conditions of the power tool (for example, the torque and rotational speed of the coring tool), the characteristics of the base material (such as hardness and abrasivity) and the wear rate of the segment metal matrix.

An experiment was performed to determine the relationship between the coring conditions in reinforced concrete and insert behaviour. Core bits of the same specification (102mm diameter segments designed for coring non-abrasive concrete with steel reinforcement bars) were tested in a small-aggregate concrete containing one 26mm reinforcement bar every 250mm. A 3.6kW drill motor was used for wet coring, and a range of different gears, each with different torques and rotational speeds, were used. For reasons of sensitivity the absolute results are not given here. However, the diamond wear state distributions under different conditions are shown in Fig. 2.

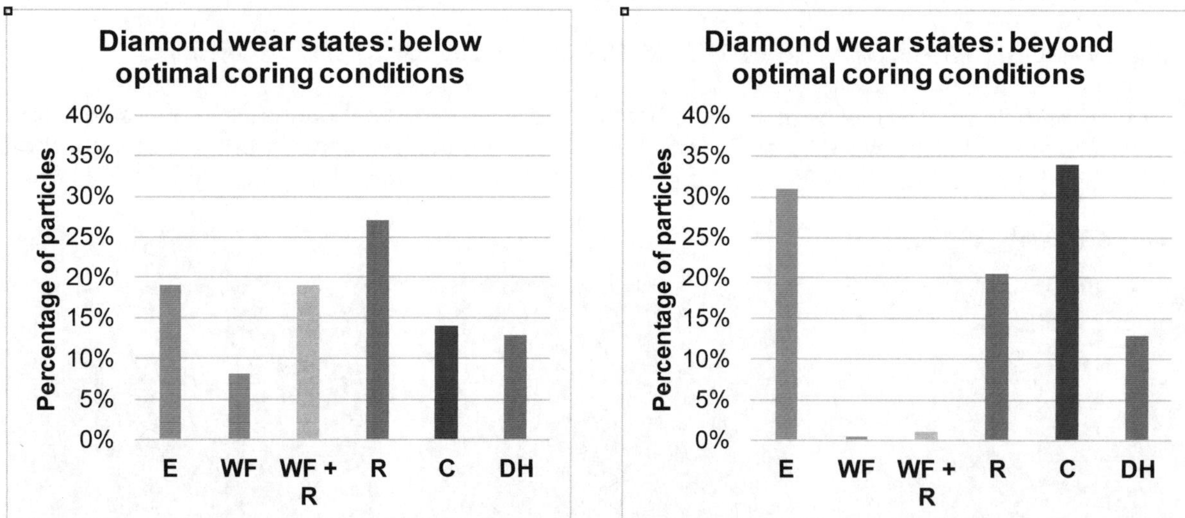

Fig. 2. The percentages of diamond particles in the six different wear states in two conditions of reinforced concrete coring: with insufficient torque (left) and with excessive torque (right).

The wear state distribution in coring conditions below optimal are shown in Fig. 2 (left). Here, the tool gear used had a torque lower than that of the recommended gear, and a rotational speed higher than that of the recommended gear. This low torque was insufficient for optimal coring speed, and therefore a relatively high degree of wear-flattening (WF and WF+R states) was seen.

Conversely, when a tool gear was used that provided a higher torque and lower rotational speed than the recommended gear for this core bit, an excessive proportion of rough and cleaved diamonds were observed (see Fig. 2 (right)). Under both of these coring conditions ('below optimal' and 'beyond optimal'), the coring speeds were below that possible in the optimal gear (20% slower and 16% slower respectively), but as can be seen from this study, the two underlying reasons for these slower speeds were very different. This shows the benefit of studying the diamond wear states to determine the factors influencing the performance of construction inserts.

2.2. DIAMOND PLACEMENT IN SEGMENTS FOR CONSTRUCTION APPLICATIONS

Recent developments in segment assembly technology allow diamond particles to be positioned at specific locations in the segment. As well as eliminating the possibility of diamond clustering, such 'arranged-diamond segment' technologies enable the diamond positioning to be designed for the application conditions (such as the expected drilling/sawing performance, drive parameters of the tool gear, and the base material to be drilled/sawn). Images of two types of segments for coring reinforced concrete are presented. Fig. 3 (left) shows a conventional segment with randomly-distributed diamonds and Fig. 3 (right) shows a segment with arranged diamonds.

Fig. 3. Photographs of grit-impregnated segments for the coring of reinforced concrete: with conventional randomly-distributed diamonds (left) and with arranged diamonds (right).

Observations on Hilti reinforced-concrete core bit segments from a previous generation (with randomly-distributed diamonds) and the current generation (with arranged diamonds) found that, when the two segment types had the same segment geometry, metal matrix, diamond type and diamond concentration, the segments containing arranged diamonds had a more consistent number of active diamonds working on the segment surface. Over the course of the segment lives, the segments with randomly-distributed diamonds had at times more active diamonds and had at other times fewer active diamonds than were observed with the arranged-diamond segments.

This greater consistency in the number of active particles on the surfaces of arranged-diamond segments can lead to improvements in coring performance. Fig. 4 shows the relative coring speeds and lifetimes in reinforced concrete of core bits with and without arranged diamonds. This graph summarises the observations from a suite of segment development activities at Hilti.

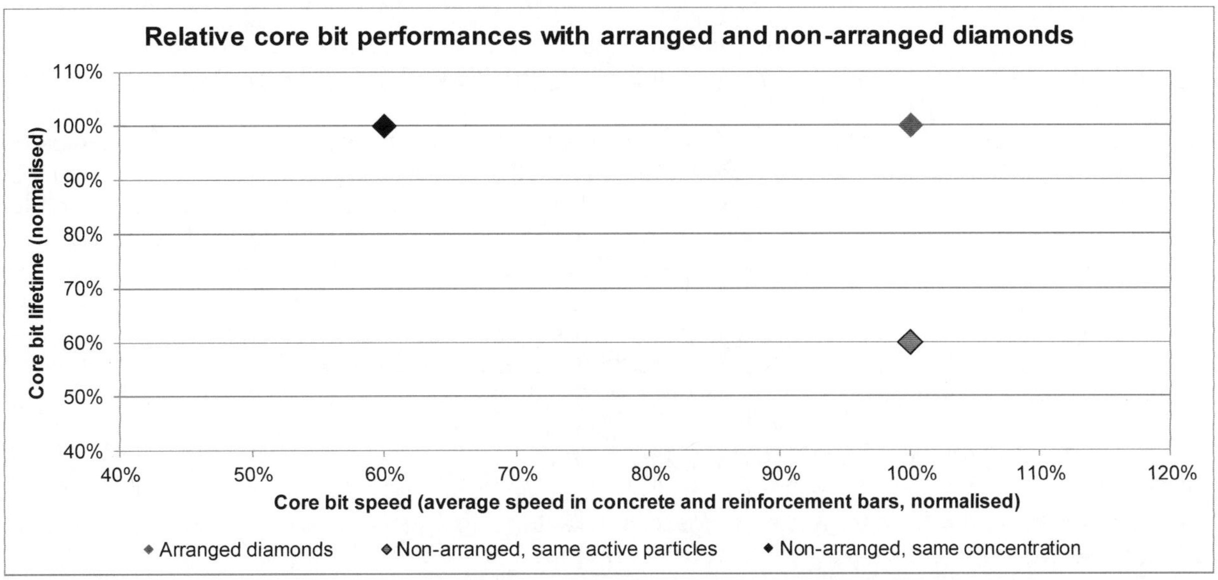

Fig. 4. Graph of the relative performances (coring speed and lifetime) of core bits with arranged and randomly-distributed diamonds, when coring reinforced concrete with 2.6kW and 3.6kW tools.

In summary, it has been found that when comparing arranged and non-arranged diamond segments with the same segment matrix and diamond concentration in the same application, the total segment lifetime is the same but the non-arranged diamond segments offer on average only 60% of the coring speed of the arranged diamond segments. The reason for this is that, while the total number of diamonds in the segment is the same (making the same overall contribution to the wear resistance of the segment and therefore the same lifetime), the more varying number of active diamonds in the non-arranged segment leads to diamond clustering where there is a reduced applied force per particle. The higher applied force per particle in the arranged-diamond segment leads to greater particle penetration into the base material, and thus faster coring.

It has been found to be possible to reduce the diamond concentration of non-arranged diamond segments in order to achieve the same number of active particles as arranged-diamond segments, and therefore the same coring speed. However, reducing the diamond concentration in this way also reduces the overall segment wear resistance, and so bit lifetime is lower (typically by 40%).

Conceptually, the ability of arranged-diamond segment technology to improve the effectiveness of the diamonds contained within the segment places additional importance on selecting the optimal diamond for the application requirements. The following sections of the paper present the key characteristics of diamond 'saw grit' for construction applications, and the suitable measurement techniques and statistical methods for studying them.

3. DIAMOND SAW GRIT AND ITS CHARACTERISTICS

Synthetic diamond grit products (produced by high-pressure, high-temperature conversion from graphite) are normally graded and selected according to the key characteristics of size and strength. The size band of the product is selected according to whether the priority is surface finish (as in 80μm diamond for polishing) or material removal rate (as in 600μm diamond for drilling).

The strength (or 'grade') of the product is then selected according to the characteristics of the base material to be cut and the drive parameters of the tool to be used. As was demonstrated earlier, diamond grit works most effectively by maintaining sharp cutting edges. The diamond strength should be high enough to prevent premature fracture and thus short working life, without being too high that the particles polish to a smooth surface and therefore lose their cutting ability.

The term 'diamond characterisation' may be defined as the laboratory measurement of the properties of diamond believed to play important roles in behaviour in application.

Two of the most important behaviours of inserts for construction applications are speed (typically the depth drilled or area sawn per unit time) and lifetime (the total depth drilled or area sawn before the segments are fully worn out). As previously implied, speed may be considered to be mostly dependent on particle size, whereas lifetime may be considered to be mostly dependent on particle strength (and its contributing factors, particle shape and metallic inclusion content).

The following sections describe in greater detail these key diamond characteristics, together with laboratory techniques for their quantification and the mathematical/statistical forms of their results.

4. CHARACTERISATION OF PARTICLE SIZE

There are perhaps three important methods of characterising or expressing the size of diamond particles: sieving, image analysis, and particles per carat.

Particles per carat ('PPC') describes the number of particles per unit mass (one carat being 0.2g), and is valuable to the segment manufacturer for expressing how a certain mass of diamond put into a segment translates into the number of available particles. PPC is most conveniently measured by using a light beam-based grain counting machine, dividing the number of particles counted by their measured mass.

4.1. SIEVING

Sieving uses sieves of sizes defined by international standards [2] to physically separate diamond particles into size fractions. Sieves are traditionally defined by the number of lines per inch, with a corresponding aperture size specification in microns. Due to its principle of physical separation, sieving is used for the creation of diamond size bands (as well as their measurement).

US mesh sizes may take the form of 'half sizes' or 'full sizes'. A half size is defined by a pair of sieves where the coarser ('upper defining') sieve has an aperture size approximately 25-30% larger than that of the finer ('lower defining') sieve. For example, a size 30/35 US mesh half size diamond product is that which sits between a 30 mesh (645μm) sieve and a 35 mesh (505μm) sieve. Adjacent half sizes may be blended together to form a full size.

Whilst the particle size distributions of the graded half sizes are approximately Normal (Gaussian) in form, the blended full sizes tend to exhibit 'bimodal' particle size distributions containing the two modes (peaks) of the component half sizes. A graph showing the theoretical particle size distribution of a 30/40 US mesh full size is shown in Fig. 5.

The green curve in Fig. 5 represents the particle size distribution from 30/35 and 35/40 blended together in a 50:50 ratio. However, such blend ratios are not specified by international standards, and any blend of the two half sizes is permitted, provided that the 'oversize' and 'undersize' fractions (in this case, coarser than 30 mesh and finer than 40 mesh) do not exceed a few percent.

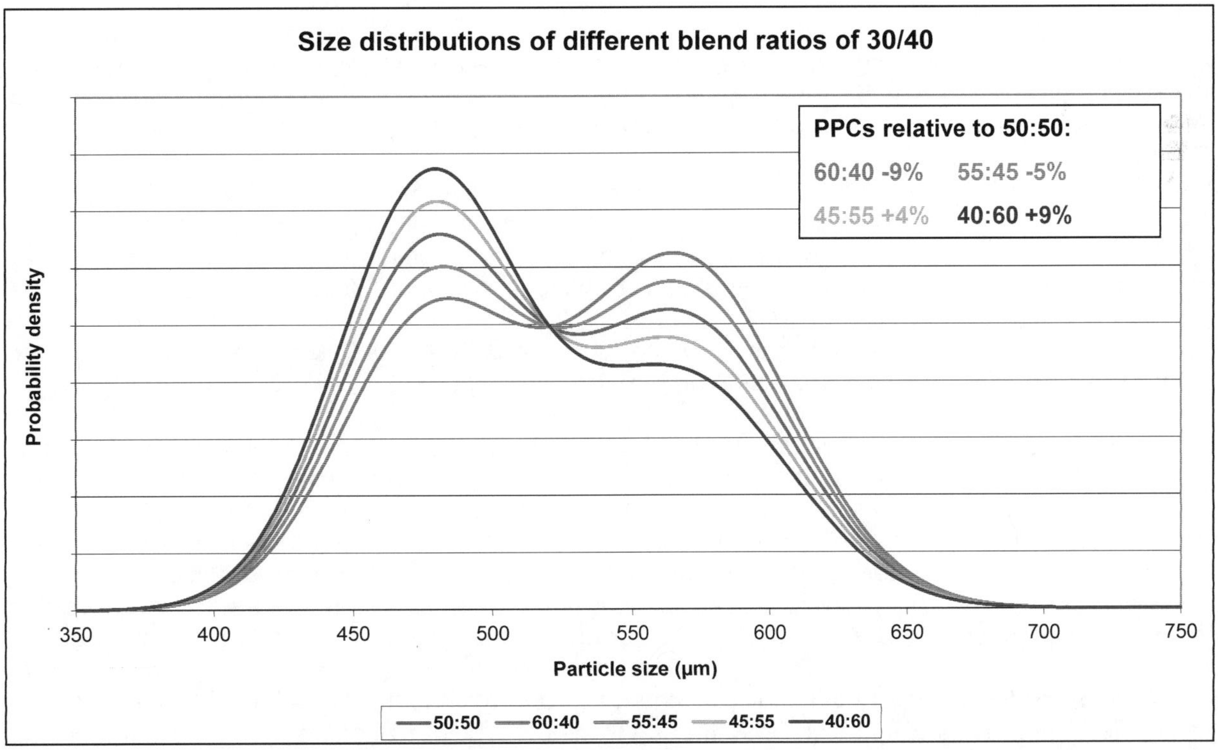

Fig. 5. Particle size distributions of 30/40 mesh diamond with different blend ratios.

Five curves representing five different blend ratios are shown in Fig. 5. It can be seen that, even within this fairly narrow range of blend ratios (60:40 to 40:60), the size distributions appear quite different. More significantly, there is a difference of around 18% in the particles per carat (PPC) of the 60:40 and 40:60 blend ratios, which could result in marked differences in application between the two blends even though they would both be correctly classified as '30/40 mesh'. From this it is clear that understanding the underlying size distribution (and not just the nominal mesh size) is important for controlling diamond behaviour in application.

4.2. IMAGE ANALYSIS

Image analysis uses a computerised optical microscope to capture digital images of individual particles, which are then measured using various size and shape parameters. Measurement of individual particles enables full particle size distributions to be recorded. However, it should be noted that image analysis generates two-dimensional images, the third dimension being not visible.

A commonly-favoured image analysis size parameter is 'equivalent circle diameter' (illustrated schematically in Fig. 6). The equivalent circle diameter is calculated from the projected area of the particle, and is defined as the diameter of a circle with the same projected area as the particle.

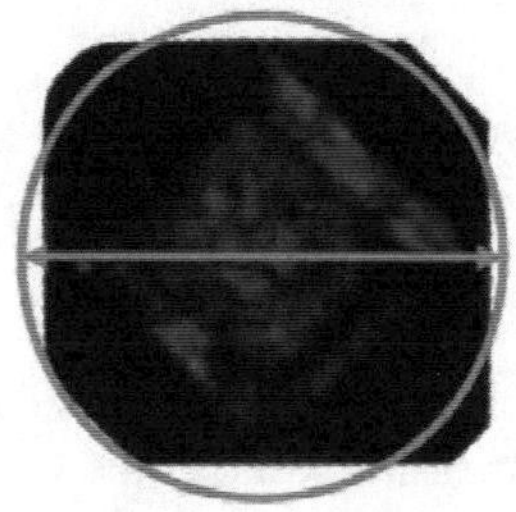

Fig. 6. Schematic illustration of equivalent circle diameter.

Equivalent circle diameter is sensitive to the longer visible dimension of the particle, and so the measured particle size distributions are larger than might be expected from sieving results. However, equivalent circle diameter, being derived from the area, has a good measurement resolution and gives smooth size distributions where small size differences are easily detected.

Particle size distributions of US mesh half sizes are typically Normal (Gaussian), and can be described by the well-known distribution statistics of the 'mean' (average) and 'standard deviation' (spread). However, when a US mesh full size is created by blending two half sizes, the resulting size distribution is 'bimodal' (it has two peaks, as previously shown in Fig. 5).

Equivalent circle diameter distributions of typical 30/35 and 35/40 half sizes are shown in Fig. 7, together with the 30/40 full size (from a 50:50 blend of the half sizes).

These distributions are now shown in cumulative form with the 'Raw' lines showing the as-measured data and the 'Normal fit' lines showing Normal distributions with the same means and standard deviations as the as-measured data. It can be seen that the Normal fits match the as-measured distributions quite well in the half sizes, but the Normal fit for the 30/40 full size deviates substantially from the as-measured data. This is because means and standard deviations always assume a Normal (Gaussian) symmetrical distribution shape, and are therefore inappropriate for describing distributions that have multiple peaks or are excessively peaked, flattened or skewed.

A more appropriate method of describing distributions of any shapes or number of peaks is to use 'percentiles', where the Xth percentile is the value below which X% of the data points lie (when ordered smallest to largest). The most commonly-used percentile is the 50th percentile ('median'), which represents the middle value in the distribution. Using additionally the 10th, 25th, 75th and 90th percentiles, a distribution shape can be conveniently described by five statistics.

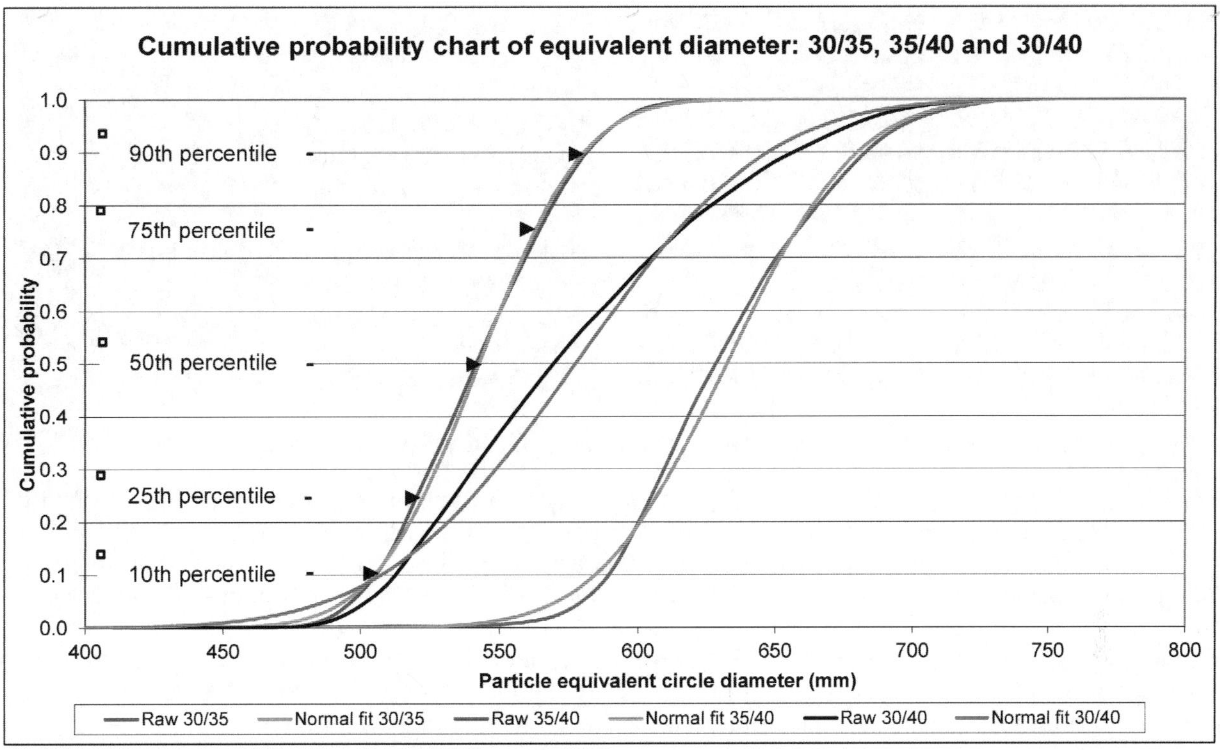

Fig. 7. Cumulative probability chart of equivalent circle diameter for US mesh half and full sizes.

5. CHARACTERISATION OF PARTICLE SHAPE

The particle images used to measure particle size can also be used to measure particle shape. A useful shape parameter for saw grit diamond is compactness, defined as the ratio between the ratio of the actual perimeter and the perimeter of a circle of the same area as the particle. A particle that appears circular (such as a highly crystalline cubo-octahedral diamond) will have a compactness near to 1, whereas particles which appear less round (for example, diamonds with elongation or lower crystallinity) will have compactness values higher than 1. Please see Fig. 8.

Fig. 8. Schematic illustration of compactness, with highly crystalline (left), elongated (centre) and less crystalline (right) diamond particles.

Compactness measurements on four adjacent saw grit product grades in size 30/35 are now demonstrated. Fig. 9 presents example images of these products, ranging from a product called here Grade 1 (a top-grade product) through to Grade 4 (a medium-grade product).

It can be seen from the small selections of particle images in Fig. 9 that there are no dramatic differences in particle shape from Grade 1 to Grade 4. More noticeable are the increased populations of darker particles in the images to the right.

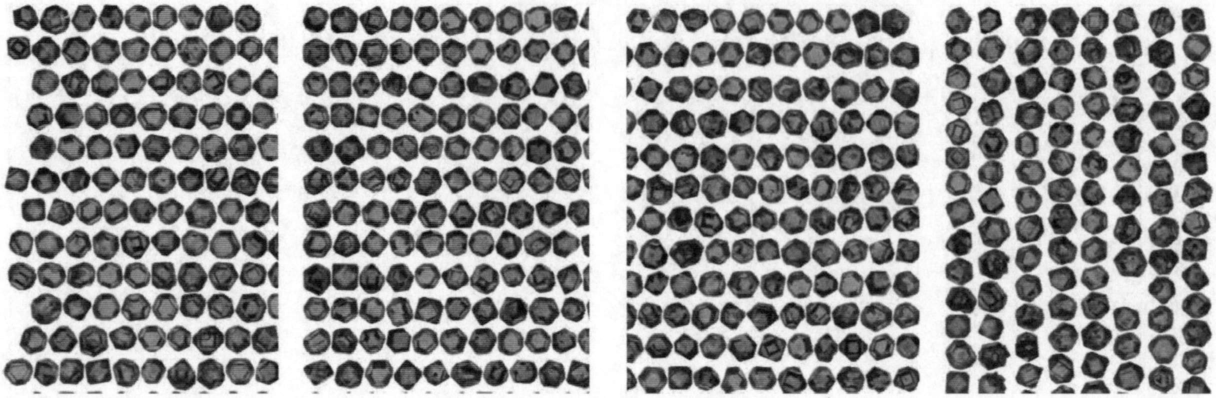

Fig. 9. Images of four 30/35 saw grit products: Grade 1 (left) to Grade 4 (right).

These observations of particle shape are confirmed by the distributions of compactness shown in Fig. 10. The compactness distributions of the four grades are close to one another, and furthermore, Grade 4 appears to have a 'better' shape than Grade 3 (the Grade 4 compactness distributions being further to the left, and therefore closer to the minimum compactness value 1).

The inset graph in Fig. 10 shows (with Grade 4 an example) how poorly a compactness distribution fits a Normal (Gaussian) curve. Even more so here than with particle size, 'percentiles' are the most appropriate way of statistically summarising the particle shape distribution.

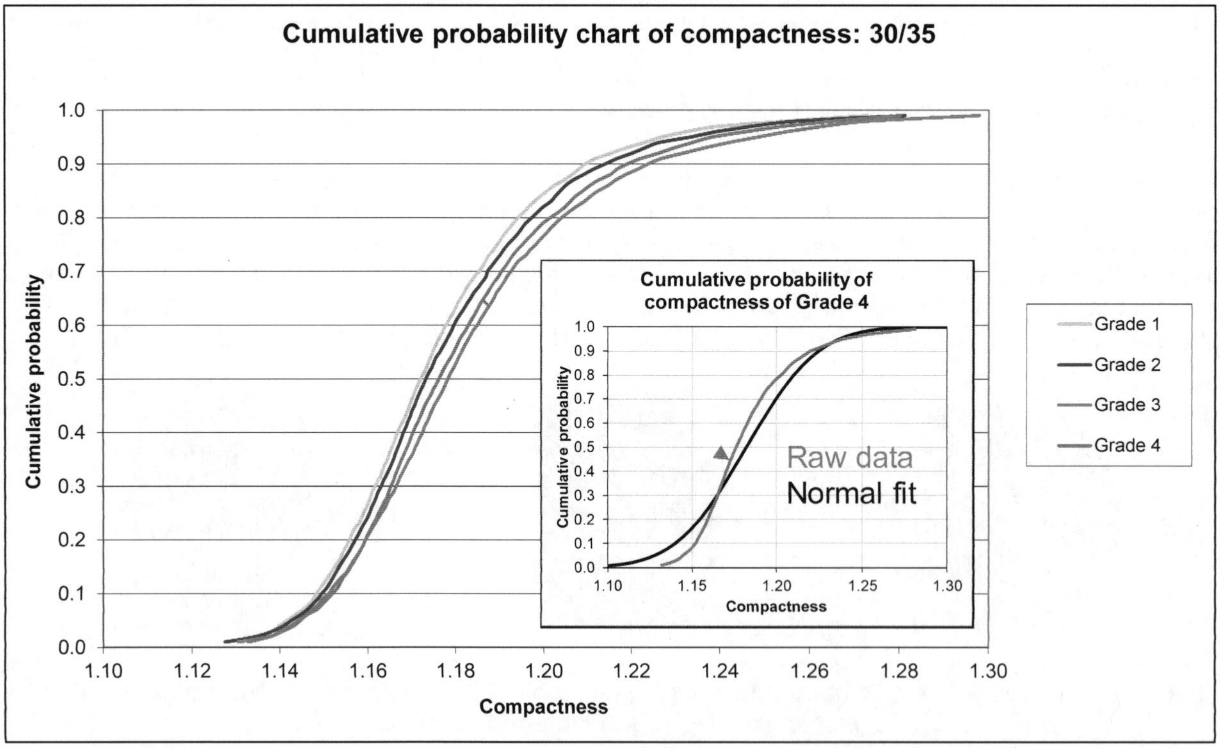

Fig. 10. Cumulative probability distributions of compactness of four saw grit grades in size 30/35.

From this graph it may be concluded that adjacent saw grit products are not substantially different in their particle shapes (they are differentiated only by small differences in the amounts of well-shaped or poorly-shaped particles), and also that particle shape does not always rank correctly with the perceived product grade.

6. CHARACTERISATION OF PARTICLE STRENGTH

Diamond product strength is typically characterised by two techniques: friability (or 'toughness') testing, and compressive strength testing.

6.1. FRIABILITY (TOUGHNESS) TESTING

Friability testing [3] measures the resistance of a bulk diamond product to cyclical impacts. The product is sieved to remove 'oversize' and 'undersize' particles, and the remaining 'on-size' particles are put into a steel capsule with a steel ball and subjected to a fixed number of oscillations ('cycles'). The product is then removed from the capsule and poured over a slightly finer 'post-impact' sieve. The diamond that has survived sits on this sieve, and is called the 'residue'. A stronger product registers a higher percentage residue (also called 'toughness index').

Friability testing is typically performed at one selection of cycles. For example, size 30/35 saw grit products are commonly tested using 1000 cycles. In this case, the test delivers a single percentage residue (toughness index) value – a figure-of-merit of product strength. However, all diamond products contain a distribution of particle strengths, and this distribution can be explored by testing at a range of different cycles. This provides a friability 'residue-time' curve, such as that in Fig. 11.

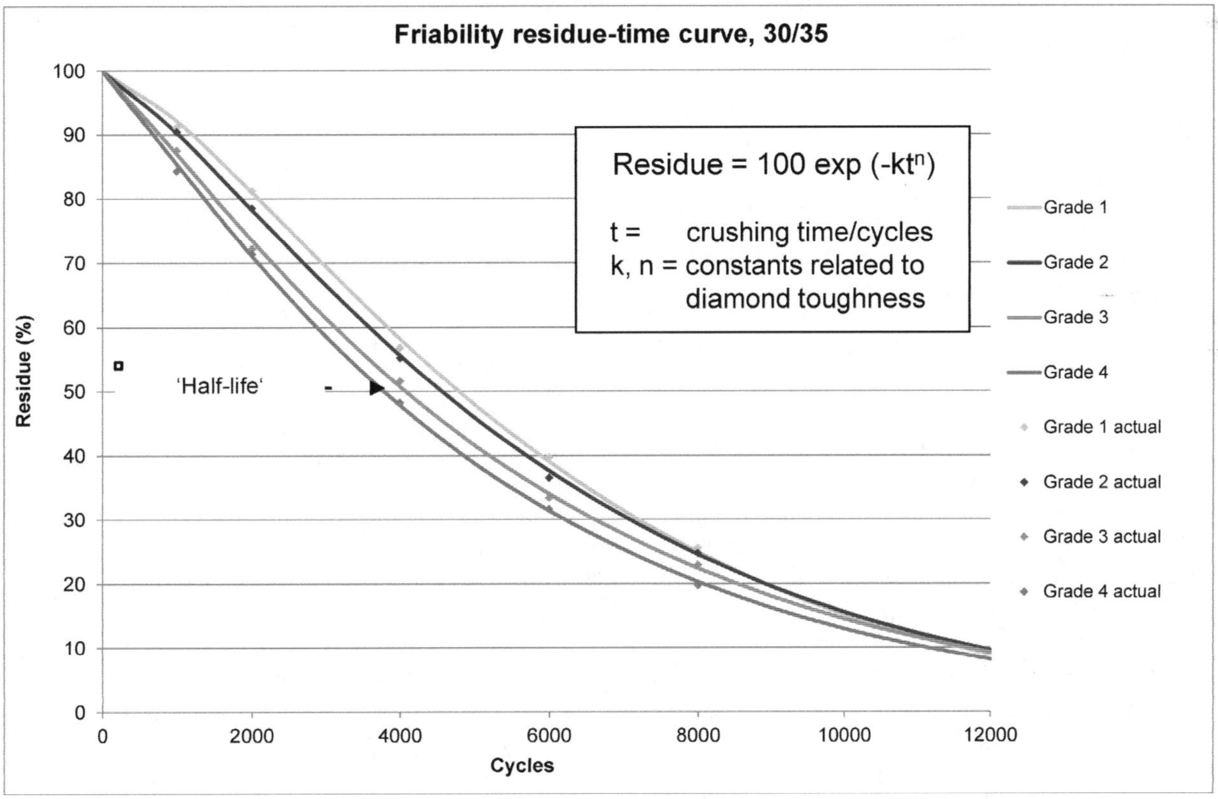

Fig. 11. Friability residue-time curves of four saw grit grades in size 30/35.

Fig. 11 shows the friability residue-time curves of the four saw grit grades in size 30/35. At all cycles, the four grades show the 'correct' ranking (Grade 1 being the strongest, Grade 4 the weakest), but their relative strengths vary as a function of cycles. The data points at different cycles may be connected by fitting a negative exponential curve according to the 'Rosin-Rammler equation' [4] (named after two scientists that studied the crushing behaviour of coal). The Rosin-Rammer equation is shown in Fig. 11, the constants k and n being characteristics of the diamond product. As well as allowing the prediction of percentage residue at any number of cycles, this equation enables the prediction of the number of cycles required for 50% residue – the 'half-life'.

6.2. COMPRESSIVE STRENGTH TESTING

Compressive strength testing of diamond measures the force required to crush individual particles. Particles are crushed between two polycrystalline diamond 'anvils', and the load cell connected to the lower anvil measures the force required to fracture each diamond particle [5].

This method provides a distribution of particle strengths ('fracture forces') for each product. Fig. 12 shows the strength distributions of the four saw grit grades. For easier comparison with the friability residue-time curves, the vertical axis shows the probability of survival (rather than failure). Furthermore, the as-measured results have been replaced by Weibull distributions, which tend to fit compressive strength distributions well [6]. In this graph, the four grades rank as expected, with Grade 1 having the highest median compressive fracture force and Grade 4 the lowest.

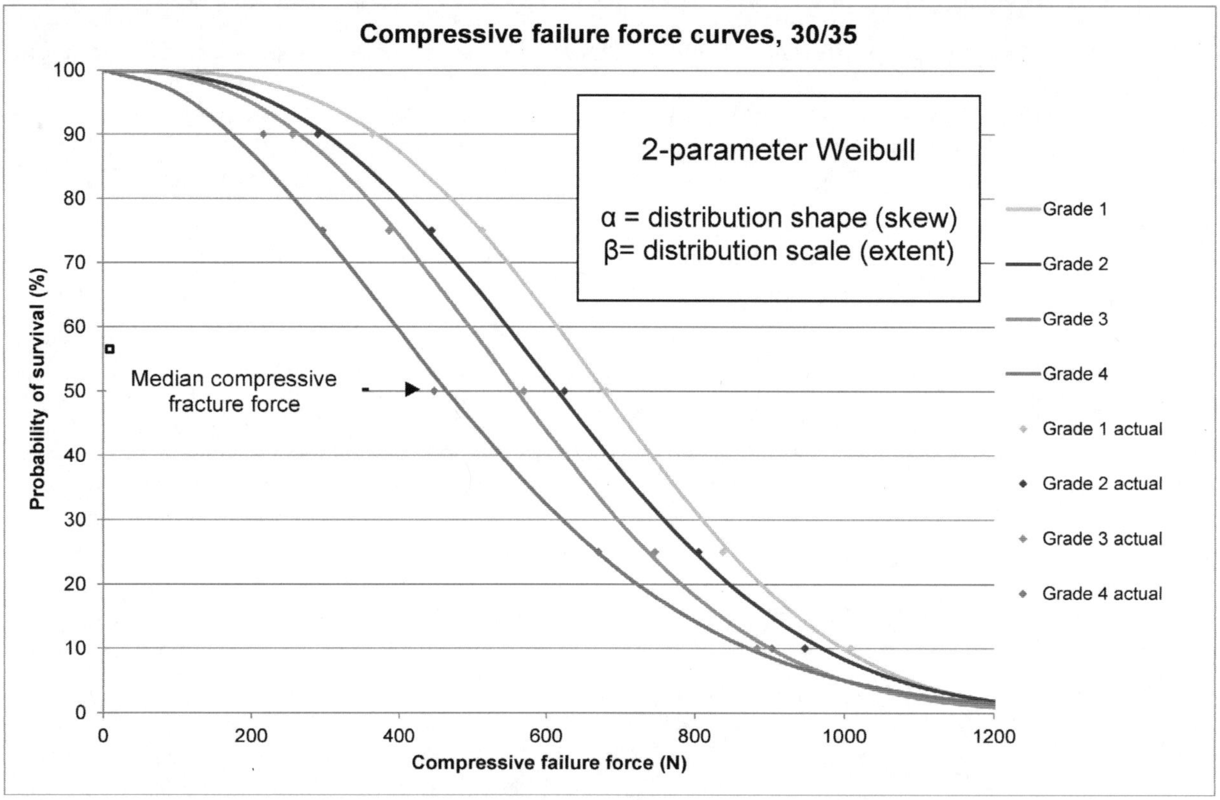

Fig. 12. Compressive strength distributions of four saw grit grades in size 30/35.

Whilst these two methods of strength testing can give complementary results, there are practical reasons for selecting one method or the other. Friability testing at one cycle selection is comparatively quick (around 10 minutes) but will only provide a single value figure-of-merit. The measurements required for a full residue-time curve would require over an hour. However, friability testing does have a good experimental error (around ±1%), and is therefore commonly used for quality control. Compressive strength testing enables a full distribution from 500 particles to be obtained in around 30 minutes, but the experimental error is usually greater due to damage of the polycrystalline diamond anvils influencing the fracture force measurements.

7. CHARACTERISATION OF METALLIC INCLUSION CONTENT

Saw grit diamonds are synthesised by the high-pressure, high-temperature conversion of graphite to diamond. Transition-metal alloys (typically iron-nickel) are used to lower the pressures and temperatures required for this conversion, and small traces of these alloys ('inclusions') can be trapped in the diamond crystals during the synthesis process (see Fig. 13 (left)). These inclusions can adversely affect diamond strength during high-temperature segment sintering (by promoting internal graphitisation), but even at room temperatures the inclusions can act as 'weak points'.

As these inclusions are ferromagnetic, their levels can be quantified in terms of 'magnetic susceptibility'. By measuring the magnetic dipole moment of the bulk diamond sample **M** in response to a magnetic field **H**, the (unitless) magnetic susceptibility χ_v can be obtained. Fig. 13 (right) shows the magnetic susceptibility values of the four saw grit grades in size 30/35.

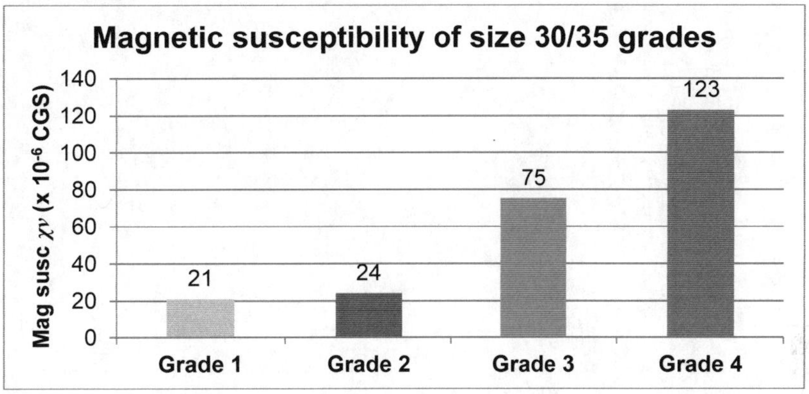

Fig. 13. Metallic inclusions in diamond (left, in red) and magnetic susceptibility results (right).

These results show that Grades 1 and 2 have similar inclusion levels, whilst Grade 4 (with the highest magnetic susceptibility) has a much higher inclusion content than Grade 3. This contributes to the darker appearance of many of the Grade 4 particles in the images shown in Fig. 9.

8. DISCUSSION AND CONCLUSIONS FROM THE STUDY OF FOUR DIAMOND GRADES

Comparison of the shape, strength and inclusion measurements from these four grades enables the influences on diamond strength to be understood. These grades were ranked by the diamond manufacturer in the order 1, 2, 3 and 4 (strongest to weakest), and the results of both friability (toughness) and compressive strength confirm this. Whilst image analyses suggested that the particle shapes of Grade 4 were slightly better than those of Grade 3, magnetic susceptibility measurements showed the metal content of Grade 4 to be much higher. Consequently, Grade 4 delivered weaker strength values than Grade 3.

The detailed results presented above focused on the mesh size 30/35. Particle shape, magnetic susceptibility and friability (toughness) results have also been obtained from the same grades in sizes 35/40 and 40/45, and compiling all these results leads to the contour plot in Fig. 14.

In Fig. 14 the vertical axis shows magnetic susceptibility and the horizontal axis the median compactness. The colours in the contour plot represent the toughness index (percentage residue), from less than 72 (dark blue) to greater than 84 (dark green). For appropriate comparison, all grades and sizes were subjected to friability testing at 2000 cycles.

It can be seen that the highest toughness comes from products with the lowest magnetic susceptibility (inclusion content) and the lowest median compactness (most circular particle shape). Furthermore, by following the border between light blue and light green (a toughness index of 78), it can be seen that the same toughness values are obtained by products with low median compactness but high magnetic susceptibility and products with low magnetic susceptibility but high median compactness.

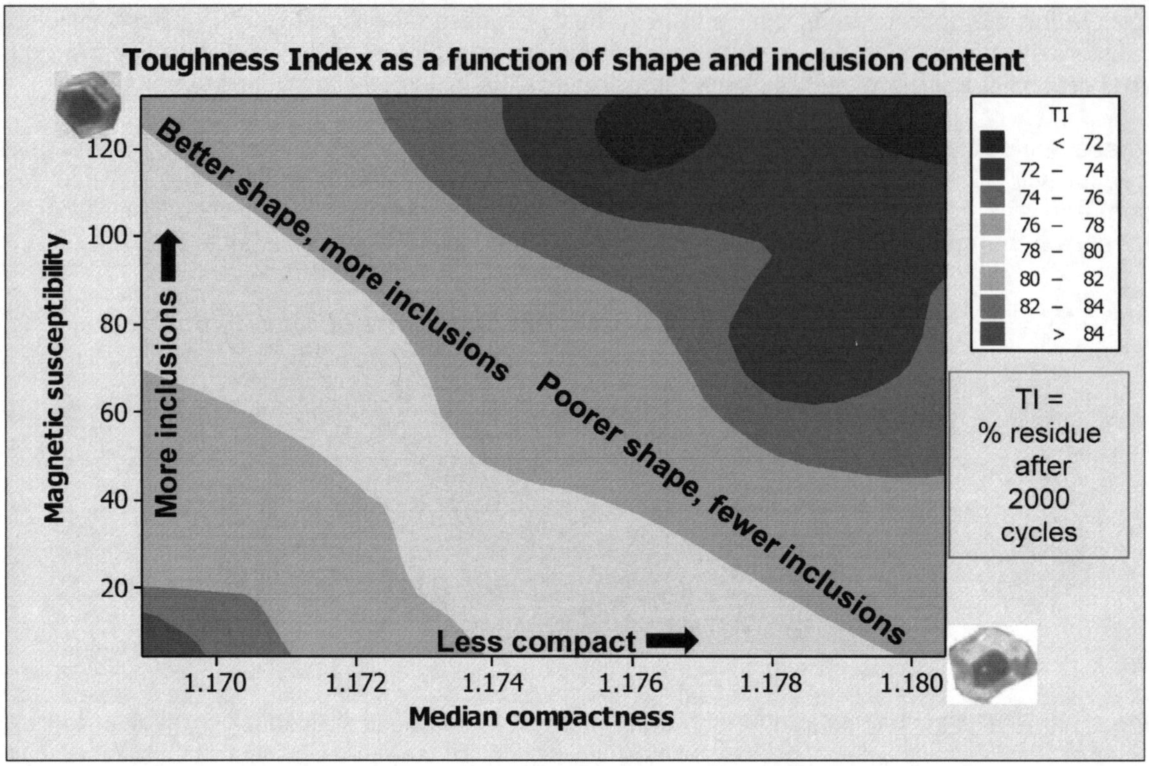

Fig. 14. Toughness index (friability residue) as a function of particle shape and inclusion content.

From this, it can be concluded that the product strength of 'saw grit' for construction applications is primarily influenced by a combination of particle shape and inclusion content.

REFERENCES

[1] Wright, D.N., 1986, 'Prediction of diamond wear in the sawing of stone', Industrial Diamond Review, 5/86, pp. 213-216.

[2] 'Abrasive products - Checking the grain size of superabrasives', ISO 6106:2013.

[3] Belling, N.G. and Dyer, H.B., 1964, 'Impact strength determination of diamond abrasive grit', Industrial Diamond Information Bureau booklet, London.

[4] Rosin, P. and Rammler, E., 1933, 'The laws governing the fineness of powdered coal', Journal of the Institute of Fuel, pp. 29-36.

[5] Vollstädt, H. and List, E., 2003, 'Controlling the stability of the properties of superabrasive powders', 4th Zhengzhou International Superhard Materials and Related Products Conference.

[6] Gallagher, J., Scanlon, P. and Nailer, S.G., 2006, 'Characterisation techniques for the study of high-strength, coarse diamond', Industrial Diamond Review, 3/06, pp. 58-65.

ACKNOWLEDGEMENTS

The authors express their gratitude to numerous Hilti colleagues. Messrs Marcel Sonderegger, Peter Jost, Roman Schneeberger, John Russell, Klaus-Peter Bohn and Dr Quentin Furnemont are thanked for their information. Messrs Stefan Fabbro, Markus Stengele and Stefan Goldener are thanked for their contribution to drilling tests, diamond characterisation and optical microscopy. Dr Matthias Lodes and Ms Nerea Gonzalez-Polvorosa are thanked for their support and advice.

The authors are also grateful to Vollstädt Diamant GmbH for the provision of equipment and further measurements.

Determining the Flowrate, Pressure and Nozzle Aperture for a Grinding Process with Consideration of Abrasive Type, Coolant Type and Wheel Structure

Dr. John A. Webster, Andreas Planakis, Chris Winkel
Worldwide Superabrasives, USA

Abstract

Grinding is a thermally dominated process where many of the operational parameters such as feeds, speeds, wheel structure, dressing and abrasive type must consider the heat generated and how to remove it from the workpiece. Optimization of coolant application is a powerful tool to suppress thermal damage and allow greater control of the process setup parameters to create the required surface topography, integrity, productivity and size.

It is well known that matching the jet speed with the grinding wheel speed is an effective way to remove the boundary layer of air, and to wet the grinding process. Mathematical models based on energy equations, such as Bernoulli's, have previously been used to predict jet velocity, however, these are generally flawed as they don't consider the kinetic energy of the coolant entering the nozzle. They are also flawed because no consideration is made of the viscosity of the coolant, which is a dramatic effect. This paper presents a much more precise analysis of the flow from a nozzle and also considers nozzle efficiency. The paper also presents the flowrate requirements based on grinding power, abrasive type (superabrasives versus conventional abrasive) and the influence of wheel structure. The models are verified by accurate flow tests performed by the author.

Finally, a case study is presented that clearly shows the economic and technological advantages of coherent-jet nozzles over a more conventional approach.

1. Introduction

Grinding is a complex chip removal process, which if done incorrectly can lead to surface damage to the work material, and unsatisfactory process economics due to inadequate removal rates and/or excessive wheel wear. The power consumed by the process is partitioned into the wheel, work, chip and coolant, and the amount that enters the workpiece must be cooled quickly to prevent high local temperatures and phase transformations from developing, and high remaining temperatures after the wheel has passed by. Phase transformations are often responsible for tensile residual stresses, white layer formation, reduced fatigue life, and surface and sub-surface cracking. Cooling of the process is achieved by the application of a cooling and lubricating fluid, as well as selecting process parameters that reduce the heat being generated.

In creep feed grinding, where the contact arc is long, the thermal input into the part is high, and the chips produced and abrasive lost from the wheel also need to be flushed away. This often results in the application of excessive flowrates in an attempt to maintain the quality and economics of the process. Such a grinding fluid application system requires a powerful pump, large settling and cooling reservoir, high worktable drain rate and an effective mist collector. The excessive heat produced by such a pump often requires refrigeration of the coolant to maintain a constant temperature throughout a production shift.

Many researchers have studied the role of grinding fluids in preventing thermal damage to the workpiece, including Brinksmeier [1]. Andrew [2] defined the critical power flux at 'burnout' to be approximately 30 W/mm^2, in the contact area between an aluminum oxide grinding wheel and the workpiece during continuous dressing creep feed grinding (CDCF). Beyond this value the workpiece surface would experience a rapid boiling of the fluid in the grinding arc, resulting in literally a dry process with its associated high temperature. Guo [3] quoted a similar power flux value at similar grinding conditions and work material. Guo [4] also explained that when the workpiece surface temperature is below the boiling point of the fluid then a liquid phase will be present in the grinding arc. As the temperature of the fluid increases, the liquid will undergo nucleate boiling, which can enhance convective cooling. As the fluid temperature increases still further a vapour film will form which blankets the heated workpiece surface suppressing heat transfer. This film boiling threshold temperature is around 130°C (266°C) for a water-based fluid. As compared to the CDCF grinding critical power flux stated above, Rowe [5] obtained values over 200 W/mm^2 , for HEDG grinding with CBN grinding wheels, in oil coolant, at 10 times the material removal rate of Andrew [2], which can be partly explained by Rowe's smaller contact length, much higher work speed and higher wheel-speed.

Yasui [6] showed that the fluid type has a large influence on burnout power flux, as the boiling point of water and oil is vastly different. Colding, et al. [7] stated that oil grinding fluids give reduced wheel wear as compared to water-based grinding fluids, more due to reduced frictional effects than chip formation effects. Vansevenant [8] showed that residual stresses improved with increasing flowrate, although he only used specific flowrates of up to 0.8 l/min·mm, much less than CDCF would use. Czenkusch [9] investigated the effect of flowrate and nozzle aperture on residual stresses and found lower tensile stresses with the smallest aperture. He found that 4 l/min·mm (25 GPM/inch) was adequate for grinding 100Cr6 with vitrified CBN wheels.

An output effect of grinding fluid application is the hydrodynamic pressure that develops just ahead of the contact arc, which tries to separate the wheel and workpiece. Klocke [10] showed that this pressure is proportional to wheel-speed and fluid viscosity. Cui [11] showed that the pressure is also related to wheel porosity. Brinksmeier [12] showed that the perpendicular force results in higher spindle power due to the drag that occurs when pumping the fluid through the grinding arc, especially so in the case of shoe or low-velocity jet nozzle applications. Malkin [13] stated that this force can be as high as the grinding forces.

Much prior work has been published on jet nozzles, shoe nozzles and chambers, with regards their overall grinding performance. Shoe nozzles have been shown to be extremely effective [11,10,14,15], but in practice are difficult to keep sealed against the wheel surface as the diameter reduces with wear (especially true with CDCF grinding).

This is one of the reasons why jet nozzles are so popular for research [16, 17], and on the shop floor. These nozzles are often designed to produce a coherent jet that allows them to be placed at a convenient distance from the grinding arc. In many industrial applications crushed-pipe or wedge shape nozzles are used, which have to be placed in very close proximity to the grinding arc to offset their dispersed flow, but just be able to clear workholding fixturing. Other advantages of coherent jets include reduced entrainment of air, more accurate velocity matching and more accurate focus into the grinding arc. Ott [18] and Bruecher [16] both recommended that the jet should be aimed just prior to the grinding arc, Mindek [19] found a horizontal aim into the grinding arc to be best with CDCF grinding and Campbell [20] recommends 5 degrees below horizontal for electroplated CBN surface grinding.

Using the process outputs of hydrodynamic pressure, critical power flux and fluid film thickness described above, several experimental tools have been developed to evaluate the effect of nozzle application parameters on the transport of fluid through the grinding arc. Inasaki [21] measured the electrical conductivity of the fluid film in the grinding arc in order to measure its thickness, Cui [11], Campbell [22] and Heinzel [14] measured the pressure in the grinding arc (since the pressure determines the fluid boiling point, and the degree of penetration of fluid into the wheel structure), and both Powell [15] and Mindek [19] measured the power flux of a heating element in the grinding arc. Apart from the more practical 'shop floor' approach taken by Campbell, these tools have not been exploited by manufacturing industry. There appears to be a need for a simple, but effective, grinding fluid flowrate model for jet nozzles that will ensure burn-free grinding in new and existing processes.

2. Existing Flowrate Models

After reviewing the abundant literature that pertains to the effect of fluid flowrate on surface quality, there is little information on estimating the flowrate based on the process conditions. For CDCF grinding in the US, a general 'rule of thumb' is to use 4 l/min·mm (25 GPM/inch), agreeing with Czenkusch [9], regardless of the material removal rate, Q'_w. Such a model is easy to apply for similar values of Q'_w , and enables the machine tool manufacturer to scale the required pump flowrate based on the maximum wheel width that can be used on the machine. However, for low values of Q'_w there will be some wasted flow and at much higher values of Q'_w there may be a tendency towards burn. Silliman [23] presented a simple flowrate model based on the spindle power consumed during grinding, as 8-10 l/min·kW (1.5-2 GPM per HP). Unlike the previous model, this model estimates the flowrate according to the process energy that is being created. A major limitation of the model is that it is independent of wheel-speed (found to be related to fluid effectiveness), fluid type, and abrasive type (CBN and aluminium oxide have different energy partition values).

Metzger [24] presented a minimum flowrate (Q_F) model, based on Ott [18], which relates to spindle power (P), but with the addition of nozzle efficiency (η), specific heat capacity of the fluid (C), the density of the fluid (ρ) and the tolerated temperature rise of the fluid ($\Delta\theta$), where:

$Q_F = P/C \cdot \rho \cdot \eta \cdot \Delta\theta$ (1)

A typical fluid temperature rise of 4°C and nozzle efficiency of 0.9 can be used.

Cui [11] developed a model that considered the influence of wheel speed when designing a fluid application for a wheel-speed (v_s) of 250 m/s, where:

$Q_F = 16.2\,(P/\Delta\theta)^{1.94}/(1.5^{1.44\ln(vs/32.2)})$ (2)

Since Cui was only considering water-based grinding fluids he does not have the fluid properties separated in his model. Using such a model Cui predicted that he required 276 l/min for a 250 m/s wheel consuming a maximum grinding power of 40 kW, with an average temperature rise of 5°C as the fluid passes through and around the grinding arc. In comparison, the Silliman model would predict 320 - 400 l/min.

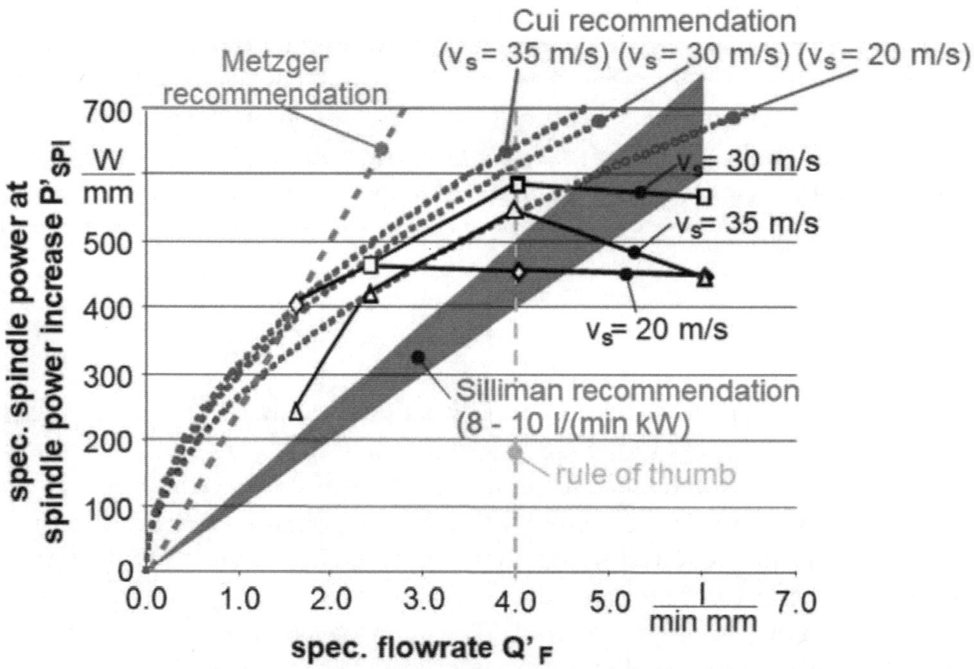

Figure 1. Comparison of flowrate predicted by Silliman, Cui and Metzger.

Figure 1 shows the relationship between specific power increase, at the moment of power surge when burnout occurs, to the specific flowrate applied by the nozzle. The result is taken from creep feed grinding research by Webster et al [25] investigating the burn limit when grinding a wedge-shaped steel workpiece, where during each grinding pass the depth of cut increases until workpiece burn occurs. Silliman's model [23] has been added as a comparison. For the conditions chosen, and the wheel specification used, there appears to be a flattening off in the graph at 4 l/min·mm, as if the maximum flowrate that the wheel can support has been reached. It can be seen that the shaded zone representing Silliman's model falls between the range of specific spindle power values measured at the onset of burn for the different specific flowrates. The specific flowrate of 4 l/min·mm is just less than the lowest flowrate predicted by his model, showing the conservative nature of his approach and is identical to the 'rule of thumb' flowrate model of 4 l/min·mm, mentioned earlier.

Using Metzger's [24] model, we would obtain a required flowrate coefficient of 4 l/min·kW. Using Cui's [11] model (equation 2), we would obtain a specific flowrate requirement of 5.5 l/min·mm at 20 m/s and 6.7 l/min·mm at 30 m/s, for the maximum critical specific power value recorded at each speed. Both results, for a 4°C temperature rise between the coolant entering the grinding machine and leaving, are shown on Figure 1. Using Silliman's model of 8 to 10 l/min·kW , the flowrate to the process is very conservative and the least likely model to produce a thermally damaged part.

3. The Cooling and Lubricating Needs of a Grinding Process

The applied flowrate discussion shown earlier cites several key researchers, all using different test platforms and different evaluation methods. For example, applying different flowrates through the same nozzle aperture by adjusting the pressure changes both the flowrate and the velocity (a function of pressure) of the jet. It is not clear if the improved cooling ability that is observed in their tests is due to the increased flowrate, or the increased velocity. In the previous work done by the author [25], he used several different nozzle apertures, at the same width, to vary the flowrate at the same velocity, and vice versa..

The applied flowrate is often much greater than the effective flowrate. The effective flowrate is dependent on wheel properties, jet quality, jet position, wheel speed, and the geometry of the grinding profile. If the grinding geometry reduces the side leakage, i.e. grinding a groove (the wheel width and grinding width are the same) versus a straddle (the wheel width is greater than the work width), the increased hydrodynamic pressure will result in greater penetration of the fluid into the structure of the wheel [25].

Spur [26] stated that for silicon nitride grinding with resin-bonded diamond wheels, only 0.4% of the 4 liters/min.mm applied was effective. Using side-shields to separate the flow going through the grinding zone from the coolant going around the grinding zone, and dividing by the applied flowrate, Krishnan measured effective flowrate of 26-55% for creepfeed grinding with high-porosity aluminium oxide wheels. Martin [27] quoted an effective flowrate of 3% in high efficiency deep grinding (HEDG) with electroplated CBN grinding of steel. Campbell achieved 9.6% with similar wheels, grinding nickel-based alloys.

As can be derived from the above discussion, conventional abrasive grinding wheels rely on highly effective flowrates in order to cool the grinding zone. The effective flowrates of superabrasive wheels are much lower than with conventional wheels, due to their much lower porosity, especially electroplated, metal-bonded, and resin-bonded wheels. Even vitrified CBN wheels are much lower porosity than vitrified bond conventional abrasive wheels. One of the reasons why it is possible to grind at much higher specific removal rates with superabrasive wheels, despite the reduced effective flowrate, is due to the much higher thermal conductivity of the superabrasive grains. Whereas aluminum oxide and silicon carbide abrasives are very good insulators, and do not rely on heat transfer into the wheel as an effective cooling mechanism, superabrasives suck a greater percentage of the energy out from the process, to be subsequently cooled by the grinding fluid. The partitioning of energy in the grinding process can be defined as in Figure 2, where:

$$q_t = e_t \cdot a_e \cdot v_w / l_c, \quad \text{or } q_t = e_t \cdot Q'_w / l_c \qquad (3)$$

where contact length (l_c) = $[a_e \cdot D]^{1/2}$, D is the wheel diameter, a_e is depth of cut, e_t is the specific grinding energy, and v_w is the workspeed.

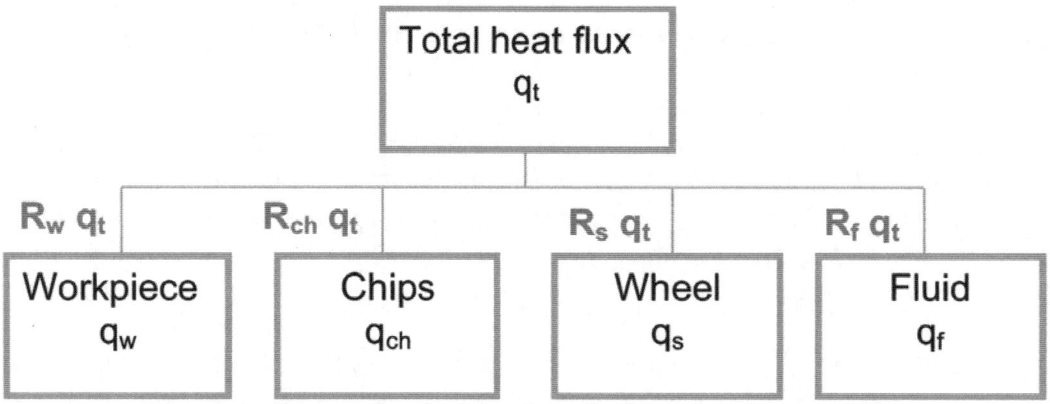

Figure 2. Energy partitioning in grinding. Stephenson [28]

The specific grinding energy, is a measure of the amount of energy required to remove one cubic millimeter of material. In a low specific removal rate process the frictional components may dominate over the chip formation components, giving a high specific grinding energy. In a high specific removal rate process, where larger chips are produced by each abrasive grain, the specific grinding energy is much lower since the frictional components become less significant. Stephenson [28] presented coefficients for the specific grinding energy, from tool steel grinding tests. The coefficients for specific grinding energy depend on the workpiece material, abrasive material, coolant type and wheelspeed. Stephenson [28] derived the coefficients for grinding different materials and an electroplated CBN wheel, in mineral oil coolant, at wheel speeds from 30 – 150 m/s. Using CBN abrasive, the specific grinding energy values tend to be much lower than with conventional abrasives. This reduced specific grinding energy means that the cooling role of the fluid is less demanding, and the potential for reduced flowrates exists with superabrasives. The much reduced pumping ability of such wheels also implies this as they are highly efficient in stock removal.

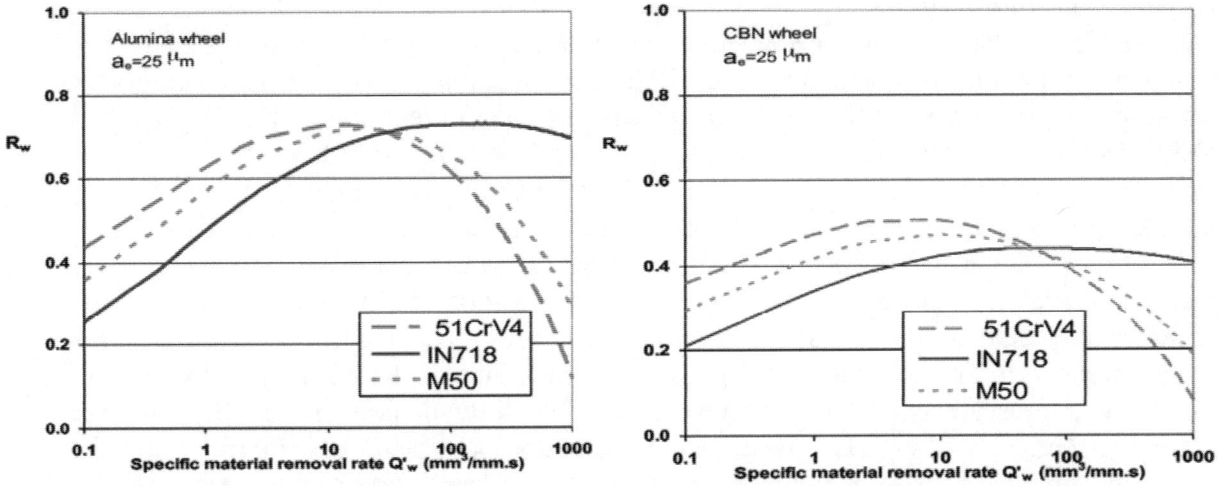

- R_w increases rapidly with Q'_w before reducing to lower values at very high material removal rates
- The largest reduction in R_w occurs for materials that exhibit high thermal diffusivity

Figure 3. Energy partitioned into workpiece for alumina vs CBN.Stephenson [28]

In addition to the reduced specific grinding energy associated with superabrasives, the energy partitioning into the workpiece is also in its' favor, especially at the very high specific material removal rates associated with HEDG. Stephenson [28] has presented a comparison between aluminum oxide and CBN grinding, for the energy partitioned into three different workpiece materials, as shown in Figure 3. The figure also shows that the cooling demands for the grinding fluid in superabrasive grinding is therefore much less than with conventional abrasives. The flowrate model of 8 to 10 liters/min.mm was established by the author for creep feed grinding with conventional abrasives [23,25], this is an indication that the flowrate can be reduced in HEDG (Figure 4).

Grinding Conditions	Nickel Superalloy (e.g. IN718 or MarM002) (low α)		Steel (e.g. 51CrV4 or M50) (high α)	
	Al_2O_3	CBN	Al_2O_3	CBN
Shallow cut (0.025mm) Q'_w 0.1 - 10mm^3/mm s	Coolant 0.2 - 0.7	Coolant 0.1 - 0.6	Coolant 0.1 - 0.5	Coolant 0.1 - 0.5
Creep (5mm) Q'_w 0.1 - 10mm^3/mm s	Coolant 0.85 - 0.95	Coolant 0.8 - 0.9	Coolant 0.7 - 0.9	Coolant 0.7 – 0.9
HEDG (5mm) Q'_w 100 - 1000mm^3/mm s	Coolant 0.3 - 0.7	Coolant 0.2 – 0.5	Coolant 0.05 - 0.4	Coolant 0.05 - 0.4

Figure 4. Energy partitioned into the fluid for different processes. Stephenson[28]

4. Coolant Nozzle Design

Coherent-jet coolant nozzles use an internal geometry that gives low dispersion, and therefore minimum entrained air within the jet. An example of coherent jet nozzle geometry is the Rouse profile [29]. In addition to an efficient internal geometry, this design requires a contraction ratio (inlet to exit diameter ratio) of at least 2:1, in order to produce a coherent jet. This means that a 12.7 mm diameter inlet pipe to the nozzle will not give a relatively coherent jet if the exit aperture is more than 6.35mm diameter. A definition of coherency has been given previously [29] as being a dispersion of 2-3 times the exit diameter, at a distance of 300 mm from the nozzle. In other words, a 6mm diameter jet would open out to 12-18mm, at a distance of 300mm from the nozzle end. Figure 5 shows a 6mm diameter coherent jet, from a nozzle with an inlet diameter of 20mm, at a pressure of 45 bar, which shows very little dispersion. In this case the contraction ratio is 20/6 = 3.33. The coolant used in this example is mineral oil, where the Reynolds number of the jet is much lower than that with water at the same pressure. For this reason, nozzles designed to be coherent in water have even better jet quality with oil.

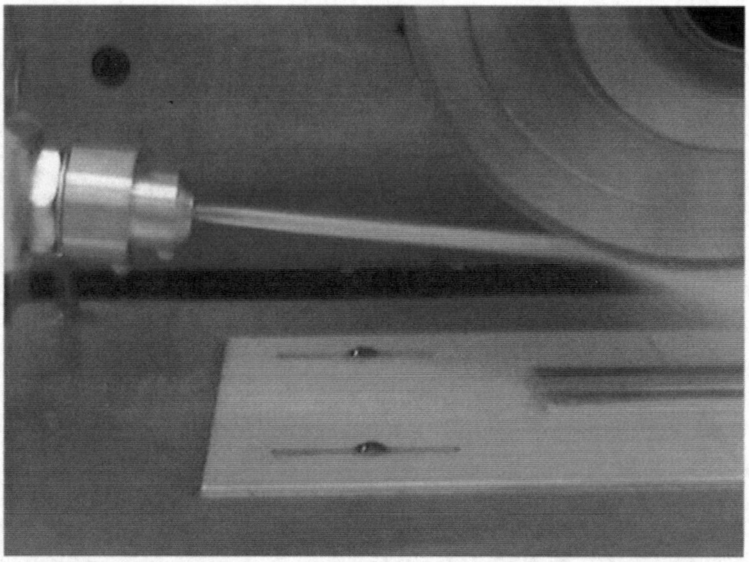

Figure 5. Coherent jet nozzle based on Rouse design [29]

Figure 6 shows an identical nozzle flowing oil and water-based coolant at the same pressure of 7bar (100 psi) and with the same jet diameter of 4mm (0.16”). The viscosity of the oil in this right side example dramatically reduces the Reynolds Number (Re) of the oil jet (Re $\approx$ 7,000) and improves the quality of the jet, as compared to the left nozzle with water as a fluid (Re$\approx$150,000).

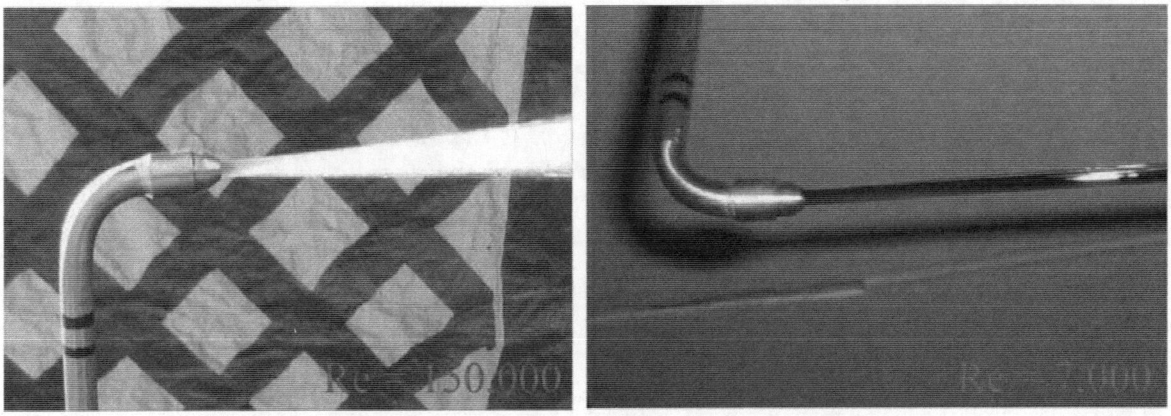

Figure 6. Influence of coolant viscosity on jet quality

Reynolds Number is defined as:

$$Re = \frac{vD}{\gamma} \qquad (4)$$

Where

D = Inside diameter (ft or m)
v = Velocity (ft/sec or m/s)
γ = Kinematic viscosity (ft^2/sec or m^2/s)
Re = Reynolds Number (dimensionless)

For the coherent-jet velocity to match wheelspeed, Bernoulli's equation is useful. Since it is based on conservation of energy, Bernoulli equates the potential and kinetic energy of the fluid entering the nozzle with the kinetic and potential energy of the jet produced. The kinetic energy of the fluid inside the nozzle can be ignored if the cross-sectional area of the pipe feeding the nozzle is much greater than the exit area of the nozzle, and therefore the velocity-squared term in the pipe is not significant as compared to the velocity squared term of the discharged jet. In the example in Figure 5, with a contraction ratio of 3.33 mentioned previously, the jet velocity is 3.33 x 3.33 = 11 times (based on area ratio) the velocity of the coolant at the inlet of the nozzle, equating to less than 1% (0.09^2) of kinetic energy (when squared) in the nozzle as compared to the 99% potential energy. The specific gravity (SG) of the fluid is also important for establishing the pressure versus velocity relationship.

The simplified version of Bernoulli's equation is:

$$P\ (\text{bar}) = \frac{SG \cdot v_j^{\,2}}{200}\text{, where } v_j \text{ is in m/s} \qquad (5)$$

Figure 7 is a simplified chart for designing the nozzle aperture based on the pressure and flowrate requirements. For example, a jet speed of 80 m/s requires a pressure of 28 bar with mineral oil coolant. At 28 bar a nozzle exit diameter of 4mm, will give a flowrate of

57 liters/minute, sufficient to quench over 6KW of grinding power. This is based on a conservative flowrate model of 8-10 liters/min.mm. In the case of steel drill-flute grinding, often at wheelspeeds between 60-80 m/s, a 4mm diameter jet will produce sufficient flowrate to cool the process, especially if the flute width is also close to 4mm. If a much lower wheelspeed is used, i.e. 30 m/s, and the grinding power is maintained at a similar level, a nozzle aperture of more than 6 mm (1/4") will be required, giving a jet much wider than the grinding wheel, and hence some wasted flow.

jet speed (m/s)	coolant nozzle pressure (bar)			flowrate (l/min) for listed nozzle diameters (mm) and areas (mm^2)								
	water	miner. oil	ester oil	0.8	3.1	7.1	13	28	50	79	113	area
	SG=1.0	SG=0.87	SG=0.93	1	2	3	4	6	8	10	12	diam
20	2	2	2	0.9	3.5	8.1	15	33	57	90	129	
30	5	4	4	1.4	5.3	12	22	49	86	134	193	
40	8	7	7	1.8	7.1	16	29	64	115	179	258	
50	13	11	12	2.2	9.0	20	36	80	144	224	322	
60	18	16	17	2.6	11	24	43	97	172	268	386	
80	32	28	30	3.6	14	32	57	129	229	358	516	
100	50	44	47	4.4	18	40	72	162	287	448	645	
120	72	63	67	5.3	21	49	86	193	344	537	774	
140	98	85	91	6.2	25	56	100	226	401	627	903	
160	128	111	119	7.1	28	64	115	259	458	716	1031	
180	162	141	151	8.0	33	73	129	290	516	805	1160	
200	200	174	186	8.9	35	81	144	323	573	895	1289	

Metric Flowrate Chart for a Nozzle with a Coefficient of Discharge of 0.95

Figure 7. Pressure vs velocity relationship, based on simplified Bernoulli's equation

The chart in Figure 7 can also be used to design a rectangular coherent-jet nozzle. The area of the nozzle exit area is shown as well as the diameter of a round jet. For a desired width of nozzle, the corresponding aperture that gives the required area, can be determined. The chart only applies to nozzles with a high coefficient of discharge, unlike the crushed and open pipes, and plastic nozzles, that dominate the market. The nozzle exit area, regardless of being round, or rectangular, is calculated as follows:

$$A_j\ (mm^2) = \frac{100}{6} \cdot \frac{Q_j\ (L/min)}{V_j\ (m/s)} \qquad (6)$$

Verifying the flowrate achieved by the nozzle(s) is difficult to do. Flowmeters can be expensive and are typically calibrated for water, not oil. A bucket test, when a container is quickly placed at the bed discharge of the grinding machine and the time measured to fill the bucket is recorded, is very difficult to do. Also, after quickly removing the bucket any foam in the coolant needs to be settled before the true volume collected can be determined during the measured time. The flow from the nozzle and bed discharge rate both have to be steady state before the bucket is inserted into the stream. After the bucket is filled and removed, the coolant supply can be turned off. Attempts to connect a bucket directly up to the coolant nozzle being investigated must consider air in the nozzle being purged as the pump is switched on and the time lag in maximum pressure being developed in the nozzle. Also, many grinding machines do not allow the front door to be open with the coolant running.

The simple equation below is used to estimate the jet velocity (V_j) from flowrate (Q_j) and nozzle exit area (A_j) :

$$V_j \text{ (m/s)} = \frac{100}{6} \cdot \frac{Q_j \text{ (L/min) } Q_j}{A_j \text{ (mm}^2\text{)}} \qquad (7)$$

The equation is used in the chart in Figure 7 and assumes that the velocity profile of the jet is constant over the entire discharge area, which is an acceptable assumption for low viscosity water-based products but very erroneous for higher viscosity coolants such as oils, which exhibit a more parabolic velocity profile.

5. Advanced Coolant Calculus

Section 4 introduced a simplified nozzle design chart based on a simplified Bernoulli's Equation assuming no kinetic energy exists prior to the nozzle, however, in order to design an ideal coolant nozzle that supplies the correct jet speed and flowrate, and considers the fluid type, process, and part profile, one must consider all of the following parameters:

- Flowrate required to cool & lubricate process (L/min)
- Fluid type: mineral oil; synthetic oil; emulsion; etc
- Fluid viscosity and specific gravity (mm^2/s
- Wheel peripheral speed (m/s)
- Coolant velocity in the pipe just before the nozzle (m/s)
- Nozzle exit area (mm^2)
- Grinding width (mm) and dressing/truing coolant needs
- Nozzle efficiency (C_d) and internal geometry

Some common challenges facing the process engineer is how the applied flowrate can be reduced, without reducing the effective flowrate supplied to the process, minimizing the electrical energy and filter size, or maximizing the number of machines connected to a central filter unit. Ensuring that the operator maintains the correct position of the jet into the process may require aiming tools, such as lasers, or changes to nozzle attachment location to reduce the influence of wheel wear on the aim of the jet to the contact zone.

Flowrate models based on grinding width and process power have been presented earlier, and the desirability to match the jet speed with the wheel speed, and matching jet width with the grinding width, have been discussed. Pressure drop between the coolant inlet at the rear of the machine and the entry to the nozzle can be huge. In the worst case a 98% pressure drop has been measured in the field. In a typical grinding machine a 20-30% drop is likely. For this reason the coolant pressure used in Bernoulli's equation must be the value just before the nozzle, before the coolant accelerates inside the contraction geometry of the nozzle. Rarely are coolant manifolds on machines tapped with a thread to fit an external pressure gauge to verify pressure drop.

Pressure drop occurs at every elbow, hose barb, and restriction (such as a valve or transition). Turbulence and cavitation also exists if the velocity of the coolant inside the plumbing exceeds the well accepted 6 m/s used by designers. In the case of the coherent jet nozzle in Figure 5, the area ratio between the inlet to the nozzle and the exit of the nozzle was 11:1, which with an efficient nozzle design would lead to an 11 times increase in jet velocity as compared to the coolant velocity entering the nozzle. At the 45 bar pressure noted for Figure 5 this would represent a jet velocity of 100 m/s according to the simplified Bernoulli equation, therefore based on the area ratio, the velocity of the coolant entering the nozzle would be 100/11 = 9 m/s, above the 6 m/s recommendation. The flow into the nozzle is therefore becoming turbulent. However, since the coolant is oil the higher viscosity will lead to a much lower Reynolds Number than the equivalent scenario with water.

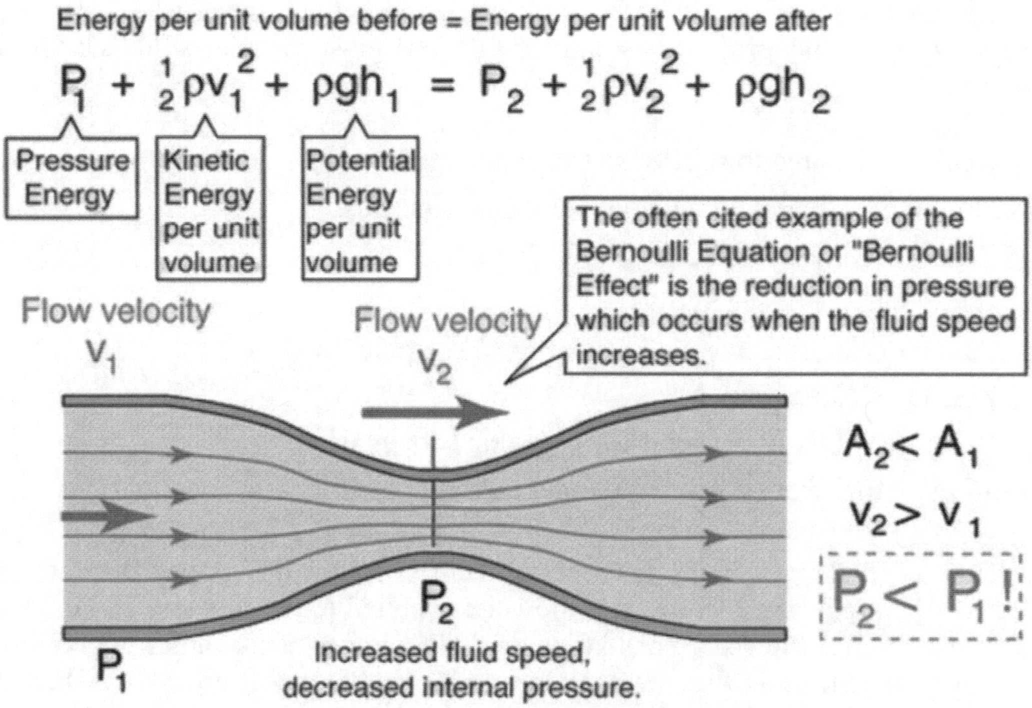

Figure 8. Bernoulli's energy equation (8)

In Equation 8 in Figure 8, consider the nozzle entry to be on the left side and the flow contracts to the center part. At this point the jet is formed and the coolant exits the nozzle, unlike the figure where the flow then expands again. The Pressure Energy P_1 is created by the pump, the kinetic energy in the nozzle is due to the coolant velocity entering the nozzle, and the potential energy is due to the height above the ground, or other reference point. At position P_2, when the jet exits the nozzle there is almost zero pressure energy, maximum kinetic energy, and typically the same potential energy as the P_1 position, cancelling that term out. In the case of the nozzle in Figure 5, the velocity v_1 entering the nozzle is 9 m/s, which is insignificant as compared to the 100 m/s jet velocity at exit, however this nozzle has a relatively high contraction ratio of 3.33. If the exit aperture was increased from 6mm to 10mm, in an effort to get more flowrate to the grinding process, the contraction ratio would now be 20/10 = 2 giving a velocity ratio of 0.25 and hence a kinetic energy at the nozzle entry of 0.25^2 = 6% of the kinetic energy at the nozzle exit, which cannot be ignored.

The design of the coolant nozzle internal geometry, and the length of the stabilizer after the geometry, both affect the discharge from the nozzle. An internal geometry with dramatic changes in cross-section can disrupt the boundary layer and create eddies and cavitation, leading to a jet that is effectively smaller than the exit area of the nozzle, which is called a Vena Contracta, shown in Figure 9 below. For example, the 10mm diameter exit of a round nozzle may exhibit a solid core of coolant of 8mm diameter, surrounded by 1mm of foam or entrained air. Any calculation of flowrate should be based on 8mm exit aperture not 10mm. This nozzle would have a discharge coefficient (C_d) of ≈0.8. The Rouse nozzle profile in Figure 10 has a typical C_d of 0.95. The traditional wedge nozzle, also in Figure 9, has a C_d of ≈0.7.

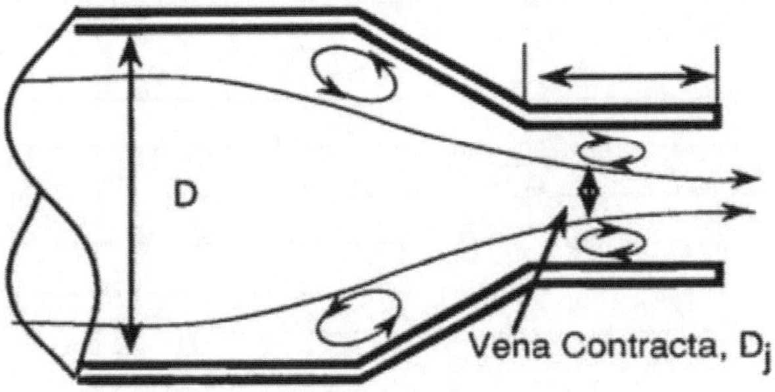

Traditional Nozzle (with eddy zones shown)

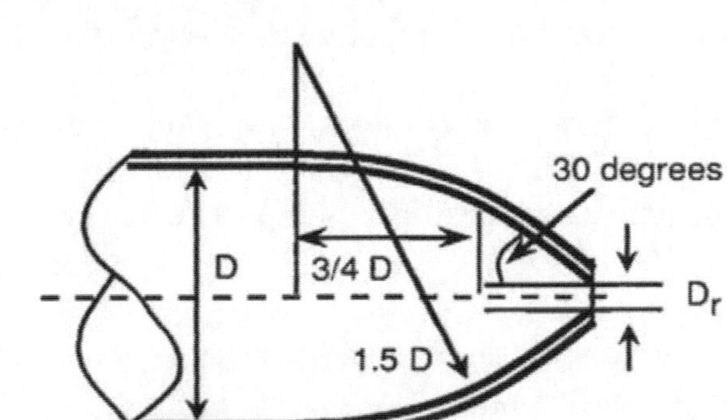

Improved Nozzle

Figure 9. Nozzle design, wedge nozzle versus Rouse nozzle

6. The Influence of Coolant Viscosity on Nozzle Performance

As previously mentioned, the coolant flow through the exit of a nozzle is not uniform, especially as the viscosity is increased. Coolant in contact with the inner surface of the nozzle is travelling at zero velocity since a boundary layer exists. The fluid further from the nozzle inner surface increases in velocity as it approaches the center of the nozzle. The velocity anywhere with the nozzle is defined by the velocity profile. When the kinematic viscosity of the coolant is increased by the end user to suit different process requirements the velocity profile becomes more parabolic, as shown in Figure 11. In most cases the center of the jet is travelling at close to the velocity that Bernoulli's equation estimates, however, with oil coolants the velocity slows down as it approaches the inside surface of the nozzle therefore the flowrate from the nozzle is based on the average velocity not the peak velocity. Whereas Bernoulli's equation predicts less pressure for oil coolants to attain the same velocity as water-based coolants, in practice the flowrate using oil coolant is lower than that of a water-based coolant using the same nozzle, even if the peak velocity on the center of the jet is the same in both cases.

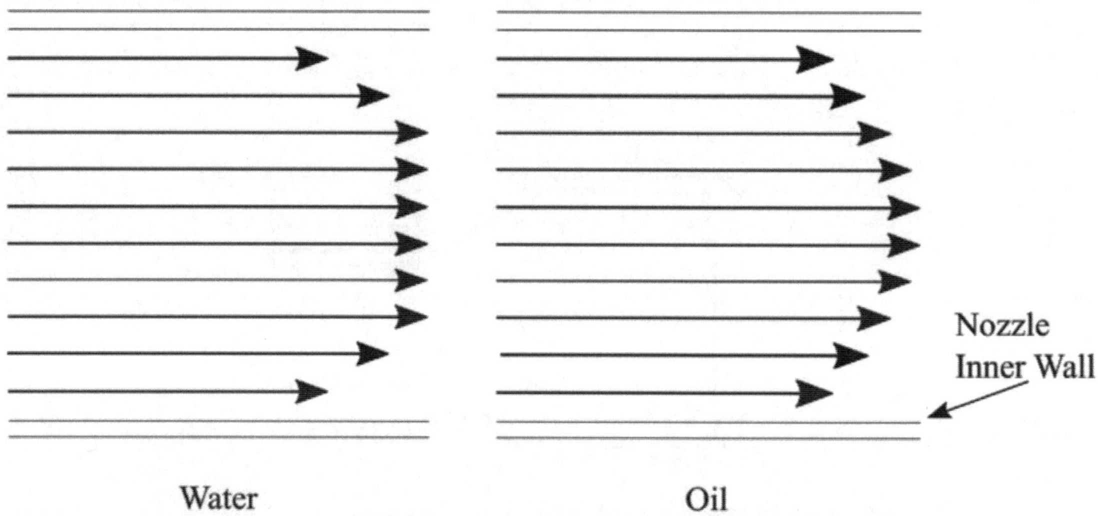

Figure 10 Shows the more parabolic velocity profile with oil vs water

A constant pressure nozzle flowtester (Figure 11) was designed which can test any nozzle up to 7 bar pressure with water-based and oil coolants, without the flow/pressure interaction that occurs with a mechanical pump. The tester was used to evaluate the following:

- Effect of aperture size on flowrate at different pressures
- Effect of stabilizer length on flowrate
- Influence of adjacent needles with needle nozzle systems
- Open-tube flowrate versus coherent-jet nozzle flowrate
- Influence of viscosity on flowrate
- Round versus rectangular nozzles
- Real data to enhance the velocity profile prediction in the coolant calculator

Figure 11. Constant pressure flowtester

The flowtester consists of an upper 30 liter cylinder which is filled with 25 liters of the coolant being tested. On the right side of the upper cylinder is a sight-gage to measure the height of the coolant inside the cylinder, and be able to estimate the volume flowed during the test. At the bottom of the upper tank is a large unrestrictive ball-valve that adjusts the flow to the much small test nozzle. The top tank is pressurized with compressed air, using a precision air regulator fed by a large compressor, and displayed by the large pressure gauge throughout each test. The test nozzle is then connected to the ball valve using a quick-disconnect, and the jet from the test nozzle aimed into the lower cylinder. Once the ball valve is opened the coolant flows into the lower cylinder and time measured from when the sight-glass level passes the top mark and then reaches the lower mark. The time taken to flow this 20 liters of coolant is recorded with a stopwatch. The time is correlated to the time shown in a video recording used as verification of time and pressure. The flowrate of each nozzle is determined for three different coolant pressures (2 bar, 4 bar, and 6 bar), and averaged over three tests at each pressure. The coolants tested were water, and three different viscosity oils. After each flow test the lower cylinder is plugged, the air is released from the upper cylinder, and the lower cylinder pressurized to return the fluid back to the top tank using a bypass loop and check valve.

The stainless steel test nozzle designs tested are shown in Figure 12, below. The large fitting is the male part of the quick disconnect system.

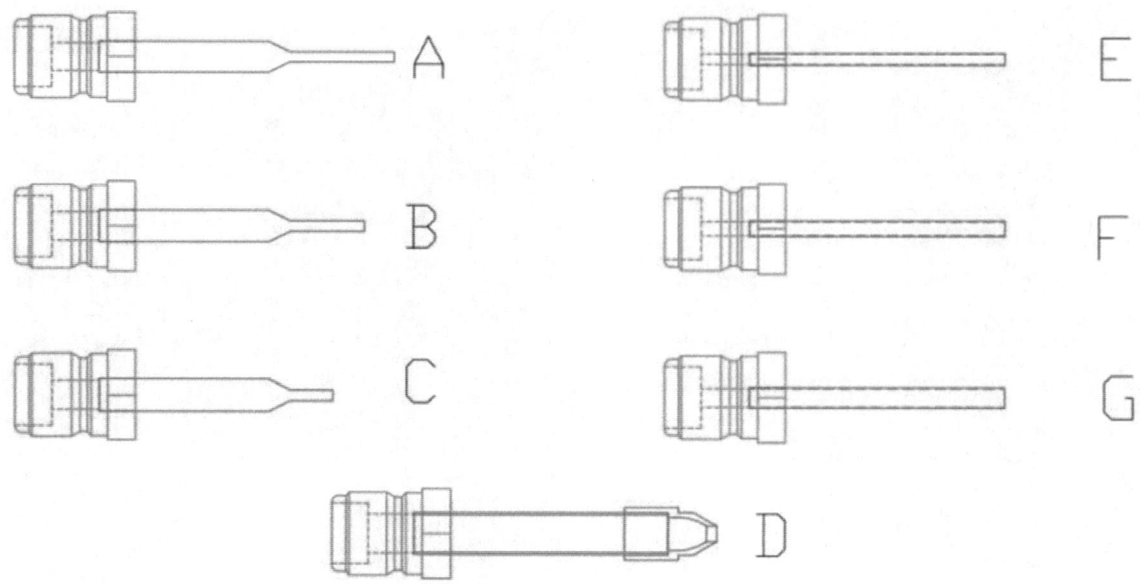

Figure 12. Seven Different Nozzle Designs Tested

Flat nozzles A, B, and C, from 75mm long and 7.8 mm ID tubing, 13 mm width of jet, three different stabilizer lengths (A=32mm, B=22mm, and C=13mm), and three different jet thickness (0.75mm, 1.25mm, and 1.8mm) for each stabilizer length. A total of 9 different nozzles

Round nozzles D, with Rouse coherent-jet inner profile, from 75mm long and 7.8mm ID tubing, 1mm long stabilizer before exit, round exit apertures of 3.18mm, 4mm, and 4.75mm. A total of 3 different nozzles

Straight tube nozzles E, F, and G, with burr free ends, from 75mm long tube at three different tube internal diameters (3.25mm, 4.22mm, and 5.54mm). A total of 3 different nozzles

In total 45 flow tests were done for each coolant type and viscosity, representing 180 tests.

7. Results

Flow rates from the constant nozzle pressure flow tests (2, 4, or 6 bar) were divided by the nozzle exit area to determine the normalized flow rate. No attempt was made to measure actual jet velocity using anemometry, although this is planned in the future.

Results from the flat nozzles showed a positive logarithmic relationship between nozzle aperture and flowrate (Figure 13.), a positive logarithmic relationship between pressure and flowrate (Figure 14.), and a negative logarithmic relationship between coolant viscosity and flowrate (Figure 15.) Tests on round nozzles also showed a negative logarithmic relationship between stabilizer length and flow rate (Figure 16). Results from the round nozzles have not been include here, but will be used for the calculator program.

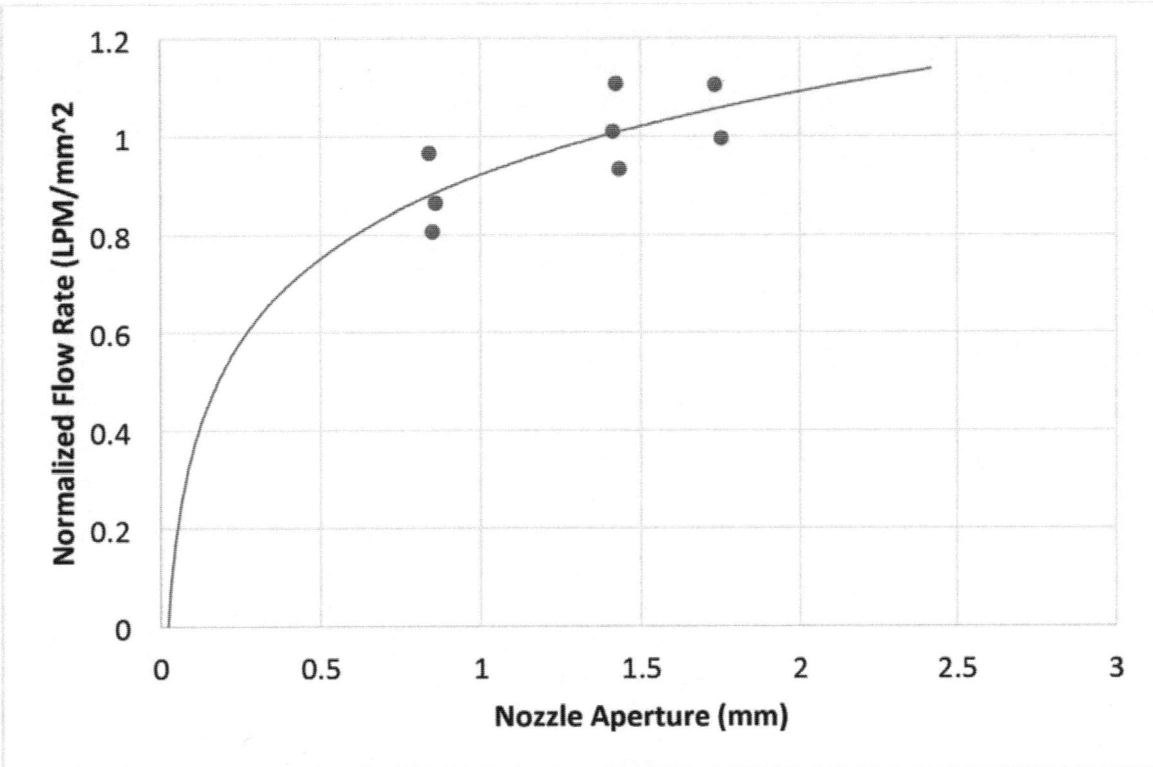

Figure 13. Effect of flat nozzle aperture on normalized flow rate

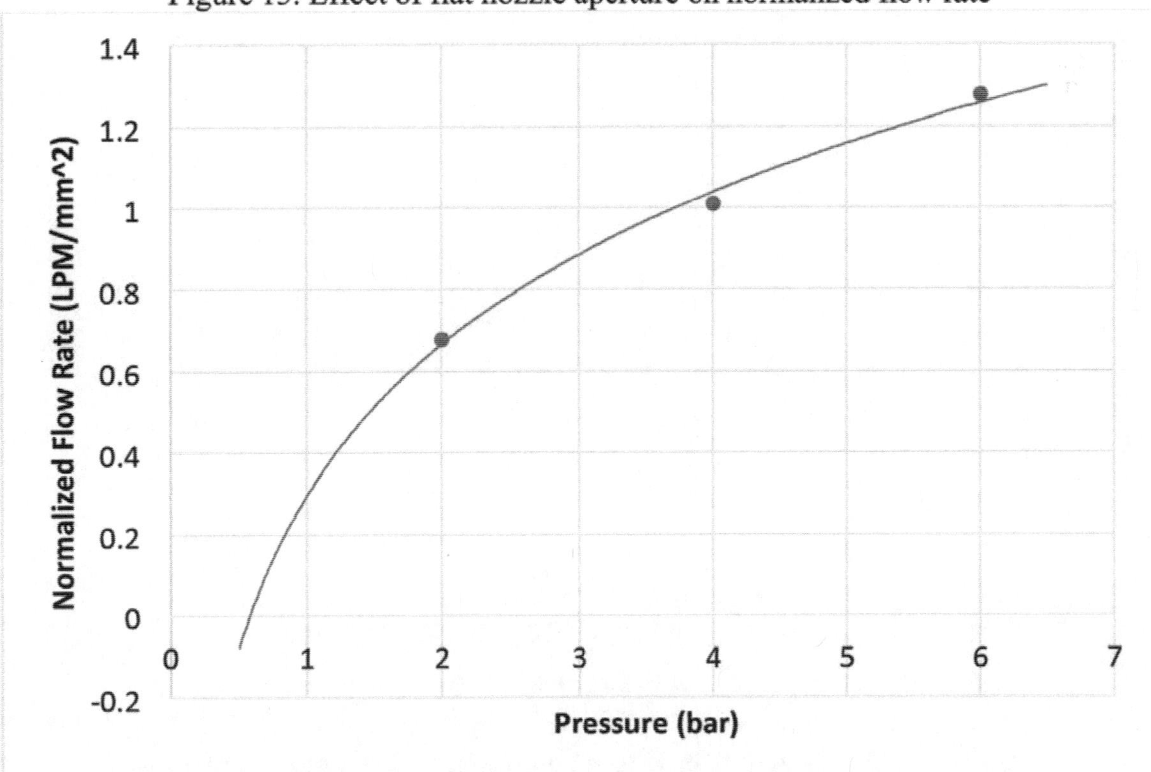

Figure 14. Effect of pressure on normalized flat nozzle flow rate

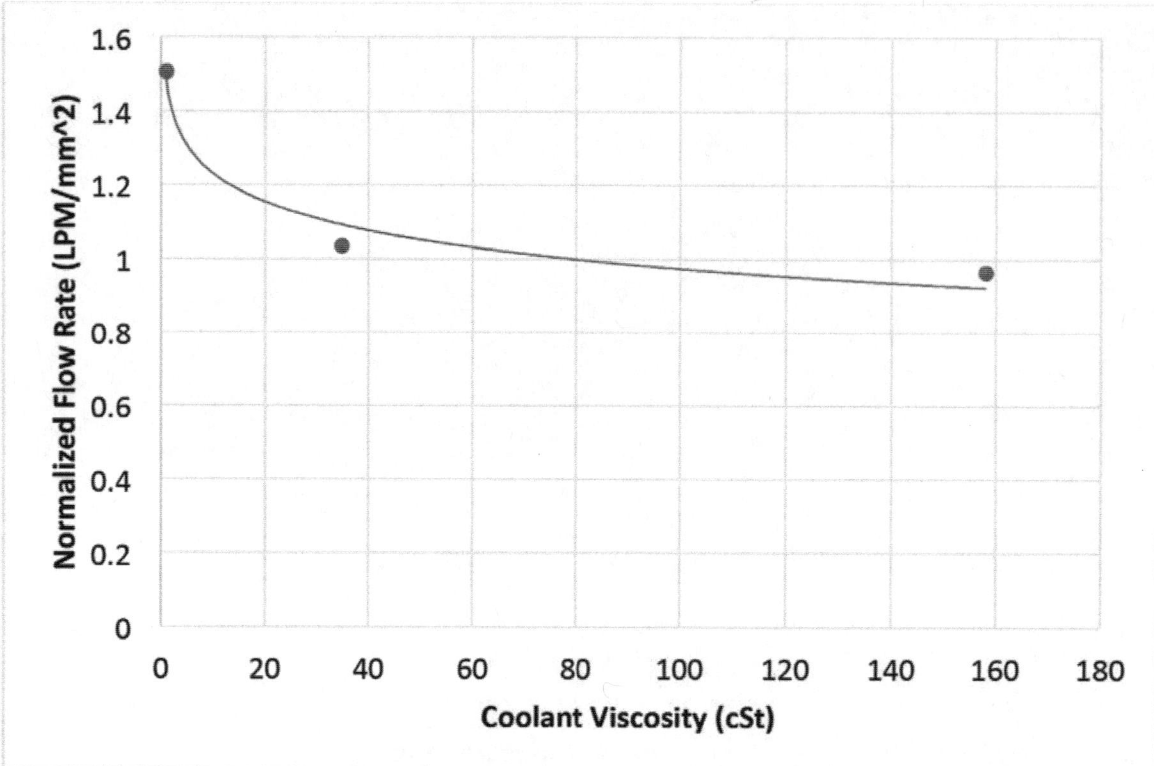

Figure 15. Effect of coolant viscosity on normalized flat nozzle flow rate

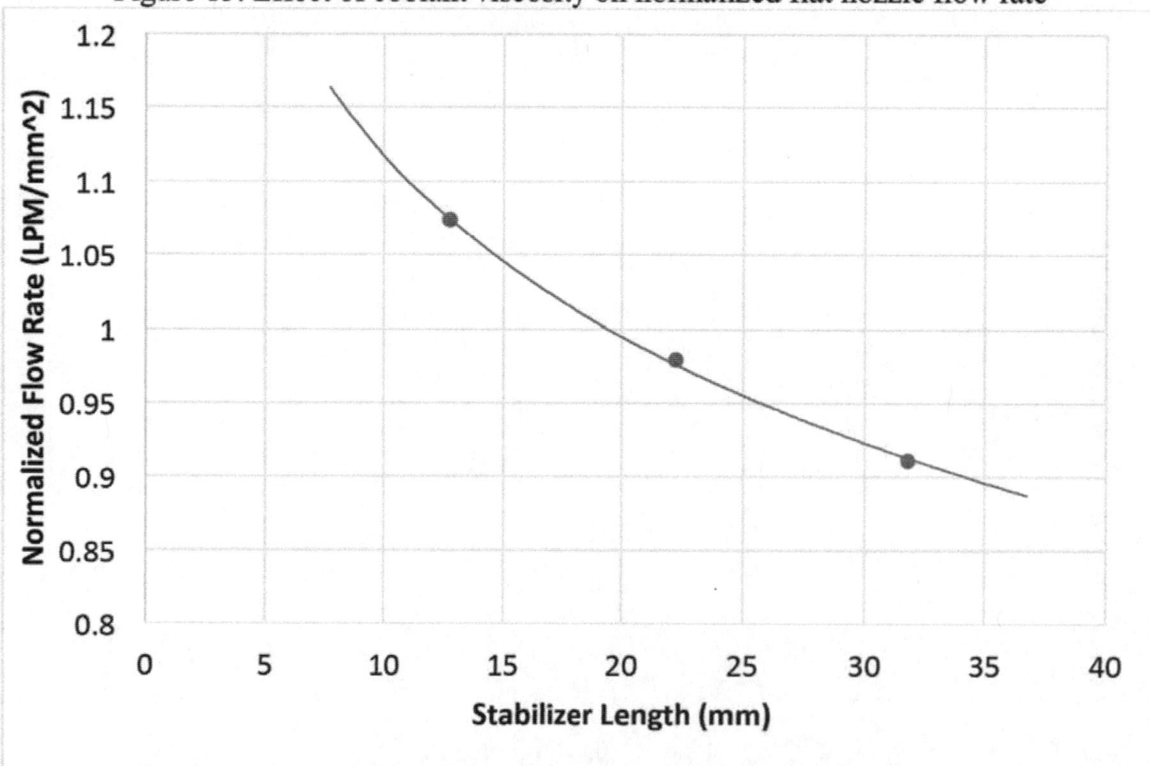

Figure 16. Effect of stabilizer length on normalized flat nozzle flow rate

The coefficient of discharge (C_d) for each nozzle can be estimated using a nozzle's exit area (A) and the slope of the plot of flow rate and $\sqrt{head}$ from the nozzle tests (*m*).

$$C_d = \frac{m}{A \times \sqrt{2 \times g}} \qquad (9)$$

A nozzle's discharge coefficient varies with coolant viscosity, decreasing as viscosity increases. As such, discharge coefficient was estimated for each combination of nozzle aperture, stabilizer length and viscosity. Multivariate logarithmic regression on the effect of nozzle aperture, stabilizer length and coolant viscosity on the coefficient of discharge was used to determine a general equation to estimate the discharge coefficient for other scenarios:

$$C_d = a(\log_{10} w) + b(\log_{10} x) + c(\log_{10} y) + d \qquad (10)$$

Where: *a, b,* and *c* are the coefficients for nozzle aperture, stabilizer length, and viscosity, respectively; *w, x,* y are novel input values for nozzle aperture, stabilizer length, and viscosity, respectively. *d* is the intercept for C_d, which can be assumed to be 0 (flow would be impossible with a nozzle aperture of 0 mm). The coefficients show the magnitude of the effect of nozzle aperture, stabilizer length, and viscosity on the efficiency of a particular nozzle.

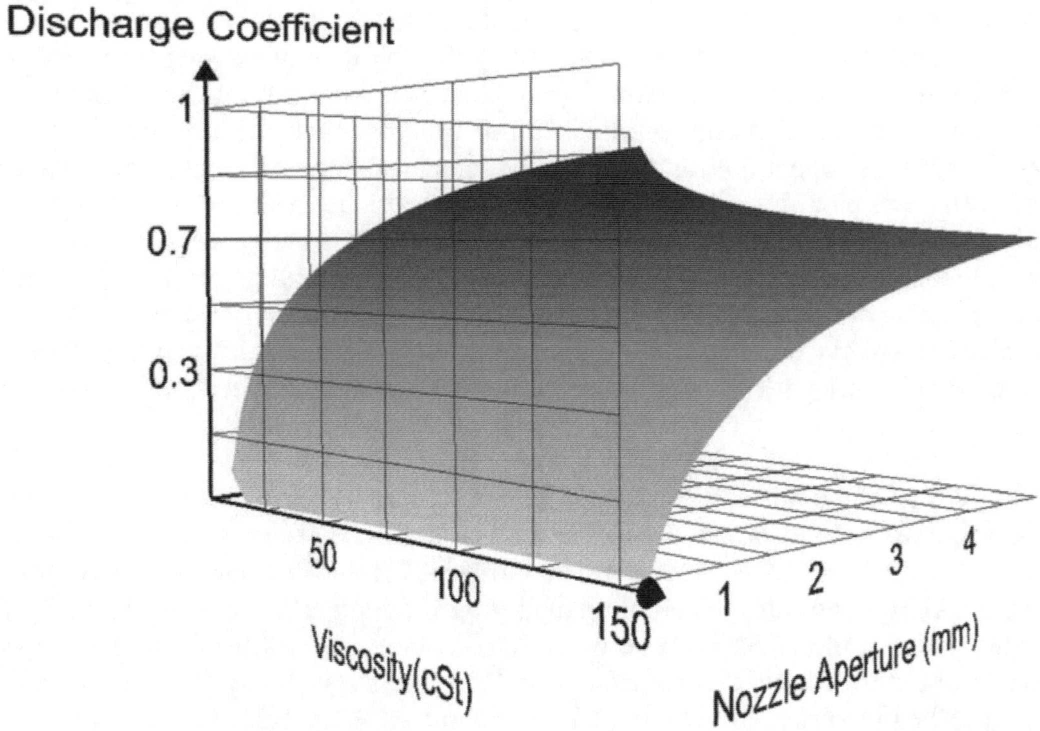

Figure 17. Effect of nozzle aperture and viscosity on discharge coefficient

These regression coefficients are dependent on nozzle geometry, meaning that the closer a nozzle's geometry is to the nozzles that were used to determine these coefficients via flow tests, the more accurate its discharge coefficient prediction will be. In terms of optimization, the ideal application of these findings would be to determine these coefficients experimentally for every nozzle type and coolant viscosity being used. That is generally not feasible but determining the coefficients for common nozzle types and incorporating the resulting equations into a coolant system optimization calculator can dramatically decrease the error in calculating the pressure necessary to achieve a flow rate that will adequately cover a grinding wheel. For example, one flat nozzle tested (12.95 mm width, 1.75 mm aperture, 31.75 mm stabilizer length) at 6 bar pressure had a flow rate of 34.1 LPM with a 34.68 cSt coolant and a flow rate of 15.75 LPM with a 158.04 cSt coolant. This is one of the more dramatic examples that the simplified Bernoulli equation would have ignored. Changes in flow rate are most extreme with high viscosity coolants, small nozzle apertures, and long stabilizers. This is especially true with the interaction of two or more of those factors.

8. Example Applications of Coherent-Jet Nozzles

After 25 years of optimizing the coolant application in a variety of grinding modes and applications, the author has learned that there is no single solution that fits all. The requirements of the customer can range from reducing the reject rate, to increased productivity, to reduced consumable wheel cost. The benefits of improved coolant application is often trivialized and as long as the coolant is going in approximately the right direction, the focus will be on wheel specification, grit type, process feeds and speeds or a new machine tool. Numerous case studies by the author have shown that the cost associated with optimized coolant application, even after paying for an audit of the current process and coolant application method, can pay back the investment within a matter of weeks. A robust nozzle system may cost $1000 for each machine, whereas a vitrified CBN wheel will cost a lot more and need to be replaced at more frequent intervals than necessary. The testing of many vitrified CBN wheel specifications can also be very expensive and often pointless if the basic coolant application set up is flawed. The author is also disappointed at the level of coolant application hardware supplied with such expensive machines, especially the open pipe and plastic nozzles that are often setup the day of the run-off with the customer, without the perceived need for an engineered product.

Figure 18 shows a coolant nozzle setup where the original open pipes in the left view, were very numerous and in some cases were crushed to conserve pressure. The setup included both a plated CBN wheel and a vitrified CBN wheel on the same spindle, grinding a nickel-based alloy. The pump supplying the mineral oil coolant had a 37 KW motor and was capable of 34 bar at a potential flowrate of 570 liters/min. The pressure measured at the vitrified wheel manifold was 6.2 bar, and at the plated wheel manifold was 21 bar. The pressure required to match the 60 m/s wheelspeed is 17 bar.

The combined area of all the nozzles on each zone allowed the flowrate for each zone to be estimated as 514 liters/min for the plated wheel and 401 liters/min for the vitrified

CBN wheel. The objective of this optimization work was primarily to reduce the flowrate of the machine to allow more machines to run simultaneously using the undersized central coolant system. The percentage spindle power load for the 23 kW motor was typically 40-45%(≈10 kW) for the plated CBN wheel, and 60-70% (≈15 kW) with the vitrified CBN wheel. These values include the idle running power without grinding.

Figure 18. Minimization of flowrate using reduced number of nozzles

The revised coolant application setup, in the right view, was based around coherent-jet nozzle tubes. As can be seen on the photo, one 4.8 mm and two 3.2 mm diameter nozzles were fitted to supply the center and each corner of the plated CBN wheel, and two 4.8 mm diameter nozzles were fitted to supply the vitrified CBN wheel. An additional nozzle was fitted underneath the vitrified wheel to lubricate the dressing operation on the vitrified wheel. The pump speed was adjusted to give 17 bar in the manifolds, to match wheelspeed. The calculated flowrate from the nozzles was 117 liters/min for the plated wheel and 148 liters/min for the vitrified wheel, a reduction of 71% for the plated wheel and a reduction of 63% for the vitrified wheel. Due to much lower viscous drag associated with the much lower flowrates, the spindle power reduced down to 30% (6 kW) on the plated wheel, and 40% (9 kW) on the vitrified wheel, for no change in the cycle parameters. Based on the spindle power recorded before the nozzle conversion, the plated wheel was applied with a specific flowrate of 12 liters/min.mm (including the deflected fluid on the two corners), and the vitrified wheel was applied with a specific flowrate of 10 liters/min.mm (including the wasted flowrate to the truer during grinding). No HP cleaning jets were deemed necessary since wheel loading was not evident.

Other benefits of the new setup for the customer were:

1. Machine more repeatable, giving less rework due to lower hydrodynamic forces
2. Easier setup using laser alignment of jets
3. First part after setup was perfect

4. Only one bend per pipe
5. Solid coolant stream with no entrained air, resulting in foam
6. No need to adjust nozzle for entire life of vitrified CBN wheel
7. No need to replace pump-back pumps due to excessive flowrate
8. No need to replace, or upgrade, the central filtration system
9. One pump can now supply 3 machines, saving considerable energy costs of at least 15 KW per machine over 3 shifts?
10. Reduced chiller energy demand
11. Nozzles are placed further away from grinding area, with less risk of damage
12. Plated CBN wheel life doubled
13. 30% more parts/dress for the vitrified CBN wheel

1. Conclusions

The application of coolant into the grinding zone needs to be taken more seriously by end-users, machine builders, coolant suppliers, wheel suppliers, etc. Carefully designed engineered nozzle products can give huge economic benefits and can be easily accepted by the shop floor. Optimized flowrate can result in large energy savings, higher productivity, better quality, and more.

As compared to dispersed jets, the use of coherent jets has several merits that were described in the paper, but they require more care to set them up. The use of laser aligning tools ensures that the operator can make quick changes to the setup without bending the pipe numerous times. Re-configurability is key with small batch sizes, and wheel diameter compensation must be eliminated by referencing the part not the wheel, or adjustment made as easy as possible.

The effect of coolant viscosity on flow rate is one of the most difficult sources of error to predict. Even before considering the complex interaction between coolant viscosity and nozzle geometry, the effect of coolant temperature on viscosity must also be considered. Most coolant manufacturers only provide one or two values for a coolant's viscosity. This makes predictions difficult because it's common for increases in temperature from 20□ to 40□ to cut an oil's viscosity in half. When manufacturers provide two or more viscosity values (temperature pairs) the viscosity can be estimated at a third temperature using ASTM D341/ Viscosity-temperature extrapolation.

This research will lead to the development of an advanced coolant delivery software for more precise optimization based on a more accurate energy equation, plus compensation for fluid viscosity when sizing the nozzle exit aperture.

Superabrasive grinding wheels require different cooling setups than conventional abrasive wheels, and potentially require less flowrate. The higher wheelspeed of CBN requires stable jets, that can work at the high pressures required to match the wheelspeed.

The author brings some real-world experience into this paper, with more than 25 years of fluid optimization research and implementation, for all common modes of grinding

Bibliography

[1] Brinksmeier, E.; Heinzel, C.; and Wittmann, M., Friction, Cooling and Lubrication in Grinding, Annals of the CIRP, 1999, 48/2: 581-598

[2] Andrew, C.; Howes, T.; Pearce, T., Creep Feed Grinding, 1985, Published by Holt, Rinehart and Winston, pp.80-90

[3] Guo, C.; Malkin, S., Analytical and Experimental Investigation of Burnout in Creep Feed Grinding, Annals of the CIRP, 1994, 43/1:283-286

[4] Guo, C., Energy Partition and Cooling During Grinding, Proc. of 3rd Int. Machining and Grinding Conference, SME, Ohio, 1999, pp.1-12

[5] Rowe, W. B.; and Jin, T., Temperatures in High Efficiency Deep Grinding, Annals of the CIRP, 2001, 50/1:205-208

[6] Yasui, H.; and Tsukada, S., Influence of Fluid Type on Wet Grinding Temperature, JSPE, 1983, Vol 17/2:133-134

[7] Colding, B.; et al., Recent Research and Developments in Grinding, Annals of the CIRP, 1972, 21/2: 157-166, Keynote STC C

[8] Vansevenant, I., 1987, A Subsurface Integrity Model in Grinding, Dr.-Ing Diss., Katholieke Universteit te Leuven, 1972

[9] Czenkusch, C., Technologie und Prozessmodelle beim Rundschleifen, Dr.-Ing Diss., IFW, University of Hannover, 2000,

[10] Klocke, F.; Baus, A.; and Beck, T., Coolant Induced Forces in CBN Grinding with Shoe Nozzles, Annals of the CIRP, 2000, 49/1: 241-244

[11] Cui. C., Experimental Investigation of Thermofluids in the Grinding Zone, PhD Diss., University of Connecticut, 1995

[12] Brinksmeier, E.; Minke, E., High Performance Surface Grinding - the Influence of Coolant on the Abrasive Process, Annals of the CIRP, 1993, 42/1: 367-370

[13] Malkin, S., Grinding Technology - Theory and Applications of Machining with Abrasives, John Wiley and Sons, 1989, ISBN 0-85312-756-5

[14] Heinzel, C., Methoden zur Untersuchung und Optimierung der Kuehlschmierung beim Schleifen, Dr.-Ing. Diss., Univ. of Bremen, 1999.

[15] Powell, J., The Application of Grinding Fluid in Creep Feed Grinding, PhD Diss., Univ. of Bristol, 1979.

[16] Bruecher, T., Kuhlschmierung beim Schleifen Keramische Werkstoff, Dr.-Ing. Diss., TU Berlin, 1996.

[17] Webster, J.; Cui, C.; and Mindek, R., Grinding Fluid Application System Design, Annals of the CIRP, 1995, 44/1: 333-338

[18] Ott, H. W., Zufuhrsysteme fuer Kuhlschmierstoff beim Schleifen, Seminar " Kuehlschmierstoffe in der Spanenden Fertigung", Deutsches Industrieforum fuer Technologie (DIF), Frankfurt, 1991.

[19] Mindek, R.; Webster, J.; and Cui, C., Minimizing Thermal Damage of Aerospace Components using Coolant Nozzle and Coolant System Optimization, ASME, IGTI-Cogen Turbo Power, 1994, 9/94: 97-104

[20] Campbell, J., Optimized Coolant Application, Proc. of 1st Int. Machining and Grinding Conference, 1995, SME: pp61-66, Detroit

[21] Inasaki, I., Fluid Film in the Grinding Arc of Contact, contribution to Paris CIRP meeting, 1998.

[22] Campbell, J., Tools to Measure and Improve Coolant Application Effectiveness, Proc. of 2nd Int. Machining and Grinding Conference, SME, Detroit, 1997, pp.657-676
[23] Silliman, J., Cutting and Grinding Fluids: Selection and Application, Published by SME, 1992, pp.181
[24] Metzger, J., 1986, Superabrasive Grinding, Butterworth Press, ISBN 0-48-01586-1
[25] Webster, J. A., Brinksmeier, E., Heinzel, C., Wittmann, M., and Thoens, K., Assessment of Grinding Fluid Effectiveness in Continuous Dress Creep-Feed Grinding, Annals of CIRP, 2002, 51/1: 235-241
[26] Spur, G., and Brucher, T., Optimization of Cutting Fluid System for Grinding of Advanced Ceramics, 1st Int. Machining and Grinding Conference, SME, 1995, pp. 933-948
[27] Martin, K. and Yegenoglu, K., HSG Technologie – Handbuch zur praktischen Andwendung, Guehring Automation GmbH, book, 1992, pp.94.
[28] Stephenson, D., Physical Basics in Grinding Technology, 1st European Conference on Grinding, Aachen, 2003, pp13.1 – 13.21
[29] Webster, J. A., Cui, C., and Mindek, R., Grinding Fluid Application System Design, Annals of CIRP, 1995, 44/1:333-338

INTERTECH 2019

- **Advances in the Processing of Semiconductor Materials using Diamond Technologies**
- Hyperion Materials & Technologies
- Timothy Dumm, Technology Manger
  - B.S. Mining Eng, PhD Mineral Processing, Penn State Univ.
  - 30 years with GE Superabrasives, Sandvik and Hyperion Materials & Technologies
  - Currently R&D Manager for Abrasives products at HYP M&T.

HYPERION Materials & Technologies

INTERTECH 2019

Outline

- Introduction
- Applications Driving Demand for Semiconductor Materials
- Diamond's Unique Role
- Basic Process Flow
- Sawing
- Lapping/Polishing
- Back Grinding
- CMP
- Trends/Summary

A historical perspective – 25 years

Then - 1993

Characterization of CBN Micron Powders *T.F. Dumm - GE Superabrasives*

DIAMOND & CBN
Ultrahard Materials Symposium
'93

Now - 2019

Advances in the Processing of Semiconductor Materials using Diamond Technologies ***T.F. Dumm – Hyperion Materials & Technologies***

INTERTECH 2019

Lighting Then

Lighting Now

Solar Power Then

Solar Power Now

Electronics Then

Electronics Now

Data Storage Then

10MB

Data Storage Now

+100,000MB

Transportation Then

Transportation Now

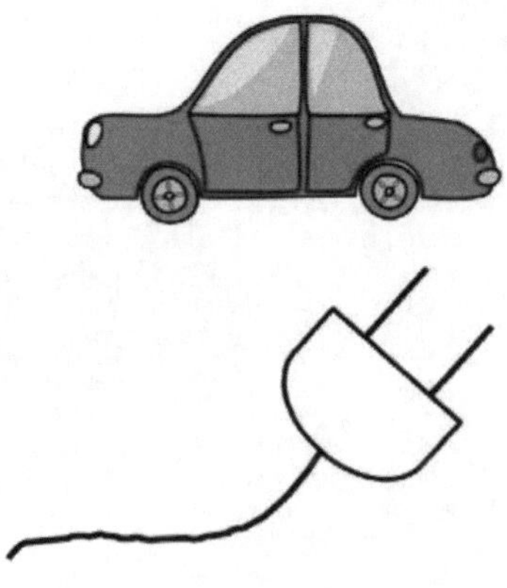

What do these new applications mean for diamond?

Silicon Market Growth

Electronics Grade

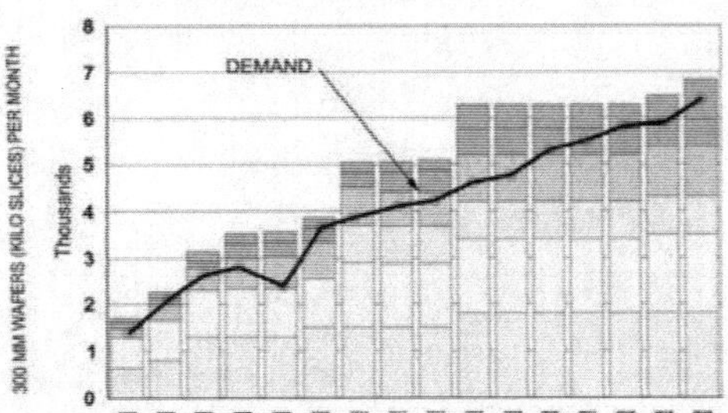

300mm supply and demand forecast (Source: Sage Concepts)

Solar Applications

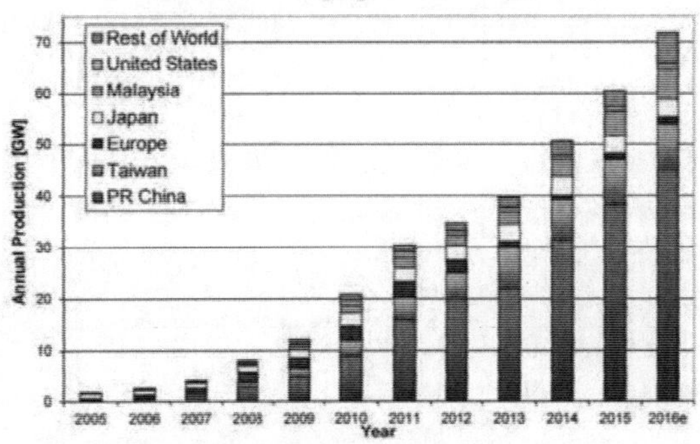

Figure 1: World PV cell/module production from 2005 to 2016 (*data source*: Photon Magazine [Pho 2012], PV Activities in Japan [Ikk 2016], PV News [Pvn 2015] and own analysis)

Sapphire Market Size for Lighting and Consumer Electronics

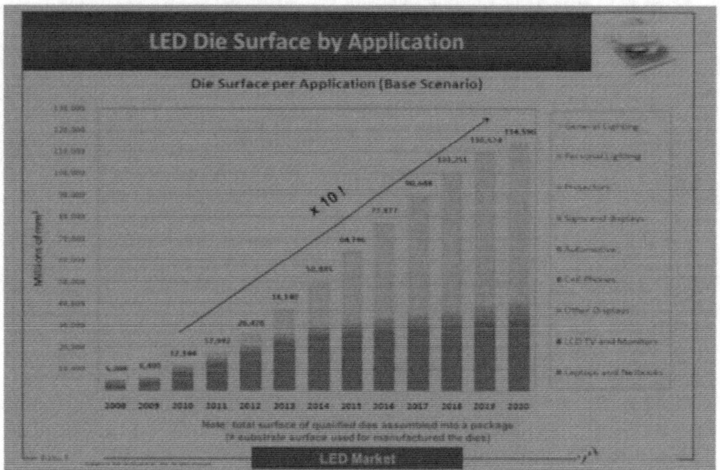

Yole Development

Sapphire market for LED applications remains strong with good growth projections.

-Current projected annual growth rate is 22.3%

-Sapphire wafer and LED manufacturing has shifted to China.

-Traditional flat lapping and polishing with slurry is prevalent, but brush lapping and fixed abrasive pads also being used.

-Wafer sizes increasing from 2-in to 6-inches.

Silicon Carbide Market Size for Lighting and Power Supplys

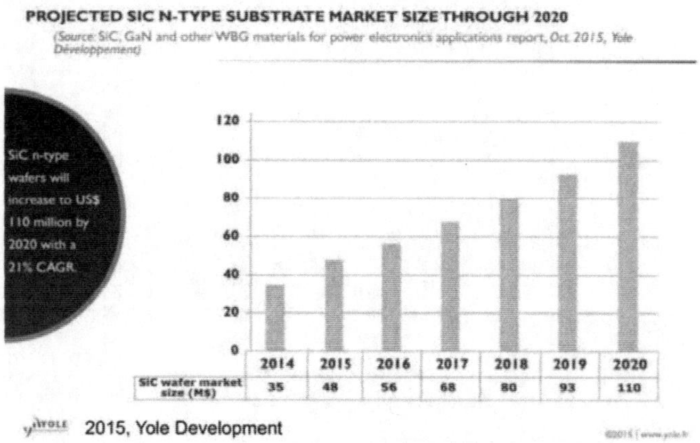

2015, Yole Development

Why SiC Products Are Gaining Traction

•Silicon carbide semiconductors offer multiple advantages over traditional silicon-based semiconductors:

- higher switching efficiency and the ability to handle higher voltages.
- allow **Electric Vehicle (EV)** mfg. to maximize battery range, reduce cooling requirements and minimize the footprint of the inverter and battery in the vehicle.

• Although SiC-based components are more expensive, the value proposition is strong for vendors.

INTERTECH 2019

Properties of Selected Ultra-Hard Materials

	Density g/cm³	Hardness Knoop	Compressive Strength Gpa	Coefficient of Thermal Expansion mm/mm/°C x10⁻⁶	Thermal Conductivity W/m K
Diamond (C)	3.52	10,000	10	4.8	2,100
Cubic Boron Nitride (cBN)	3.48	4,500	7	5.6	1,400
Boron Carbide (B_4C)	2.60	3,200	4.6	3.1	90
Silicon Carbide (SiC)	3.21	2,700	1.3	4.5	42
Aluminum Oxide (Al_2O_3)	3.92	2,100	3	8.6	33
Tungsten Carbide (WC-Co, 6%)	15.0	1,700	5.4	4.5	105

Diamond is the only practical abrasive for processing SiC and Sapphire

INTERTECH 2019

Basic Semiconductor Process Flow

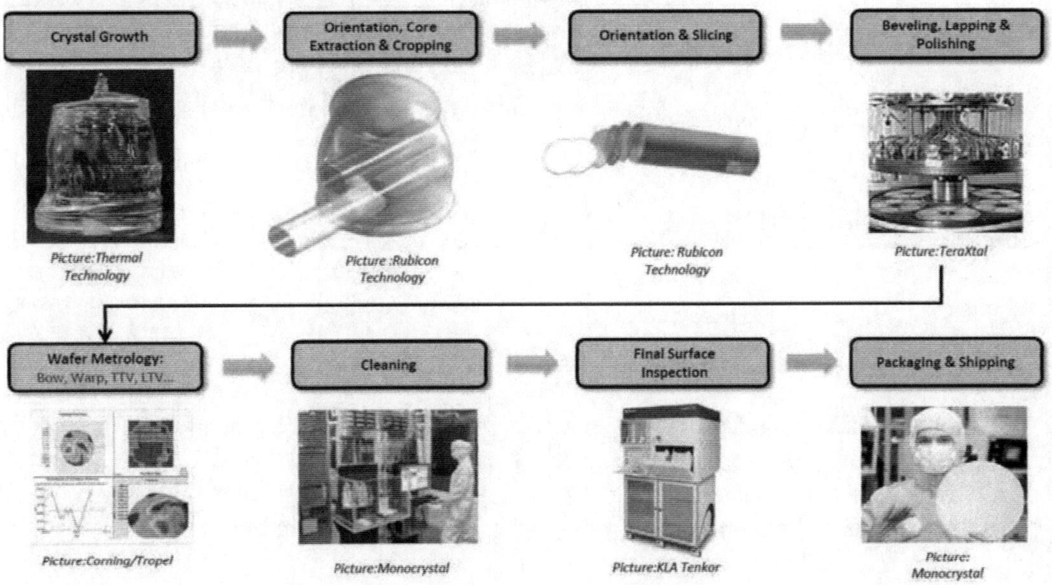

Multi-Wire Precision Saws

Wire Sawing Then

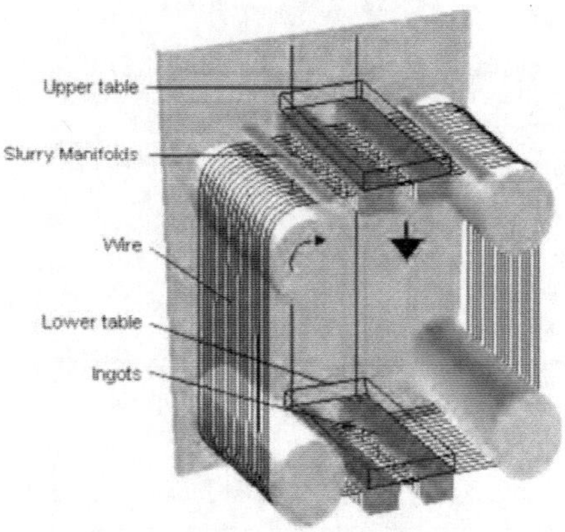

A single strand of bare steel wire
– 50-250km long, 100-300um dia
- Wafer thickness: 100um to 250um

Can be looped 1000 times.
Self contained recycle slurry system.
Generally SiC, B_4C or Diamond abrasives
Considerable sludge is generated.

Wire Sawing Now

- Fixed abrasive wire instead of slurry
- Wire diameters down to 50um
- Diamond electroplated to wire
- Diamond size down to 5-10um
- Thinner kerfs
- No abrasives in sludge

Fixed Abrasive Wire

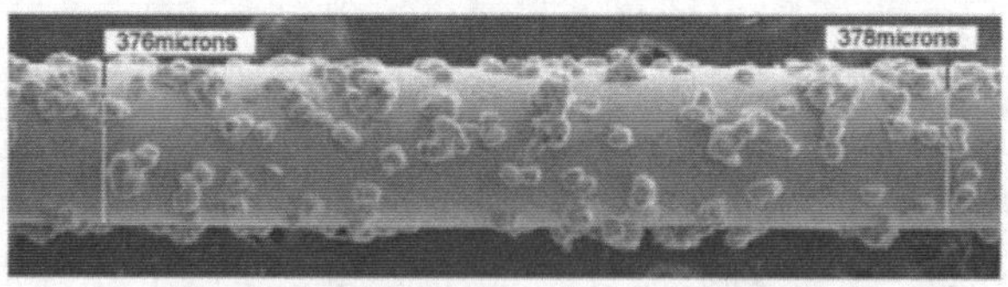

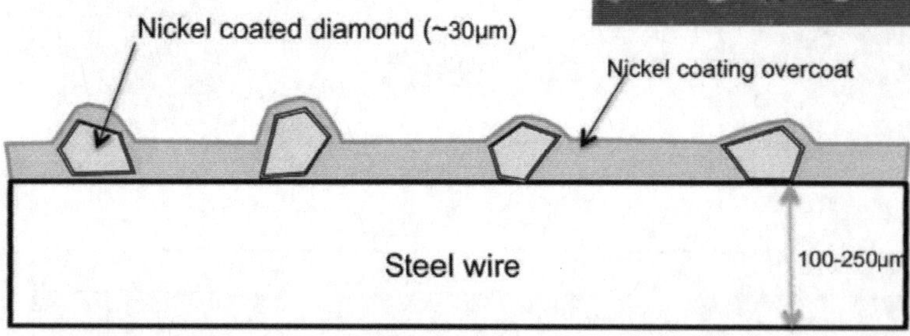

FAW vs. Slurry:
- Higher cut rates
- No slurry or sludge
- Lower cost of ownership

- Diamond is attached to fixed abrasive wire is made in electrolytic plating bath.
- The nickel coating on the diamond improves attachment rate to the wire.
- However, the nickel coating on the diamond can dissolve in the plating solution.
- Challenge is to make a **corrosion resistant** coating on the diamond.

INTERTECH 2019

Fixed Abrasive Wire Product Requirements

EP-wire manufacturer CTQ' s

- High platability rate (improves throughput)
- Lower crystal pull-outs: improved adhesion of diamond to the core wire
- Optimum concentration and uniformity of diamond on the wire
- Minimal post processing

Diamond CTQ' s

- Corrosion resistant coating in Ni bath.
- Coating Adhesion- both mech. & chem.
- Non-magnetic coating
- Coating assay- 20 or 30wt%
- Coating coverage and texture
- Absence of agglomeration
- Particle Size Distribution

17

INTERTECH 2019

Typical Sapphire Lapping and Polishing Process Details

Typical process used for making sapphire surface ready for epitaxial layer.

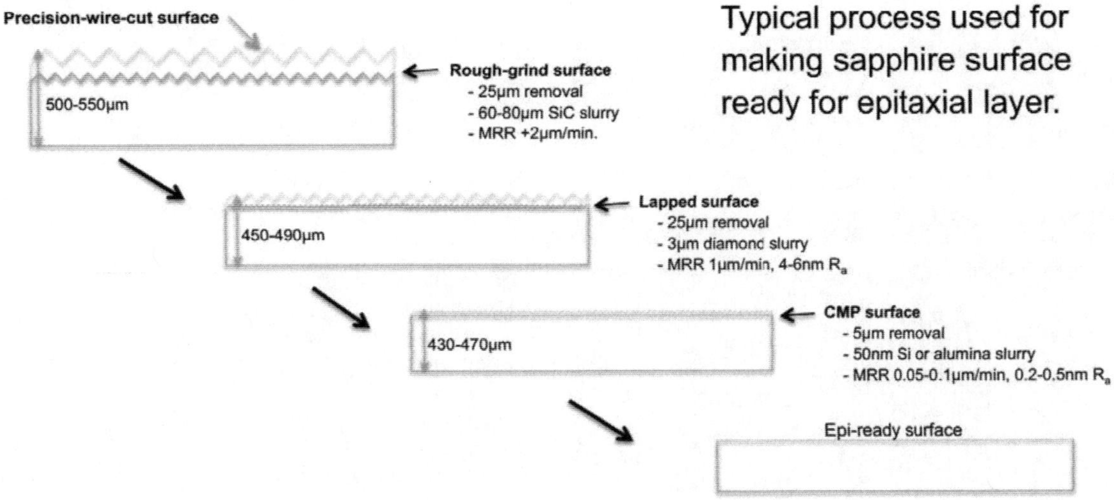

INTERTECH 2019

Lapping and Polishing

Abrasives used then:

"Mono-crystalline" Micron Diamond
-Made from Mesh mono-crystalline diamond
-Few cutting points
- Suitable for general performance applications
- High volume availability

"Shock" Diamond
-Made from explosion process
-Few cutting points
- Size becomes smaller as particle crushes
- Size availability can be limited, esp. in coarse sizes (>10um)

INTERTECH 2019

Lapping and Polishing

Abrasives used now:

"Surface Modified" Diamond
- Many micro-cutting points
- Does not crush easily
- Retains size and effectiveness
- Highest performance in sapphire lapping applications
- Can be supplied in higher volume and coarser sizes than "Shock" diamond

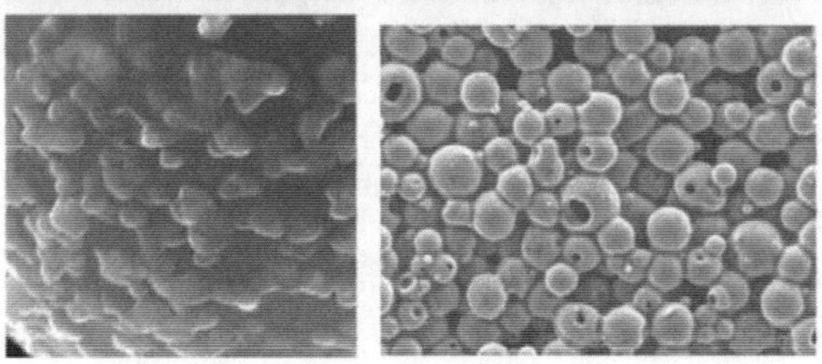

Diamond Clusters
- Many micro-cutting points
- Crush strength varies
- Performance relies on breakdown and diamond exposure
- Highest performance in sapphire lapping applications
- Often incorporated into a lapping pad or film.

INTERTECH 2019

Slurry Product Performance on Sapphire

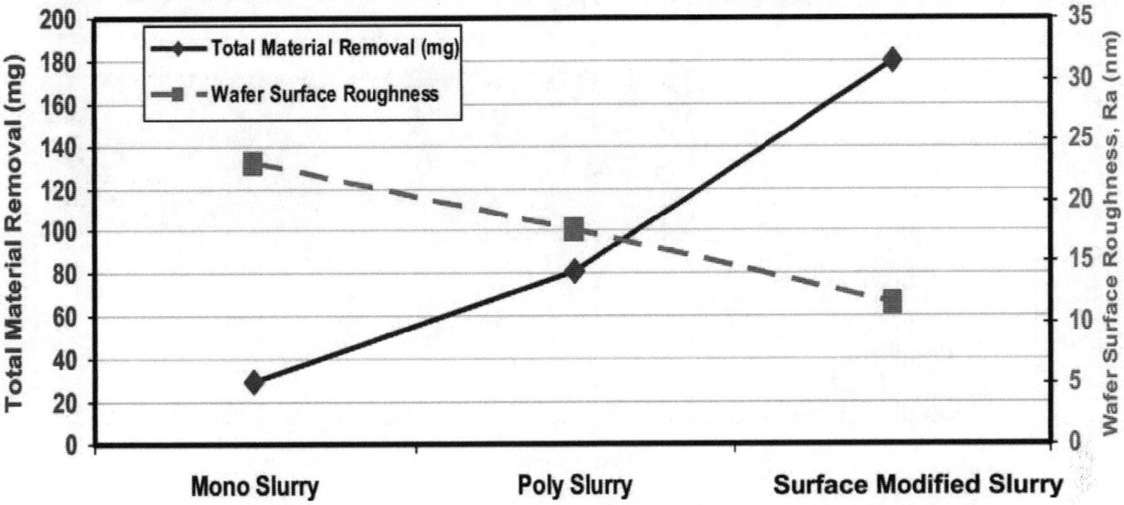

Surface Modified slurry product provides 2X to 5X material removal rate and 33% reduction in surface roughness in lab tests vs. poly diamond.

INTERTECH 2019

Semiconductor Backgrinding

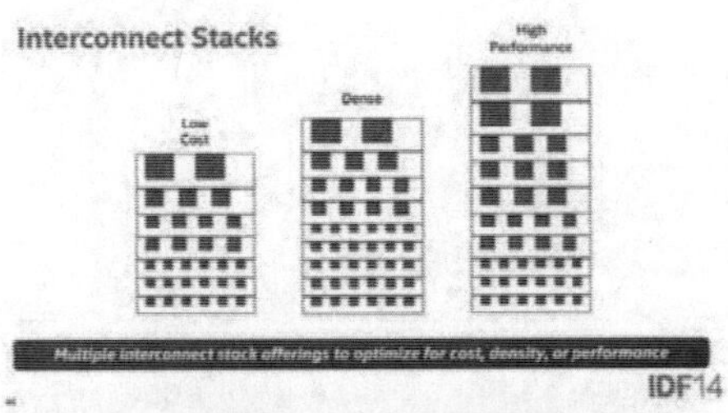

Chip-stacking for 3-D architecture requires **thin** chips that are obtained by backgrinding/polishing.

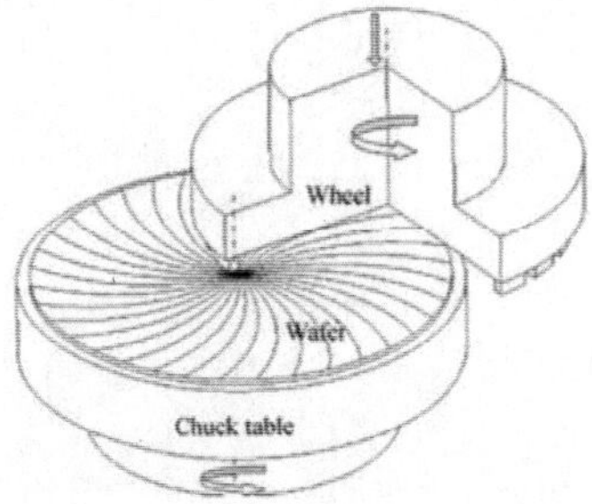

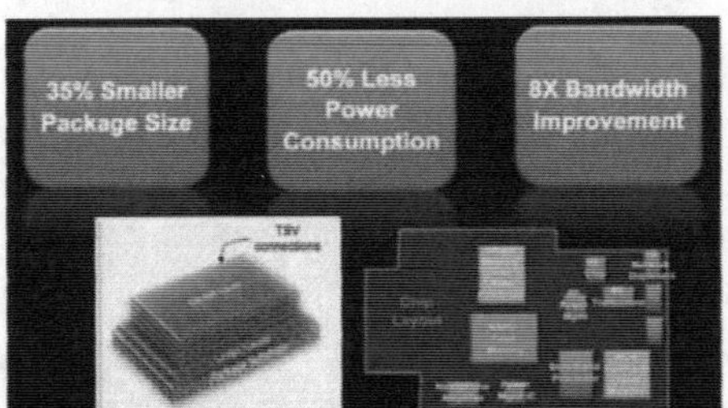

Defects like scratches on thinned wafers are stress risers that can fracture the wafers and chips.

Semiconductor Backgrinding

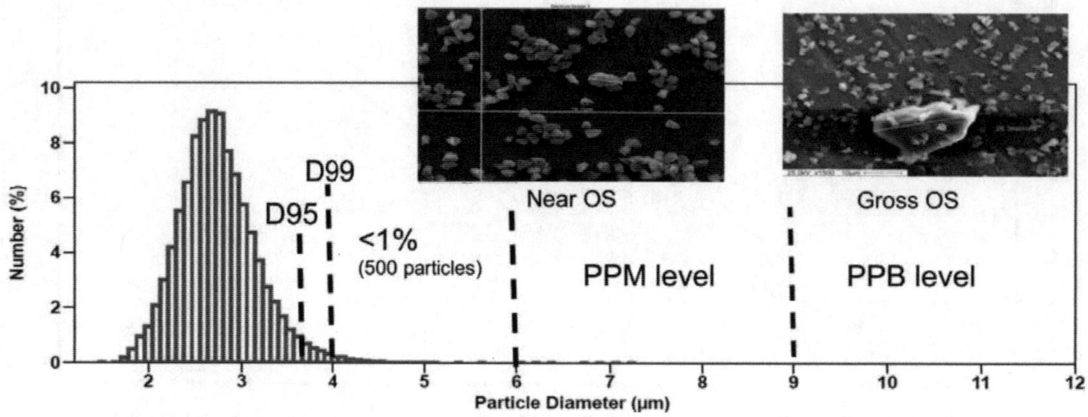

Ever more demands on quality of diamond powders used in back grinding applications

CMP Pad Conditioning

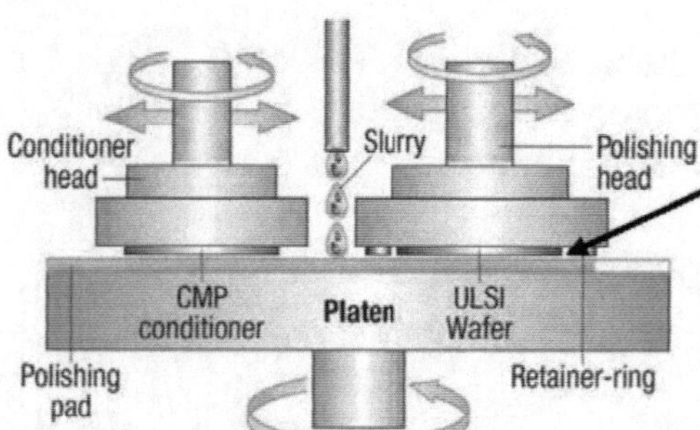

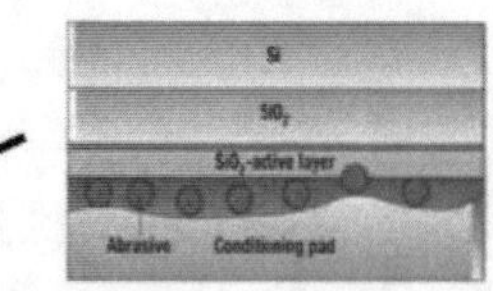

Functions of CMP Pad Conditioner

a) To machine a pad to create flat surface and remove glazing.
b) To remove debris from polishing and restore pores

Major requirements for CMP Pad Conditioners	Equivalent Diamond Requirements
No Diamond pop out	No defects and impurities on diamond surface
Constant removal rate	Uniform shape and size distribution throughout the diamond matrix
Consistency of pad among batches	Lot to lot consistency in terms of shapes, sizes and other properties
Long tool life	Low cost and high yield diamond

CMP Pad Conditioning

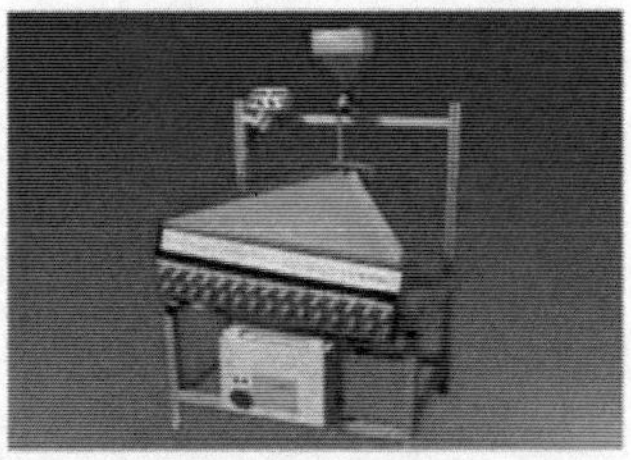

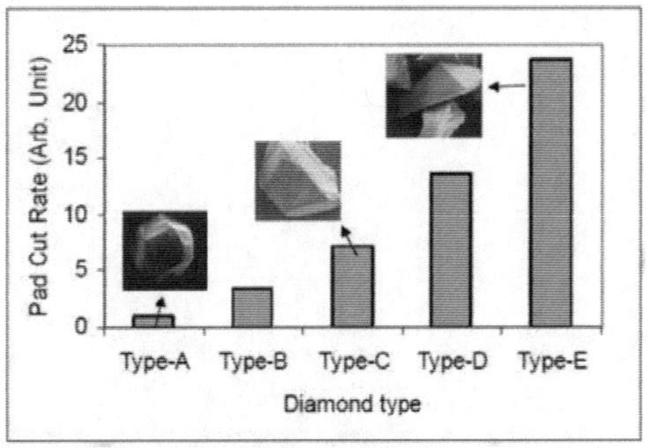

CHALLENGES

- Each customer has varying and different demands – ability to respond rapidly being established & investments in characterization on-going
- Customers ultimately desire very high level of specification to ensure repeatability
- Implementation of more complex processing and testing in operations:
  - ✓ Narrow shape distribution requirements
  - ✓ Testing for shape parameters
  - ✓ Magnetic Susceptibility

INTERTECH 2019
An International Technical Conference on Diamond, Cubic Boron Nitride and their applications
www.intertechconference.com
Hyatt Regency New Orleans
NEW ORLEANS, LOUISIANA USA
May 29, 30 & 31, 2019

Summary

•Evolving technology using semiconductors relies on the use of diamond for high quality, low-cost and efficient processing.

•In general, diamond products have kept pace with the need of industry:

- Tighter size, shape, coating control
- Improved material removal rates and surface finishes
- Improved consistency and certification

•Consumer demand and strong competition for market share is driving electronics manufacturers to develop new materials and requirements that will continue to challenge the diamond industry to keep up with their needs.

Thank You

HYPERION
Materials & Technologies

Laser Technology and Machining Brittle Materials with PCD

There is industry-wide pressure on manufacturing machining sites to continuously increase productivity, improve surface quality and reduce cost per part. This pressure is the trigger for an increasing need to adopt new tooling technologies especially in cost sensitive, high volume environments. In the ever-changing consumer electronics landscape, this highly innovative space which is often overlooked from a novel tooling perspective, can reap significant rewards from developing creative tooling solutions.

Traditionally, nickel electro-plated (EP) tools have been used in the manufacture of glass components for consumer electronics products such as watches and smart phones. Despite being inexpensive, EP tools have numerous disadvantages including limited tool life due to the single diamond abrasive layer, limited productivity with average machining speeds plus time wasted during tool changing and an environmentally unfriendly nickel electroplating process.

At the same time, increasingly difficult-to-machine brittle materials such as toughened engineered glass, zirconia, sapphire and onyx, are being incorporated by design engineers into new smart devices. Not only are these being used more, but the demand for stronger, higher performing products means that they are also ever more difficult to machine. This means the manufacture of components such as product displays and housings is becoming more challenging using the traditional nickel EP approach. Additionally, these brittle materials must provide suitable part strength when finished. To achieve this, we must successfully control surface finish and residual stress, which are all areas of attention and concern for OEMs and consumer electronics companies.

Ensuring successful and cost-effective machining of these brittle materials in the future depends on developing a combined understanding of materials science, machining theory and manufacturing technology. This article will discuss how Element Six developed key expertise in all three areas, combining the results to develop a new manufacturing solution for brittle materials in the consumer electronics space.

Brittle materials machining theory

In developing a solution to more effectively machine brittle materials, a better understanding of the fundamentals of the tool-workpiece interaction is required. Due to the brittleness of these materials, any machining process can lead to subsurface damage such as micro-cracks, scratches and residual stress. To avoid those defects, brittle materials must be machined in the **ductile mode or regime**, which is characterised by plastically deformed or a combination of ductile-brittle.

The critical undeformed chip thickness (h_m) is the most important parameter to characterize ductile regime, as illustrated in Fig. 1. Generally, h_m is affected by several parameters: the mechanical properties of workpiece material, operating conditions, process operation, just to name a few.

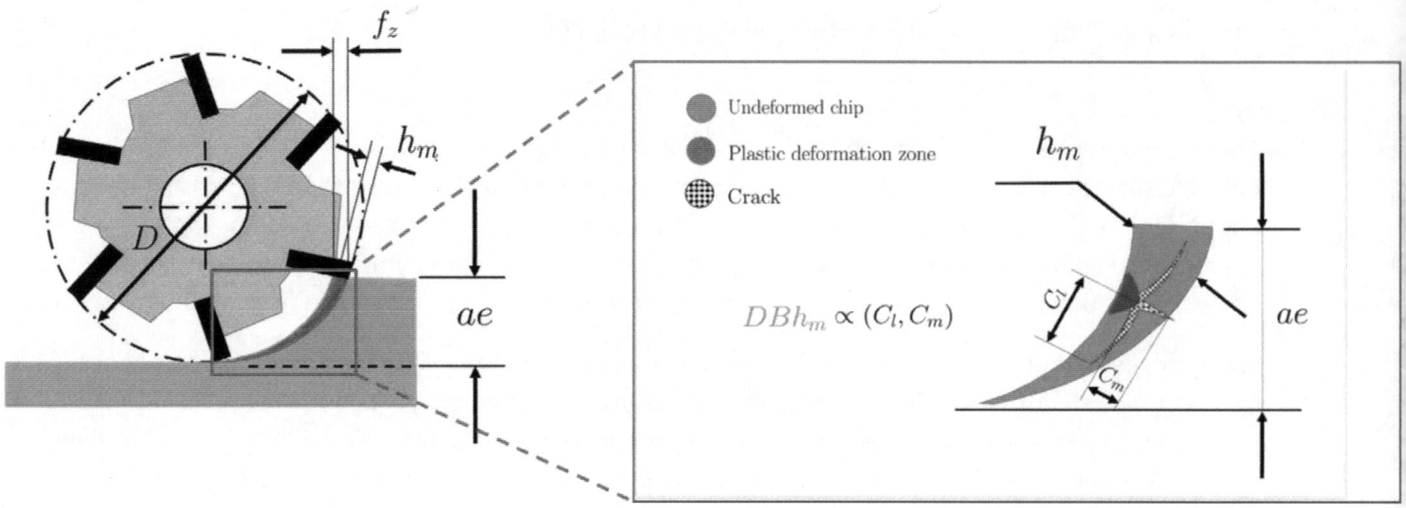

Illustration 1 – The undeformed chip thickness (h_m) in the combined ductile-brittle mode, during milling operation. Notice that the micro-cracks are constrained to the chip, after Marshall et al. (1982)

For the vast majority of studies, the ductile mode is defined when h_m is smaller than the brittle-to-ductile transition chip thickness (DBh_m) that can be estimated analytically. Based on the indentation method, Bifano et al. (1991) proposed a simple power–law model to identify DBh_m:

$h_m < DBh_m$;

and

$$DBh_m = \alpha \frac{E}{H}\left(\frac{K_c}{H}\right)^2:$$

where α is a dimensionless tool factor [0-1], H is the workpiece hardness, E the Young's modulus and K_c the fracture toughness. A schematic representation of h_m-H space of the Bifano et al. model is shown in Fig. 2. In the case of brittle materials, which are generally characterised by large values for H, the typical values for h_m is quite small, i.e., the machining conditions are constrained in a very narrow band.

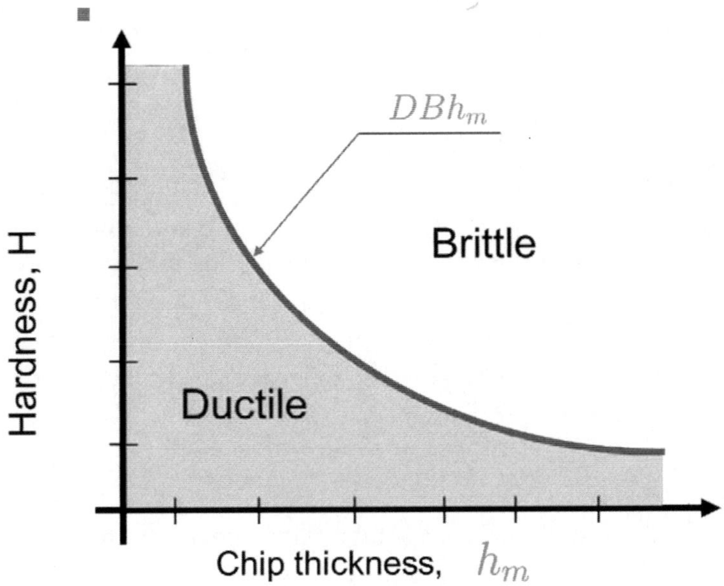

Illustration 2 – Theoretical schematic of the h_m-H diagram based on Bifano et al. (1991) phenomenological model.

An example of one such brittle material is BK7 glass, a high quality borosilicate glass from Schott typically used in consumer electronics. The experimental value of DBh_m = 0.34 μm obtained by Arif et al. (2013) indicates a good agreement with the Bifano's indentation model. This result suggests that this model can be indeed used to estimate h_m, and consequently, machining conditions and tool geometry (e.g. number of flutes) to ensure we remain in a ductile machining regime.

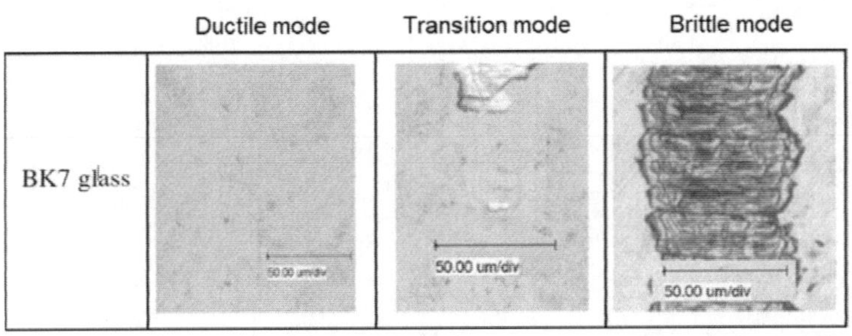

Illustration 3 – The optical images of the experimentally machined grooves conducted in BK7 glass at different h_m (0.20; 0,34 and 0.43 μm), after Arif et al. (2013).

Typical DBh_m values for Zirconia, Glass and Sapphire are shown in Table 1.

Table 1 - Typical values for DBh_m of materials used in consume electronic application.

Material	DBh_m(μm)
Zirconia	0.15-0.25
Glass	0.10-0.30

Sapphire	0.05-0.20

Considering these typical values for h_m for brittle materials, conventional EP (micro-grinding) tools have been frequently used. Unlike the conventional sharp-edge tools, the random distribution of grits on the cutting edge of EP tools can have a tremendous effect on surface roughness, forces (side effect and sub-surface damage), tool wear and thermal interactions. Nevertheless, the latest advances in laser ablation technology has lead the development of micro-milling PCD tools that can provide relatively high material removal rate with minimum wear. Despite the advantages of micro-milling, there are still challenges to overcome, especially in the case of dealing with micro edge processing.

References:

1) D. B. Marshall, B. R. Lawn, and A. G. Evans, "Elastic/plastic indentation damage in ceramics: the lateral crack system," J. Am. Ceram. Soc. 65(11), 561–566 (1982).
2)
3) T.G. Bifano, T.A. Dow, R.O. Scattergood, Ductile-mode grinding: a new technology for machining brittle materials, J. Eng. Ind. 113 (2) (1991) 184–189.

4) Muhammad Arif n, Zhang Xinquan, Mustafizur Rahman, Senthil Kumar, A predictive model of the critical undeformed chip thickness for ductile–brittle transition in nano-machining of brittle materials, International Journal of Machine Tools & Manufacture 64 (2013) 114–122

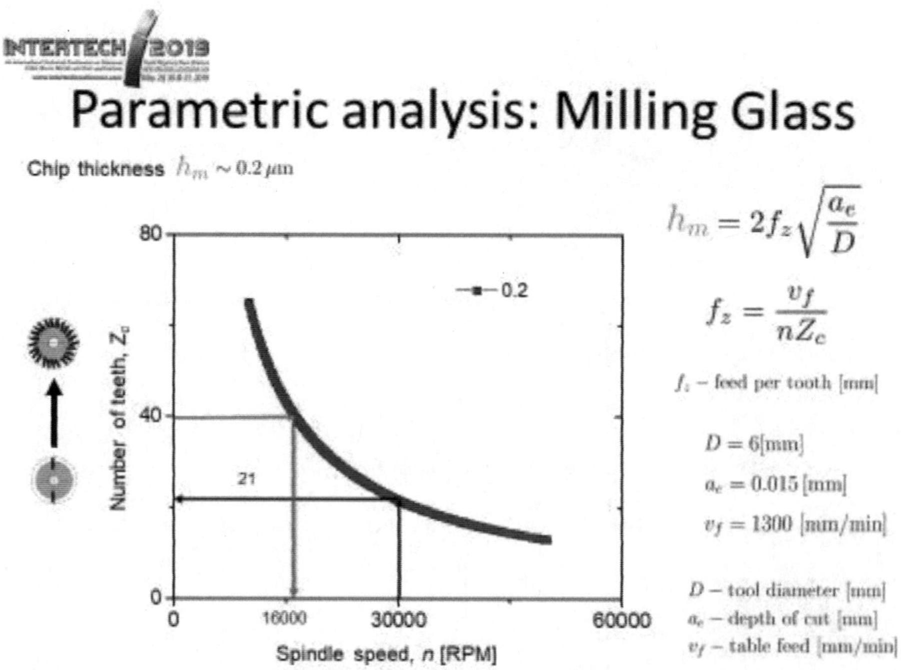

Laser Ablation of PCD

Laser ablation is a blanket term for the process of material removal by laser. The energy from the laser beam is transferred to the atoms of the work piece, giving them sufficient energy to be ejected from the surface, by mechanisms such as ionisation, vaporisation, melting and chemical breakdown.

In the case of diamond, laser energy is only absorbed weakly and heats up the surface, causing the diamond to graphitise. Graphite absorbs light more strongly than diamond, causing more energy transfer to the material. The material is essentially vaporised by this well-coupled energy, thus removing a layer of material but leaving a self-sustaining layer of graphite behind. A well-defined boundary exists between the underlying diamond and the graphite layer, meaning there is very little structural damage to the diamond underneath. The vaporised material is extracted as a gas or redeposited as a solid on the surface. Excess energy is transferred to the bulk as heat, which can induce stress in the bulk.

Laser ablation of PCD is more complicated than the case of diamond, because of the presenceof the binder. Again the diamond is graphitised then vaporised, but the binder may melt or react with the surrounding carbon because of excess heat. The different thermal expansion co-efficients of the diamond grains and the binder can lead to stress build up and material weakening.

Within academia, pulse duration has been explored as a method of reducing heating and therefore damage. Femtosecond (fs) lasers transfer energy to the material in a shorter time frame than energy can be transferred to the bonds of the material, so minimise heat induced laser damage. However fs lasers have historically been prohibitively expensive for a manufacturing environment. Nanosecond (ns) lasers are typically used in manufacturing operations since they are stable, reliable and cost effective. Nanosecond lasers have been used to produce all the tools in the following case studies.

Laser finishing is an enabling technology which allows fine tool geometries to be produced. While laser ablation is not necessary for all tool designs or machining applications, and has struggled to match grinding and erosion for removal rates in the past, its fine geometry capability is vital to the kind of tools we are developing for brittle material machining.

PCD Tool Produced Using Laser Ablation

Test Results

Milling of Zirconia – Case Study 1

Zirconia is challenging to machine with manufacturing sites reporting poor surface finish generated through brittle mode of material removal and surface cracking on parts. Both of these contribute to higher rates of scrap and weaker parts as tested in drop testing and final product strength testing. By understanding the ductile machining of brittle materials, we can select a particular feed rate to ensure the maximum undeformed chip thickness is in the recommended zone thus giving ductile machining and reducing any subsurface damage and micro cracks in the brittle material. In this test we show the affect changing the undeformed chip thickness has on surface finish. We improve surface finish and process efficiency through the correct choice of cutting conditions using PCD tools. Different proprietary laser technologies have been used to produce the tools in each of the case studies in this document.

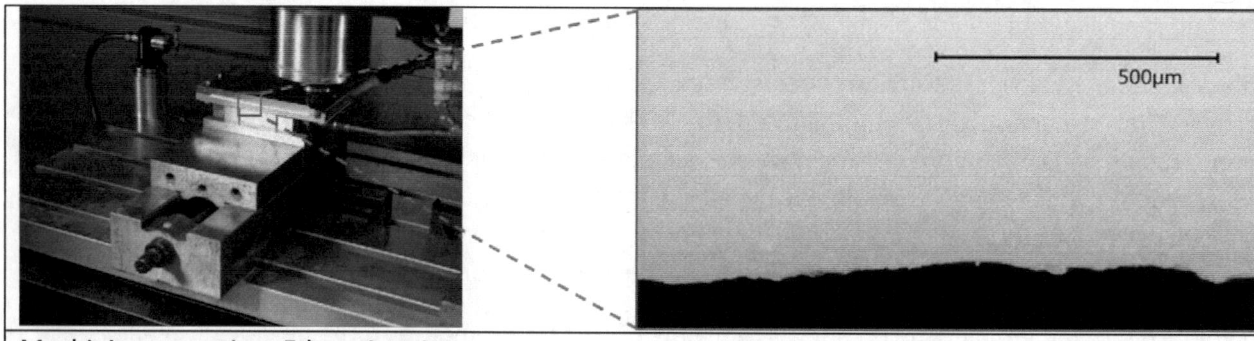

Machining operation: Edge trimming
Workpiece: Zirconia
Tool: 4 mm diameter brazed PCD endmill
Coolant: Water
Cutting parameters:

- $n = 12500$ RPM
- $v_f = 25\text{-}200$ mm/min
- $a_e = 0.05$ mm
- $Z_c = 4$ flutes

@Luiz – please can you confirm details of the tests (equipment, conditions, etc.). I need your support here. Thanks.

Hurco VM 20i

We vary the feed rate from 25 to 200 mm/min to show the effect of varying the undeformed chip thickness. The undeformed chip thickness is 0.11 – 0.90 microns. By understanding ductile machining of brittle materials, we can select a particular feed rate to ensure the maximum undeformed chip thickness is in the recommended zone. Thus controlling the surface finish through an engineered fixed cutting edge tool.

Additionally, if we have a target feed rate for the ductile regime, we can select the number of flutes to machining the brittle material in a ductile regime.

By increasing the feed rates, the chip thickness increases and the average surface roughness also increases. By understanding the machining conditions needed for ductile mode machining, we can control the surface finish and limit cracking and sub-surface damage.

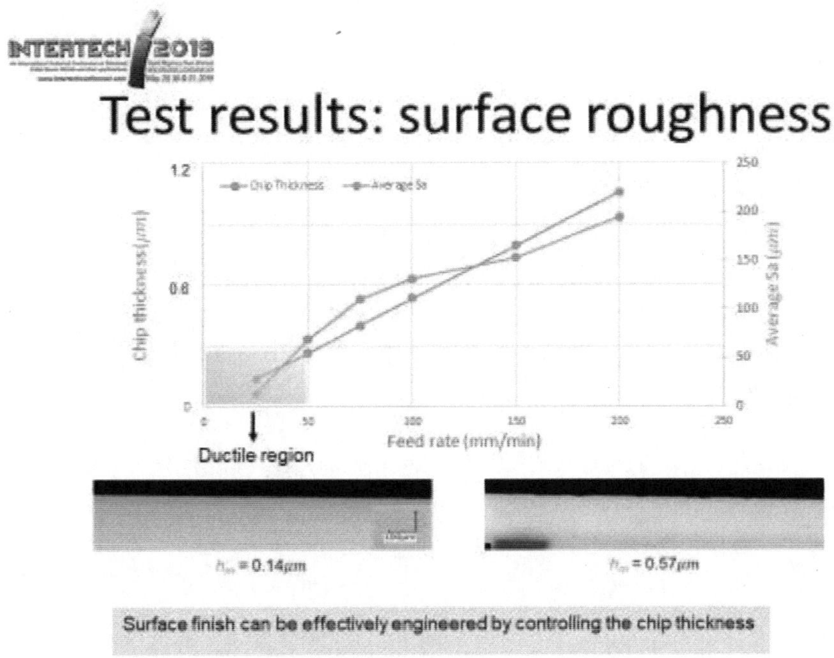

We see from the graph below that the number of flutes required to maintain an undeformed chip thickness of 0.14 microns at 400 m/min is 64 flutes. With a lot of the worlds' largest manufacturing companies searching for ways to use existing equipment in a flexible demand plan for future products, this understanding enables us to use tool design to meet productivity needs rather than spending hundreds of millions on new equipment.

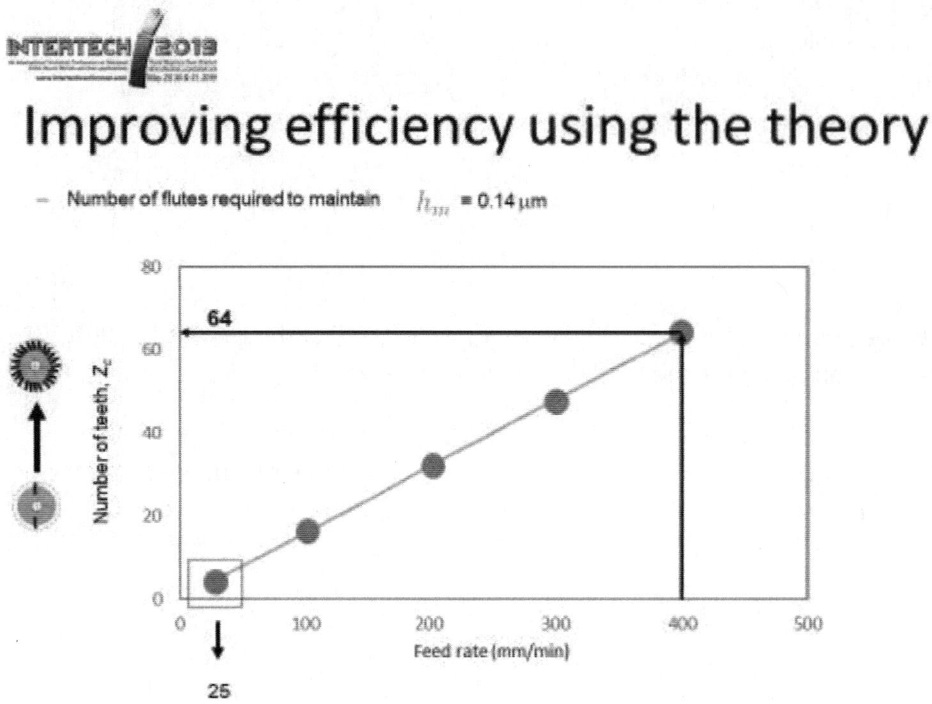

Milling of Sapphire – Case Study 2

With the same tool, equipment and machining conditions as used in zirconia machining, we now look at machining sapphire in a ductile regime. With a tighter band for maximum undeformed chip thickness, we need to be more careful with sapphire.

The following cutting conditions are used;

Cutting Operation: Edge trimming

Workpiece: Sapphire

Tool: 4 mm diameter brazed PCD endmill

Coolant: Water

Cutting parameters:

n = 12500 RPM

v_f = 25-100 mm/min

a_e = 0.05 mm

Z_c = 4 flutes

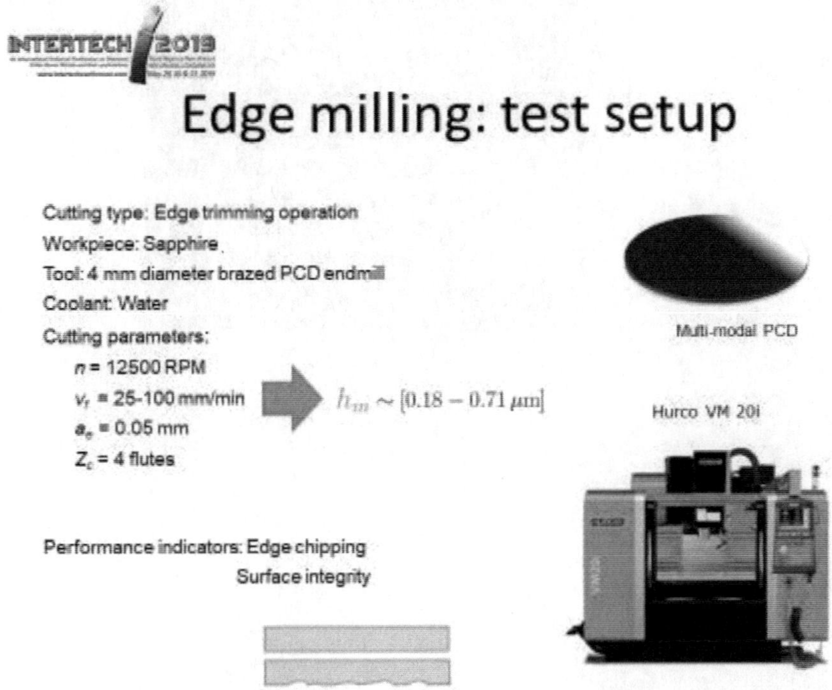

By looking at opposite ends of the recommended maximum undeformed chip thickness, we visibly see a rougher surface with greater h_m values approaching the upper limit. At the upper limit, surface cracks begin to show as well as surface pitting.

Milling of Glass – Case Study 3

The glass for phones and watches is currently processed using EP tools made from diamond grits. However, the customer is always looking to improve the performance of the manufacturing process and ultimately to reduce the cost. This is particularly true as the glass used in these applications are consistently being upgraded. Having to machine harder glass usually means higher cost and lower productivity. Another challenge is that we need to produce enough products during the ramping up for the new product launch season. Therefore, there is a high requirement on the productivity and yield of the manufacturing.

With these challenges, EP tools are gradually losing their advantage. The EP tool life is limited which means much time is wasted during tool changing. Therefore, our approach is to machine glass using a PCD tool. The advantages PCD tools have over EP tools are the longer tool life, more control of workpiece surface integrity by using an engineering approach than random grit orientation. This will be shown in the following slides.

The foundation of the proposed approach is the brittle to ductile transition region of the brittle materials. For this idea to work, a defined edge tool is needed and for each of the cutting edge, the load per tooth needs to be controlled below the critical load per tooth. h_m of 0.29 microns was selected at the upper end of the recommended target. By working closely with capable tool making partners a multi-fluted tool processed by laser was designed. The methodology behind the design is that due to the critical load per tooth being maintained below the recommended value for glass, the amount of flutes was increased so that the glass can be machined without reducing the productivity.

Cutting type: Profile milling operation Workpiece: 131Q Glass Tool: EP and PCD Coolant: Water Cutting parameters: n = 24000 RPM v_f = 1300 mm/min a_e = 0.015 mm Z_c = 50 flutes	

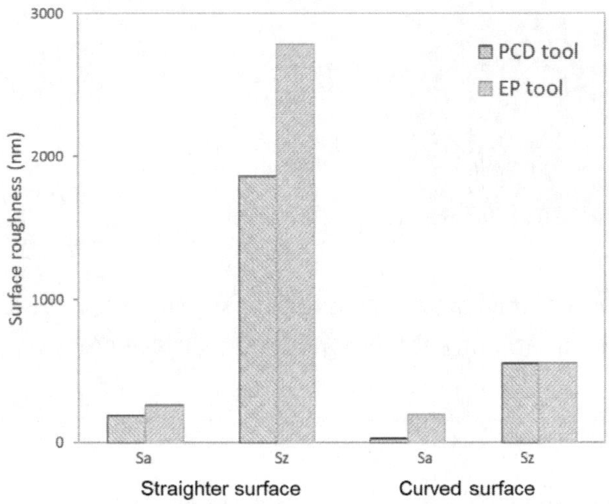

50% improvement in the S_z in the straighter section of the glass using defined edge tools

Milling of Glass – Case Study 4

The surface roughness target for the phone ear slot is below 150 nm after machining with the PCD tool.

For machining the ear slot in the phone screen, the following details describe the conditions used:

- Cutting operation: End milling
- Workpiece: 131Q Glass
- Tool: 1 mm diameter
- Coolant: Water
- Cutting parameters:
  n = 28000 RPM
  v_f = 25 - 200 mm/min
  a_e = 0.010 and 0.020 mm
  Z_c = 18 flutes

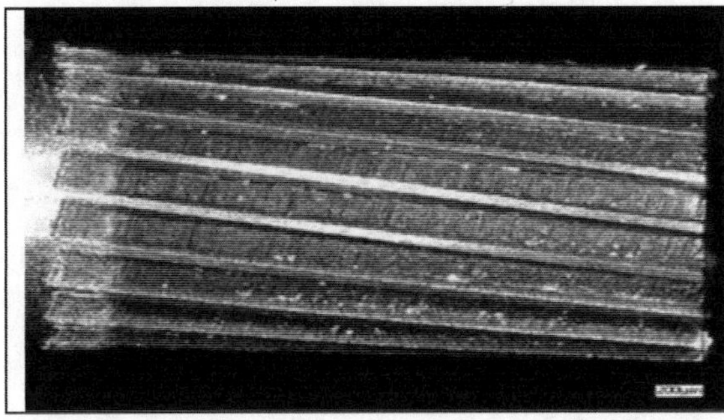

Shown here is a 1 mm diameter laser ablated tool produced using a carefully selected PCD from Element Six. There are 18 flutes which when used with the selected machining conditions ensure the glass in the ear slot is cut in a ductile manner. The undeformed chip thickness was between 0.013 – 0.12 microns.

We proved here that we can actually achieve the required surface finish of 150 nm at feed rates above 200 mm/min and bigger depths of cut than initially planned. This gives potential productivity improvements. At smaller depths of cut, we can achieve even better surface finish at higher feed rates meeting stringent specifications for surface roughness if required.

Summary

Understanding brittle materials machining theory and developing new PCD materials/formats are an enabler. Laser processing enables intricate tool geometries, which have not been possible through traditional processes. By laser ablation, edge roughness and cutting edge sharpness are independent of grain size.

Title	**Milling Glass**	**Milling Zirconia**	**Milling Sapphire**
h_m (mm)	0.10-0.20	0.15-0.25	0.05-0.2
PCD attributes	Bi-modal & Thick PCD	Bi-modal & Thick PCD	Bi-modal & Thick PCD
Result	Improved surface finish, productivity and tool life compared to EP tool	Good control of surface roughness via chip thickness	Keeping the chip thickness below 0.18mm eliminates edge chipping

Laser processing allows new 3D geometries of tools, textured surfaces to enhance cutting, and exploitation of new grades which would be expensive to process by EDM or grinding. Element Six is working on new platforms and products to allow efficient processing by laser ablation. We find laser efficiently processes our PCD with a smooth, sharp cutting edge, regardless of grain size, by cutting grains rather than removing whole grains at a time.

Thanks to Matthew Lay and Nastja Macerol for their support, help and collection of experimental data.

TARIFFS AND TRADE ACTIONS AFFECTING THE SUPERABRASIVES INDUSTRY

By Daniel B. Pickard, Wiley Rein LLP

Numerous international trade developments over the past two of years should be of the highest concern to the superabrasives industry. The current Administration has responded to ongoing challenges with an array of new and traditional measures. This white paper provides an overview of traditional and non-traditional actions taken by the Trump Administration in regard to international trade issues, as well as providing a case study involving diamond sawblades.

The Trump Administration Team and Strategy

Three key officials in the current Administration who are responsible for developing and implementing trade policy are: U.S. Trade Representative Robert Lighthizer; White House Office of Trade and Manufacturing Policy Director Peter Navarro; and Secretary of Commerce Wilbur Ross.

U.S. Trade Representative Robert Lighthizer is a Georgetown University Law Center graduate and was a practicing international trade attorney for several years before and between periods of government service. Ambassador Lighthizer previously served as the Chief of Staff for the Senate Committee on Finance under Senator Bob Dole, and then as Deputy U.S. Trade Representative under President Ronald Reagan. President Donald Trump appointed Ambassador Lighthizer to his current post as United States Trade Representative in 2017.

Director of the White House Office of Trade and Manufacturing Policy (OTMP), Dr. Peter Navarro, holds a PhD in Economics from Harvard and is the author of several books including *The Coming China Wars* (2006) and *Death by China* (2011). Dr. Navarro's Office of Trade and Manufacturing Policy was created by President Trump in 2017. The OTMP's mission is to "defend and serve American workers and domestic manufacturers while advising the President on policies to increase economic growth, decrease the trade deficit, and strengthen the United States manufacturing and defense industrial bases." In his capacity as OTMP Director, Dr. Navarro has advocated reducing or eliminating the trade deficit and repatriating global supply chains.

Secretary of Commerce Wilbur Ross, a Harvard MBA, was an investment banker specializing in corporate restructuring. Secretary Ross is both a policy advisor and a spokesman for the Administration's trade initiatives and the Secretary is an influential voice both within the Administration and in the public square.

The Trump Administration's trade strategy can be viewed on three distinct bases, namely: unilateral actions, bilateral negotiations, and multilateral disruption. *First*, the Trump administration is acting unilaterally to remedy certain perceived problems where years of negotiations have failed. *Second*, this unilateral action is being used as leverage in bilateral negotiations to force changes to trade behavior. *Third*, both the unilateral actions and bilateral negotiations are conducted in a context of potential changes to the multilateral system.

The Current Challenge

China's state-led industrial expansion has been viewed by many as harmful to U.S. interests and the global trading system. The Administration has voiced concerns regarding the fact that some significant portion of China's GDP growth is due to perceived violations of international treaty obligations, including the protection of intellectual property rights.

Alternative Remedies in Action

The Administration has "dusted off" two tools in the U.S. trade policy toolbox to address growing concerns about over capacity outside of the United States, failure to protect U.S intellectual property rights, and/or the effects of imports on U.S. national security. Both tools are unilateral measures that were more common in the years predating the World Trade Organization (WTO). The first is Section 232 of the Trade Expansion Act of 1964; the second is Section 301 of the Trade Act of 1974.

The first tool, Section 232 of the Trade Expansion Act, allows the President to adjust imports if the Commerce Department finds that the imports threaten to impair U.S. national security. In the steel industry, the Administration has used section 232 to impose a 25% tariff on iron and steel imports from most countries, but with certain exceptions. For example, Australia has a blanket exemption as compared to Argentina, Brazil and South Korea that are subject to tariff rate quotas. In the aluminum industry, the Administration imposed a 10% tariff on imports from all countries except Argentina and Australia with a similar quota for Argentina but none for Australia.

The second tool, Section 301 of the Trade Act of 1974, authorizes investigations into foreign trade policies and practices that (1) violate trade agreements, or (2) are unjustifiable and burden or restrict U.S. commerce, or (3) are unreasonable or discriminatory and burden or restrict U.S. commerce. The Act authorizes the President, at the recommendation of the U.S. Trade Representative, to take "appropriate and feasible" action to enforce agreements and respond to such policies or practices.

Section 301 has been used by this Administration to address Chinese intellectual property violations, among other concerns. In a Section 301 investigation conducted under the current Administration, USTR found that China was using unfair trade practices to dominate access to important raw materials. Key findings of the investigation included that, "A longstanding focus of China's foreign investment has been the acquisition of mineral deposits ... with a particular aim to mitigate China's reliance on resource imports." The investigation also revealed that, "China pursues an outbound industrial policy with government capital and highly opaque investor networks to facilitate high-tech acquisitions abroad."

On March 22, 2018, USTR found that Chinese intellectual property and technology transfer policies were violations within the meaning of Section 301. On June 20, 2018, USTR announced a 25% tariff on 818 tariff lines covering $34 billion in Chinese imports effective July 6, 2018. On the day the tariff took effect, China retaliated in kind. A month later, on July 17, 2018, USTR proposed a 10% tariff on an additional $200 billion in Chinese imports. Then, on August 16, 2018, USTR announced a 25% tariff on 279 additional tariff lines covering $16 billion in Chinese imports effective August 23, 2018.

Subsequently, President Trump announced via tweet that the current "Tranche 3" import duties on $200 billion in Chinese goods would rise from 10% to 25% effective Friday, May 10, 2019. On May 8th, the Chinese delegation arrived in Washington, D.C. Two days later, Ambassador Lighthizer released the following statement regarding additional action under Section 301: "Earlier today, at the direction of the President, the United States increased the level of tariffs from 10 percent to 25 percent on approximately $200 billion worth of Chinese imports. The President also ordered us to begin the process of raising tariffs on essentially all remaining imports from China, which are valued at approximately $300 billion."

On August 13, 2019 the United States Trade Representative (USTR) issued two lists of Chinese-origin goods that would be subject to 10% tariffs. The first list, List 4A, identified tariff lines that would be subject to additional duties of 10% effective September 1, 2019. The second list, List 4B, identified certain tariff lines for which duties would not go into effect until December 15, 2019. Products on List 4B include, among other things, cell phones, laptop computers, video game consoles, certain toys, computer monitors, and certain items of footwear and clothing. The announcement came after the Trump Administration indicated on August 1, 2019 that it intended to impose additional Section 301 tariffs covering $300 billion in consumer and industrial items, nearly all Chinese imports not subject to the prior three rounds of tariffs.

On Friday, August 23, 2019, after the Chinese Government unveiled its plan to impose additional tariffs on approximately $75 billion in annual U.S. imports, the Administration made two announcements of its own. First, the Administration announced that it would increase the existing and planned Section 301 tariffs on China by 5%. Specifically:

- Chinese-origin goods subject to the first three rounds of Section 301 tariffs, i.e., Tranches 1, 2, and 3, would be subject to 30% duties effective October 1, 2019. Currently, the Section 301 tariff rate applicable to such goods is 25%.
- Chinese-origin goods subject to the forthcoming Tranche 4A and Tranche 4B tariffs would additional duties of 15%, rather than the previously announced 10%.

Recent reporting has indicated that US-China negotiations, including the Section 301 tariffs, will begin again shortly. However, this is far from a foregone conclusion.

Traditional Actions

In addition to the recently "dusted off" tools, the Administration continues to employ traditional tools to counteract what they perceive to be unfairly priced and/or injurious imports. These traditional remedies include Section 201 investigations, antidumping and countervailing duty investigations.

Section 201 allows the President to impose relief—usually tariffs or quotas—if the International Trade Commission (ITC) determines that a surge in imports is a substantial cause of serious injury or threat thereof. Invoking this section does not require a showing of "unfair trade," *i.e.*, dumping or subsidies. Section 201 is generally a global measure, and it has heightened requirements for relief on NAFTA imports.

Antidumping and countervailing duties (AD/CVD) are imposed by the U.S. Department of Commerce (DOC) to remedy unfair trade practices that injure or threaten a U.S. industry. Notably, the Trump Administration has broken with previous administrations by "self-initiating" antidumping and countervailing duty cases against aluminum sheet from China.

The DOC will impose antidumping duties when a foreign producer sells a product in the United States at prices below the product's normal value. Normal value is defined as either a foreign producer's sales price in its home market, a foreign producer's sale price in a third country, or a foreign producer's cost of production plus their profit.

The DOC will impose countervailing duties on imports that it finds to be improperly subsidized. Examples of potentially countervailable subsidies include grants, loans at preferential rates, preferential relief from taxes and fees, and provision of inputs at artificially low prices.

Antidumping and countervailing duties can be imposed only if the U.S. International Trade Commission finds that the unfairly traded imports are materially injuring the domestic industry or threaten to injure the domestic industry. To make this determination, the ITC traditionally examines the volume, price effects, and impact of the imports. Material injury is defined as "harm that is more than inconsequential, immaterial, or unimportant." Importantly, imports do not have to be the only cause of harm to the industry.

Case Study

One well known example of a trade remedy investigation involving the superabrasives industry is the antidumping order on diamond sawblades. An ad hoc coalition of mostly small, family-owned U.S. diamond sawblade manufactures filed petitions that alleged that Chinese sawblades were being sold below normal value and that such imports were injuring domestic production operations. As a result, for the past several years, the majority of imports of such diamond sawblades have been subject to antidumping duty deposits of 82.12% and as a result there has been a significant decrease in the volume of such imports (see graph below).

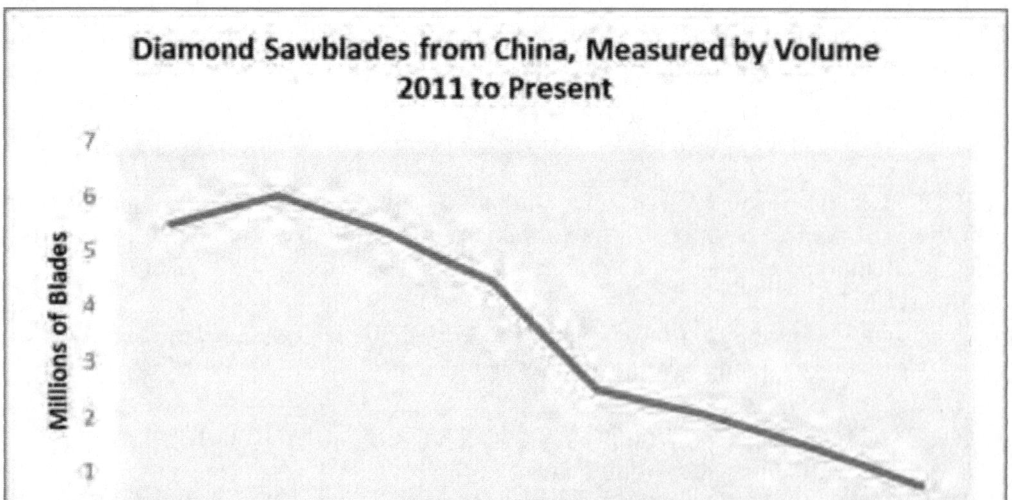

Conclusion

Trade issues are a high priority for the current Administration. Part of this Administration's strategy is to use unilateral tools like Section 232 and Section 301 that were more common in the pre-WTO years. But traditional options like AD/CVD and

safeguards investigations are still in play and have been used with demonstrable success. The situation remains fluid and to a certain degree unconventional. It will be imperative for the foreseeable future for both domestic producers and importers to closely monitor this extraordinarily dynamic scenario.
For Questions Contact: Daniel B. Pickard, Wiley Rein LLP, 202.719.7285, dpickard@wileyrein.com

International Journal of Refractory Metals & Hard Materials 85 (2019) 105051

Contents lists available at ScienceDirect

International Journal of Refractory Metals & Hard Materials

journal homepage: www.elsevier.com/locate/IJRMHM

Characterization of the residual stresses introduced by a new joining method in diamond and tungsten carbide composites

Olivier Lavigne[a,*], Vladimir Luzin[b], Marco Mendez[a], Abds-Sami Malik[c], Oscar Carrasco[a], Floriana Salvemini[b]

[a] Hyperion Materials & Technologies, Pol. Ind. Roca - C/ Verneda s/n, Martorelles, Barcelona 08107, Spain
[b] Australian Nuclear Science & Technology Organisation (ANSTO), Lucas Heights, NSW 2234, Australia
[c] Hyperion Materials & Technologies, 6325 Huntley Road, Worthington, OH 43085, USA

ARTICLE INFO

Keywords:
Diamond composite
Cemented carbide composite
Co-sintering
Bond strength
Neutron diffraction
Residual stresses
Neutron tomography

ABSTRACT

In this work, a co-sintering method was used to attach diamond to cemented carbide composites. The joining method consists of sintering a green part (ring) of cemented carbide (CC) around a thermally stable diamond composite (TSDC) part (plug) to radially contain it. During the sintering step, the green body shrinks to a controlled level and therefore forms interference fit between the two parts (mismatch between the inner diameter (ID) of the CC ring and the outer diameter (OD) of the TSDC cylinder). The residual stresses induced by this process as well as the bond strength between the CC and the TSDC parts were quantitatively evaluated. It is shown that the interface pressure between the two parts, and the level of residual stresses, increased with increase in the designed interference fit, as well as with the increase of the ID/OD ratios of the CC ring.

For the chosen material combination (cemented carbide ring comprised 90 vol% WC and 10 vol% Co with medium coarse WC grains; diamond composite plug comprised 84 vol% diamond and 16 vol% SiC), the values of the hoop stresses at the interface in the CC ring measured by neutron diffraction was determined to be between 150 MPa and 550 MPa, depending on the ID/OD ratio. It was also found that for a given ID/OD ratio, the increase of the designed interference fit had little effect on that attained due to the plastic deformation of the cemented carbide material at the interface during the sintering (dynamic) process. A mechanical bond around 60 MPa was nevertheless achieved.

1. Introduction

Polycrystalline diamond (PCD) cutters are extensively used in the oil, gas, geological drilling, mining, engineering and manufacturing industries all around the world [1,2]. PCD materials present superior wear resistance and hardness properties. Cobalt or other transition metals are traditionally used as sintering aids in forming the PCD at high pressure and temperature. Since these metals remain in the PCD, its operational temperature is limited to approximately 800 °C, due to the coefficient of thermal expansion mismatch between the diamond and metal, as well as the graphitization of diamond in the presence of the metal. The most modern type of thermally stable diamond composites (TSDC) do not contain transition metals, but instead employ reactive sintering with Si to form a diamond/SiC composite, that enables much higher operational temperatures (up to 1400 °C) [3]. The reactive sintering is accomplished either within the diamond stability field, ~1500 °C and ~5.5 GPa, or in the graphite phase field at ~1550 °C and ~2–3.5 GPa, [4,5].

TSDC parts are generally produced with small dimensions to limit costs and subsequently have to be attached to another material such as cemented carbides (CC) for example, typically by brazing. To avoid this joining step and to allow higher tool operational temperatures, a co-sintering method has recently been developed [6]. This method consists of sintering a green cemented carbide part around the TSDC part to radially contain it. During the sintering step, the green body shrinks to a controlled level and therefore forms interference fit between the two parts (mismatch between the inner diameter of the CC ring and the outer diameter of the TSDC cylinder). Notwithstanding this new process providing a mechanical bond between the two parts, residual macro-stresses are introduced close to the interface with the CC part in tension and the diamond composite in compression. This can potentially affect the tool service behaviour as during operational use the diamond composite is subjected to external fluctuating, or cyclic, and rapidly applied loads. Such fluctuating loads induce cyclic stresses that often

* Corresponding author.
E-mail address: Olivier.lavigne@hyperiommt.com (O. Lavigne).

https://doi.org/10.1016/j.ijrmhm.2019.105051
Received 24 May 2019; Received in revised form 5 August 2019; Accepted 6 August 2019
Available online 07 August 2019
0263-4368/

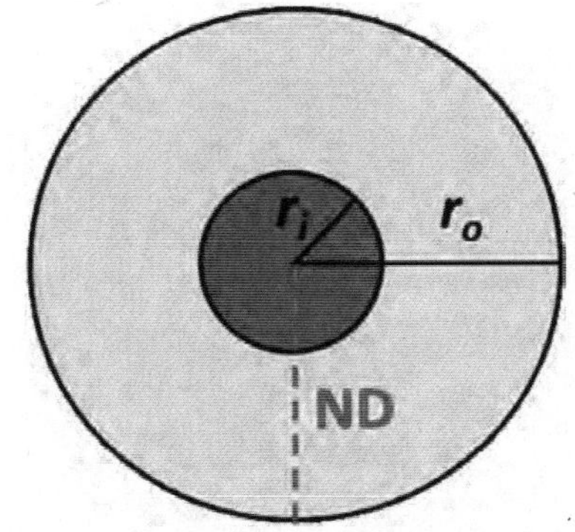

Fig. 1. Cross sectional view of the co-sintered TSDC/CC sample assembly showing the location of the neutron diffraction measurements (ND) along a radius of the part (r). r_o and r_i are respectively the outer and inner radii of the CC ring.

result in failure by fatigue.

The quantitative characterization of the CC and TSDC materials in terms of macro-stress is therefore an essential element in stress control and mitigation in the high-demanding targeted applications. In this study the level of residual stresses introduced by the new joining process in the CC and the TSDC parts were measured using the neutron diffractometer, KOWARI, at the OPAL research reactor (ANSTO, Sydney) [7]. The results are presented as function of the TSDC diameter and the level of interference fit between the two parts.

2. Material and methods

2.1. Samples

The samples are 5 mm high cylinders of CC (90 vol% WC and 10 vol % Co, medium-coarse [8] WC grains) containing a diamond composite cylinder (84 vol% diamond and 16 vol% SiC) in its centre (Fig. 1). Two different external diameters were used for the CC, 17.8 mm and 24 mm, and three for the diamond composites, 6.4 mm, 8.6 mm and 12.4 mm.

The green cemented carbide rings were sintered around the diamond composite cylinders at 1400 °C using the hot isostatic pressing process and to achieve different final interference fit levels (between 3% to 10%) between the two composites.

The interference δ is defined as:

$$\delta\,(\%) = \frac{r_i - r_{i\ stand\ alone\ CC}}{r_i} \tag{1}$$

where r_i is the inner diameter of the CC ring co-sintered with the diamond composite cylinder (TSDC) at its centre (Fig. 1) and $r_{i\ stand\ alone\ CC}$ is the inner diameter of the cemented carbide ring sintered without the diamond composite part.

2.2. Neutron tomography

The neutron tomography investigations were performed on the neutron imaging station DINGO [9] at the Australian nuclear research reactor OPAL using a Photometrics 5 Megapixel (5056 × 2968) IRIS CMOS camera. The instrument was set to yield an image at a pixel size of ~13 µm by coupling a 100 mm lens to a 30 µm thick Gadox scintillation screen. The tomographic scan consisted of 1800 angular projections covering 180°. As the counting statistic is low at this configuration, 3 accumulations with an exposure time of 20 s each were acquired at each step angle and integrated to improve the image quality. With the field of view being 750 × 450 mm^2, 5 samples were stacked vertically and scanned simultaneously.

The data were processed using the Octopus code for tomographic reconstruction [10]. The measured slices were recomposed and evaluated using ImageJ [11] and Avizo 9.4 [12] software.

In Avizo, the tomographic reconstruction of each sample was first aligned to a reference xyz coordinate system and resampled perpendicular to the axis of the cylinder, cropped and exported. ImageJ was used to compute the image projection of the average attenuation coefficient along the sample axis. Based on histogram thresholding, the region occupied by the sample was determined and selected. The profile plot of the attenuation coefficient integrated around concentric circles as a function of distance from the centroid of the selection (approximate to a circle) was performed. The integrated attenuation coefficient was divided by the number of pixels in the circle that is also part of the image, yielding normalized comparable values [13].

2.3. Mechanical characterizations

2.3.1. Hardness measurements

Hardness measurements at a load of 30 kgf were carried out on the cemented carbide part in accordance with ISO 28079 using a semi-automated Emco-Test DuraScan 50 G5 apparatus equipped with a diamond Vickers indenter.

2.3.2. Bond strength measurements

The bond strength at the interface between the two composites was evaluated with a push-out test using a Shimadzu Autograph AG-X plus series universal tensile tester. The CC ring part of the sample was held on a hardened steel support, while a compressive load was applied at a rate of 2 mm/min to the diamond composite plug using a hardened steel rod. The maximum force required to push-out the TSDC was recorded for different levels of interference (between 3% and 10%), as well as for different ID/OD ratios (between 0.17 and 0.5). Four samples per condition were tested for consistency.

2.4. Residual stress measurements

Residual stress profiles were measured with neutron diffraction (ND) using the KOWARI strain scanning diffractometer at the OPAL nuclear research reactor. The measurements were done in their assembled (under stress) conditions across the CC and the diamond composite (TSDC) parts for five samples (Table 1) and compared against stand-alone (unstressed) data to derive macro-strain. Using a gauge volume of 1 × 1 × 5 mm^3, measurements were taken in the CC part along a radius at steps of 0.5 mm using the WC(201) reflection. Three strain components hoop, radial and axial were measured at each position. Since the macro-strain and macro–stress in the diamond composite plugs were uniform and equal-biaxial (plane-stress approximation of a thin ring-and-plug system), only one measurement was needed. The diamond (311) and SiC (220) reflections were used to measure the diamond composite part with a 2 × 2 × 5 mm^3 gauge volume from the centre of the cylinder.

2.5. Residual stress analytic solution

The overall macro-stress elasticity problem for such ring-and-plug systems is well known [14] and an analytical solution can be fit to the experimental data to fully resolve the stress problem.

Table 1
Sample matrix.[a]

CC outer diameter (OD, mm)	CC inner diameter (ID, mm)	ID/OD	Interference (δ, %)		
			3	6	9
17.8	6.4	0.36	a		b
17.8	8.6	0.47		c	
17.8	12.4	0.70	d		e

[a] Note: The CC inner diameter is equal to the TSDC outer diameter.

The pressure at the interface, p, is written:

$$p = \frac{\delta_r}{\frac{r_i}{E_{CC}}\left(\frac{r_0^2 + r_i^2}{r_0^2 - r_i^2} + \nu_{CC}\right) + \frac{r_i}{E_{TSDC}}(1 - \nu_{TSDC})} \tag{2}$$

where δ_r is the radial interference between the two parts (in mm), E_{CC} and E_{TSDC} are the elastic modulus of the cemented carbide (625 GPa) and the diamond composite (890 GPa) respectively, and ν_{CC} and ν_{TSDC} are the Poisson's ratio of the cemented carbide (0.25) and the diamond composite (0.25) respectively.

In the CC ring the hoop stress, $\sigma_{\theta,\ CC}$, is tensile:

$$\sigma_{\theta,CC} = p\frac{r_i^2\left(1 + \frac{r_0^2}{r^2}\right)}{r_0^2 - r_i^2} \tag{3}$$

and the radial stress, $\sigma_{r,\ CC}$, is compressive:

$$\sigma_{r,CC} = p\frac{r_i^2\left(1 - \frac{r_0^2}{r^2}\right)}{r_0^2 - r_i^2} \tag{4}$$

In the diamond composite plug, the hoop stress, $\sigma_{\theta,\ TSDC}$, is equal to the radial stress, $\sigma_{r,\ TSDC}$, and is compressive and uniform across the plug:

$$\sigma_{\theta,TSDC} = \sigma_{r,TSDC} = -p \tag{5}$$

The axial stress in the CC ring and diamond composite (TSDC) plug is zero (plane-stress approximation of a thin ring-and-plug system).

3. Results

3.1. Neutron tomography characterizations

The neutron tomography measurements were conducted on samples a to e (see Table 1). No macroscopic defects or cracking were observed in any samples at a spatial resolution of 50 μm. Fig. 2 shows examples of both xy and xz cross sections obtained for sample a.

The profile of the neutron attenuation coefficient, μ, averaged on the whole sample volume, as a function of the distance from the centroid of the samples is shown in Fig. 3. A gradient for this coefficient can be observed in the CC rings, with values increasing from the TSDC/CC interface towards the outer diameter of the CC ring. At the same time, no gradient in the through-thickness (axial direction) was detected for any sample.

3.2. Mechanical characterizations

The nominal hardness value measured for the cemented carbide grade under consideration was 1460 HV30. Fig. 4 presents the HV30 profiles measured in the carbide ring for a sample with the TSDC contained and a sample after the TSDC extraction (ID/OD ratio = 0.36 and δ = 6%). Both profiles are similar considering the standard deviation on the measurements. Close to the TSDC/CC interface the hardness values are higher (1490 HV30) than the nominal measured hardness. The hardness values decrease with increase of the part radius to reach approximately 1400 HV30 close to the outer diameter of the cemented carbide ring.

The bond strength between the two composite parts for TSDC/CC assemblies with an ID/OD ratio of 0.27 was measured for interference fits of 3%, 3.75%, 5.25%, 6.5%, 7.75%, 9% and 10.25% (Fig. 5a). Considering the standard deviation on the measurements, this figure shows that the bond strength increases moderately with increase in the interference fit. However, as shown by Fig. 5b, increasing the ID/OD ratio of the CC ring significantly increases the bond strength values.

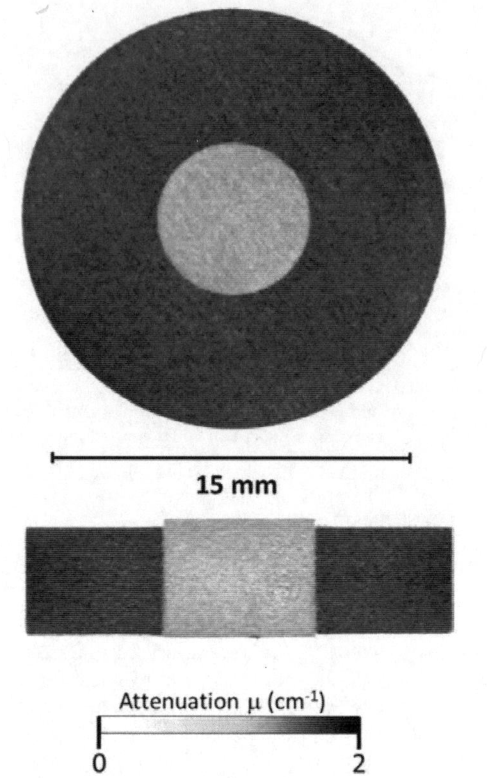

Fig. 2. Example of a tomographic reconstructed view for sample a, Table 1. The top view shows the xy cross section with the xz cross section (side view) below it. The scale at the bottom shows the extent of the attenuation.

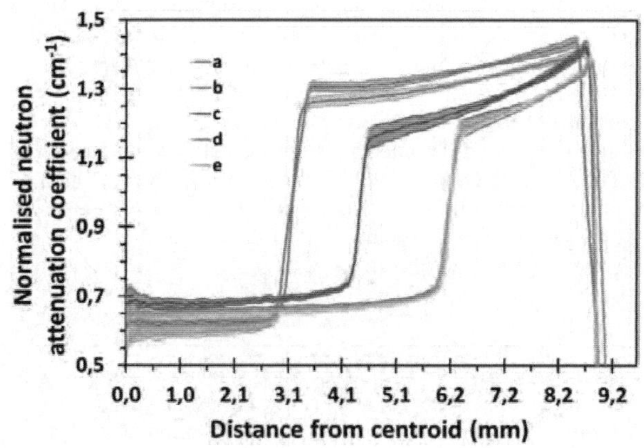

Fig. 3. Normalized neutron attenuation coefficient profiles for TSDC/CC assemblies a to e (*see* Table 1).

3.3. Residual stresses measurements

Fig. 6a–e present the residual stress profiles along the radius of the assembled TSDC/CC parts for the five selected samples. The analytical solution is plotted as continuous lines superposed over the experimental data. These figures show that there is scatter in the experimental data, especially in the CC ring, because of smaller gauge volume and WC being a weaker neutron scatterer than diamond. However, considering the experimental errors bars (~100 MPa in WC), the model appears to be in good statistical agreement with the measurements. The interface pressure, p, between the CC and TSDC parts can be estimated from fitting the experimental data with the analytical solution. This is presented in Fig. 6a–e which show that the interface pressure and the level

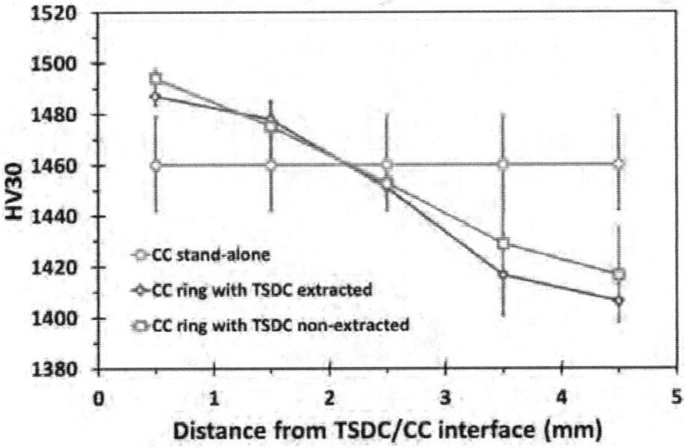

Fig. 4. HV30 profiles in the cemented carbide ring for a TSDC/CC assembly (ID/OD = 0.36, δ = 6%) before and after the TSDC extraction. The nominal hardness of the cemented carbide is also indicated.

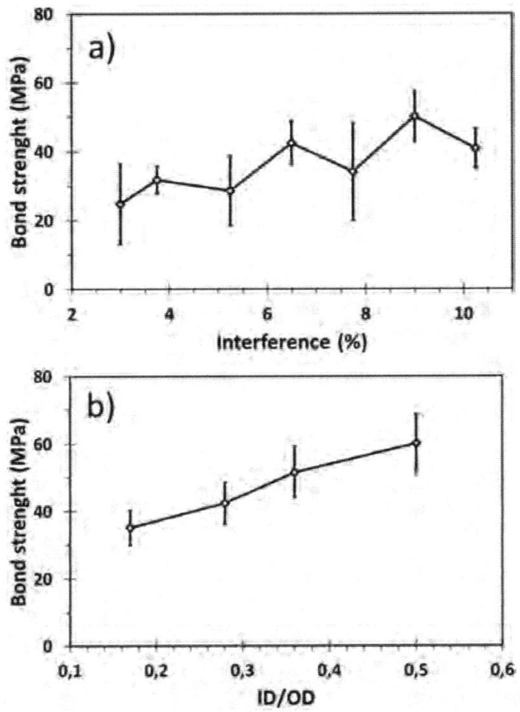

Fig. 5. Bond strength between the two composite parts as function of: a) the level of interference fit for TSDC/CC assemblies with an ID/OD ratio of 0.27; b) the variation of ID/OD ratio for TSDC/CC assemblies with a 6% interference fit between the parts.

of residual stresses increase with the increase of the designed interference fit, as well as with the increase of the ID/OD ratio of the CC ring. Tensile hoop stress in the CC part close to the interface was measured to be up to 600 MPa for the highest ID/OD ratio (Fig. 6d and e).

From the experimentally obtained values of p, the interference fit values, δ, can be estimated (Eq. (2)), and are presented in Fig. 6a–e. It appears that the measured values of δ, δm, are significantly lower than the intended designed levels, between 30 and 190 times, depending on the considered sample.

4. Discussion

The high neutron absorption in the CC rings (Fig. 3) comes from the Co matrix, as in comparison WC is neutron transparent ($\mu_{Co} = 2.53\,cm^{-1}$, $\mu_{WC} = 1.03\,cm^{-1}$). Subsequently the profiles measured in Fig. 3 imply the presence of a Co gradient in the CC rings, with the depletion zone being at the TSDC/CC interface. This gradient can be explained by the mechanical effect of the pressure applied by the solid TSDC on the CC during the liquid phase sintering that would push the Co away from the TSDC/CC interface [15]. Additionally, some level of graphitization occurred on the surface of the TSDC during the sintering cycle, since carbon could diffuse in the carbide part from the TSDC/CC interface and consequently generate a carbon gradient in the CC ring that would promote Co migration towards the outer diameter of the ring [16]. Some level of WC grain growth on a layer thickness < 100 μm was also observed close to the TSDC/CC interface, suggesting C intake in the CC. The gradient of Co explains the higher hardness values than the nominal one measured close to the TSDC/CC interface, as well as the decrease of the hardness values across the CC rings (Fig. 4).

The neutron diffraction measurements (Fig. 6) show that the attained level of interference fit between the two parts is generally significantly lower than the designed criterion. Moreover, for a given ID/OD ratio, increasing the designed interference fit has very little effect. For example, for the ID/OD ratio of 0.36, increasing the designed interference fit from 3% to 9%, actually only led to a real increase of 0.005% (from 0.042% to 0.047%). This explains the moderate increase of bond strength (Fig. 5a) from 28 MPa to 44 MPa by increasing the designed interference fit between the two parts from 3% to 10.25%.

The loss of interference between the two parts is due to the microstructural changes and the plastic deformation of the CC material at the interface during the dynamic sintering process. The green cemented carbide ring cannot shrink to the designed level during sintering due to the presence of the diamond composite plug and the material flows from the interface towards the outer areas of the part. For higher ID/OD ratios, this results in higher level of residual stresses in the cemented carbides rings. For example, from the ID/OD ratios of 0.36 to 0.70 the residual hoop stresses increase from approximately 150 MPa to 550 MPa at the interface, and from approximately 50 MPa to 250 MPa at the outer surface of the CC ring. The higher levels of residual stresses, for higher ID/OD ratio, induced by higher level of deformation in the CC rings, increase the interface pressure on the TSDC and therefore the bond strength between the two parts (Fig. 5b).

A point to note is that the whole system is in fact characterized by a single parameter, i.e. the interface pressure (Eq. (2)). This study shows that in principle, a single measurement (plus one reference) in the diamond composite part would be sufficient to reconstruct all stress profiles of all components and would define the whole stress field to adequate accuracy. Moreover, the TSDC part can be measured with higher accuracy (as a combination of diamond and SiC) of 10 MPa than the CC with 100 MPa error bars and the necessity to measure three stress components and three profiles. This scatter in the experimental neutron diffraction data is related to the fact that both the cemented carbide and the diamond composite materials are elastically hard and difficult to measure accurately with the required high resolution due to the small sample size.

5. Conclusions

It is shown that the co-sintering joining method provided a defect-free mechanical fit between the cemented carbide and the diamond composite materials with bond strength values up to 60 MPa for the material combination and ID/OD ratios considered in this study.

The interface pressure between the two parts, and the level of residual stresses, increased with the increase of the designed interference fit, as well as with the increase of the ID/OD ratios of the cemented carbide ring.

For a given ID/OD ratio, the increase of the designed interference fit had little effect on the actually achieved values due to the plastic deformation of the cemented carbide material at the interface during the

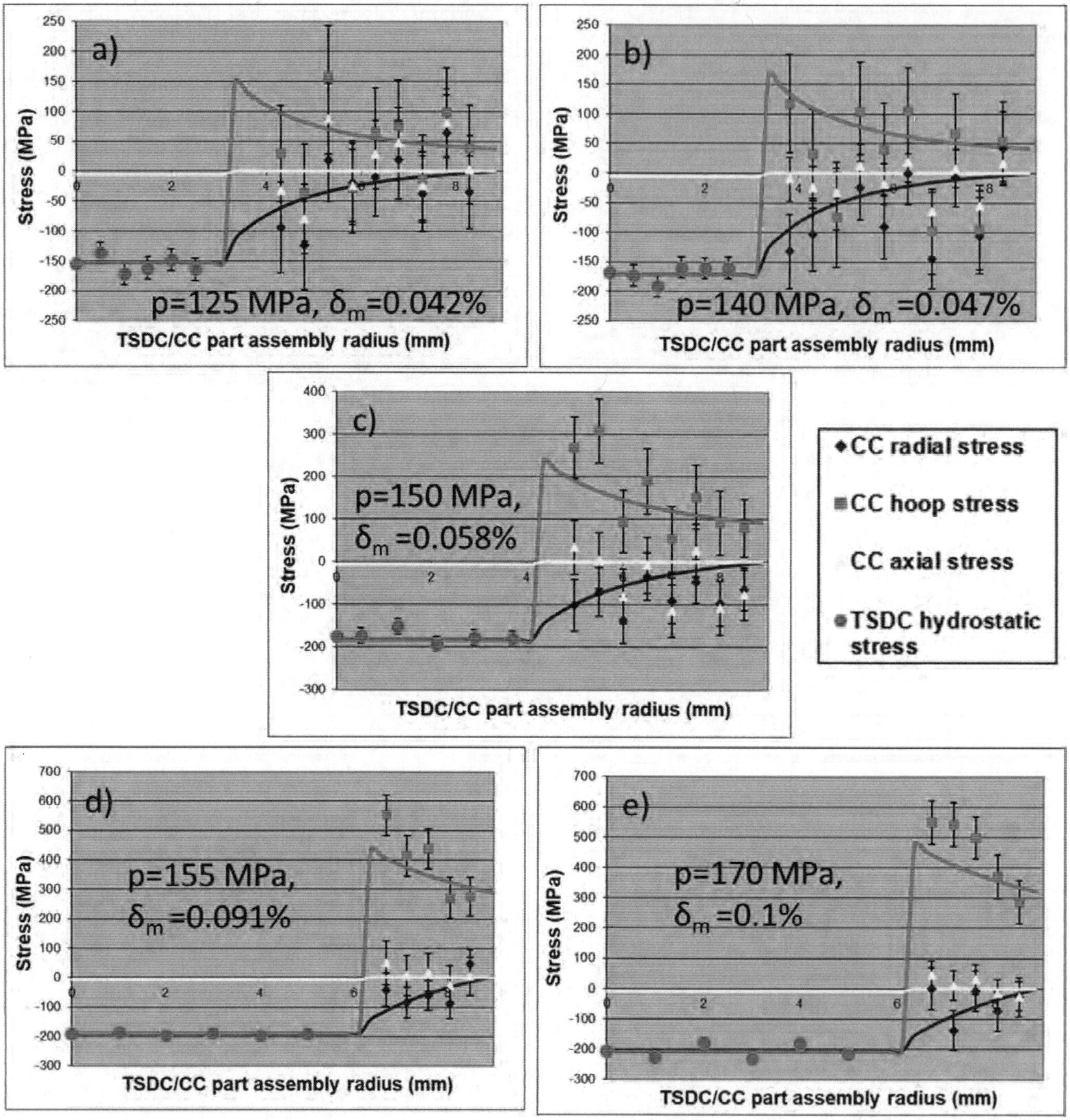

Fig. 6. a) to e) Residual stress profiles for the TSDC/CC assembly parts (*see* Table 1). Solid lines are the analytical solution. p and δm are respectively the interface pressure and the level of interference fit estimated from the model.

sintering (dynamic) process.

On an experimental point of view, it is shown that in principle, a single measurement (plus one reference) in the diamond composite part would be sufficient to reconstruct all stress profiles of all components and would define the whole stress field accurately enough.

References

[1] D. Belnap, A. Griffo, Homogeneous and structured PCD/WC–Co materials for drilling, Diam. Relat. Mater. 13 (2004) 1914–1922.

[2] F. Bellin, A. Dourfaye, W. King, M. Thigpen, The Current State of PDC Bit Technology, World Oil, 2010, pp. 41–46.

[3] J.N. Boland, X.S. Li, Microstructural characterisation and wear behaviour of diamond composite materials, Materials 3 (2010) 1390–1419.

[4] V. Luzin, G. Voronin, M. Avdeev, J. Boland, Microstresses in thermally stable diamond composites made by high pressure infiltration technique, Mater. Res. Proc. 4 (2018) 65–70.

[5] A.E. Ringwood, Diamond Compacts and Process for Making Same. Patent 4,948, 388, (1989).

[6] A.-S. Malik, T.C. Easley, S.A. Kaiser, T. Selinder, Diamond Composite Cutting Tool Assembled with Tungsten Carbide, US Patent 9,827,611 B2, Nov. 28, 2017.

[7] V. Luzin, J. Boland, M. Avdeev, X. Li, Characterization of thermally stable diamond composite material, Mater. Sci. Forum 777 (2014) 165–170.

[8] Cemented Carbide, Sandvik New Developments and Applications, Ref. No. H-9116

ENG (2005).
[9] U. Garbe, T. Randall, C. Hughes, G. Davidson, S. Pangelis, S.J. Kennedy, A new neutron radiography / tomography / imaging station DINGO at OPAL, Phys. Procedia 69 (2015) 27–32.
[10] M. Dierick, B. Masschaele, L. Van Hoorebeke, Octopus, a fast and user-friendly tomographic reconstruction package developed in LabView®, Meas. Sci. Technol. 15 (7) (2004) 1366–1370.
[11] C.T. Rueden, J. Schindelin, M.C. Hiner, B.E. DeZonia, A.E. Walter, E.T. Arena, K.W. Eliceiri, ImageJ2: ImageJ for the next generation of scientific image data, BMC Bioinformatics 18 (2017) 529, https://doi.org/10.1186/s12859-017-1934-z.
[12] https://www.thermofisher.com/au/en/home/industrial/electron-microscopy/electron-microscopy-instruments-workflow-solutions/3d-visualization-analysis-software.html.
[13] https://imagej.nih.gov/ij/plugins/radial-profile.html.
[14] S.R. Schmid, B.J. Hamrock, B.O. Jacobson, Fundamentals of Machine Elements, CRC Press, Taylor & Francis Group, Boca Raton, FL, USA, 2014, p. 626.
[15] O. Ther, Élaboration de carbures cémentés à gradient de propriétés par procédé d'imbibition réactive - Application aux inserts WC-Co et aux taillants en diamant polycristallin pour le forage pétrolier en conditions sévères, PhD Thesis Mines ParisTech, France, 2015 (In French).
[16] P. Fan, Z.Z. Fang, J. Guo, A review of liquid phase migration and methods for fabrication of functionally graded cemented tungsten carbide, Int. J. Refract. Met. Hard Mater. 36 (2013) 2–9.

*** Formatting Issue - Best Available Paper/Graphic ***

Paper #1938

Evaluation of the Influence of Impurities in Single Crystal Diamond on Wear Characteristics

Minori Teramoto
Sumitomo Electric Industries, Ltd.
Itami-shi, Hyogo, Japan

1. INTRODUCTION

Diamond differs in the type and content of impurities, and the form of its presence in the crystal, due to differences in the crystal growth conditions and environment. In particular, it is known that there is a remarkable difference in the amount and form of nitrogen impurities between synthetic diamond and natural diamond [1]. While nitrogen contained in synthetic diamond included as substitutional impurity, in natural diamond, nitrogen is present in aggregated forms. In some cases, multiple nitrogen atoms are bonded to a vacancy. The differences due to the content, existence form of such impurities affect the properties of the diamond material and may significantly change the tool performance. In addition, single crystal diamond has a large difference in mechanical properties depending on the crystal orientation. Therefore, it is important to recognize these effects in detail when we used diamond crystals as cutting tools or wear resistant tools. We established a method to evaluate the mechanical wear due to the sliding movement along a specific crystal orientation using a small diameter diamond wheel and a machining center. We report the results of evaluating the difference in wear characteristics using single crystal diamond with different types and contents of impurities.

2. EXPERIMENTAL METHOD

Focusing on the fact that natural diamonds used for tools are identified not only by impurities but also by color as an important factor, three kinds of natural diamonds with different colors shown in Fig. 1 were prepared. There are some types of nitrogen aggregation. In each sample, the presence of "A-center (2N)" composed of two neighboring substitutional nitrogen atoms, "B-center (4N-V)" which is a cluster of 4 nitrogen atoms surrounding a vacancy and planar aggregation "Platelets" was observed by infrared spectroscopy. On the other hand, HPHT synthetic diamonds do not have these aggregates but contain some isolated substitutional nitrogen atoms "C-center". The nitrogen content in the crystal can be estimated from the absorption characteristic of 1130cm-1 of "C-center" by infrared spectroscopy. The nitrogen content of the 7 synthetic crystals prepared for evaluation is from 100 to 700 ppm.

Using these HPHT synthetic diamonds and three kinds of natural diamonds were subjected to the wear tests using metal bonded diamond wheel. A schematic diagram of the wear test method is shown in Fig.2. The test pins was prepared as below. The test pieces of φ 2 × 2 mm prepared from each diamond specimens were brazed to sample holders, and processed into a conical shape with a tapered angle of 120 °. And a test surface of φ0.3 ± 0.005 mm was formed by scaife polishing at the tip of the cone of each sample. A φ30 mm metal bond grinding wheel diamond disk (grit size 3000) was clamped on the main axis of the machining center so that the diamond disk is vertical. From the lower part of the grinding wheel surface rotated at high speed at 6000 rpm (sliding speed 260-340 m / min), the test pin was pressed and slid with a constant load of 2.83N. After sliding for 60 minutes, the change in diameter of the test surface was measured to derive the wear rate.

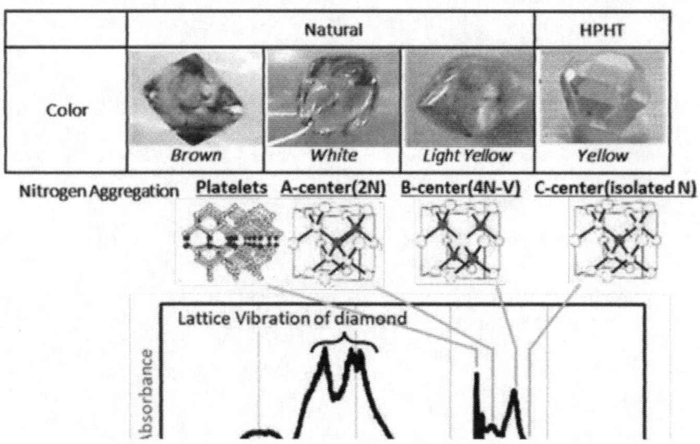

*** Formatting Issue - Best Available Paper/Graphic ***

Fig.1 Nitrogen existance form in diamond sample and an example of infrared absorption spectra pattern of diamond

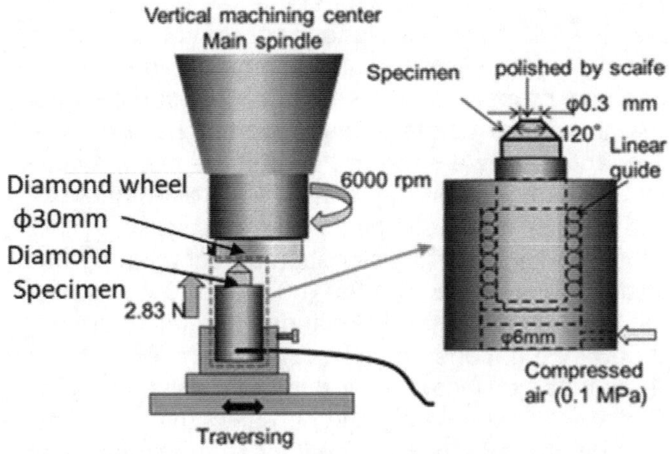

Fig.2 Schematic diagram of evaluation method

3. RESULTS AND DISCUSSION

3-1. WEAR RESISTANCE EVALUATION

The results of the evaluation of wear rate for grinding wheel are shown in Fig.3. For both single crystal diamonds of HPHT synthetic diamond and natural diamond, the wear rate of the (001) <100> direction (Fig. 3-a), which is the easy wear direction, is at least 100 times higher than that of the (001) <100> direction (Fig.3-b), which is the wear resistance direction. In the easy wear direction (001) <100>, the wear rate of HPHT synthetic diamond increases rapidly with nitrogen content of 400 ppm or more. In addition, the wear rate of *White* is at least ten times higher than *Brown* or *Light Yellow*. On the other hand, in the wear resistance direction (001) <110>, the wear rate of the samples of the nitrogen content 700ppm of HPHT synthetic diamond and *White* of natural diamond are significantly high.

In each of HPHT synthetic diamond and natural diamond, the difference ratio of the maximum wear rate to the others of the wear resistance direction was smaller than the easy wear direction.

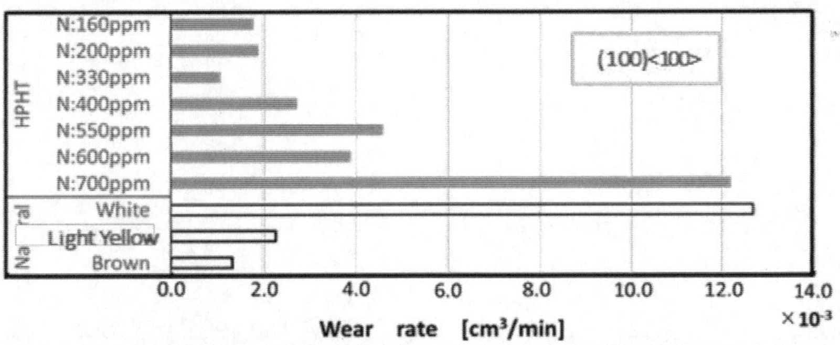

Fig.3-a Wear rate of each sample for diamond grinding wheel – easy wear direction (100)<100>

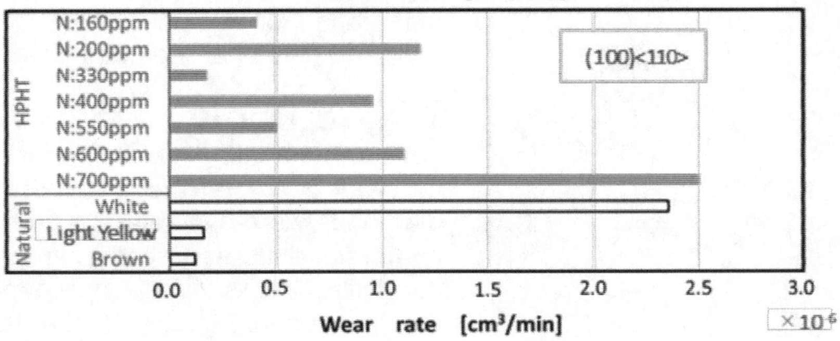

Fig.3-b Wear rate of each sample for diamond grinding wheel – wear resistance direction (100)<110>

3-2. NITROGEN AGGREGATION MORPHOLOGY AND MECHANICAL PROPERTIES IN NATURAL DIAMOND

The absorption spectra of three natural diamonds used in the experiment were analyzed, and the height of the peak from each aggregation form of nitrogen were compared. The results are shown in Fig.4. *Light Yellow* and *Brown* crystals have 2N, 4N-V, and Platelet at almost the same ratio, whereas *White* crystals has very few few Platelets. The Platelets is a planar aggregation of nitrogen atoms in an oval shape of about 30 to 60 nm in diameter parallel to the (100) plane. In order to confirm the influence of the Platelets on the mechanical properties, thin sections were sampled from the end of the Knoop indentation with FIB, and the propagation state of the micro cracks and dislocations were observed by TEM. The Knoop indentations were formed so that the major axis was parallel to the (001) <100> direction. By forming the Knoop indentation in the crystal orientation, it will be possible to observe the dislocations along the (111) <110> slip plane and micro cracks [2].

Fig. 5 shows a TEM image observed from the [100] direction. Oval dislocation defects generated around Platelets on the (100) plane and similar dislocations present on the (001) and (010) planes are observed as horizontal lines and vertical lines. In addition, many dislocations and micro cracks have developed in the [011] direction from the Knoop indentations, and propagate parallel to the cleavage plane (111). It was confirmed that the tip of these dislocations and micro cracks are pinned down or changed the propagation directions at Platelets. Thus, Platelets suppresses the propagation of dislocations and cracks. These phenomena are thought to produces an effect of improving wear resistance.

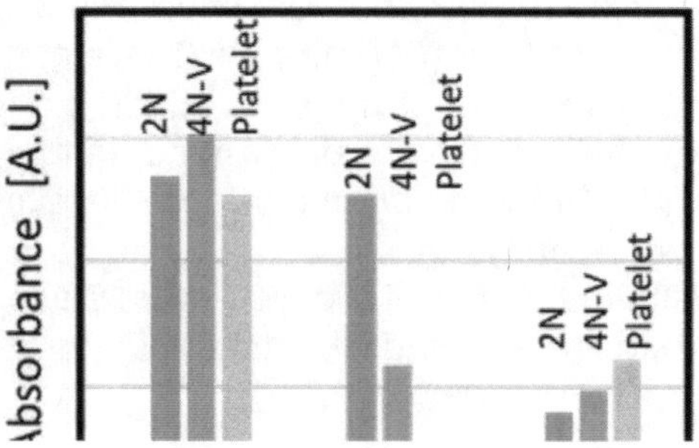

Fig.4 FT-IR absorbance peal heights due to each aggregated nitrogen impurity.

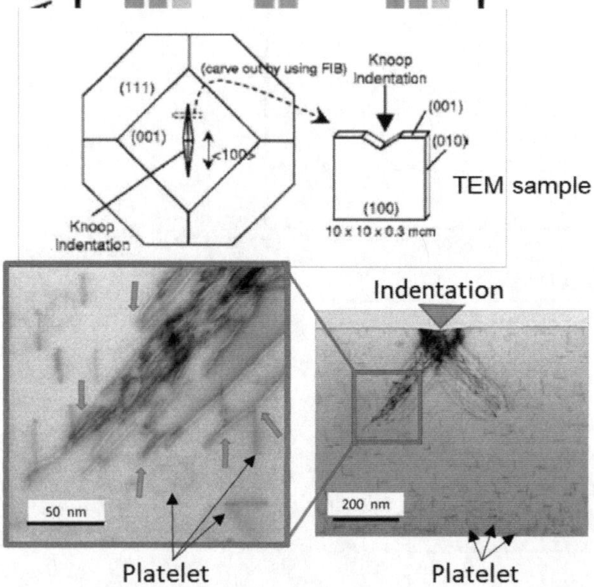

Fig.5 TEM image of microstructure under Knoop indentation formed on (100)<100> of Natural diamond crystal at room temperature.

4. CONCLUSION

In wear test using metal bonded diamond wheel, the wear characteristics of synthetic and natural diamonds were evaluated by comparing the content and the aggregation form of nitrogen impurities. It was found that the mechanical wear resistance decreases with increasing nitrogen content in the easy wear direction, but the effect is low in the wear resistant direction. It is suggested that Platelets in natural diamond may suppress the propagation of dislocations and micro cracks and improve the wear resistance.

***** Formatting Issue - Best Available Paper/Graphic *****

5. REFERENCES

[1] P.B.Hirsch, P. Pirouz, J.C. Barry: Proc. R. Soc. London, 407 (1986) 239.
[2] H. Sumiya, Appl. Phys. Lett, 88, (2006) 161904-161904-3.

Understanding sources of residual stress in PDC by X-Ray diffraction

D.K. Mukhopadhyay and K.E. Bertagnolli
US Synthetic Corporation, 1260 South 1600 West, Orem, Utah, 84058, USA

Abstract:

An attempt had been made to identify the sources of residual stress in polycrystalline diamond cutters (PDC) by X-ray diffraction (XRD). Residual stress was measured in a finished 16mm PDC insert. Changes in residual stress were observed as the substrate was removed by grinding. Residual stress was also measured in the remaining diamond disc after leaching out the cobalt. Comparison between the as-pressed PDC, after removing the substrate, and after removing the cobalt show the relative contribution of geometry, cobalt, and high pressure-high temperature processing to overall residual stress. An additional measurement was made of diamond powder compacted at HTHP without sintering to evaluate the influence of diamond particle deformation on residual stress.

Keywords: Polycrystalline Diamond Cutter, XRD, Residual stress, HTHP

1. Introduction

Since their introduction over 30 years ago, polycrystalline diamond cutters (PDC) have made a significant impact on the oil and gas drilling industry [1]. High penetration rate and long life are some of the attributes of PDC bits. However, PDC cutters may fail prematurely during hard rock drilling. Fig. 1. shows a typical PDC failure mode which may be attributed to residual stress.

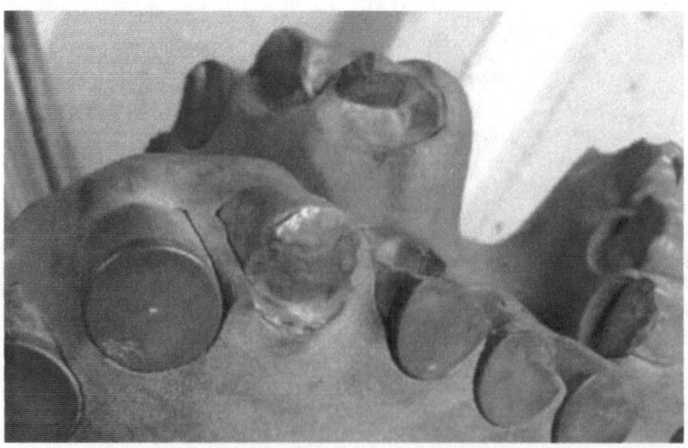

Fig.1. Typical PDC failure mode experienced in hard rock drilling

Conventional PDC inserts are made by sintering diamond powder on a cobalt–cemented tungsten carbide (WC–Co) substrate under high pressure–high temperature (HPHT) conditions around 6 GPa and 1400°C. During sintering, cobalt from the substrate melts and infiltrates through the pores among the diamond crystals, producing a zone of depleted cobalt in the substrate near the interface. The molten cobalt dissolves carbon atoms from the diamond crystals and quickly saturates. The dissolved carbon precipitates and forms diamond–diamond bonding among the diamond crystals [2]. As the pressure is released and the PDC cools from sintering, the diamond layer and the carbide substrate shrink at different rates, thus giving rise to residual stress in the PDC.

There are several techniques to measure residual stress in PDC [3]. Diffraction methods include X-ray diffraction (XRD), neutron diffraction, synchrotron X-ray, and Raman scattering. Diffraction techniques are generally non-destructive. Both neutron and synchrotron techniques have high penetration depth in comparison to XRD or Raman but they need special facilities. Shallow penetration depth of XRD and Raman techniques allows measurement of stress on the surface of the PDC. Additionally, stress can be measured using relaxation methods including hole drilling, ring coring, slitting, and grinding [4]. Relaxation methods are destructive and often difficult given the super hard material properties of PDC.

2. Experimental Procedure

High pressure sintering was carried out using a cubic-type HPHT apparatus. Temperature was measured with a W5%Re-W26%Re thermocouple without correction for the pressure effect. Pressure was calibrated at room temperature using the known transitions of Bi, Tl, and PbTe [5]. Diamond powder (30 μm grain size) and a WC-Co substrate (13 wt.% Co, 2–4 μm WC grain size) were enclosed in a niobium capsule, surrounded by sodium chloride, placed inside a graphite resistance heater, and enclosed in a pyrophyllite cube as shown in Fig. 2. The samples were first subjected to a pressure of 8 GPa. The temperature was then raised to approximately 1400 °C by passing electric current through the heater. The sample was maintained at temperature for approximately 2 minutes. Power was then turned off and the cell was cooled before pressure was released. Typical run times were on the order of 6 minutes. Once sintered, the samples were removed from the cube assembly, and the diamond table was lapped flat. The parts were then ground to a diameter of 15.9 mm and an overall height of 8.38 mm. The diamond table thickness was 1.52 mm.

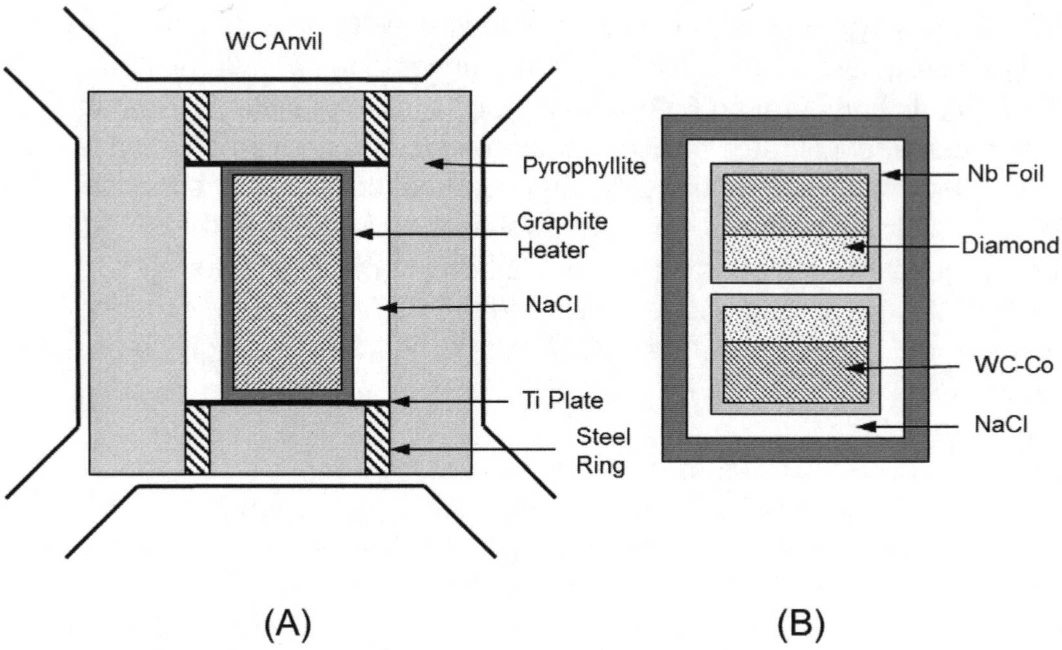

Fig. 2. Sample assembly for high pressure-high temperature (HPHT) sintering of diamond. (A) Overall view; (B) enlarged view of graphite heater

A Rigaku Ultima IV powder diffractometer was used for the XRD residual stress measurements. Copper K_{α} beam was used. The (311) diamond peak at a two-theta value of 91.5 degree was used for the measurement. The $\sin^2\psi$ method was used where ψ is the tilt angle. Stress measurements were made at the top center of the cutter with a spot size of 3 mm x 1 mm.

The substrate was ground to different thicknesses until it was fully removed, and stress measurements were made at every thickness. Stress was also measured in the diamond disc after removing the substrate. The diamond disc was then fully leached by acid to remove the cobalt phase, and stress measurements were made of this leached discs. Finally, to evaluate the contribution of HTHP processing on residual stress, diamond powder was compacted without cobalt at 1400°C at 8 GPa. Residual stress in the compacted diamond powder was also measured.

3. Results and discussion

A crystalline material is defined as a solid that consists of definite periodic arrangement of atoms in three dimensions. These planes of atoms can create constructive or destructive interference of X-ray beam, and diffraction of an X-ray beam happens by constructive interference when the beam obeys Bragg's law [6],

$$\lambda = 2d \sin\theta$$

where λ = X-ray wavelength, d = interatomic spacing, and θ = diffraction angle. If the residual stress is present in the sample, then the interatomic spacing (d) will be different than in an unstressed state. If the lattice is under under compression, the interatomic spacing (d) will be smaller, and the peak will shift to higher angles according to Bragg's law. This change in the lattice spacing can be measured from the peak shift. Thus, the lattice spacing acts as a strain gauge in this XRD stress measurement technique.

In our present experiment the (311) diamond peak at 91.5^0 was chosen due its relatively higher intensity and high angle position. A high-angle peak is chosen because the error in Sin θ for a given error in θ decreases as the value of θ increases [6]. The diffraction data was collected at three psi angles. The Young's modulus (E) of the diamond table used in this calculation is 850,000 MPa. The Poisson's ratio (ν) used was 0.2.

Fig. 3. shows the residual stress change in the PDC with decrease in the substrate length. Residual stress of the PDC sample was measured at the center of diamond table. XRD measurements show the diamond surface is in compression when attached to the full substrate. As the substrate is removed, the diamond begins to expand outward, relieving some of this compression. In addition, the relatively stiff carbide support is being removed, allowing the diamond surface to bend slightly. Eventually enough carbide is removed to allow a significant amount of bending of the diamond surface, nearly eliminating the compressive residual stress on the diamond surface. When the substrate is completely removed, the diamond table is no longer bent and the surface is again in compression. These measurements show that the diamond table is not stress free even after removal of the whole substrate.

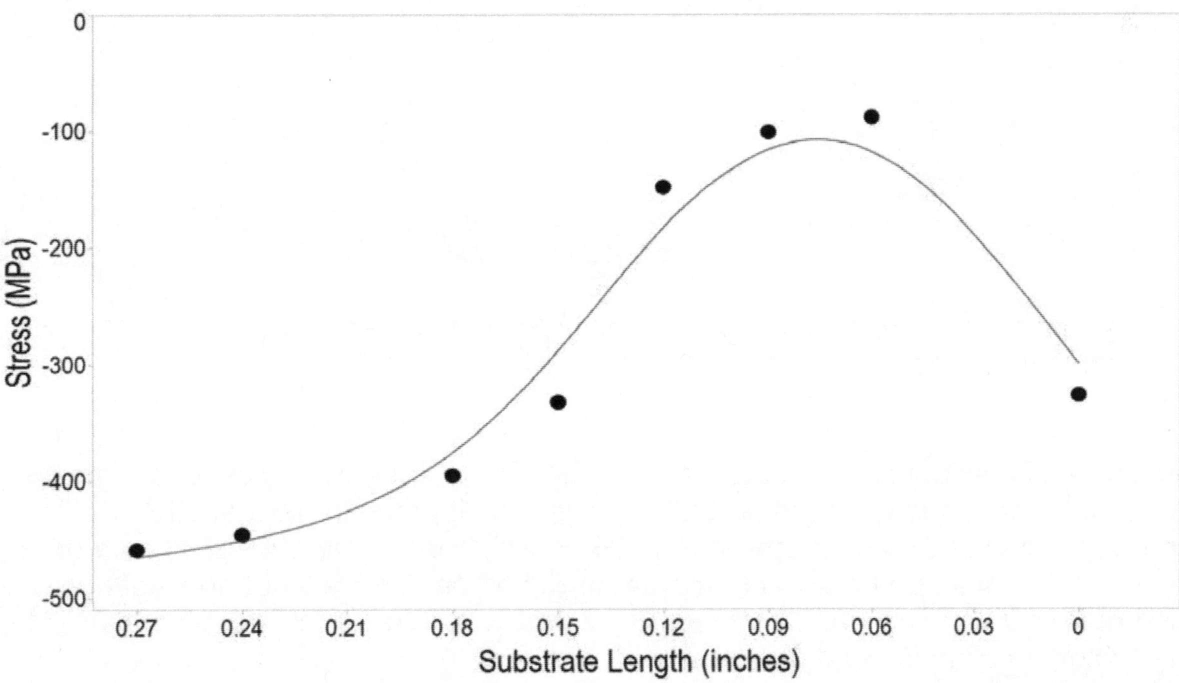

Fig 3: Residual stress change in the PDC with decrease in the substrate length

The presence of stress in the diamond table after complete removal of substrate could be due to the interstitial cobalt. The cobalt in the diamond table tends to pull the diamond into compression even in the absence of the substrate. To test this hypothesis, we measured the diameter change of the diamond table before and after removal of the cobalt phase by acid leaching. The diameter increased an average of 0.0051 mm (.0042-.0060, 95%, n=18) after leaching. This indicates that the cobalt phase pulls the diamond table into compression. Stress measurements of the fully leached disc by XRD show that the diamond disc is not stress free even after complete removal of cobalt (see Fig. 4). In fact, the diamond disc appears to still be in compression after complete leaching of cobalt, but the magnitude of the compressive stress is less than that observed in the un-leached diamond disc.

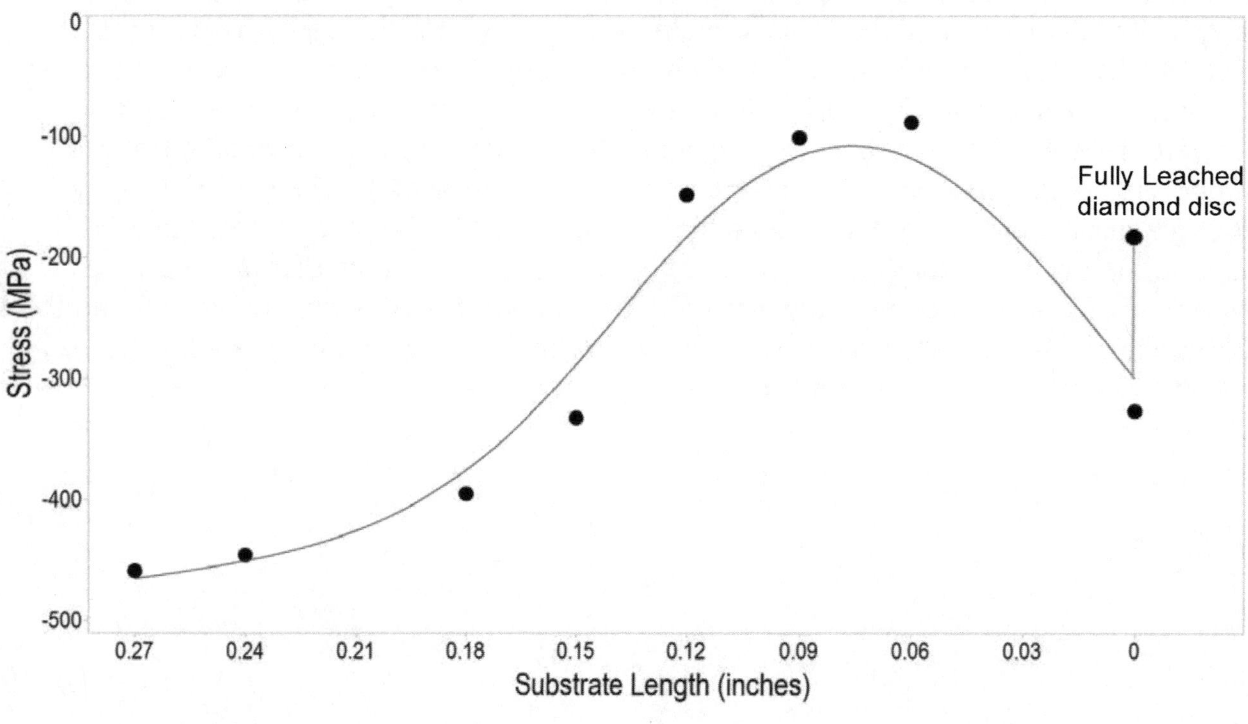

Fig. 4. Residual stress map shows presence of stress in the diamond disc after complete removal of cobalt

The presence of residual stress in the diamond table after full leaching of cobalt may be due to deformation of the diamond particles by the HTHP process. To test this hypothesis, we pressed diamond powder at 8 GPa and1400°C without a substrate or added cobalt. The diamond powder formed a hard, compacted disc, but no evidence of diamond-diamond bonding was observed. Fig. 5. shows the residual stress measured in the disc of compacted diamond particles.

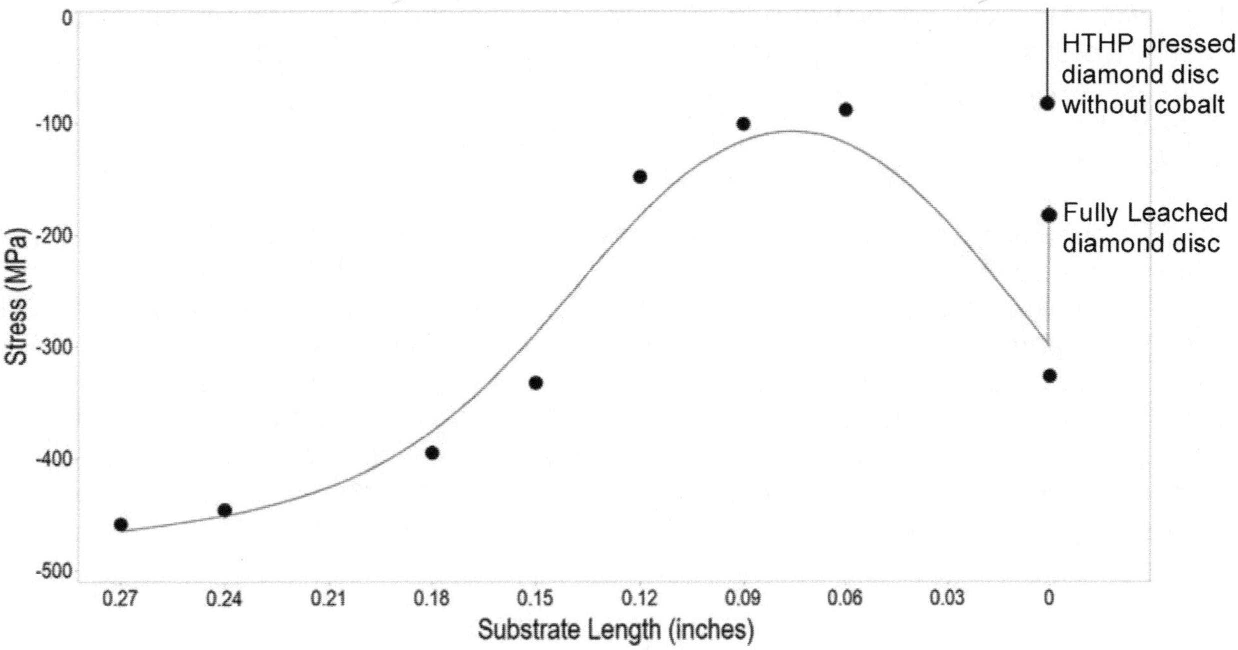

Fig. 5. Residual stress map shows presence of stress in diamond particles after HTHP process in the absence of cobalt

The disc of compacted diamond particles was then evaluated in a high resolution TEM to look for the evidence of deformation. The TEM image in Fig. 6. shows the presence of {111} micro twins and dislocations in a single diamond grain. The presence of dislocations and micro twining in plastically deformed diamond particles has been observed previously [7].

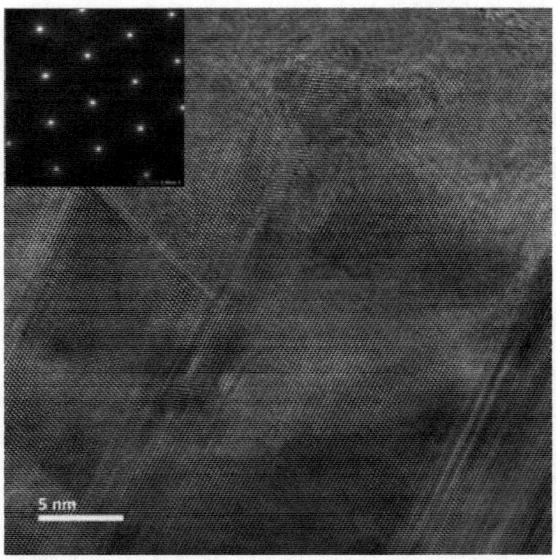

Fig. 6. Presence of micro twins in the diamond particle after HTHP compaction in the absence of cobalt

From Fig.5. the stress gap between HTHP pressed diamond disc without cobalt and a fully leached diamond disc suggests that diamond-diamond bond is also responsible for residual stress in the diamond table in PDC.

4. Conclusions

The XRD technique was used to identify the sources of residual stress in polycrystalline diamond cutters (PDC). Measurements show that the diamond surface is in compression when attached to the full substrate. As the substrate is removed by grinding, the diamond begins to expand outward, relieving some of this compression. When the substrate is completely removed, the diamond table is not stress free but remains in a state of compression. After leaching the cobalt from the diamond table, the compressive stress is reduced but is still not zero. TEM evaluation of HTHP pressed diamond particles showed the presence of dislocations and micro twins. We speculate this deformation is partially responsible for the remaining compressive stress.

References

[1]. Clark, I. E., & Bex, P. A. (1999). The use of PCD for petroleum and mining drilling. IDR. Industrial diamond review, 59(580).

[2]. Wentorf, R. H., DeVries, R. C., & Bundy, F. P. (1980). Sintered superhard materials. Science, 208(4446), 873-880.

[3] Schajer, G. S. (Ed.). (2013). Practical residual stress measurement methods. John Wiley & Sons.

[4] Lin, T. P., Hood, M., Cooper, G. A., & Smith, R. H. (1994). Residual stresses in polycrystalline diamond compacts. Journal of the American Ceramic Society, 77(6), 1562-1568.

[5] Qian, J., McMurray, C. E., Mukhopadhyay, D. K., Wiggins, J. K., Vail, M. A., & Bertagnolli, K. E. (2012). Polycrystalline diamond cutters sintered with magnesium carbonate in cubic anvil press. International Journal of Refractory Metals and Hard Materials, 31, 71-75.5

[6] Cullity, B.D. (1967). Elements of X-ray Diffraction, Addison-Wesley Pub. Co.

[7] Walmsley. J.C. & Lang. A.R., Journal of Materials Science 23, (1998), 1829-1834

Inspecting Ultra-Hard Materials and Processes using Acoustic Micro Imaging (AMI)

*Jack H. Richtsmeier**
Nordson SONOSCAN
Elk Grove Village, Illinois, USA

Abstract

Acoustic Micro Imaging (AMI) is a nondestructive inspection technique that utilizes high frequency ultrasound to examine and characterize a wide variety of high-tech materials and processes. The technique can be used to 1) measure empirical material properties such as velocity, layer thickness, porosity and attenuation as well as, 2) detect a wide variety of defects and/or failures within materials such as cracks, voids, dis-bonds and/or delamination's. With source frequencies ranging from 5 to 300 MHz, the technology has the capability to image features as small as 3 to 5 microns in size. Applications range from inspecting ultra-hard materials to ceramics, composites, metals and plastics as well as epoxies, adhesives, solders and pastes. This paper will review the basic theory and principles behind the technology of AMI and cover specific applications within hard materials including the inspection and characterization of synthetic diamond or poly-crystalline diamond (PCD), bond evaluation of PCD to tungsten carbide (WC) cutters and flaw detection in related bonding, cutting and polishing processes.

Key words: Acoustic Micro Imaging (AMI), bonding, non-destructive testing & poly-crystalline diamond (PCD).

Introduction

Ultrasound is an elastic disturbance or mechanical wave motion that propagates through materials at frequencies above 20 KHz. It is highly sensitive to the elastic properties of materials and can be transmitted, absorbed, scattered and/or reflected by various material changes (Figure 1). By observing those changes in the ultrasound, we can make certain observations about changes in the material. At high source frequencies (5-300 MHz), air or a vacuum does not have the required elastic properties to support ultrasound and is detected as an anomaly relative to the surrounding material(s). This relationship makes the technology extremely sensitive in detecting air-gap type features such as porosity, voids, cracks, delamination's and/or disbonds.

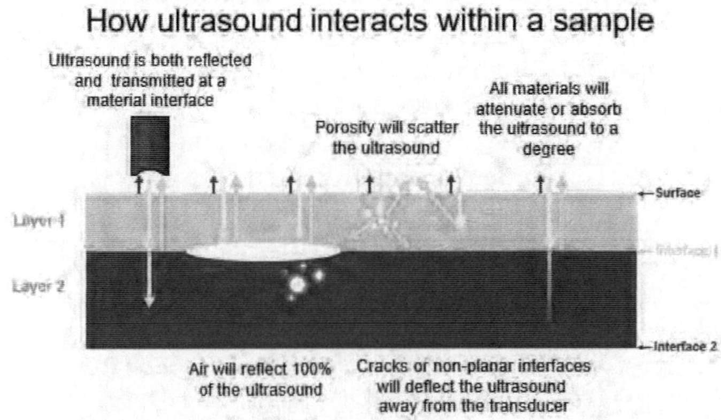

Figure 1: The figure displays a block diagram showing how ultrasound can be transmitted, absorbed, scattered and/or reflected within materials.

The standard piece of equipment used in AMI is the C-Mode Scanning Acoustic Microscope (C-SAM^{TM}). The system is a pulse-echo or reflection type microscope that generates images by mechanically scanning a piezoelectric transducer back and forth, in a raster pattern, over the sample (Figure 2).

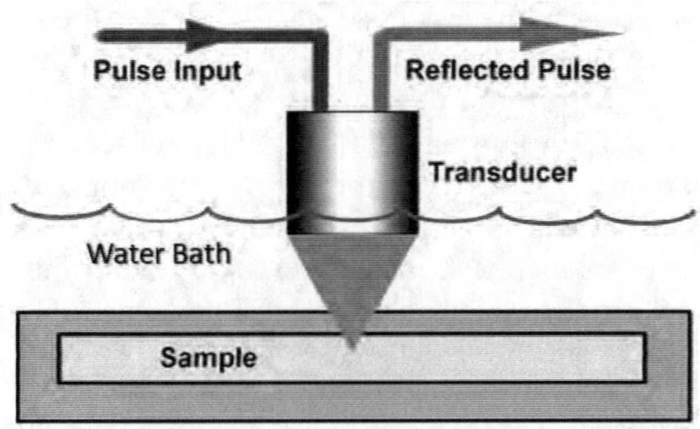

Figure 2: The figure displays a block diagram of the tranducer relative to the sample in a refection mode (pulse-echo) AMI system.

The source transducer is coupled to the sample by a fluid medium (bath), typically de-ionized water or some type of inert fluid. The transducer has piezoelectric properties, which allow it to act as both a sender and receiver being electronically switched back and forth between transmit (pulse) and receive (echo) modes.

As the acoustic pulse penetrates the sample, return echoes are produced from material changes (or interfaces) at specific depths within the sample. The echo return times are based on the thickness and acoustic velocity of the respective material(s). A digital oscilloscope, known as the A-Scan, displays these echoes along with their relative time/distance relationships, amplitude, and polarity information *(Figure 3)*. These time/distance relationships within the A-scan provide the basis for investigating features at specific levels or depths within a part.

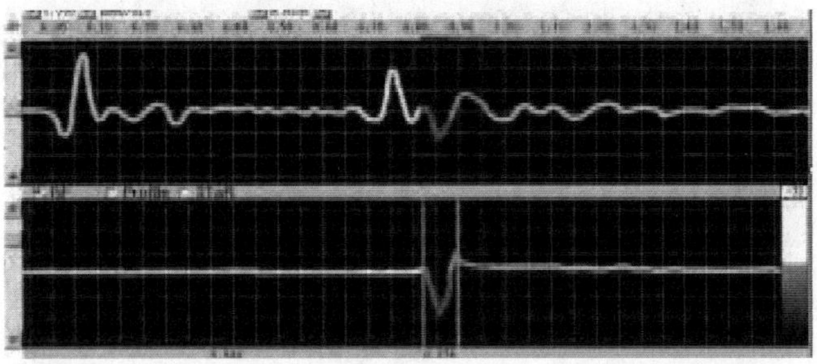

Figure 3: The figue displays the A-scan oscilloscope (top waveform) as the mechanical signature that corresponds to the position of the transducer over the sample. Also displayed is the "gated" signal (bottom waveform), which determines the depth or level of the resulting image.

Once the echo or depth of interest has been identified, an electronic timing gate is positioned at that specific level *(reference the lower trace in Figure 3)*. Only information from the designated gate level is gathered. The "gated" information is digitally processed and displayed on a video monitor that is synchronized with the transducer's position over the sample. As the transducer scans back and forth over the sample an image is produced of the x-y plane known as a C-mode image or c-scan *(Figure 4a)*.

The ultrasonic signal contains time, distance (depth), velocity, amplitude, polarity (phase) and frequency information. Any one of these characteristics can be captured, measured and displayed - many of them simultaneously. As acoustic signals are received and processed, changes in the signal (due to changes in the material) are displayed as important or relevant information. In all cases air-gap type flaws such as voids, cracks, disbonds and delamination's provide the highest contrast and are easily distinguishable.

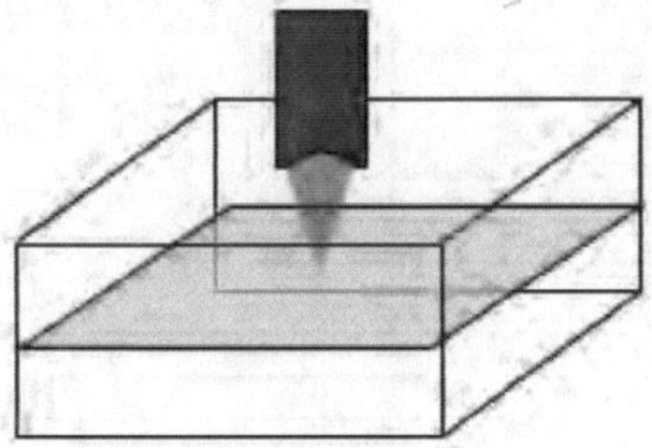

Figure 4a: The figure displays a block diagram of the "gated" region or level of interest in the x-y plane know as a c-scan or C-Mode image.

Another example includes monitoring changes in time that can be displayed to map a materials profile and/or layer thickness (Figure 4b).

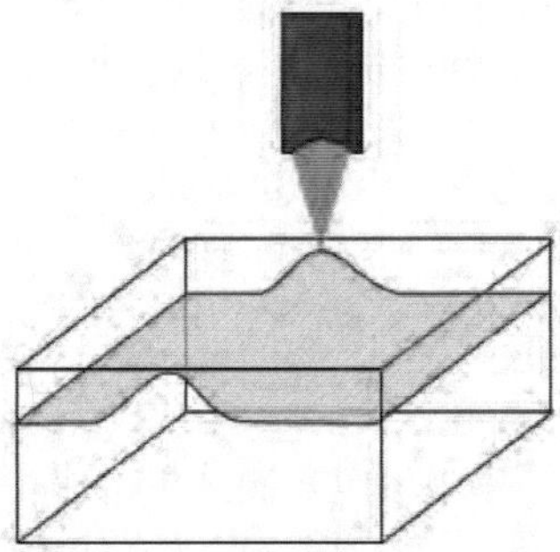

Figure 4b: The figure displays a block diagram tracking changes in time to map a materials profile and/or thickness.

Polycrystalline Diamond (PCD) or Synthetic Diamond – wafers, rounds and blanks

Polycrystalline Diamond (PCD) or Synthetic Diamond was engineered in the early 1970's and has been in continuous development over the past 45 years. It is widely used in a variety of high-tech cutting, grinding and drilling applications and covers numerous market sectors including, but not limited to, the aerospace, automotive, energy, mining and petroleum industries.

PCD is typically manufactured as a full round and/or wafer – which is then cut and shaped into smaller blanks for various tooling applications (*Figure 5*). The material is composed of a top layer called the diamond table along with a bottom substrate of Tungsten Carbide (WC). It is manufactured under very high temperatures and pressures (2800° F and 1,000,000 psi) along with a metallic binder (cobalt alloy) to help bond diamond and carbide during the sintering process. The WC substrate provides 1) mechanical support to the PDC layer, 2) increased impact strength and 3) a larger footprint making it easier to mount for various tooling applications. Both the PCD and WC layer thickness specifications vary depending on the intended use or application.

(Figure 5 PCD full rounds or wafers and a variety of cut tooling blanks)

When manufacturing, both the top PCD layer and the lower WC layer are supposed to maintain a consistent thickness across the whole wafer. The sintering process is not easy to control and variations in thickness do occur (*Figure 6*). Maximizing production (yield) is based on the number of good blanks that can be cut from the source wafer.

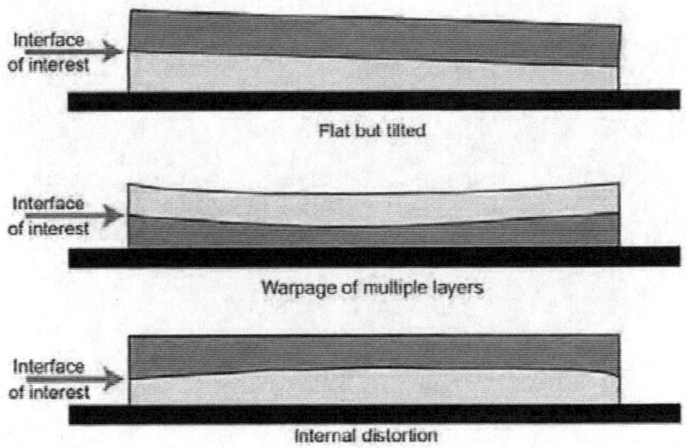

Figure 6; Both the top PCD layer and the lower WC layer are supposed to maintain a consistent thickness across the whole wafer. A variety of inconsistencies can occur including tilting, warpage and/or internal distortion.

Ultrasound travels at a relatively high velocity through both the PCD (17,500 m/s) and WC (6,655 m/s) materials while the C-SAM™ scans at a rate 1000 mm/s. This combination allows data to be gathered from thousands of x-y locations per second, providing data and feedback in a relatively short amount of time. By measuring the travel time and velocity of the material(s), AMI can measure the depth and/or thickness of the PCD layer over the entire wafer - without need for physical cross-sectioning.

Figure 7 is an acoustic image of a 52 mm diameter PCD/WC wafer. The color scale to the right references thickness ranges from 0.70 mm (dark green) to 1.20 mm (magenta). Each color in the scale represents a thickness change of 0.05 mm, or 50 um. Assigning colors to each distance makes it relatively easy to identify thickness variations in the wafer. As a result, the data can immediately be used to help determine production yields for specified blanks. In this example, the thickest region of the PCD material is at the center of the wafer (>1.00 mm - identified by its pale blue color). In contrast, the thinnest region is at the lower left edge of the wafer (<0.75 mm - identified by its dark green color). The total thickness ranges ~300 um across the wafer.

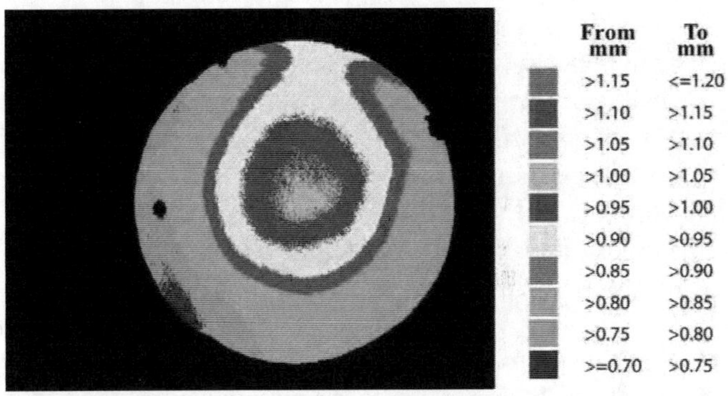

Figure 7: AMI profile mode image of a 52mm diameter wafer showing PDC thickness (color) variations represented in 50 um steps.

Figure 8 represents an acoustic image of four (4) 52 mm diameter wafers imaged in one field of view (note the color and thickness scale differs from those used in Figure 7). The specified thickness for these wafers is 0.60 mm. Note the upper left wafer has a good amount of area that falls within specification (displaying the three adjacent colors that best meet the desired thickness - pale blue, red and brown). Alternately the lower right wafer has a large region that is very thin - between 0.30 and 0.35 mm - yet still showing areas that fall within specification. A wafer may have areas of PCD that are too thin or too thick to be used for the specified purpose however areas of the wafer may still yield acceptable blanks for alternate applications.

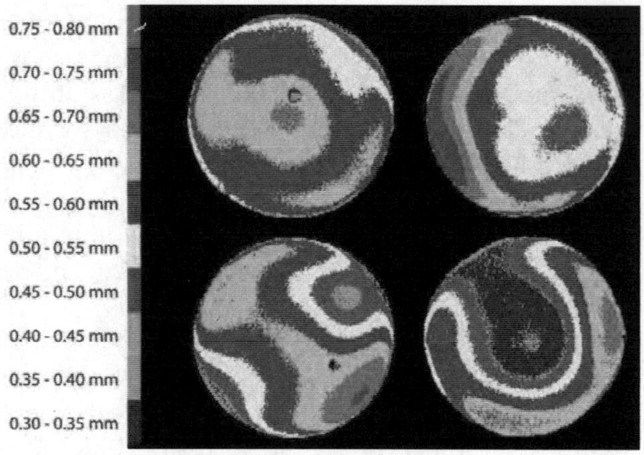

Figure 8: AMI profile mode image of a four (4) 52mm diameter wafers showing PDC thickness (color) variations represented in 50 um steps.

Polycrystelline Diamond Compact (PDC) Cutters

Polycrystalline Diamond Compact (PDC) cutters or inserts are widely used in all types of drilling operations. Areas include gas and oilfield, geological exploration, geothermal energy, water well, coal mining and many other drilling, milling and tooling applications. High rates of penetration, long life, and design versatility are some of the benefits offered by PDC cutters. Their impact on the drilling industry has been dramatic with today's operations taking a fraction of the time to drill compared with 20-30 years ago.

PDC cutters or inserts consist of a layer of synthetic polycrystalline diamond bonded (sintered) to a layer of cemented tungsten carbide (PCD/WC). The blanks are then bonded to tungsten carbide studs, which are then brazed or fitted into the head of a drill bit. PDC cutters or inserts range in size from 5 to 25 mm in diameter and 3 to 25 mm in length. They can be manufactured in many different shapes and sizes from basic planer cylindrical cutters to inserts with customized angles, chamfers and diamond table thickness (Figure 9).

Figure 9. Polycrystalline Diamond Compact (PDC) cutters or inserts designed and developed specifically for the mining, drilling and exploration markets.

Due to the extreme forces and overall stressful environment of downhole drilling, the PDC cutter interface has been modified and designed to aid in both bond strength as well as stress relief to help prolong tool usage (Figure 10). Increasing bond strength and stress relief has proven to be critical in the cutters overall performance and eventual longevity. High stress concentrations in the PCD (diamond table) can result in residual delamination failures at the PDC/WC interface or in the diamond table itself causing fracture and edge chipping. Poor bonds between grains of diamond grit can also lead to cracking in the diamond table leading to similar failures. A single drill head may include dozens of cutters so failures can be quite costly, especially when retracting tooling from thousands of feet below the earth's surface.

Figure 10. Schematic cutaway diagram of a representative interface and joint design on a PDC cutter.

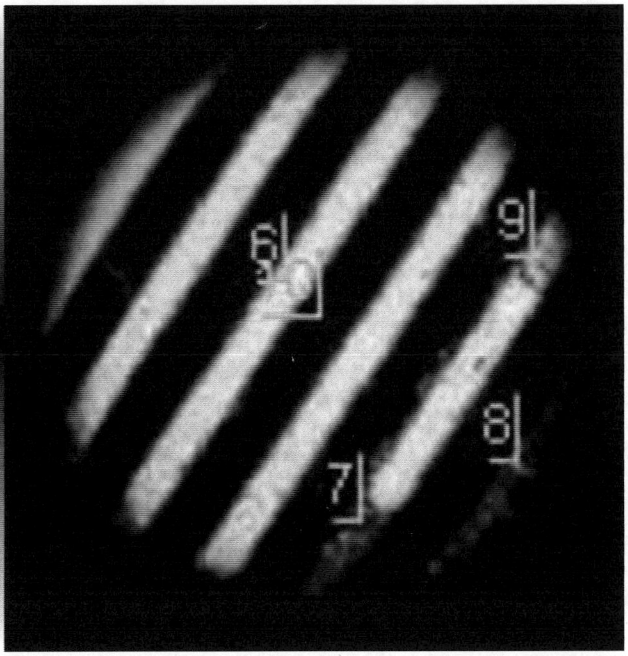

Figure 11. Acoustic image of a PDC cutter focused and gated at the PCD/WC interface. The numbers 6-10 correspond to the transducer (pixel) position of the respective waveforms represented in Figure 12.

Figure 11 is an acoustic image of a single mounted PDC cutter. The ultrasound is focused thru the diamond table and gated at the PCD/WC interface (note the "tongue & groove" joint design is apparent as a series of parallel lines where the two materials meet). The numbers 6-10 reference the transducer or pixel locations over the sample corresponding to the A-scans represented in Figure 12. In Figure 12, each A-scan includes the time/distance, echo amplitude and polarity information for that specific pixel position. The A-scan represents a time (or depth) profile displaying increasing time (or depth) as it is read from left to right. In other words, echoes to the left are closest in time (or nearer to the surface) while echoes to the right are further in time (or deeper in the profile. The first echo on the left side of each A-Scan represents the top surface of the sample (or water to PCD surface). The two vertical lines near the center of the A-scan represent the "gate" or the time/depth location designated to receive data and generate the image. The gate has been positioned at the PCD/WC interface. As echoes are returned from the interface, the strongest or largest (amplitude) echoes within the gate come from regions where there is no bonding (air) as represented in A-scans 6, 8 & 9. In contrast, smaller amplitude echoes within the gate generally represent well bonded areas as represented in A-scans 7 & 10. The gray scale or color map used in Figure 11 (located along the left margin of the image) designates the strongest or largest echoes as bright white while smaller echoes appear gray or black. Much of the interface shows very strong (bright white) reflections indicating regions of poor bonding while the lower right corner shows nominal (gray) reflections indicating regions of good bonding.

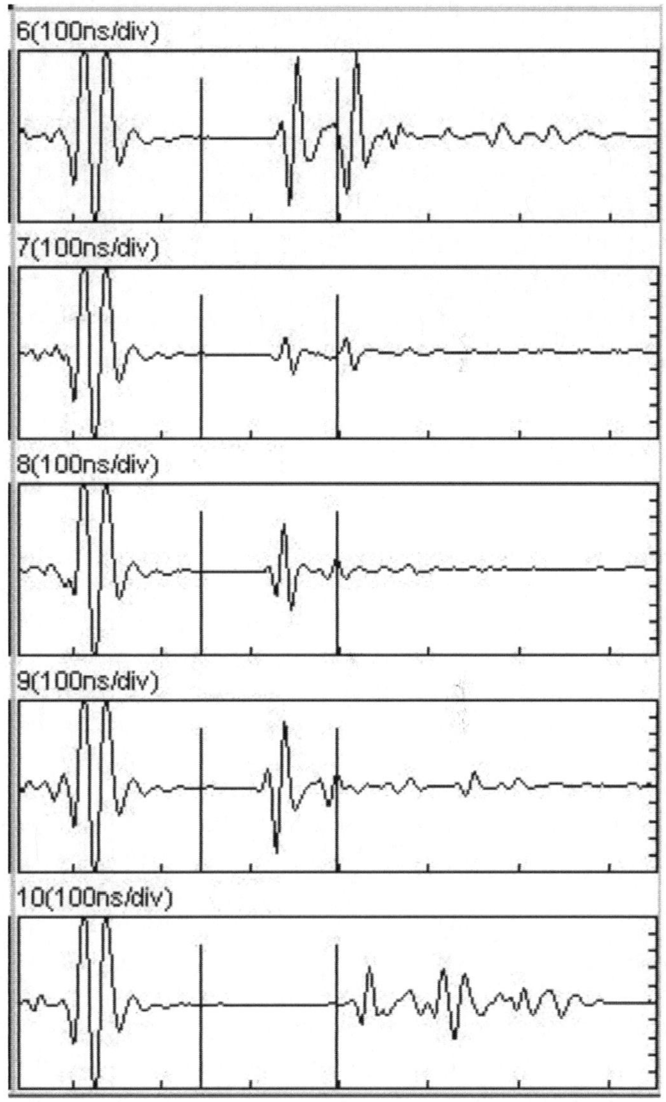

Figure 12. A-scan waveforms corresponding to the transducer positions represented in Figure 11.

Figure 13. Acoustic image of two PDC cutters focused and gated at the PCD/WC interface.

Figure 13 is an acoustic image of two individual mounted cutters. The ultrasound is focused thru the diamond table and gated at the PCD/WC interface (note the "tongue & groove" style joint design is apparent as a series of parallel lines where the two materials meet). As echoes are returned from the interface, the strongest (amplitude) echoes come from regions where there is no bonding (air). The cutter in the upper left corner shows very strong reflections (bright white) coming from numerous non-bonded regions. The cutter in the lower right corner shows nominal reflections (gray) indicating overall good bonding.

Conclusions

With the ability to gain information at specific levels, AMI is a powerful non-destructive inspection tool for analyzing various materials and processes. Over the past couple of decades AMI has proved to be a cost effective, nondestructive means of inspecting PCD blanks and PDC cutters. With the ability to capture information at a specific depth, ultrasound can be focused and gated specifically at the PCD/WC interface to help measure PCD layer thickness as well as detect flaws and/or defects associated with the manufacturing process. Today AMI technology is being actively applied to 1) design and development, 2) failure analysis, 3) manufacturing process control and 4) final QA/QC inspection.

Bibliography

1. Adams T., Acoustically Measuring Polycrystalline Diamond Thickness – Ceramics Industry, August 2014

2. Adams T., Acoustic Imaging Techniques Effectively Map Buried Layer Contours - Advanced Materials & Processes, June 2015

3. Adams T., Acoustically Imaging Diamond Drill Bit Inserts for Oil Wells - Ceramics Industry, June 2018

4. Richtsmeier J., Kessler L., Inspecting Bond Quality and Seal Integrity in Medical Devices using Acoustic Micro Imaging (AMI) – American Society of Metals, September 2003

* john.richtsmeier@nordson.com Nordson Sonoscan 2149 East Pratt Blvd., Elk Grove Village, IL 60007 phone (847) 437-6400: fax (847) 437-1550: http://www.sonoscan.com

Theoretical research on the Uniformity of Planetary Double－Side Lapping trajectory

Yaoyu Wu

Zhengzhou Institute of Technology， Zhengzhou 450044， China

Abstract: The reduction to absurdity is adopted to analyze equation of planetary double－side lapping motion built by kinematics, and study the important parameter in the process of lapping－ I (the ratio of grinding disc and the inner pin ring in rotary speed) satisfies the condition of uniformity of lapping trajectory of a point in workpiece relative to surface of grinding disc. The results show the ratio I that meet uniformity of lapping trajectory is a function of time t. On the other hand, for a given planetary double－side lapping machine, the I of constant ratio (rational or irrational) that satisfies uniformity of lapping trajectory does not exist unless it continues to change over the time. This research has laid a theoretical foundation for the further studying in uniformity of lapping trajectory, and also provided a new idea and method for the analysis of relative velocity between workpiece and grinding wheel, the wear uniformity of grinding wheel, and the study of workpiece surface quality.

Key words: planetary double－side lapping; trajectory; wear ; uniformity

0. Introduction

Using abrasive precision grinding and polishing technology is an important method of the ultra precision machining , to obtain the shape of the limit of accuracy, dimensional accuracy, surface roughness, surface integrity as the goal on the premise of not changing the workpiece material physical properties[1] ,have a wide application in machinery, automobile, electronics, communications, computer, laser, aerospace and other fields , is the key technology of precision machining parts of a mechanical seal ring, gauge block, hydraulic and pneumatic components, precision ceramics, monocrystalline silicon slice, sapphire, etc [2-3]. The grinding motion trajectories reflects the relative motion of the grinding disc and the workpiece in the grinding process, which is the basis for an in-depth understanding of the grinding mechanism [4].It can be divided into two categories: one is the trajectories of movement of a point on the workpiece relative to the grinding disc, and the other is the trajectories of movement of a point on the grinding disc relative to the workpiece. The former is of great significance for revealing the grinding uniformity of the grinding disc, while the latter is directly related to the morphological features and texture distribution of the workpiece processed surface. The significance of studying the corresponding trajectory uniformity is reflected in the following aspects: ensuring that all points on the workpiece surface have the same material removal probability and ensuring uniform wear of the tool disc [5].Therefore, domestic and foreign scholars have conducted a lot of researches on the motion trajectories distribution and grinding mechanism of plane grinding and polishing [6-15].The results show that the driving mode and speed ratio are the main factors affecting the grinding uniformity [6].Yang changming et al. [7] studied the relationship between the rotational speed of planet gear and solar wheel and the motion trajectory of abrasive particles and the quality of grinding plane. Zhao wenhong et al. [8] studied the plane grinding uniformity by means of grinding with definite eccentric and indefinite eccentric. Wen donghui et al. [9-10] carried out theoretical analysis on the influence of rotation speed ratio and eccentricity of grinding disc and workpiece on the trajectories uniformity. The results showed that increasing the rotation speed ratio made the processing uniformity worse .Liu qing et al. [11] showed that x-y linkage of the workbench has an important impact on the trajectory ; The theoretical research results of Yan zhen et al. [12] show that the introduction of yaw motion is conducive to improving the trajectories non-uniformity and improving the surface quality of the processed workpiece. Adam Barylski et al. [13] analyzed the influence of coupling relationship between velocity ratio K1 of workpiece disc and grinding disc and cycle ratio K2 of reciprocating cycle of workpiece disc and rotation cycle on trajectories distribution under linear swing drive, and optimized kinematic parameters to obtain better grinding trajectories uniformity.

No matter which kind of motion trajectory, when the grinding machine with a given driving mode is used for grinding processing, which motion parameters are closely related to the grinding motion trajectory, and what are the conditions for the uniformity of motion trajectory? Based on the kinematics analysis to establish the trajectories of movement of a point on the workpiece relative to the grinding disc in planetary double—side lapping , the reduction to absurdity is adopted to deduce mathematically the equation of motion trajectory , studied the important parameters in the process of grinding - I (the ratio of grinding disc and the inner pin ring in rotary speed) satisfying the conditions for the uniformity of motion trajectory , for the uniformity of planetary plane lapping movement laid a theoretical foundation for further study. It also provides a new idea and method for the analysis of the relative velocity between workpiece and grinding disc, the wear uniformity of grinding disc and the surface quality of workpiece. The method can also be used to evaluate the uniformity of the trajectory simulation of single-side grinding, single-side polishing and double-side polishing.

1.Grinding machine principle and motion trajectory equation

Figure 1 is a common structure schematic diagram of planetary double—side lapping machine. The inner pin ring is driven by the motor through the reducer, and the outer pin ring is fixed on the machine base (there is also a grinding machine with the outer pin ring rotation), which can load multiple planet gear (workpieceholder) at the same time to drive multiple workpieces to move together, the upper and lower grinding disc is the consolidated abrasive grinding disc. The planet gear is driven by the inner pin ring to go around the center of the inner pin ring and rotate. The workpiece is placed in the hole of the planet gear, and the movement of the workpiece is driven by the planet gear. At the same time, under the action of the grinding force of the lower grinding disc and the upper grinding disc, rotation will also occur. Therefore, the movement of the workpiece is the synthesis of the planetary movement and rotation.

According to the motion principle of the grinding machine, considering that the influence of the workpiece motion on the motion of the planet gear train is far less than that of the planet gear train on the motion of the workpiece, namely ignoring the rotation of the workpiece ; At the same time, considering the similarity of the workpiece's relative motion with the upper grinding disc and the lower grinding disc, only the relative motion of the workpiece and the lower grinding disc is studied here. Therefore, it can be simplified into the relative motion schematic diagram of the workpiece and the lower grinding disc as shown in figure 2. In the plane perpendicular to the spindle of the machine, the rotation axis of the grinding disc and the inner pin ring is O_1, and the rotation axis of the planet gear is O_2.Take O_1 and O_2 as the origin of coordinates respectively, and set up the coordinate systems $O_1X_1Y_1$ and $O_2X_2Y_2$ in the plane, and fix them on the grinding disc and workpiece respectively. The O_1X_1 axis and O_2X_2 axis of the two coordinate systems are along the line direction of O_1O_2.The distance between O_1O_2 is the sum of the radius r of the inner pin ring pitch circle and the radius r of the planet wheel pitch circle, namely r+ R

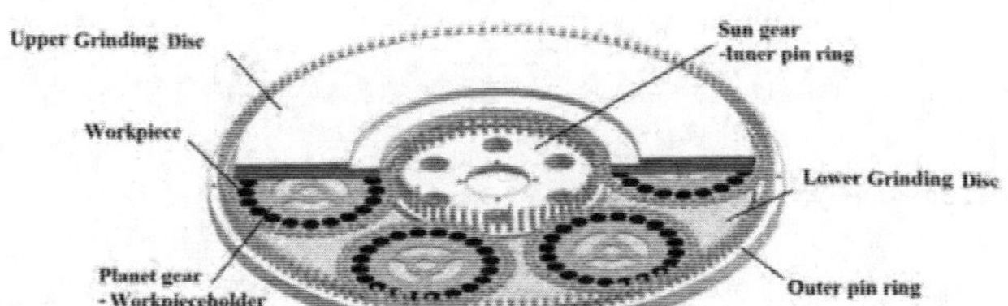

Figure 1 Principle of double side grinding operation

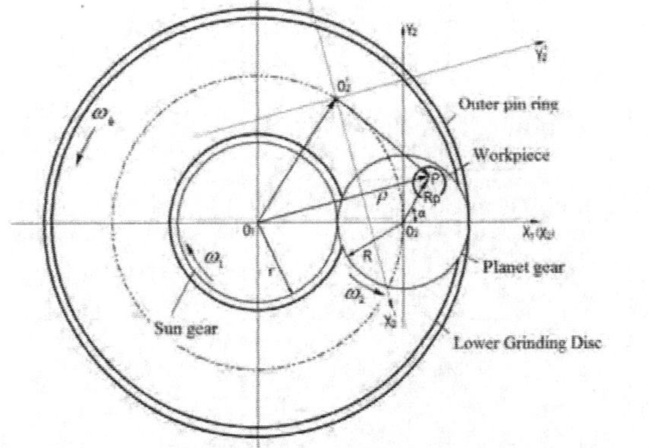

Figure 2 Schematic diagram of relative motion of workpiece and lower grinding disc

The motion path equation of workpiece relative to the lower grinding disc is established by means of graph transformation. If any point P is selected on the workpiece with diameter d, the initial distance between point P and point O_2 is R_P, and the Angle between the vector O_2P and the axis O_1X_1 is α , then the homogeneous coordinates of point P (an n+1-dimensional vector to represent an n-dimensional vector method) can be expressed by the vector as:

$$L = \left[R_p \cos\alpha, R_p \sin\alpha, 1\right]$$

Let the revolvation tation angular velocity of the planet gear be ω_H, rotation angular velocity ω_2, and the lower grinding disc angular velocity ω_0. After time t, the rotation Angle $\theta_1 = \omega_H t$ of the planet gear around O_1 and the rotation Angle $\theta_2 = \omega_2 t$ of the vehicle around O_2 can be expressed by matrix as follows:

$$T_1 = \begin{bmatrix} 1 & 0 & 0 \\ 0 & 1 & 0 \\ (r+R) & 0 & 1 \end{bmatrix} \times \begin{bmatrix} \cos\theta_1 & \sin\theta_1 & 0 \\ -\sin\theta_1 & \cos\theta_1 & 0 \\ 0 & 0 & 1 \end{bmatrix} \times \begin{bmatrix} 1 & 0 & 0 \\ 0 & 1 & 0 \\ -(r+R)\cos\theta_1 & -(r+R)\sin\theta_1 & 1 \end{bmatrix} \times$$
$$\begin{bmatrix} \cos(-\theta_2) & \sin(-\theta_2) & 0 \\ -\sin(-\theta_2) & \cos(-\theta_2) & 0 \\ 0 & 0 & 1 \end{bmatrix} \times \begin{bmatrix} 1 & 0 & 0 \\ 0 & 1 & 0 \\ (r+R)\cos\theta_1 & (r+R)\sin\theta_1 & 1 \end{bmatrix}$$

Similarly, the grinding disc is rotated by an Angle $\theta_0 = \omega_0 t$ in time t. If the grinding disc is regarded as fixed, it is equivalent to the re-rotation Angle θ_0 of the planet gear, in the direction opposite to the revolution of the planet gear at point O_1, that is, the coordinate system $O_2X_2Y_2$ is finally rotated to the position of coordinate system $O'_2X'_2Y'_2$. The corresponding transformation matrix is:

$$T_2 = \begin{bmatrix} 1 & 0 & 0 \\ 0 & 1 & 0 \\ -(r+R) & 0 & 1 \end{bmatrix} \times \begin{bmatrix} \cos\theta_0 & \sin\theta_0 & 0 \\ -\sin\theta_0 & \cos\theta_0 & 0 \\ 0 & 0 & 1 \end{bmatrix}$$

Therefore, the motion trajectory equation of any point P of the workpiece relative to the grinding disc is the motion of any point P of the workpiece relative to the grinding disc. It comes down to the coordinate XpYp of point P in the coordinate system $O_1X_1Y_1$ after the transformation, which is expressed by the homogeneous coordinate $(X_P, Y_P, 1)$. Therefore, the coordinates $(X_P, Y_P, 1)$ of any point P on the workpiece on the planet gear in $O_1X_1Y_1$ are:

$$(X_P, Y_P, 1) = L\times T1\times T2 = [R_P\cos\alpha, R_P\sin\alpha, 1]\times$$

$$\begin{bmatrix} 1 & 0 & 0 \\ 0 & 1 & 0 \\ (r+R) & 0 & 1 \end{bmatrix}\times\begin{bmatrix} \cos\theta_1 & \sin\theta_1 & 0 \\ -\sin\theta_1 & \cos\theta_1 & 0 \\ 0 & 0 & 1 \end{bmatrix}\times\begin{bmatrix} 1 & 0 & 0 \\ 0 & 1 & 0 \\ -(r+R)\cos\theta_1 & -(r+R)\sin\theta_1 & 1 \end{bmatrix}\times$$

$$\begin{bmatrix} \cos(-\theta_2) & \sin(-\theta_2) & 0 \\ -\sin(-\theta_2) & \cos(-\theta_2) & 0 \\ 0 & 0 & 1 \end{bmatrix}\times\begin{bmatrix} 1 & 0 & 0 \\ 0 & 1 & 0 \\ (r+R)\cos\theta_1 & (r+R)\sin\theta_1 & 1 \end{bmatrix}\times$$

$$\begin{bmatrix} 1 & 0 & 0 \\ 0 & 1 & 0 \\ -(r+R) & 0 & 1 \end{bmatrix}\times\begin{bmatrix} \cos\theta_0 & \sin\theta_0 & 0 \\ -\sin\theta_0 & \cos\theta_0 & 0 \\ 0 & 0 & 1 \end{bmatrix}$$

$$= [R_P\cos(\theta_1-\theta_2-\theta_0+\alpha)+(r+R)\cos(\theta_1-\theta_0),\ R_P\sin(\theta_1-\theta_2-\theta_0+\alpha)-(r+R)\sin(\theta_1-\theta_0), 1]$$

The trajectory equation of point P, that is, the motion equation of the workpiece at any point P relative to the grinding disc is:

$$\begin{cases} X_P = R_P\cos(\theta_1-\theta_2-\theta_0+\alpha)+(r+R)\cos(\theta_1-\theta_0) \\ Y_P = R_P\sin(\theta_1-\theta_2-\theta_0+\alpha)-(r+R)\sin(\theta_1-\theta_0) \end{cases}$$

That is:

$$\begin{cases} X_P = R_P\cos[(\omega_H-\omega_2-\omega_0)\ t+\alpha]+(r+R)\cos(\omega_H-\omega_0)t \\ Y_P = R_P\sin[(\omega_H-\omega_2-\omega_0)\ t+\alpha]+(r+R)\sin(\omega_H-\omega_0)t \end{cases} \quad (1)$$

Relative to the center radius ρ of grinding disc:

$$\rho = \sqrt{X_p^2+Y_P^2} = \sqrt{R_P^2+(r+R)^2+2R_P(r+R)\cos\left(\frac{r}{2R}\omega_1 t+\alpha\right)} \quad (2)$$

As can be seen from equation (1), the trajectory of point P relative to the grinding disc is a cluster of cycloid. At this point, the distance that any point P of the workpiece moves relative to the grinding disc in dt time:

$$\begin{cases} dX_P = \{-R_P(\omega_H-\omega_2-\omega_0)\sin[(\omega_H-\omega_2-\omega_0)\ t+\alpha]-(r+R)(\omega_H-\omega_0)\sin(\omega_H-\omega_0)t\}dt \\ dY_P = \{R_P(\omega_H-\omega_2-\omega_0)\cos[(\omega_H-\omega_2-\omega_0)\ t+\alpha]+(r+R)(\omega_H-\omega_0)\cos(\omega_H-\omega_0)t\}dt \\ dS = \sqrt{dX_P^2+dY_P^2}dt \end{cases} \quad (3)$$

In order to discuss the influence of grinding parameters on the motion trajectory, the revolution speed of the planet gear and the rotation speed of itself must be converted into the actual grinding parameters, such as the sun gear revolution speed and the grinding disk revolution speed. The relationship between the revolution speed of the planet gear and the sun gear rotation speed is as follows:

$$\omega_H = \frac{z_1}{z_1+z_3}\omega_1 \quad (4)$$

The relation between the rotation speed of the planet gear and the rotation speed of the sun gear is as follows:

$$\omega_2 = \frac{z_1(z_2 - z_3)}{z_2(z_1 + z_3)}\omega_1 \tag{5}$$

Where, z1 is the number of sun gear teeth, z2 is the number of the planet gear teeth, and z3 is the number of out pin ring teeth. Since the sun gear, the planet gear and the inner pin ring are cylindrical pin gears engaged with each other, their modulus are equal. Let the modulus be m, then:

$$\begin{cases} z_1 = 2\mathrm{mr} \\ z_2 = 2\mathrm{mR} \\ z_3 = 2\mathrm{m(r+2R)} \end{cases} \tag{6}$$

Substitute equations (4), (5) and (6) above into equation (3), and $I = \frac{\omega_0}{\omega_1}$ get:

$$\begin{aligned} \mathrm{d}S &= \sqrt{\mathrm{d}X_P^2 + \mathrm{d}Y_P^2}\,\mathrm{dt} \\ &= \omega_1\sqrt{\left[\frac{r(r+2R)}{2R(r+R)} - I\right]^2 R_p^2 + \left[\frac{r}{2(r+R)} - I\right]^2 (r+R)^2 + 2R_P(r+R)\left[\frac{r(r+2R)}{2R(r+R)} - I\right]\left(\frac{r}{2(r+R)} - I\right)\cos\left(\frac{r}{2\mathrm{R}}\omega_1\mathrm{t}+\alpha\right)}\,\mathrm{dt} \end{aligned} \tag{7}$$

2.Analysis of conditional of uniform distribution of grinding trajectory

At any point P of the workpiece, the trajectory density relative to the grinding disc is uniformly distributed, that is, the arc appearance of the trajectory is completely uniform for the radius of the grinding disc. In other words, the derivative of the trajectory arc length S with respect to the radius ρ of the grinding disc (that is, the change rate of the track arc length with the radius) should be linearly related to ρ [14]:

According to formula (2) :

$$\frac{\mathrm{d}S}{\mathrm{d}\rho} = K\rho \tag{8}$$

$$d\rho = -\frac{R_P(r+R)r\omega_1}{2R} \times \frac{\sin\left(\frac{r}{2R}\omega_1\mathrm{t}+\alpha\right)}{\sqrt{R_P^2 + (r+R)^2 + 2R_P(r+R)\cos\left(\frac{r}{2R}\omega_1\mathrm{t}+\alpha\right)}}dt \tag{9}$$

Replace: $\mathrm{a} = \frac{\mathrm{r}}{2R}$

$\mathrm{b} = \mathrm{r} + R$

$i_1 = \frac{r(r+2R)}{2R(r+R)} - I$

$i_2 = \frac{r}{2(r+R)} - I$

Substitute equations (7) and (9) into equation (8), and get:

$$K=\frac{ds}{\rho d\rho}=\frac{\omega_1\sqrt{i_1^2R_p^2+i_2^2b^2+2R_pbi_1i_2\cos(a\omega_1 t+\alpha)}dt}{\sqrt{R_P^2+b^2+2R_pb\cos(a\omega_1 t+\alpha)}\times\left[-abR_p\omega_1\times\frac{\sin(a\omega_1 t+\alpha)}{\sqrt{R_P^2+b^2+2R_pb\cos(a\omega_1 t+\alpha)}}dt\right]}$$

$$=-\frac{\sqrt{i_1^2R_p^2+i_2^2b^2+2R_pbi_1i_2\cos(a\omega_1 t+\alpha)}}{abR_p\sin(a\omega_1 t+\alpha)}$$

According to equation (8), if the linear relation is true, K should be a constant, i.e: $\frac{dK}{dt}=0$

be: $$\frac{dK}{dt}=\omega_1\frac{1+\frac{i_1^2R_P^2+i_2^2b^2}{i_1i_2R_Pb}\cos(a\omega_1 t+\alpha)+\cos^2(a\omega_1 t+\alpha)}{i_1i_2\sin^2(a\omega_1 t+\alpha)\sqrt{i_1^2R_P^2+i_2^2b^2+2R_pbi_1i_2\cos(a\omega_1 t+\alpha)}}=0 \quad (10)$$

$$1+\frac{i_1^2R_P^2+i_2^2b^2}{R_pbi_1i_2}\cos(a\omega_1 t+\alpha)+\cos^2(a\omega_1 t+\alpha)=0 \quad (11)$$

It can be seen that equation (11) is a quadratic equation with one unknown of $\cos(a\omega_1 t+\alpha)$, and the solution is:

Solution 1:

$$\cos(a\omega_1 t+\alpha)=\frac{-2i_1^2R_P^2}{2i_1i_2R_pb}=\frac{-i_1R_P}{bi_2} \quad (12)$$

Solution 2:

$$\cos(a\omega_1 t+\alpha)=\frac{-2b^2i_2^2}{2i_1i_2R_pb}=\frac{-bi_2}{i_1R_P} \quad (13)$$

As can be seen from equations (12) and (13), the two solutions are reciprocal to each other, so only one of them can be solved.

Replace:

$$i_1=\frac{r(r+2R)}{2R(r+R)}-I$$

$$i_2=\frac{r}{2(r+R)}-I$$

$$I=\frac{\omega_0}{\omega_1}$$

$$a=\frac{r}{2R}$$

$$b=r+R$$

After substituting back into equation (12) and arranging, we can get:

$$I_1=\frac{(r+R)+\cos(\frac{r}{2R}\omega_1 t+\alpha)\ R_P}{\frac{r}{2}+(\frac{r}{2R}+\frac{r}{2(r+R)})\cos(\frac{r}{2R}\omega_1 t+\alpha)\ R_P}=\frac{e+\cos(\frac{r}{2R}\omega_1 t+\alpha)\ R_P}{\frac{r}{2}+(\frac{r}{2R}+\frac{r}{2e})\cos(\frac{r}{2R}\omega_1 t+\alpha)\ R_P} \quad (14)$$

The formula,

r-- the radius of the sun gear pitch circle

R -- the radius of the planet gear pitch circle

R_P -- the radius of any point of the workpiece relative to the center of the planet gear

ω_1-- the rotation speed of the sun gear

e -- eccentricity, that is, the distance between the center of the planet gear and the grinding disc.

3.Analysis of uniform distribution condition of grinding trajectory

Equation (14) is derived by using the reductive method -- first, it is assumed that the track density of any point of the workpiece relative to the grinding disc is uniformly distributed -- that is, the trajectory arc length S of any point of the workpiece relative to any radius ρ of the grinding disc is linearly related to its center radius of the grinding disc through mathematical derivation. This means that in order to achieve uniform distribution of the workpiece's trajectory density relative to the grinding disc at any point, equation (14) must be true. The equation (14) shows that meet of uniformity of lapping trajectory the rotating speed ratio I (grinding disc angular velocity ω 0 and the ratio of the sun gear angular velocity ω 1)is the function of r, R, R_P and t. For a given grinding machine and workpiece processing, r、R、R_P has been determined, that is, constant, only the sun gear angular velocity ω_1 and grinding disc angular velocity ω_0 is the operating parameters can be changed. More importantly, the rotating speed ratio I is a function of the grinding time t. This indicates that in order to satisfy the uniformity of lapping trajectory of the workpiece relative to the grinding disc, the rotating speed ratio must change with time. On the other hand, corresponding to a constant speed ratio (whether rational or irrational), the uniformity of lapping trajectory the workpiece relative to the grinding disc is impossible to achieve a linear uniform distribution. The effect of time t can be ignored only if and only if the $\cos(\frac{r}{2R}\omega_1 t+\alpha)$infinitesimal in $\frac{r}{2R}\omega_1$ is close to zero and the amplitude $\cos(\frac{r}{2R}\omega_1 t+\alpha)$is close to a straight line, but this is impossible in reality.

In order to discuss the influence of rotation speed ratio on the motion trajectory uniformity, formula (1) can be converted into the following formula:

$$\begin{cases} X_p = R_p\cos\left[(\frac{r(r+2R)}{2R(r+R)}-I)\omega_1 t+\alpha\right]+(r+R)\cos\left(\frac{r}{2(r+R)}-I\right)\omega_1 t \\ Y_p = R_p\sin\left[(\frac{r(r+2R)}{2R(r+R)}-I)\omega_1 t+\alpha\right]+(r+R)\sin\left(\frac{r}{2(r+R)}-I\right)\omega_1 t \end{cases} \quad (15)$$

Taking the AC700F planar CNC grinding machine of planetary pin gear of Peter Wolters company in Germany as an example, this paper discusses and analyzes the influence of the rotational speed ratio I of grinding disc and sun gear on the workpiece's motion trajectory relative to the grinding disc. Its parameters are as follows: sun gear pitch radius is $r = 137.85$ mm; The pitch radius of the planet gear is $R = 122.15$ mm; For the sake of representativeness and convenience of calculation, the workpiece center point is taken as the starting point of motion, That is, R_p is the pitch circle radius of the workpiece hole distribution on the planet gear $R_p = 90.5$ mm. Considering that the influence of initial Angle on the trajectory distribution uniformity can be ignored, we can set $\alpha=0$.Take the sun gear speed n1=100rpm, t=0.42s and substitute the above parameters into (14) and (15) to obtain the workpiece center point relative to the grinding disc motion trajectory distribution in FIG. 3. It can be judged from the figure that the track point distribution is uniform, but if it exceeds t=0.45s, the track will be very chaotic, because the rotational speed ratio is close to infinity. Rotational speed ratio of the corresponding is changing with the time as shown in figure 4, there are the actual value of the mutation (in theory to plus or minus up) in the change cycle,is clearly impossible, but you can take the middle period (t=0.62 s to 1.52 s) by getting rid of the extremum and access to basic meet the requirements of trajectory distribution

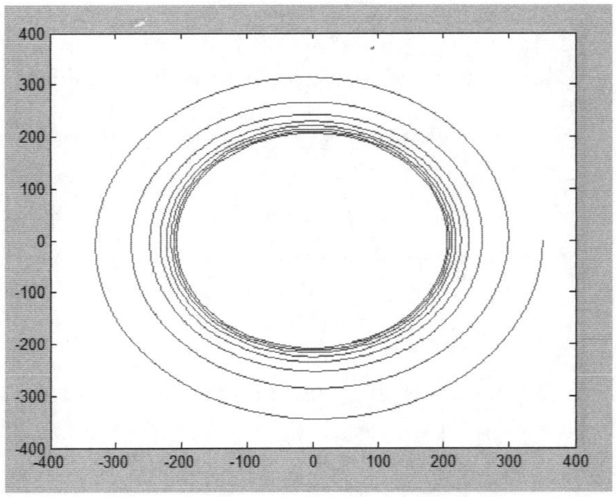

Figure. 3 Workpiece center point relative to grinding disc motion track distribution

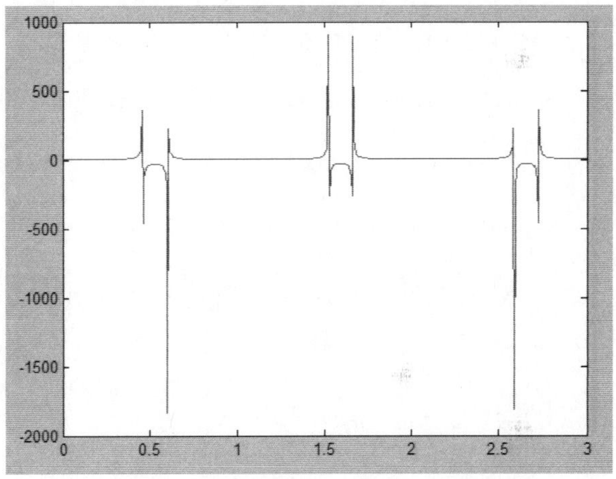

Figure.4 variation of rotational speed ratio satisfying the unformity of motion trajectory

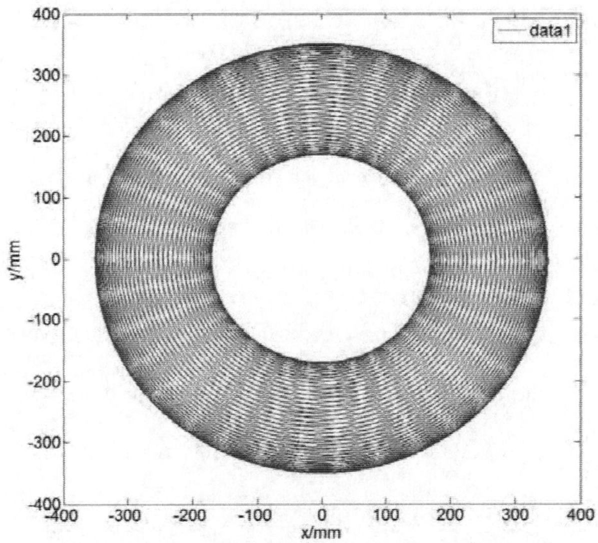

Figure.5 movement point distribution of workpiece center

uniformity, shown in figure 5 to ignore the extreme value point and the simulation time t = 18 s track points distribution, subject to further research on this aspect.

By inverting (14), equation (16) can be obtained:

$$I_2 = \frac{\frac{r}{2} + \left(\frac{r}{2R} + \frac{r}{2(r+R)}\right)\cos\left(\frac{r}{2R}\omega_1 t + \alpha\right) R_P}{(r+R) + \cos\left(\frac{r}{2R}\omega_1 t + \alpha\right) R_P} \quad (16)$$

Similarly, by substituting the above parameters into equation (16), the change curve of rotational speed ratio with time as shown in FIG. 6 can be obtained. It can be seen from the curve that it is a time-dependent trig function with zero and negative values. Similarly, the rotational speed ratio (16) was substituted into the trajectory equation (15), and the above related parameters were substituted. The grinding parameters were taken as n1=100rpm of the sun gear, and the grinding time t=20s to obtain the workpiece center's motion trajectory curve relative to the grinding disc as shown in FIG. 7.It can be judged from the figure that the distribution of its trajectory points is uniform, but there is a ring blank belt between them. What causes it remains to be studied in the next step.

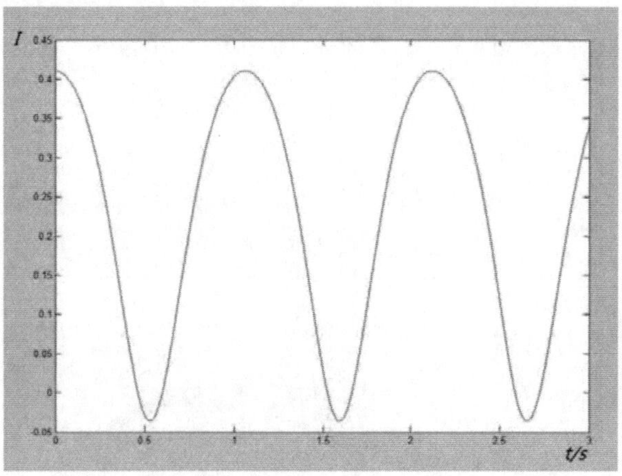

Figure. 6 Movement point distribution of workpiece center relative to grinding disc

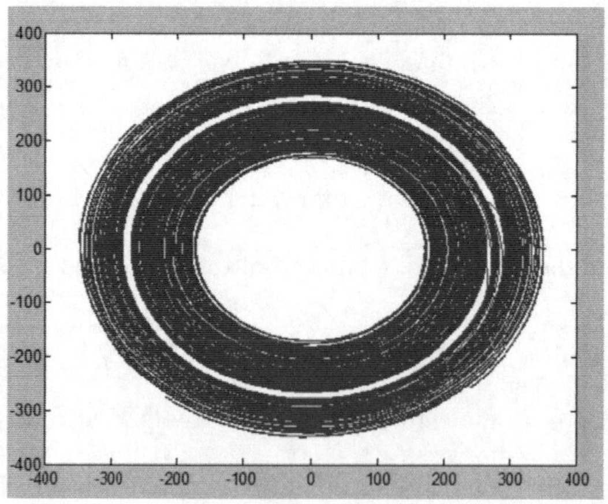

Figure. 7 Workpiece center point relative to grinding disc motion track distribution

The uniformity of the grinding trajectory is the premise to ensure the uniform wear of the grinding disc or the improvement of the flatness of the workpiece, but it is not enough to ensure the uniform wear of the grinding disc or the removal of the workpiece by the uniform material of the machining surface. The decisive factor should be the relative speed between the workpiece and the grinding disc. According to Preston equation, the grinding disc wear or workpiece material removal rate is a function of the relative velocity. In the next step, according to the above ideas, the condition that the relative velocity density of the workpiece at any point P relative to the grinding disc is uniformly distributed can be studied. In other words, the derivative of the synthetic relative velocity with respect to the radius of the grinding disc (that is, the rate of change of the synthetic relative velocity with the radius) should be linearly related to the radius of the grinding disc.

4.Conclusion

Based on the reductionist method, the motion equation of planetary plane grinding established by kinematics is analyzed, and the important parameter in the grinding process -- rotational speed ratio I(grinding disc rotational speed and sun gear rotational speed ratio) is studied to meet the condition that the motion trajectory uniformity of any point on the workpiece relative to the grinding disc is satisfied, and the following conclusions are drawn:

(1) the rotational speed ratio I, which satisfies the motion trajectory uniformity of any point on the workpiece relative to the grinding disc, is not only related to the structural parameters of the grinding mechanism (eccentricity, radius of the sun gear, radius of the planet gear, coordinates of the point on the workpiece), but also a function of time t.

(2) for a given planetary plane grinding grinder, the fixed speed ratio I (whether rational or irrational) that satisfies the motion trajectory uniformity of the planetary grinding disc does not exist. If the rotational speed ratio changes periodically with time, the possibility of motion trajectory uniformity distribution exists.

(3) this study lays a theoretical foundation for the further study on the uniformity of the motion trajectory of planar planar grinding, and also provides a new idea and method for the analysis of the relative velocity between workpiece and grinding disc, the wear uniformity of grinding disc, and the study on the surface quality of workpiece.

References

[1] Yuan julong. Ultra-precision machining technology of functional ceramics [M].Harbin: Harbin Institute of Technology press, 2000. (in Chinese)

[2] Zhou zhibin, Xiao shali, Zhou yan, etc. General situation and application of ultraprecise machining technique [J].Modern manufacturing engineering, 2005(1):121-123. (in Chinese)

[3] Wu yaoyu, Liu dongmin, Xie jixun. High efficiency lapping technology for support plate of fuel injector with vitrified bond CBN lapping plate [J].Diamond & abrasive engineering, 2013, 33(4):68-70(in Chinese)

[4] Wu hongji, Cao lixin, Liu jian. Analysis of motion geometry based on the grinding process of planar grinding machine [J].Chinese Journal of mechanical engineering, 2002,38 (6) : 144-147. (in Chinese)

[5]Zhao ping, Tao li, Wang zhiwei, Yuan julong. Review on Trace of Plane Lapping/polishing [J].Aviation precision manufacturing technology, 2009,45 (2) : 1-6. (in Chinese)

[6] Yang jie, hong tao, wen donghui, etc. Influence of lapping drive mode and rotational speed ratio on the track of abrasive particles [J]. Journal of Mechanical & Electrical Engineering, 2018, 35(5):453-458. (in Chinese)

[7] Yang changming, Zhu li, Zhang mao.Analysis of grinding motion trajectory of grinding machine [J].Machine tools and hydraulics, 2012,40 (15):7-9. (in Chinese)

[8] Zhao wenhong, Zhou zhaozhong, Wen donghui, etc.. Research on the uniformity of certain and uncertain eccentricity plane lapping processes [J]. Diamond & abrasive engineering, 2006(3) : 46-49. (in Chinese)

[9] Sheng jisheng, Wen donghui, Timing-ming. Uniformity of kinematic trajectory with wafer driving Initiatively in plane lapping process[J]. Transactions of the Chinese society of the Agricultural Machinery, 2010,41 (4) : 209-212. (in Chinese)

[10] Ji mengtuo, Hong tao, Wen donghui, etc. Analytical study on uniformity of path distribution with irrational rotational speed ratio in plan lapping process [J]. Journal of Mechanical & Electrical Engineering,2016,33 (5) : 532-536. (in Chinese)

[11] Liu qing, Zhang guangpeng, Gao zhixue etc. Experimental Study of X-Y Linkage Plane Lapping Track [J]. Mechanical Science and Technology for Aerospace Engineering, 2008, 37(6) : 903-909. (in Chinese)

[12] Yan zhen, Fang congfu, Liu chong.Theoretical study on trajectory of swinging plane lapping and polishing [J].Diamond & abrasive engineering, 2008,38(4) : 77-82. (in Chinese)

[13] Adam Barylski & Norbert Piotrowski. Non-conventional approach in single-sided lapping process：kinematic analysis and parameters optimization［J］. The International Journal of Advanced Manufacturing Technology (2019) 100:589–598

[14] Yang j d, Tian c l. High-speed Lapping technology [M]. Beijing: National Defense Industry Press, 2003. (in Chinese)

Volume Production of Machining Tools Using Laser MicroJet® Technology

Volume production of high precision tools with new 5-axis Laser Machining System LCS 305

Jacques Coderre, Rémi Laure, Amédée Zryd

Introduction

Industrial diamond materials are replacing tungsten carbide and ceramic composites in cutting tools where high surface finish quality is required. The choice of materials ranges from lower end polycrystalline diamond (PCD) to high-end single crystalline diamond (SCD). As an example, SCD tools have proven their value-add in the super-finishing of ultra-precision optical grade surfaces. In many cases, the use of PCD tool inserts can eliminate grinding operations, thereby improving machining process times. PCD tools have also shown a longer tool life.

Such materials being as hard as natural diamond, lasers are most suitable for machining them. Synova's patented Laser MicroJet® (LMJ) technology delivers better results than conventional dry lasers and has become a benchmark in the processing of diamond materials. In this paper, results achieved in machining diamond inserts are presented. The novel LCS 305 Laser MicroJet system with 5 axes is introduced as a production-ready system to manufacture such tools.

The LMJ Process

In the Laser MicroJet system, a laser beam, passing through a pressurized water chamber, is focused into a nozzle.

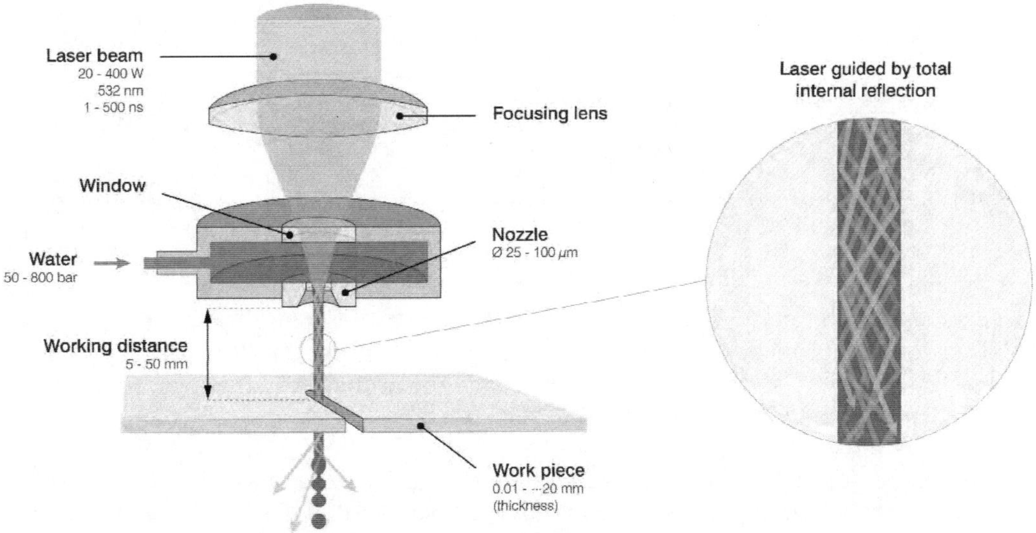

Figure 1: LMJ technology

The low-pressure water jet emitted from the nozzle guides the laser beam by means of total internal reflection at the water/air interface. The water jet diameter is usually 50 microns and the laser power required is between 25 and 30 watts. While the principle is simple, years of experimentation and optimization have gone into the industrialization of the process and in the design and manufacturing of advanced machining systems.

The LMJ process works in two stages. The energy of the laser pulses vaporizes the workpiece material by heating, while the water cools and cleans the surface in the interval between the pulses. Through a scanning process, a trench is formed that becomes deeper with each pass.

As compared to conventional dry lasers, the LMJ 'wet laser' technology has many advantages.

The most important advantage is that the Laser MicroJet cuts with a cylindrical beam (no V-shape) and the cutting depth can reach up to several centimeters.

This is not the case with conventional lasers where the focused laser beam has a limited working distance of just a few millimeters due to beam divergence. The beam converges at a focal point and then diverges. Therefore, a focus distance control is required, and the working distance is short.

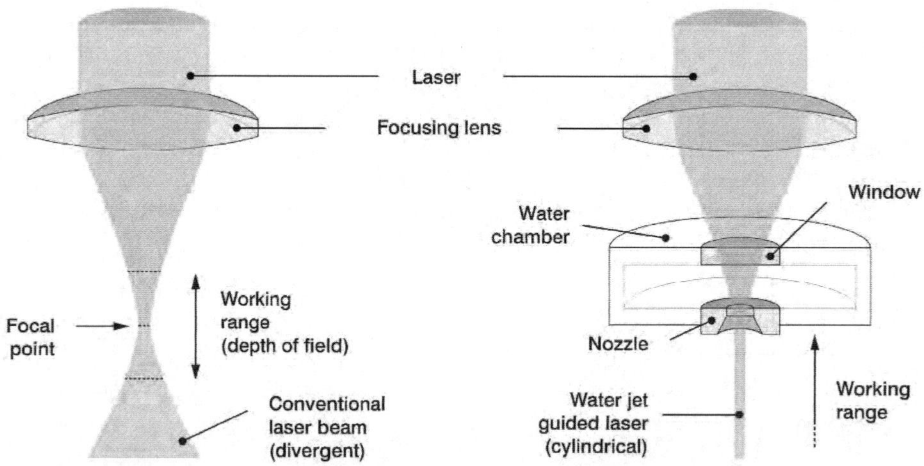

Figure 2: Difference between conventional laser beam (left) and water jet guided laser (right)

The technology behind the Laser MicroJet is based on creating a laser beam that is completely reflected at the air-water interface, using the difference in the refractive indices of air and water. The laser is, therefore, entirely contained within the water jet as a cylindrical beam, similar in principle to an optical fibre.

There is no need for focal adjustment, and one obtains parallel kerf sides. There is a minimum heat affected zone thanks to the cooling effect of the water. Finally, there is a high removal rate with debris washed from the kerf.

The LCS 305 Laser Cutting System

The LCS 305 laser cutting system can cut extremely hard materials with high precision and complex shapes. It is a 5-axis machine with precision components for processing accurate tools.

Figure 3: Overview of LCS 305

Linear and rotative axes are mounted on separate carrier frames to increase stability at high speed. This results in a dynamic accuracy lower than ± 3 µm and repeatability of ± 1 µm with angular precision of ± 0.9 sec.

Figure 4: Overview of LCS 305 axis system

The LCS 305 is typically configured with an HSK-63 holder, enabling a positioning radial accuracy lower than 3 µm, and an axial accuracy lower than 1µm. Combined with a unique axis system, it allows a cylindrical working envelope of 130 mm in diameter, with a height of 260 mm. It was designed with high-accuracy machining of hard materials in mind, from small inserts to large rotative tools.

Figure 5: HSK63 holder

Volume Production Capabilities

To ensure high-volume manufacturability of hard material tools, the LMJ technology leverages geometrical precision with high-quality finished cutting surfaces, ensuring the manufacturing of a tool that shows a long tool life and optimal performance. The LCS 305 is designed and manufactured in a way to get the full potential out of the LMJ technology, thanks to its axis and holding system. In addition of cutting quality and speed, it is possible to produce parts in volume by eliminating the need for probing or pre-cutting procedures.

To generate the data presented in this paper, a production run of 20 samples was done. After one calibration cut, all 20 parts were machined without any additional setup.

Figure 6: 3D view of processed sample

Results pertaining to the 20-sample run are presented in the table below. Using the LCS 305, a geometrical tolerance of profile lower than ± 0.1 mm was obtained, versus drawing nominal targets. The achieved surface quality eliminates the need for a subsequent grinding operation.

Edge radius	Roughness	Overall cutting speed
5 µm	< 0.300 µm	4.5 mm/min

Figure 7: 3D view of cutting edge (Alicona Infinite Focus SL)

This test also demonstrated how high precision in primary and secondary clearance angles can be achieved, as shown in the graph and table below. The results are consistent all along the profile path, even if there are different angles for each face.

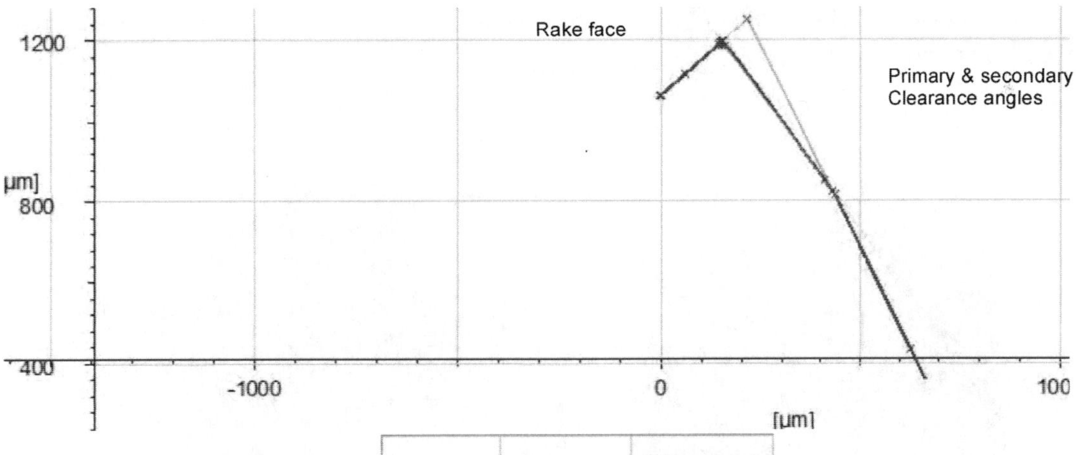

	Angle [°]	Nominal Angle [°]
Angle 1	84.937	85
Angle 2	75.467	75

Figure 8: Measure of cutting surface (Alicona Infinite Focus SL)

The same test was performed on rotative tools composed of several inserts. Similar high accuracies were obtained on all inserts of a given tool. The use of an HSK 63 holder guarantees a highly precise positioning. The availability of a fully automated calibration of the

LCS 305, along with an optimal repeatability in the process, enables a production with minimum operator coverage and maximum machine availability.

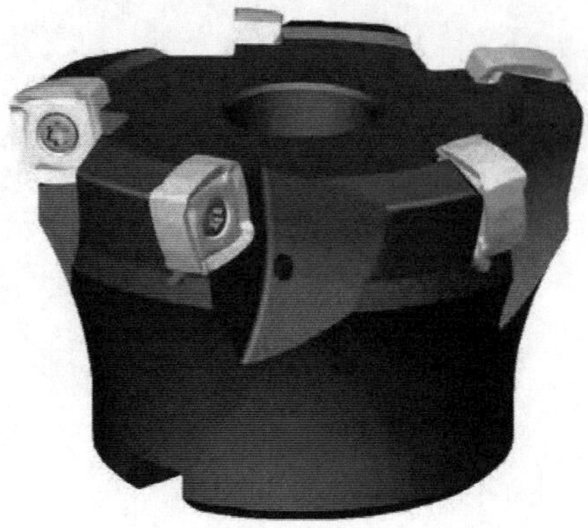

Figure 9: Example of processed rotative tool

Complex 3D Shaping

The LCS 305 allows complex 3D form-cutting or form-engraving. To demonstrate such capability, small pyramids with a 1 mm square base and 1 mm in height, were cut out of a SiC blank. A production run of 40 pieces was completed in order to demonstrate the repeatability of the process, as well as the uniformity of the cutting profile along the 4 faces. As depicted below, a well-defined peak pyramid, with a roughness of 500 nm on all the faces, was achieved.

Figure 10: Alicona measure of pyramidal 3D form

In addition to 3D cutting, 3D engraving trials were conducted, where blind holes (diameter 0.25 and depth 0.9) and conic shapes (diameter 1.20 and depth 0.85) were ablated (see Figure 11). First, the ablation rate per pass is determined. The number of repeated passes is then adjusted to achieve the desired depth. Repeatability and shape control of this test demonstrate the ability to produce a wide range of high-precision forms by using the LCS 305.

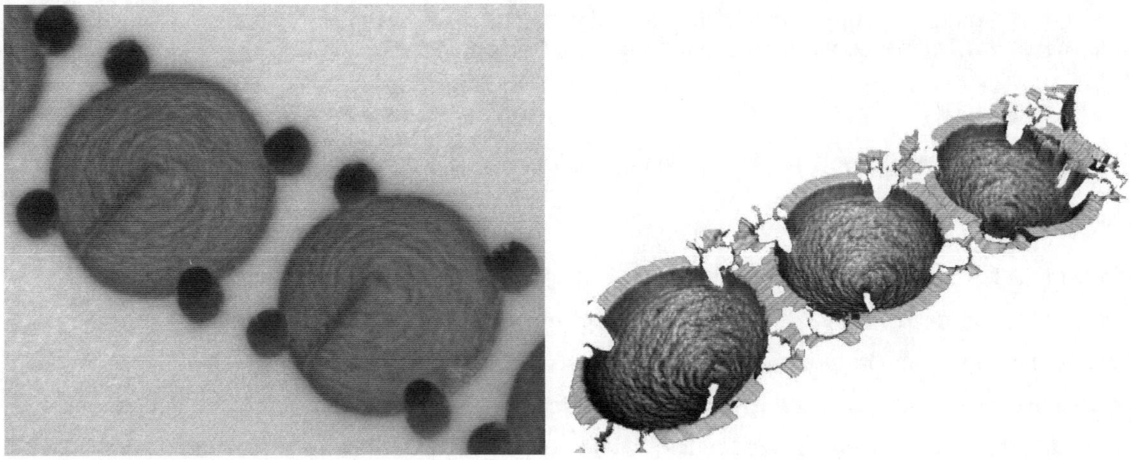

Figure 11: Example of engraving with LCS 305

Conclusion

The results shown in this paper highlight the capabilities of the LCS 305. With its 5-axis capability and intuitive CAM software, this advanced system is ideally suited for the machining of super hard materials, able to cut tools of different geometries and size.

By using automated setup procedures, a high-precision axis system and holders with high repeatability in positioning, the LCS 305 provides low cost of ownership by reducing the need for operator interventions during volume production. Once calibrated, parts are produced without the need for probing or pre-cutting procedures.

The LCS 305 produces parts with high cutting quality that ensure geometrical and surface tolerances. Combined with an overall cutting speed of up to 5 mm/min in PCD, it delivers finished parts at a high rate.

In cases where manufacturers need a high cutting-tool edge finish, a hybrid solution can be used whereby an LCS 305 is combined with a CNC polisher. In such cases, a sub - 0.1 um Ra roughness is achievable.

Differentiating performance of Polycrystalline Cubic Boron Nitride (PCBN) composite with nanostructured binder

Authors: Chris Graham, Atsuko Tomoda, Myriam Motchelaho, Åke Andersin, Antionette Can, Miriam Miranda, Alexandre Thomasson, Andreas Malmquist and Santonu Ghosh

Keyword: PCBN, nanostructured, finish machining, productivity

Abstract

A large part of ferrous machining, primarily finishing and semi-finishing operation, is carried out by polycrystalline cubic boron nitride (PCBN) cutting tool material. Although PCBN cutting tool material operates under high temperature machining conditions (>800°C) in hardened steel machining, an increased demand in productivity is pushing PCBN tool material to perform at even higher temperatures for longer durations. This requires PCBN cutting tools to exhibit higher crater wear resistance and delay failure.

PCBN is a composite material where binder phase plays a critical role along with CBN grain size and distribution. It has been shown that a stronger bonding between the binder particles and CBN-binder particles enhances the wear resistance of the PCBN and reduces crack propagation through the material. This paper looks at the influence of nano TiC/TiN material in PCBN (PureCut™ platform from Element Six) as a means to achieve a stronger binder to CBN particle bonding, and discusses improvements seen to the microstructure, mechanical behaviour and machining performance in interrupted finish machining. The paper also analyses the influence of impurities on the sintering of nano TiC/TiN PCBN.

1 INTRODUCTION

High speed machining of hardened steel using polycrystalline cubic boron nitride (PCBN) has converted several grinding operations into machining [1] .Hard part turning using PCBN equipped tools is a single step operation which offers surface finish and integrity that is similar to or better than finish grinding, with an energy consumption five times lower than grinding [2]. The introduction of PCBN in machining has not only resulted in higher production efficiency and higher quality of the finished parts but also a reduction in cost per machined part. In one case study, Whitney et al. showed a reduction in cost per part of 38% compared to silicon nitride cutting tools [3]. In a separate study, Pu and Sing demonstrated that the cost per part could be reduced by more than 10% and cycle time reduced by more than 40% despite the cost of the tool being 5 times higher [4].

Improvement in productivity depends on several factors including tool life, tool changing time, cutting conditions, capacity of the machines and the velocity of the assembly and supply line. There are different optimisation strategies for extracting higher productivity in metal cutting [5]. An increase in productivity can be achieved mainly through an increase in the cutting speed or through the use of higher feed rate. The former strategy has the added advantage of improved workpiece surface roughness and has been the focus of Element Six in this paper. Of critical importance when pursuing productivity improvements through higher cutting speeds is the crater and flank wear resistance of the PCBN [6,7].

For the design of a PCBN material, much research has been done to understand the influence of three design elements: (i) CBN content; (ii) CBN grain size and size distribution and (iii) the chemical composition of the binder. CBN content has a significant influence on chemical wear resistance and toughness of the PCBN whereas CBN grain size and distribution mainly dictate the abrasive wear resistance of the material [8] but also influences toughness.

On the chemical composition of the binder there are three main ceramics binders that are used in low content PCBN material. Titanium nitride (TiN), titanium carbide (TiC) and titanium carbonitride (TiCN) based binders each provide different property advantages. TiC provides better abrasion resistance than TiN, but TiN offers improved thermal conductivity over TiC to better manage thermal shock resistance seen in interrupted machining. TiCN provides a balance of properties based on the application area.

Equally important to chemical composition is the level of reaction and bonding between binder particles and the CBN grains. It has been demonstrated in other composite materials that bonding can be improved through the use of ultrafine binder particles. For example, in TiCN cermets use of ultrafine grain TiCN has been shown to improve the strength and toughness of the material [9]. Experiments with ultrafine binder particles in PCBN have shown similar improvements to reaction and bonding but also emphasise the importance of reducing impurities to achieve such benefits. [10].

This paper has studied the influence on wear resistance and physical properties in machining from using a PCBN with a nanocrystalline binder with minimised impurities. This paper also assesses the likely improvements to productivity of such a PCBN material.

2 EXPERIMENTAL

2.1 Synthesising nanostructured PCBN

A comparative study was carried out to investigate performance and characteristics of different PCBN materials. Different compositions were produced through two different powder manufacturing technologies to reduce impurities in the sintered material. Compositions are summarised in Table 1. CBN binder mixed powder was sintered under a High Pressure High Temperature process with pressure between 4-6 GPa and temperatures between 1200-1400°C. Two synthesis processes produced PCBN with submicron (PBA) and nanosized binder <100 nm (PBC). Phase analysis of the material was carried out using XRD (Pananalytical). The microstructure of the PCBN was also analysed using FIB-SEM (Auriga, Zeiss) and STEM (JEOL 2100F) to analyse the presence of impurities in the material.

A Vickers micro-indenter (weber 572) with diamond pyramidal indenter was used for measuring the micro-hardness using a 500 g load for 15 seconds. Indentation size was measured using an optical microscope. Micro-hardness Hv was calculated using the expression [11]

$$Hv = \frac{1.8544 P_l}{d_l^2} \quad (1)$$

Where, P_l is the force, kgf and d_l mean diagonal length of the indentations.

Table 1: Composition of the PCBN tested for measurement of properties and application testing

Sample ID	CBN content (vol.%)	CBN grain size (µm)	Binder type	Binder size
PAB-1	50	4	TiCN	Submicron
PAB-2	55	1	TiCN	Submicron
PAB-3	65	1-3 (Bimodal)	TiCN	Submicron
PAB-4	70	1-3 (Bimodal)	TiCN	Submicron
PBC-1	50	4	TiCN	Nanostructured
PBC-2	55	1	TiCN	Nanostructured
PBC-3	65	1-3 (Bimodal)	TiCN	Nanostructured
PBC-4	70	1-3 (Bimodal)	TiCN	Nanostructured

Along with these four compositions, four more compositions with a different level of inclusions were produced to investigate the influence of impurities in the material.

2.2 Wear testing

Wear and fracture behaviour of the PCBN grades was investigated through machining of hardened steel bars (SAE 8620). Chemical composition of the workpiece material is shown in Table 1. Hardness of the workpiece was measured between 60-62 HRC and the depth of the case hardened layer was 1.5 mm. Length on the bar was 120 mm with 39 mm diameter, with two slots to introduce interruption in the material (Fig. 1). Tools that were used for the testing had the ISO geometry SNNM090804S with 150±20 µm chamfer width, 25±1° Chamfer angle and 20±5 µm hone and 0.8 mm nose radius. Tool life was characterised as failure through fracture. Test conditions comprised finish turning at a cutting speed (V_c) of 150 and 180 m/min, feed (*f*) of 0.1 mm/rev and depth of cut of 0.15 mm and testing was performed on a CNC lathe (Glidemeister, CTX 410). Flank wear was measured using an optical microscope and crater wear was measured using a 3D microscope (Alicona). The PCBN grade used for this application test contained 65% CBN. A further test matrix with 65% CBN content was tested to study the influence of impurity content in PCBN.

Table 2: Chemical composition of SAE 8620 case-hardened steel

Workpiece	C (%)	Si (%)	Mn (%)	Ni (%)	Cr (%)	Mo (%)	S (%)	P(%)
SAE820	0.18-0.23	0.15-0.35	0.7-0.9	0.4-0.7	0.4-0.6	0.15-0.25	0.04	≤ 0.035

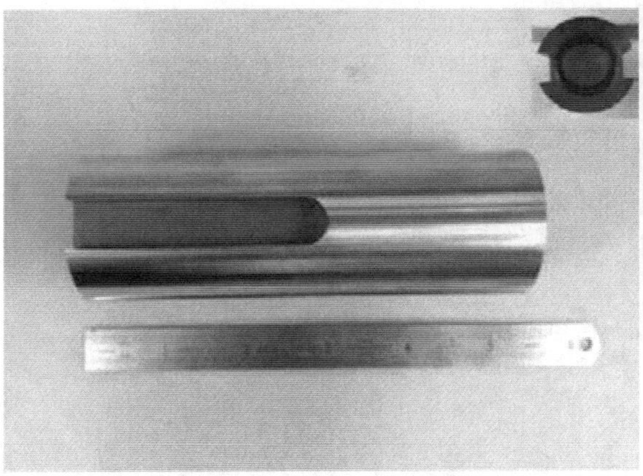

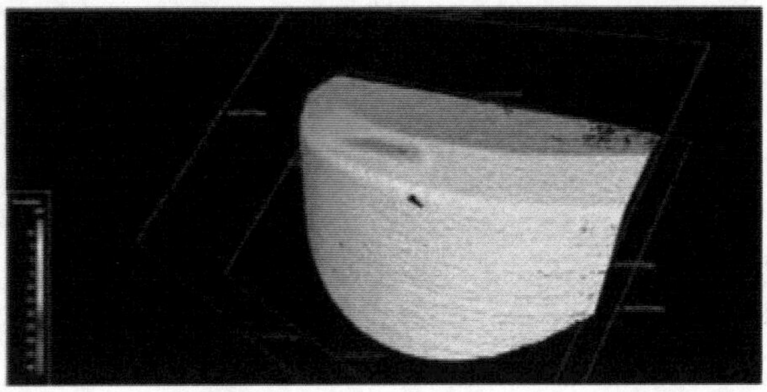

Fig 1. (a) SAE 8620 Workpiece for machining experiment. (b) 3D image of the crater development, deepest point of the crater was measured by overlapping the original uncut model with the machined insert.

3 Results and Discussion

3.1 Nanostructured PCBN

Figure 2a is the TEM micrograph of the PAB-2 with sub-micron grain-sized binder phase. The dark large particles in the micrographs are the CBN particles in the submicron grained binder phase. The distinction is much clearer in Fig. 2b, which is the elemental map, red grains in the micrographs are CBN grains and the binder phase shows a colour distribution which is a composite map, showing the presence of Ti, C, N, and Al in the binder phase. Submicron structure of the binder grains could be seen from the micrograph. In Fig. 2c, the nanostructured binder of PBC-2 could be seen in the micrograph with an average grain size less than 100 nm like Fig 2b. Figure 2d showed the elemental map which indicated the presence of the same elements as for sub-micron structured material.

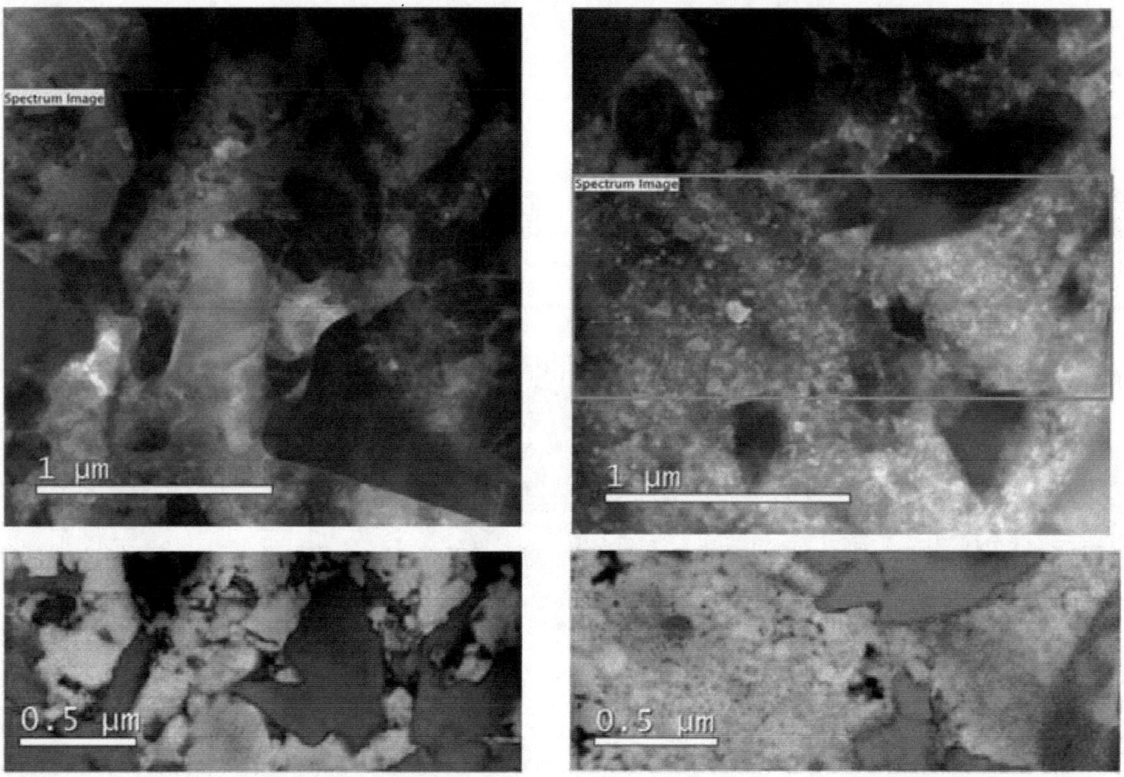

Fig 2. (a) Microstructure of PAB-2 with submicron grain (b) (d) Elemental map of the binder phase showing CBN (in red) and the binder phase in green and blue a composite of Ti, N, C and Al. **(c)** Microstructure of PBC-2 with nanostructure binder

3.2 Low impurity containing material

Achieving a nanostructured binder will help to improve reaction and bonding of binder to CBN in the PCBN material. However, the presence of impurities could minimise the sintering of the particles, which is a common concern for the non-oxide nanostructured material [11]. In PCBN, impurities can be introduced in the system at various stages of the powder manufacturing process and can be both oxide and non-oxide in nature. The nanostructured material used in this study included various levels of impurities as well as materials largely devoid of impurities. Oxide impurities are difficult to investigate because of the layer thickness and they alter the wetting behaviour of the binder particle. Figure 3 shows the EDS map of the binder phase of the nanostructured material.

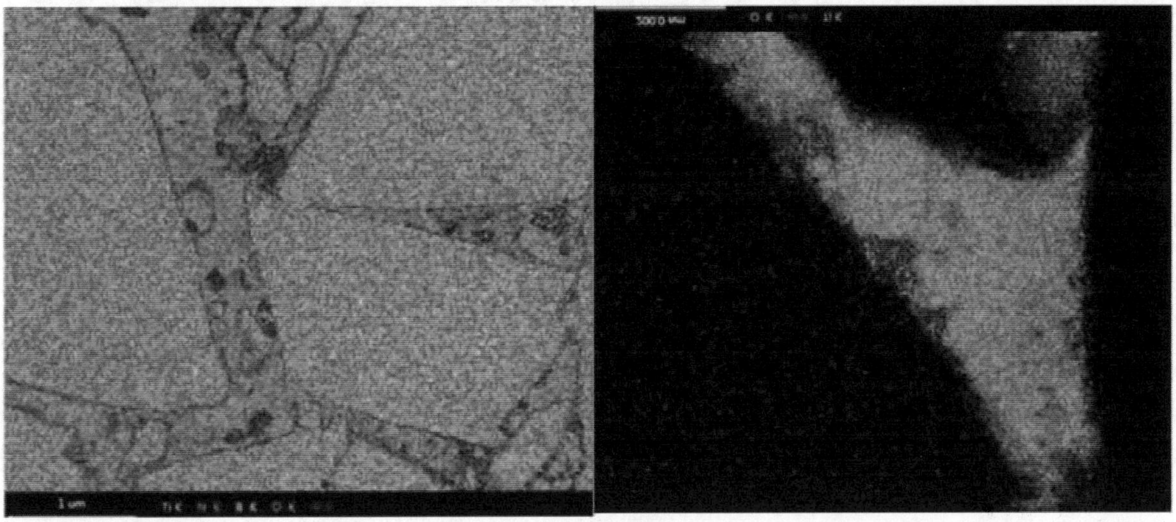

Fig 3. (a) (b) EDS Elemental map of the binder phase of the material PBC-3, showing formation of Aluminium Oxide as the main oxygen source

In Fig. 3, the green grains are CBN grains whereas red and blue spots are aluminium and oxygen respectively. Oxygen stays in the material as aluminium oxide. Oxygen content was less than 3wt% in the powder before sintering which is considered low for any low content PCBN material.

To explore the role of non-oxide impurities in nanostructured PCBN E6 designed two PCBN materials both based on 70% CBN content, 2 μm CBN grain size and TiCN binder composition. For the first PCBN material (PAB-4) Element Six added impurities (for this experiment carbide particles were used) to the powder manufacturing process whereas the second PCBN material had these removed (PBC-4).

Fig 4a. shows SEM scan images of the two materials where the white particles in the microstructure are the carbide particles whereas Fig 4b shows the impurities free PCBN material. A higher purity of the binder resulted in not only improved sintering between the binder particles but also improves both thermal and mechanical properties of the material.

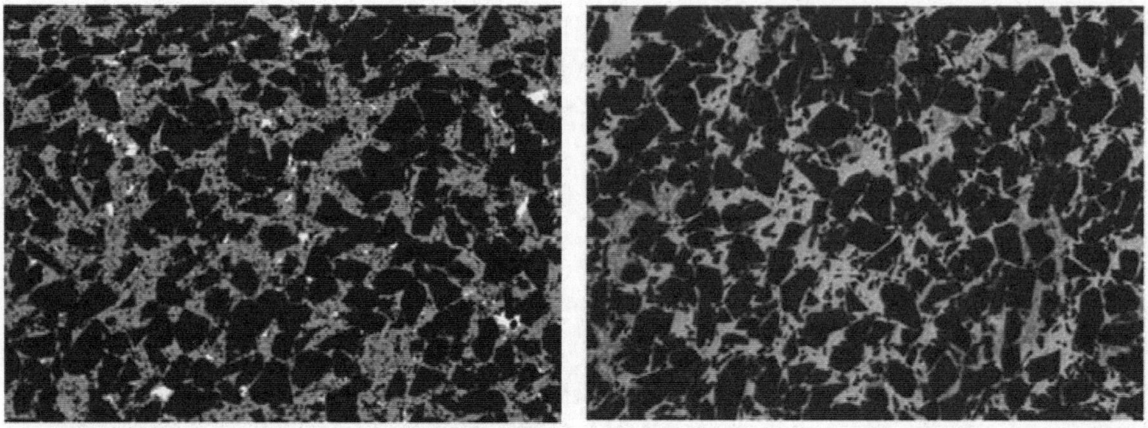

Fig 4 (a). Microstructure of 70% CBN containing PCBN showing the white carbide inclusion which in absent in (b) with same composition as a, containing nanostructured pure binder

3.3 Hardness measurement.

Improved mechanical properties of the binder require properties of the binder to be measured. Vickers indentation under a load of 1 N was carried out on the binder region of 2 PCBN materials (PAB-1 and PBC-1). One with submicron grain size (PAB-1) and the other with the nanostructured grain size (PBC-1). Fig 5 a shows the impression of the indent in the binder area of a submicron structured PCBN material and compared with a similar indentation on nanostructured PCBN material. Fig 5.c shows the difference in the hardness of the binder where it is evident the nanostructured binder is providing higher hardness. Although the original Hall-Petch relationship was developed for metallic materials, hardness of PCBN material can also be approximated by the Hall-Petch relationship [12]. Hence, it is not surprising that hardness of the nanostructured phase is higher than the submicron grained binder phase.

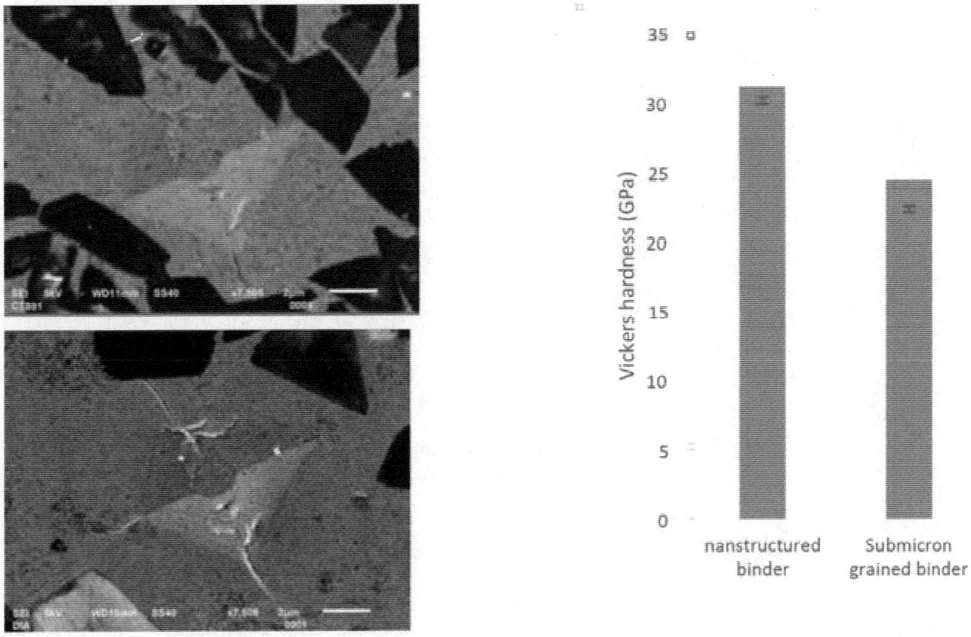

Fig 5 (a). Micro-indentation on the binder with submicron grain structure **(b)** Indentation on the nanostructured grained PCBN (c) Measured hardness of the binder phase

3.4 Wear tests

Superior sintering manifests in improved wear resistance of the cutting tool. Inclusions and impurities in PCBN play a key role in tool material performance. , To investigate this, five different PCBN tool materials with 65l% CBN and with different level of carbide inclusions were machined in the interrupted test as described in the previous section. Figure 6 is a plot for crater depth of the cutting tool after machining for approximately 560 m under a cutting speed of 180 m/min, feed rate of 0.1 mm/rev and a depth of cut of 0.1 mm against the area of the carbide particles measured using image analysis technique. It is quite clear that an improvement of crater depth could be correlated with the reduction in the carbide inclusion in the matrix. Relationship with carbide inclusion and the wear result is complex and different [9, 13], it was reported that above a certain level of WC, both abrasive and crater wear start enhancing. When the oxide protective layer from the tool surface is removed, it opens up large number of interface boundaries which are potential sites for grain pull out thereby aggravating the wear process. Minimizing the carbide inclusion in the PCBN could enhance wear properties of the material

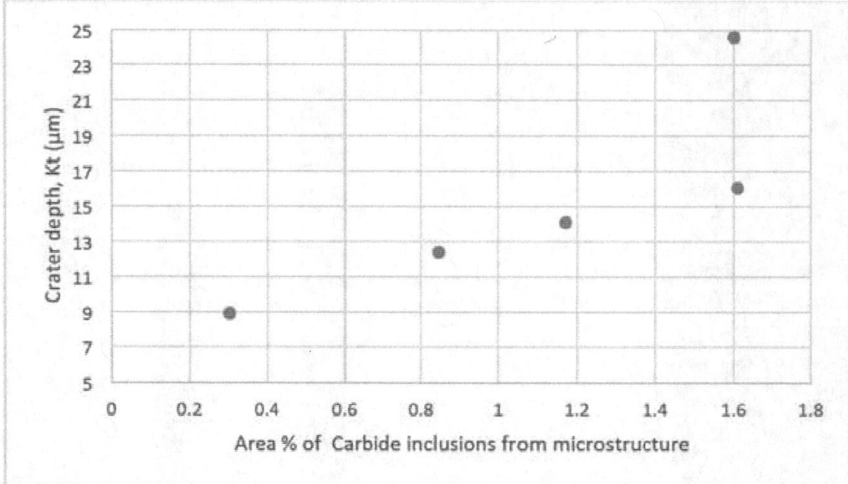

Fig. 6. Improvement in crater wear resistance with decreasing WC content in the PCBN cutting tool material

Machining tests were carried out comparing two 65% CBN material PAB-3 and PBC-3. Tool life test was carried out until the test insert fails through catastrophic failure. For each grade 3 corners were tested and the average tool life for the 3 corners were shown in Figure 7.

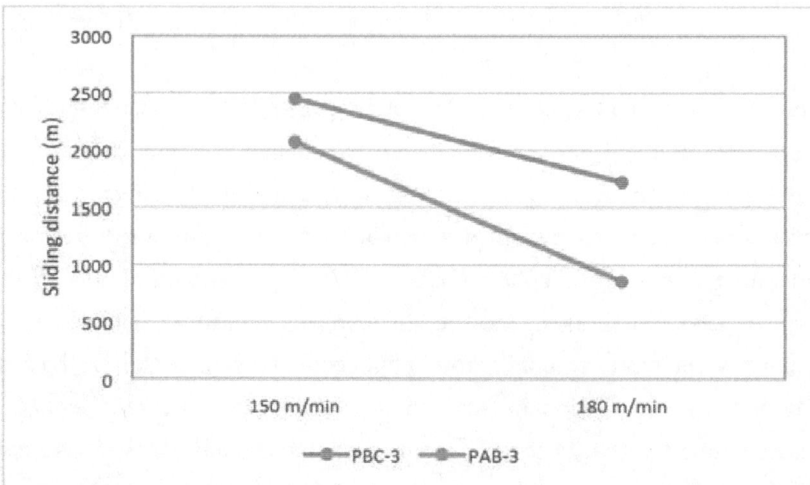

Fig 7. Tool life of the two grades showing drop in tool life with increasing cutting speed

At both cutting speeds PCBN with submicron grain material showed substantially higher tool life. It was also noticed that the drop in the tool life with increasing speed is less pronounced in PBC-3 than in PAB-3 which indicate that PBC-3 is less sensitive towards increasing speed. Flank wear rate and crater depth after first interval was plotted in Figure 8 a and 8 b respectively.

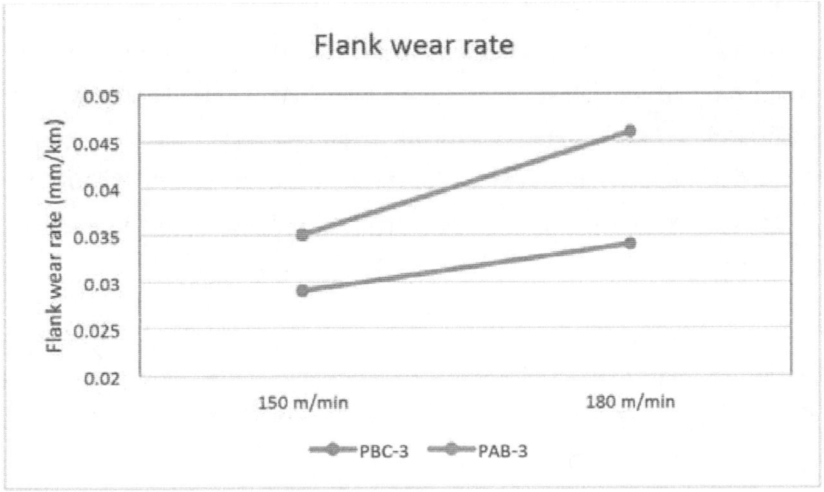

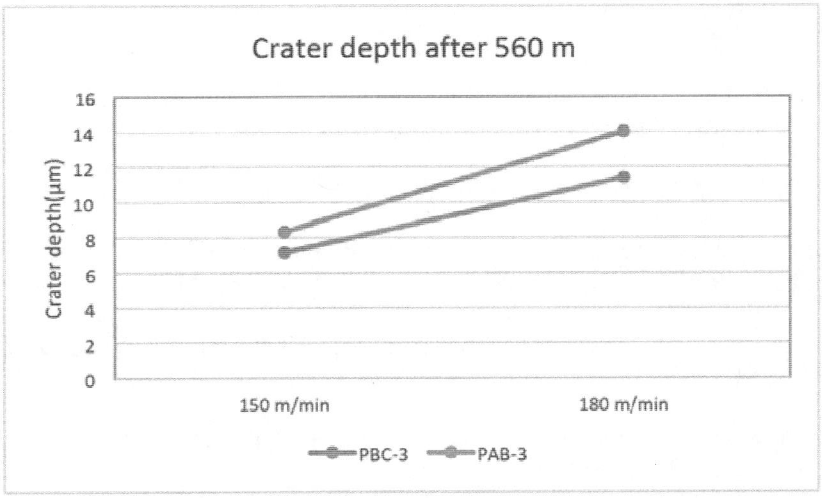

Fig 8 (a) Flank wear rate at 2 different cutting speed **(b)** Crater wear after 560 m of machining

From the flank and crater depth measurement it is evident that the trend in the both wear pattern followed that of the tool life trend. With improvement in the impurity and hardness of the binder, both flank and crater wear could be improved along with tool life. These positive performance characteristics all have a significant impact in the productivity.

4 CONCLUSION

Improvement in productivity for low content PCBN requires a cutting tool that performs at high speeds and requires higher abrasion and chemical wear resistance. This study demonstrated that an improvement could be achieved through using nanostructure binder systems free from impurities. High chemical and flank wear resistance resulted in higher tool life. It was also noticed the pure fine binder material is less sensitive towards change in cutting speed than the submicron binder product. Element Six has launched its new PureCut™ product range based on this novel technology.

5 REFERENCES

1. H.K. Tönshoff, B. Karpuschewski, C. Borbe, Comparison of basic mechanisms in cutting and grinding of hardened steel, Prod. Eng. IV (2) (1997) 5–8.

2. S. R. Agha and C. R. Liu, 'Experimental study on the performance of super finish hard turned surfaces in rolling contact', Wear, (244) (2000) 52–59.

3. E. D. Whitney, *Ceramic Cutting Tools,* Noyes Publication,(1994), 300-303,

4. Z. Pu and A. Sing, 'High speed end nose ball milling of hardened AISI A2 tool steel with PCBN and coated carbide tools,' J. Manufact. Proces. (15) (2013), 467-473.

5. S.S.K. Deepak, 'Applications of Different Optimization Methods for Metal Cutting Operation – A Review' Res. J. Engg. Sci., (1) (2012) 52-58

6. A. R. Motorcu, ' Tool wear, chip formation and workpiece surface issues in CBN hard turning: a review', Ind. J. Eng. Mat. Sci.,(18) (2011) 137-146.

7. S. S. Das, A. Panda and D. Dhupal, Experimental investigation of surface roughness, flank wear, chip morphology and cost estimation during machining of hardened AISI 4340 steel with coated carbide insert' Mech. Adv. Mat. and Mod. Proc. (3) (2017) 1-14.

8. Y. K Chou and C. J. Evans, Tool wear mechanism in continuous cutting of hardened tool steels, (212), (1997) 59-65

9. Y. Peng, H. Miao and Z. Peng, 'Development of TiCN based cermets: Mechanical properties and wear mechanism' Int. J. of Ref. Met. and Hard Mater. (39) (2013) 78-89

10. Cemented Carbide (2007), 24, 129-33 (ref 127 in 8)

11. ASTM384: Standard test method for Knoop and Vickers hardness of materials,

12. B. V. Manoj Kumar, J. Ram Kumar, B. Basu, 'Crater wear mechanisms of TiCN-Ni-WC cermets during dry machining,' (25) (2007) 392-399.

*** Formatting Issue - Best Available Paper/Graphic ***

Diamond synthesis "The science"
A theoretical and experimental study

Ion C. Benea, Ph.D.
Engis Corporation
Wheeling IL 60090, USA

Abstract

Sixty-five years ago, on December 16, 1954, the first repeatable diamond synthesis was successfully achieved by H. T. Hall of General Electric.
Nowadays, billions of carats of diamond powders are produced annually worldwide, based on the high pressure-high temperature catalytic diamond synthesis process.
Although, the pressure-temperature conditions needed for diamond synthesis can be easily established, the mechanism of high pressure-high temperature catalytic transformation of graphite to diamond, remains insufficiently understood. Over time, both published and, unpublished theoretical and experimental studies on graphite-diamond polymorphic transformation have been performed by many.
This work, one of many, proposes a theoretical approach to catalytic diamond synthesis, by way of a mathematical model, with the purpose of understanding the mechanism of the catalytic transformation of graphite to diamond in the binary Carbon-Metal (C-Me) systems with eutectic, at high pressure and high temperature (HP-HT).
The results of mathematical calculations were verified against published experimental data for C-Ni system, and experimental results acquired through diamond synthesis experiments designed to: a) ascertain the pressure-temperature conditions required for diamond formation in C-Co system, and b) establish the effect of pressure and temperature on the nucleation and growth rates that control the physical and chemical properties of synthesized diamond crystals.
Validity of the mathematical model developed was substantiated by the agreement between theoretical calculations and experimental results.
Understanding the mechanism that governs the HP-HT catalytic transformation of graphite to diamond, represents a valuable approach toward an efficient control of the diamond synthesis processes.

1. Phase diagram of carbon

Phase diagram of carbon has risen continuous interest from worldwide scientists, especially after successful production of synthetic diamond by General Electric, USA in 1955. (1)
The phase diagram of carbon, presented in the approximate diagram of Fig. 1.1, represents the summation of all known data pertaining to elemental carbon over a wide range of pressures and temperatures.
The main boundaries between different carbon phases are *graphite/diamond equilibrium line; graphite/liquid equilibrium line* and *diamond/liquid equilibrium line*.

2. Proposed mechanism for HP-HT catalytic transformation of graphite to diamond

It is well known that HP-HT conversion of graphite to diamond involves the presence of a metal in molten state.

There are a number of hypothesis regarding the role metal plays in the conversion process of graphite to diamond (2, 3, 4, 5 – 24).

Of those, solvent-catalyst hypothesis (3, 10, 16, 17, 20) was taken into consideration for the development of a mathematical model for the high pressure-high temperature catalytic transformation of graphite to diamond.

This model assumes that high pressure-high temperature transformation of graphite to diamond is a two-step process: ***nucleation*** *– formation of diamond nuclei, as a result of a solid-solid catalytic transformation process, and* ***growth*** *– growth of diamond crystals by carbon deposition onto existing diamond nuclei, through a diffusion process.*

Under favorable pressure-temperature conditions for thermodynamic stability of diamond, graphite dissolves in metal catalyst with formation of carbon-metal solution. Metal atoms and clusters penetrate graphite lattice and, when in close proximity with carbon atoms, partially filled electronic sub-shell *d* of metal will receive delocalized electrons from the weak π bond of graphite. As a result of loss of electrons from π bond, σ bond becomes weaker and the structure will break into free carbon atoms and carbon clusters with graphitic structure. Graphite dissolution in molten metal continues until carbon reaches equilibrium concentration corresponding to pressure-temperature conditions at which carbon-metal system is submitted to. As a result of energetic (thermodynamic) fluctuations in the carbon-metal solution, carbon atoms and carbon clusters exist in dynamic transitional states – sp^2 hybridization (trigonal-planar geometry) $\rightleftarrows$ sp^3 hybridization (tetrahedral geometry). Some of the carbon clusters with graphitic structure experience a solid-solid catalytic transformation to diamond structure - homogeneous nucleation. This transformation is associated with transition of carbon atoms from sp^2 hybridization to sp^3 hybridization, with rearrangement of the weak graphite structure into diamond structure, through formation of covalent bonding on sp^3 hybrid orbitals between carbon atoms (σ bonds).

Furthermore, carbon-metal intermediate compounds (interstitial solid solutions and/or interstitial compounds) can also be formed between graphite mono-layers and metal atoms.

In addition to spontaneous formation of diamond nuclei, formation of nonspontaneous diamond nuclei in the carbon-metal solution is also possible (heterogenous nucleation). Solid particles of carbon-metal intermediate compounds, high melting point impurities, etc., can become diamond nucleation sites.

If energetic equilibrium in the carbon-metal solution is favorable to diamond growth (pressure-temperature conditions inside the thermodynamic stability domain of diamond), carbon atoms and clusters head toward existing diamond nuclei, driven by the difference in solubility limits between diamond + liquid/liquid boundary and graphite + liquid/liquid boundary.

Given that under these conditions the solubility of graphite in molten metal is higher than that of diamond, carbon-metal solution becomes oversaturated with carbon, prompting deposition of carbon atoms and clusters with sp^3 hybridization onto diamond nuclei. As a result, diamond crystal starts to develop from diamond nuclei. Following precipitation of carbon atoms and clusters onto diamond nuclei, carbon concentration in carbon-metal solution around newly grown diamond crystal, is reduced. Hence, more carbon atoms and clusters with graphitic structure are dissolved into carbon-metal solution until carbon concentration reaches oversaturation limit again. At the same time, more carbon atoms and clusters in sp^2 hybridization state are activated to sp^3 hybridization state.

Crystal growth process continues for as long as carbon atoms and clusters in sp^3 hybridization state are present in the carbon-metal solution surrounding the newly grown diamond crystal.

If, on the other hand, the energetic equilibrium changes and becomes favorable to graphite growth (pressure-temperature conditions outside the thermodynamic stability domain of diamond), crystal growth process is reversed into diamond dissolution process. Diamond crystal dissolves in metal catalyst simultaneously with deactivation of carbon atoms and clusters from sp^3 hybridization state to sp^2 hybridization state, rearrangement in graphite structure, and formation of graphite nuclei. Under these conditions, the solubility of diamond in molten metal is higher than that of graphite, and carbon-metal solution becomes oversaturated with carbon, prompting deposition of carbon atoms and clusters with sp^2 hybridization onto graphite nuclei.

According to this mechanism, in the HP-HT catalytic transformation of graphite to diamond, the function of metal is preponderantly catalytic, in the process of $sp^2 \rightarrow sp^3$ activation of carbon atoms and clusters and formation of diamond nuclei, based on spontaneous graphite-diamond polymorphic transformation, while in the dissolution of graphite and diamond growth processes, the function of the metal is preponderantly that of carbon solvent.

Thus, mathematical modeling of HP-HT graphite-diamond catalytic transformation in the binary systems C-Me, presumes calculation of pressure-temperature conditions required for the thermodynamic stability of diamond (thermodynamic calculations), as well as calculation of nucleation and growth rates (kinetics calculations) that are responsible for physical and chemical properties of diamond crystals.

With the exception of Fe, at atmospheric pressure, group VIII metals, known as effective catalysts for conversion of graphite to diamond, form with C simple solution systems with eutectic. In this work, mathematical calculations related to solid-liquid thermodynamic equilibrium, were limited to C-Me systems with eutectic, particularly to C-Ni and C-Co systems.

3. Mathematical model for HP-HT catalytic graphite-diamond transformation in binary carbon-metal systems with eutectic

Thermodynamics of phase equilibrium in C-Me systems at high pressure

Peters (25) and Dickinson (26), calculated phase diagrams of C-Ni, C-Co, C-Mn and C-Fe at high pressure and indicated the pressure-temperature conditions for thermodynamic stability of diamond in these systems.

Comparison between the results of theoretical calculations performed by Peters and Dickinson for C-Ni phase diagram at high pressure, with experimental data obtained by Strong (27) for same system, shows that calculated pressure-temperature values are approx. 0.5 GPa and approx. 100 K higher than experimental data. Particularly, the equilibrium temperature calculated for graphite/diamond/liquid equilibrium at a given pressure, exceeds the calculated temperature of graphite/diamond equilibrium at same given pressure, based on the mathematical relation proposed by Berman and Simon (28) for graphite/diamond equilibrium line, that is well verified experimentally:

$$P\ (GPa) = 0.6865 + 0.00266\ T\ (K) \qquad (3.1)$$

Analyzing the reasons for the erroneous calculations by Peters and Dickenson, we concluded that the mathematical relation chosen for the excess free energy, E_G, of a regular carbon-metal (C-Me) solution used in their calculations, is independent of pressure and temperature.

$$E^f_G(X) = X^f_{Me} * X^f_C * E^{(0)} \qquad (3.2)$$

where:

f: hc; vcc; fcc, or L;

X_{Me} is concentration of metal in carbon-metal solution;

X_C is concentration of carbon in carbon-metal solution;

E = constant.

To calculate the solid/liquid equilibrium diagrams of C-Me systems at high pressure, we chose a mathematical relation for the excess free energy of C-Me solution which is dependent of concentration, X, pressure, P, and temperature, T, as follows:

$$E^f_G = X^f_{Me} * X^f_C\ [E^{(0)}{}_1 - TE^{(0)})_2 + P(E^{(i)}{}_1 + E^{(i)}{}_2)] \qquad (3.3)$$

where:

$E^{(f)}{}_G$ is the excess free energy;

$E^{(0)}{}_1$ is the excess enthalpy in phase f;

$E^{(0)}{}_2$ is the excess enthropy in phase f;

$E^{(i)}{}_1 + P\ E^{(i)}{}_2$ is the excess volume in phase f and is dependent of pressure.

For a polymorphic transformation, phase equilibrium between two phases at temperature T and pressure P, implies equality between their respective free energies:

$$G_1 = G_2,\ \text{or}\ \Delta G = \Delta H - T\Delta S + PV = 0 \qquad (3.4)$$

To describe the solid/liquid phase equilibrium in a binary system C-Me with eutectic, at high pressure, the following thermodynamic relations were used in our calculations:

<u>Me^{fcc} – Liquid equilibrium line:</u>

$$\Delta G_{Me}{}^{fcc/L}(X,T,P) = \Delta^0 H_{Me}{}^{fcc/L} - T\Delta^0 S_{Me}{}^{fcc/L} + P\Delta^0 V_{Me}{}^{fcc/L} + RT\ln(X^L{}_{Me}) + (1-X_{Me}{}^L)^2[(L_1{}^{(0)} - T*L_2{}^{(0)}) + P(L_1{}^{(1)} + PL_2{}^{(1)})] = 0 \qquad (3.5.a)$$

<u>Graphite – Liquid equilibrium line:</u>

$$\Delta G_C{}^{G/L}(X,T,P) = \Delta^0 H_C{}^{G/L} - T\Delta^0 S_C{}^{G/L} + P(K_1{}^{(2)} + PK_2{}^{(2)}) + RT\ln X_C{}^L + (1-X_C{}^L)^2[L_1{}^{(0)} - T*L_2{}^{(0)} + P(L_1{}^{(2)} + P*L_2{}^{(2)})] = 0 \qquad (3.5.b)$$

<u>Diamond – Liquid equilibrium line:</u>

$$\Delta G_C{}^{D/L}(X,T,P) = \Delta^0 H_C{}^{D/L} - T\Delta^0 S_C{}^{D/L} + P(K_1{}^{(3)} + PK_2{}^{(3)}) + RT\ln X_C{}^L + (1-X_C{}^L)^2[L_1{}^{(0)} - T*L_2{}^{(0)}) + P(L_1{}^{(3)} + P*L_2{}^{(3)})] = 0 \qquad (3.5.c)$$

where:
Me: Ni, Co:
$X_{Me}^{L} + X_{C}^{L} = 1$;
$\Delta^{0}V_{Co}^{fcc/L}$ = constant;
$\Delta^{0}V_{C}^{G/L} = [K_1^{(2)} + PK_2^{(2)}]$;
$\Delta^{0}V_{C}^{D/L} = [K_1^{(3)} + PK_2^{(3)}]$;
$L^{(i)}_1\ L^{(i)}_2$ = interaction parameters for free energy in liquid phase; i = 0, 1 ,2, 3;
$K^{(2)}_1$; $K^{(3)}_1$; $K^{(2)}_2$; $K^{(3)}_2$ = constant

While for C-Ni system at high pressure some experimental data is available, experimental data for C-Co system at high pressure is practically nonexistent.

To verify that mathematical relations developed for binary C-Me systems with eutectic at high pressure correctly describe solid-liquid (S-L) equilibrium lines, calculations were first made for carbon-nickel (C-Ni) system based on mathematical relations 3.5.a, 3.5.b, and 3.5.c. Calculated data were then compared with published experimental data.
Interaction parameters $L_1^{(0)}$ and $L_2^{(0)}$ are determined at P = 0 for two known experimental points (X, T) either from f.c.c. + L/L equilibrium line or from G+L/L equilibrium line, under the condition that the respective equation be verified simultaneously by both pair values (X, T).
High pressure experimental data regarding solubility of carbon (as graphite or diamond) in molten metal, as well as regarding melting of pure components (metal, graphite, diamond) is limited. In addition, experimental errors associated to melting points at high pressure (especially for graphite and diamond) are significant. Hence, interaction parameters $L_1^{(i)}$ and $L_2^{(i)}$ are determined separately for each equilibrium line under the condition that respective equations simultaneously verify two pairs of values (X, P, T).

Even less rigorous from the physical point of view, this approach allows to differently adjust the excess free energy term of C-Me solution for each equation, such that respective equations can be simultaneously verified by reciprocal experimental points, and by individual experimental points belonging to each equilibrium line, not only at moderate pressures but also at very high pressures.

Consequently, according to proposed mathematical model, the equilibrium phase diagram of a binary C-Me system with eutectic, is described by relations 3.5.a; 3.5.b; 3.5.c.
If $L_2^{(0)} = L_1^{(i)} = L_2^{(i)} = 0$, these relations describe solid/liquid equilibrium according to regular solution model.

3.1 Thermodynamic calculations

Calculation of C-Ni phase diagram at high pressure

Thermodynamic data for S-L transformation of pure components C, Ni and Co, is presented in Tab. 3.1. Experimental data used for the calculation of interaction parameters L and constants K, is presented in Tab. 3.2.
Calculated L and K values for solid/liquid (S/L) equilibrium in C-Ni system are shown in table 3.3. Comparison between calculated data and experimental data available for C-Ni system at high pressure is exhibited in Tab. 3.4.

Based on the agreement between the result of theoretical calculations with experimental data, we concluded that the mathematical model proposed for the calculation of S/L equilibrium diagrams in C-Me binary systems with eutectic is correct.

Calculation of C-Co phase diagram at high pressure

Thermodynamic data for S-L transformation of Co is presented in Tab. 3.1, while experimental data used for the calculation of interaction parameters L and constants K, are presented in Tab. 3.5.
Thus, dependence of temperature and concentration of Co-G eutectic and Co-D eutectic with pressure is described by following mathematical relations:

$$T_{EG} = 1548 + 11.3488 * P + 0.3682 * P^2 \quad (3.1.1)$$
$$X_{EG} = 11.6 + 0.5973 * P + 0.0452 * P^2 \quad (3.1.2)$$

$$T_{ED} = 1453 + 22.1352 * P + 1.5071 * P^2 \quad (3.1.3)$$
$$X_{ED} = 15.5 + 0.4617 * P + 0.0688 * P^2 \quad (3.1.4)$$

Calculation of *"experimental points"* Co-G eutectic and Co-D eutectic was based on the assumption that their dependence of temperature and concentration is similar to that of Ni-G eutectic and Ni-D eutectic, for which experimental data at high pressure is available.

Temperature and concentration values of the points considered as *"experimental points"* in our calculations were calculated based on relations 3.1.1 to 3.1.4 and are displayed in Tab. 3.5.

The values of interaction parameters L and constants K were determined based on thermodynamic values for solid-liquid transformation of pure components in C-Co system, from table 3.1, and *"experimental data"* from Tab. 3.5. Calculated L and K values for solid/liquid (S/L) equilibrium in C-Co system are shown in Tab. 3.6.
Solid/liquid equilibrium phase diagrams of C-Co system computed based on the mathematical model developed, at 0 GPa; 5.01 GPa (pressure corresponding to invariant equilibrium Co^{fcc}/graphite/diamond/liquid); 5.6 GPa; 8.0 GPa and 12.5 GPa, are exhibited in Fig. 3.1. to 3.3.

*** Formatting Issue - Best Available Paper/Graphic ***

Taking into account temperature and concentration data for G-L and D-L, resulted from calculation of solid-liquid equilibrium at different pressures, the following mathematical relations are obtained for pressure dependence of temperature and concentration for graphite/diamond equilibrium:

$$T_{GD} = 152.4 + 270.4331 * P + 4.1144 * P^2 \quad (3.1.5)$$
$$X_{GD} = 9.03 - 1.2685 * P + 0.5109 * P^2 \quad (3.1.6)$$

Graphic representation of dependence with pressure of temperature and concentration for cobalt-graphite eutectic and cobalt-diamond eutectic (described by mathematical relations 3.1.1 to 3.1.4), as well as of temperature and concentration of graphite/diamond/liquid equilibrium (described by mathematical relations 3.1.5 and 3.1.6), are presented in Fig. 3.4 and Fig. 3.5.
As it can be observed from Fig. 3.4, conditions for thermodynamic stability of diamond in C-Co system, are represented by pressure-temperature pair values included between melting line of diamond-cobalt eutectic line and graphite/diamond/liquid equilibrium line

It is worth mentioning that, using a pressure dependent asymmetrical model, Munke (33) calculated the phase diagram of C-Co at high pressure. However, as in the case of calculations by Peters and Dickinson, temperatures associated to graphite-cobalt eutectic, diamond-cobalt eutectic, and graphite/diamond/liquid equilibrium register unjustified increases with pressure.

Analyzing the solid/liquid equilibrium diagrams calculated at different pressures, the following conclusions regarding thermodynamic stability of diamond, that is, pressure-temperature conditions required for catalytic transformation of graphite to diamond in C-Co system can be drawn:

- *Pressure-temperature minimum conditions (P_0, T_0) required for diamond synthesis, are determined by the invariant equilibrium between all four phases present in the system: f.c.c.-Co/graphite/diamond/liquid; $P_0 = 5.01$ GPa; $T_0 = 1601.5$ K;*
- *For any given pressure value $P > P_0$ minimum and maximum temperature values for diamond synthesis are represented by temperature values pertaining to diamond-cobalt eutectic and, respectively, to graphite/diamond/liquid equilibrium, corresponding to that particular pressure;*
- *For any temperature value $T > T_0$ the minimum pressure for diamond synthesis is represented by pressure value pertaining to graphite/diamond/liquid, corresponding to that particular temperature.*

3.2 Kinetic calculations

It must be mentioned that experimental data regarding kinetics of HP-HT catalytic transformation of graphite to diamond, in C-Ni and C-Co systems is scarce.

*** Formatting Issue - Best Available Paper/Graphic ***

Calculation of nucleation rate of diamond

According to this mechanism for catalytic transformation of graphite to diamond, diamond nuclei are formed as a result of statistical phase fluctuations that take place in carbon-metal solution, based on solid-solid transformation – homogeneous nucleation. Larger nuclei that survive phase fluctuations, become sites for diamond crystals growth.

To calculate the nucleation rate of diamond as a function of pressure and temperature, the following approximative relation was developed:

$$NR \sim \exp \{- C_1 / T [P - (a + bT)]^2\} \quad (3.2.1)$$

where:
$C_1 = 16\pi\sigma^3 / 3\rho_D^2 (\Delta V)^2 K$ = constant
ρ is the diamond density;
σ is the superficial tension on the interface;
a and b are the constants of the graphite/diamond equilibrium line, with a = 0.6865 GPa and b = 0.00265 GPa/K, according to relation 3.1.

Since relation 3.2.1 has physical significance only for temperatures above melting temperature of metal in the presence of carbon, but below graphite/diamond equilibrium temperature, the nucleation rate is described by the following relation:

$$NR = C_0 \exp \{- C_1 / (T-T_E) * [P_{GD} - a - b (T_{GD} - T)]^2\} \quad (3.2.2)$$

where:
T_E is the carbon-metal eutectic temperature;
P_{GD} and T_{GD} are pressure and temperature corresponding to graphite/diamond equilibrium.

The limiting pressure-temperature conditions for thermodynamic stability of diamond are derived from analysis of solid/liquid equilibrium diagrams, calculated at different pressures.
Thus, the approximate relationship 3.2.2. that described the rate of diamond nucleation as a function of pressure and temperature within the diamond thermodynamic stability domain was written as follows:

$$NR = C_0 \exp \{- C_1 / [(T_{GD} - \Delta T) - T_{ED}] * [P_{GD} + \Delta P - a - b(T_{GD} - \Delta T)]^2\} \quad (3.2.3)$$

where:
T_{ED} is the diamond-cobalt eutectic temperature corresponding to graphite/diamond equilibrium pressure P_{GD};
T_{GD} is the diamond/graphite/liquid equilibrium temperature corresponding to graphite/diamond equilibrium pressure P_{GD};

P_{GD} is the pressure along isotherm T_{GD}; $P_{GD} \geq P_0$
a and b are the constants of the graphite/diamond equilibrium line, with a = 0.6865 GPa and b = 0.00265 GPa/K, according to relation 3.1.

Following substitutions were made in relation (3.2.3)
$T = (T_{GD} - \Delta T)$, is the temperature along isobar P_{GD} and is confined between T_{ED} and T_{GD};
$P = (P_{GD} + \Delta P)$, is the pressure along isotherm T_{GD} with $P_{GD} > P_0$

By analyzing the mathematical relation 3.2.3, which describes the dependence of nucleation rate with respect to pressure-temperature conditions from within the diamond thermodynamic stability domain, following considerations can be made:

- *For $P = P_0$ and $T = T_0$, implicitly $\Delta P = 0$ ($P_{GD} = P_0$) and $\Delta T = 0$ ($T_{GD} = T_0$), nucleation rate is zero, which means that under pressure-temperature conditions pertaining to invariant equilibrium: f.c.c.-Co/graphite/diamond/liquid, probability for diamond nuclei to form is zero;*
- *For $P = P_{GD} > P_0$, implicitly $\Delta P = 0$, nucleation rate along isobar P_{GD} is zero for $\Delta T = 0$ ($T = T_{GD}$), and is different from zero for $0 < \Delta T < T_{GD} - T_{ED}$ ($T_{ED} < T < T_{GD}$), which indicates that along isobar P_{GD}, probability of diamond nuclei to form, is zero for temperature corresponding to carbon-cobalt eutectic and, respectively for temperature corresponding to graphite/diamond/liquid equilibrium, and is different from zero for temperatures between these limits;*
- *For $T = T_{GD} > T_0$, implicitly $\Delta T = 0$, nucleation rate along isotherm T_{GD} is zero for $\Delta P = 0$ ($P = P_{GD}$) and increases for $\Delta P > 0$ ($P > P_{GD}$), which indicates that along isotherm T_{GD}, the probability for diamond nuclei to form is zero for temperature corresponding to graphite/diamond/liquid equilibrium, and increases for higher temperatures.*

The results of theoretical calculations for diamond nucleation rate as a function of pressure and temperature, are represented graphically in Fig. 3.6 and Fig. 3.7.
To conclude, dependence of nucleation rate with pressure and temperature conditions from within diamond thermodynamic stability domain, is described as follows:

- *At constant pressure, $P > P_0$, diamond nucleation rate is zero for the temperature corresponding to diamond-cobalt eutectic, rises rapidly to a maximum value for a temperature $T_{max} = (T_{GD} + 2T_{ED})/2$, and decreases slowly to zero for the temperature corresponding to graphite/diamond/liquid equilibrium;*
- *At constant temperature, $T > T_0$, diamond nucleation rate is zero for the pressure corresponding to graphite/diamond/liquid equilibrium, and increases with pressure, as long as the temperature does not exceed the diamond-cobalt eutectic temperature for any pressure value.*

Calculation of growth rate of diamond

In case of crystallization from solution, diffusion mechanism takes place as a result of existence of a crystallization film around nuclei in which concentration around saturation (X_0), is lower than concentration of supersaturated solution (X).
Concentration gradient ensures a diffusion flux so that, according to Fick law, crystal growth rate can be expressed as:

$$dm/dt = \beta S(X-X_0) \tag{3.2.4}$$

where:
dm/dt is the quantity of substance deposited on crystal surface in unit time;
β is the mass transfer coefficient;
S is the surface of growing crystal;
$X - X_0$ expresses supersaturation of solution.

Since β is determined by ratio of diffusion coefficient, D, and thickness of diffusion film, d, the above expression becomes:

$$dm/dt = D/d \, {*}S \, (\text{grad } X) \tag{3.2.5}$$

where:
grad X = $(X - X_0)$ is the concentration gradient.

If a spherical shape is assumed for diamond crystals, growth rate can be expressed as:

$$GR = dr/dt = 1/\rho \, {*} \, [D(X-X_0)/d] \tag{3.2.6}$$

where:
r is diamond crystal radius;
ρ is the diamond density;
D is diffusion coefficient;
X_0 is the concentration on crystal surface;
X is the concentration in supersaturated solution.

Assuming that at constant temperature, diffusion coefficient is constant and independent of concentration, within concentration range ($X-X_0$), and taking into account the thickness of 0.01 mm of metal catalyst film around the diamond crystal (27), relation (3.2.6), becomes:

$$GR = C\,(X - X_0) \tag{3.2.7}$$

where:

$C = D/\rho d$

Taking into consideration that $(X - X_0)$ represents the concentration difference between carbon concentration in carbon-metal solution, in graphite phase and diamond phase, at pressure-temperature conditions within diamond thermodynamic stability domain, the growth rate of diamond crystals was calculated using the following approximate relation:

$$GR = C\,(X_G^{\,L} - X_D^{\,L}) \qquad (3.2.8)$$

From analysis of solid/liquid equilibrium diagrams calculated for different pressures, it becomes clear that concentration difference between graphite and diamond in carbon-metal solution is dependent of thermodynamic parameters pressure-temperature.

The results of mathematical calculations regarding dependence of growth rate of diamond crystals with temperature at different pressures and with pressure at different temperatures are presented in Fig. 3.8 and Fig. 3.9. From these graphs, following conclusions can be drawn:

- *At constant pressure, $P > P_0$, diamond growth rate registers a maximum value for a temperature close to diamond-cobalt eutectic temperature and decreases with temperature increase, becoming zero for the temperature corresponding to graphite/diamond/liquid equilibrium, for which: $X_G^{\,L} = X_D^{\,L}$;*
- *At constant temperature $T > T_0$, diamond growth rate is zero for the pressure corresponding graphite/diamond/liquid equilibrium and increases with pressure increase, provided that temperature is higher than diamond-cobalt eutectic temperature, for any pressure value.*

Finally, based on the mathematical model developed for the HP-HT catalytic transformation of graphite to diamond (thermodynamic and kinetic calculations), the diamond synthesis process in binary carbon-metal systems with eutectic, subjected at pressures and temperate within the diamond thermodynamic stability domain, can be described as follows:

- *At pressure-temperature conditions in the vicinity of diamond-metal eutectic line, both nucleation rate and growth rate are high. Due to high nucleation and growth rates, under these conditions, a large number of low quality diamond crystals with high level of crystal growth defects (CGD) and residual metal catalyst (RMC) impurities can be synthesized. Diamond crystals will exhibit low mechanical strength;*

- *At pressure-temperature conditions away from diamond-metal eutectic line and graphite/diamond/liquid equilibrium line, nucleation rate and growth rate are lower. Due to reduced nucleation and growth rates, under these conditions, a lower number of medium quality diamond crystals with lower level of crystal growth defects (CGD) and residual metal catalyst (RMC) impurities can be synthesized. Diamond crystals will exhibit medium mechanical strength;*

- *At pressure-temperature conditions in the vicinity of graphite/diamond/liquid equilibrium line, both nucleation rate and growth rate are much lower. Due to much lower nucleation and growth rates, under these conditions, a small number of high-quality diamond crystals with low level of crystal growth defects (CGD) and residual metal catalyst (RMC) impurities, can be synthesized. Near graphite/diamond/liquid equilibrium line, large size, high-quality diamond crystals with high mechanical strength can be grown with time;*

- *At pressures much higher than pressure corresponding to graphite/diamond/liquid equilibrium for any given temperature, nucleation and growth rates are very high. As a result, under these conditions, intergrown, twined (macles) and multicrystalline diamond crystals can be synthesized;*

- *At pressure-temperature conditions close to invariant equilibrium f.c.c.-metal/graphite/diamond/liquid, diamond synthesis conditions are uncertain. Thus, small pressure-temperature fluctuations will "remove" the system from diamond thermodynamic domain. Under these conditions, diamond crystals (if any) will be heavily included with graphite and metal.*

4. Experimental

High pressure apparatus

The opposed anvils high pressure apparatus, known as Bridgman anvils, is considered as ideal for high pressure generation, due to its simplicity and unlimited stroke, which allows unlimited compression of the pressure media. In its original design, the main disadvantage of Bridgman anvils apparatus is represented by the small volume of the sample subjected to compression, which makes this device unsuitable for diamond synthesis studies.

For the purpose of the study of HP-HT synthesis of diamond in carbon-cobalt system, a modified opposed anvils high pressure apparatus was design and built (34).
High pressure apparatus *"opposed anvils with spherical cavity"* employed in diamond synthesis experiments is shown schematically in Fig 4.1. a. and b.
In this design, the high pressure chamber of high pressure apparatus is enclosed between the spherical segment shaped depressions on top faces of the anvils. The total surface of the high pressure chamber is 16 cm^2. When fully compressed, high pressure chamber volume is approx. 4.0 cm^3, down from 5.0 cm^3, in uncompressed stage.
Sealing of the high pressure chamber was realized in two stages, using a combination of two gaskets: inner gasket, made of pyrophyllite, supported by an outer gasket made from polyamide resin. In compressed state, the gasket covers almost entirely the front faces of the anvils, so that high pressure generated inside the high pressure chamber, decreases gradually to atmospheric pressure, at the outer diameter of anvils. Consequently, the pressure gradient inside the reaction chamber is considerably reduced.

Diamond synthesis capsules

Diamond synthesis experiments were carried out using two types of diamond synthesis capsules: a) direct heating capsule and b) indirect heating capsule. Drawings of both types of diamond synthesis capsules are shown in Fig 4.2.

Pressure calibration at room temperature

Pressure calibration of the assembly comprised of high pressure apparatus and diamond synthesis capsule was performed using the so-called fixed pressure points technique. Pressure generated inside diamond synthesis capsule was measure by detecting the abrupt change in electrical resistivity, which accompanies the phase transitions of bismuth (Bi), thallium (Tl) and barium (Ba) with pressure.
Pressure values recommended for fixed pressure points on the international practical pressure scale are presented in Tab. 4.1.
Experimental setup for pressure calibration of both direct heating (a) and indirect heating (b) diamond synthesis capsules is shown in fig 4.3.
A number of five experiments were performed for the detection of each pressure point, and average value and variance of compressive force required for pressure generation, was calculated.
In Tab. 4.2, experimental data acquired in pressure calibration experiments are presented as pressure at transition points vs. press force. Data on pressure generation efficiency (E) at pressure points and relative error (I) in pressure generation, calculated according to relations (4.1) and (4.2) respectively, is also included in Tab. 4.2.

$$E = S*P / 2F \quad (4.1)$$

$$I = s / F \quad (4.2)$$

where:
S is the total surface of high pressure chamber;
P is the pressure generated inside the capsule;
F is the compressive force (press force) used to generate pressure P;
s is the average squared deviation (variance) calculated for compressive force.

Pressure calibration curves for direct heating capsule (a) and indirect heating capsule (b) described by mathematical relations (4.3) and (4.4), for pressures between 2.0 GPa and 8.0 GPa, are presented in Fig. 4.4.

a) $$P = 0.0995 + 0.659 F + 0.0208 F^2 \quad (4.3)$$

b) $P = 0.3751 + 1.029\ F + 0.0249\ F^2$ (4.3)

Temperature calibration at high pressure

Temperature inside diamond synthesis capsules was measured at 5.5 GPa, with the aid of Ni/CrNi thermocouple.
Experimental setup for pressure calibration of both direct heating (a) and indirect heating (b) diamond synthesis capsules is shown in Fig. 4.5.
A number of five temperature measurements were performed for each data point, and average value and standard deviation of temperature generated inside the capsule, for the same level of electrical power, were calculated.
Temperature values were corrected for pressure effect on thermocouple.
Temperature calibration curves for direct heating capsule (a) and indirect heating capsule (b) described by mathematical relations (4.4) and (4.5), for temperature range between 25°C and 1600 °C are presented in Fig. 4.6

a) $T = 22 + 1242\ W + 4.8\ W^2$ (4.4)

b) $T = 25 + 1429\ F - 155.3\ W^2$ (4.5)

Confirmation of pressure and temperature calibration of high pressure apparatus and capsule

The techniques of pressure calibration using fixed pressure points and temperature calibration with thermocouple are unanimously accepted and practiced by those involved in high pressure research.
Pressure-temperature calibration of the high pressure apparatus-capsule assembly was verified by comparing our calibration results with experimental results obtained by others.
For this purpose, *alpha iron* (αFe) – gamma iron (γFe) equilibrium line was determined experimentally for pressures up to 8.0 GPa.
It is known that at normal pressure and temperatures up to 1183 K, *alpha* phase of iron with volume centered cube (v.c.c.) structure is the stable phase. At temperatures above 1183 K, *gamma* phase of iron with face centered cube (f.c.c.) structure becomes the stable phase.
In order to detect the αFe – γFe transition, the capsule for diamond synthesis was assembled as shown in Fig. 4.7. The capsule so assembled made possible temperature measurement with the aid of Ni/CrNi thermocouple, simultaneously with measurement of the electrical resistance of an iron wire.
Transition *alpha* - *gamma* (on heating) and *gamma* - *alpha* (on cooling) are accompanied by significant variations of the electrical resistance of the iron wire. The resistance vs. temperature graph acquired at 7.0 GPa is presented in Fig. 4.8. As it can be seen from Fig. 4.8, temperature corresponding to αFe – γFe transition at this pressure is 562 °C and the transition interval is 65°C.
In Tab. 4.3, our experimental data is presented together with experimental data obtained by Clougherty and Kaufman (36).

Based on the data acquired in the calibration experiments, we concluded that experimental errors in evaluation of pressure and temperature conditions associated with diamond synthesis experiments are approx. ± 0.1 GPa and ± 20°C.

5. Experimental study of diamond synthesis in carbon-cobalt system

The experimental study of diamond synthesis in carbon-cobalt system consisted of experiments designed to ascertain the pressure-temperature conditions for diamond synthesis and the nucleation and growth rates of diamond, as a function of pressure and temperature.

High pressure phase quenching technique was used in all diamond synthesis experiments. Practically, phase quenching under high pressure was accomplished based on the algorithm, depicted in Fig 5.1:

- Pressure was increased to a pre-established value and maintained constant for a set time interval;
- Following a set time (needed for high pressure uniformization) temperature was increased to a pre-established value and was maintained constant for a set time interval (reaction time);
- Once reaction time expired, temperature was reduced rapidly under existing high pressure;
- Following a set time (needed for quenching the phases formed under high pressure), the pressure was reduced to atmospheric pressure.

High pressure phase quenching experiments to determine pressure-temperature parameters for diamond formation were performed under the following pressure-temperature-time conditions: P = 5.2 GPa – 6.0 GPa; T = 1300 °C to 1700 °C; t = 60 sec.

Diamond nucleation and growth experiments were performed at constant pressure, P = 5.6 GPa, and variable temperature, T = 1300 °C - 1550 °C, and at constant temperature, T = 1475 °C, and variable pressure, P = 5.2 GPa – 6.2 GPa, respectively. In all diamond nucleation and growth experiments the reaction time was set for 300 sec.

Following extraction from capsule, graphite-cobalt specimens reacted under different pressure-temperature-time conditions, were inspected under optical microscope, for visual detection of diamond crystals, and then chemically processed for removal of unreacted graphite and cobalt. Presence or absence of diamond was the only criteria used to evaluate the results of high pressure quenching experiments, designed to determine pressure-temperature conditions for diamond formation. Experimental results for diamond formation in the carbon-cobalt system are presented in Tab. 5.1 and displayed graphically in Fig. 5.2. Based on these experimental results, diamond synthesis in carbon-cobalt system takes place under pressure-temperature conditions comprised between line 1, which represents the high temperature limit and line 2, which represents the low temperature limit. The intersection of these two lines defines the minimum pressure-temperature conditions for diamond system: 5.0 GPa < P_0 < 5.2 GPa; 1325 < T_0 < 1350.

In Tab. 5.2, pressure-temperature conditions for thermodynamic stability of diamond in carbon-cobalt system at P = 5.6 GPa, calculated using the mathematical model developed, are compared with pressure-temperature conditions for diamond formation in carbon-cobalt system obtained in our diamond synthesis experiments at P = 5.6 GPa.
Recovered diamond crystals from nucleation and growth experiments were first weighted and then analyzed for particle size distribution. Subsequently, optical microscopy, scanning electron microscopy and inductively couple plasma spectroscopy analysis were performed to assess the quality of synthesized diamond crystals.

Optical microscopy and scanning electron microscopy images of synthesized diamond crystals, present in carbon-cobalt reaction charge extracted from diamond capsule following a diamond synthesis cycle, are presented in Fig. 5.3.

Diamond nucleation rate was calculated according to the following formula:

$$NR = N / V * t \qquad (5.1)$$

Using experimental data, the number of crystals was calculated for each diamond nucleation experiment according to the following approximate relation:

$$N = \sum_{k=1}^{n} p_k * n_k \qquad (5.2)$$

where:
p_k is the mass of crystals of size k in carats;
n_k is the approximate number of crystals of size k in unit mass (one carat).

The approximate number of particles per carat and the average diameter of diamond crystals per size grade used in these calculations is depicted in Tab. 5.3.
Experimental conditions and the number of diamond crystals calculated according to (5.2) are included in Tab. 5.4.
Since in all these experiments, graphite volume subjected to HP-HT treatment and reaction time were kept constant, $V_G = 0.4$ cm^3, t = 300 sec., the number of diamond crystals is directly proportional with nucleation rate.

Graphic representation of nucleation rate as a function of temperature at constant pressure and, as a function of pressure at constant temperature, was obtained by fitting the data calculated for the number of crystals, for each pressure-temperature pair value, from Tab. 5.3, using the following mathematical relations:

$$N(T) = C_1 \exp [- C_2 / (T - T_m) (T_M - T)^2] \qquad (5.3)$$

$$N(P) = C_3 \exp [- C_4 / (P - P_m)^2] \qquad (5.4)$$

For P = 5.6 GPa, numerical values of constants C_1 and C_2 together with T_m (low temperature limit) and T_M (high temperature limit) values, are listed below:

N (T) (nr. crystals)	NR (T) (nr. crystals/ cm^3 *s)
$C_1 = 3.334 * 10^6$	$C'_1 = 0.027 * 10^6\ cm^{-3}\ s^{-1}$
$C_2 = 0.632 * 10\ ^{\circ}C^{-3}$	
$T_m = 1325\ ^{\circ}C$	
$T_M = 1550\ ^{\circ}C$	

For T = 1475 °C, numerical values of constants C_3 and C_4 together with P_m (low pressure limit) and P_M (high pressure limit) values, are listed below:

N (P) (nr. crystals)	NR (P) (nr. crystals/ cm^3 *s)
$C_3 = 4.338 * 10^6$	$C'_3 = 0.036 * 10^6\ cm^{-3}\ s^{-1}$
$C_4 = 0.546 * 10\ GPa^{-2}$	
$P_m = 5.4\ GPa$	

Dependence of nucleation rate with temperature at constant pressure, and with pressure at constant temperature is displayed in Fig. 5.4.a and b., respectively.

Crystal growth rate, expressed as linear growth of diamond crystal diameter per unit time, was calculated as follows:

$$GR = \Delta d_{avg} / \Delta t \tag{5.5}$$

Using the data presented in Tab. 5.3, the average size (diameter) of crystals was calculated, according to following approximate relationship:

$$d_{avg} = \sum_{k=1}^{n} p_k * d_k = \frac{1}{2} \sum_{k=1}^{n} p_k * (d_k + d_{k-1}) \tag{5.6}$$

where:
d_k, d_{k-1} represent maximum and minimum diameter of crystals belonging to class size k;
n_k is the approximate number of crystals of size k in unit mass (one carat).

Experimental conditions and the growth rate of diamond crystals calculated according to (5.6) are included in Tab. 5.5.

For a set time interval, growth rate is directly proportional to the linear growth of the average diameter of diamond crystals. It is worth mentioning that, in the case of diamond synthesis experiments performed at constant pressure, due to thermal inertia of diamond capsule, the

effective reaction (synthesis) time is not rigorously constant and depends on temperature. Therefore, to avoid experimental errors that could be introduced by thermal inertia of the capsule, in our crystal growth calculations, reaction time was measured from the moment the graphite-cobalt specimen started to melt, which was detected by abrupt change of electrical resistance of the specimen. Therefore, as presented in Tab. 5.5 reaction time shorter at lower temperatures and reaches the set value at higher temperatures.
Similarly, graphic representation of growth rate as a function of temperature at constant pressure and as a function of pressure at constant temperature, was achieved by fitting data calculated for the average size of diamond crystals, for each pressure-temperature pair value shown in Tab. 5.4, using the following mathematical relations:

$$GR(T) = K_1 + K_2 * T + K_3 T^2 \tag{5.7}$$

$$GR(P) = K_4 + K_5 * P + K_6 P^2 \tag{5.8}$$

For P = 5.6 GPa, numerical values of K_1, K_2 and K_3, are listed below:

d_{avg} (T) (μm)	GR (T) (μm/s)
K_1 = -234.84 μm	K'_1 = -234.84 μm s^{-1}
$K_2 = 71.7 * 10^{-2}$ μm $°C^{-1}$	$K'_2 = 71.7 * 10^{-2}$ μm $°C^{-1}$ s^{-1}
$K_3 = -31.2 * 10^{-5}$ μm $°C^{-2}$	$K'_3 = -0.104 * 10^{-5}$ μm $°C^{-2}$ s^{-1}

For T = 1475 °C, numerical values of K_4, K_5 and K_6, are listed below:

d_{avg} (T) (μm)	GR (T) (μm/s)
K_4 = 119.36 μm	K'_4 = 2.336 μm s^{-1}
K_5 = -3.02 μm GPa^{-1}	K'_5 = -0.672 μm s^{-1} GPa^{-1}
K_6 = 1.01 μm GPa^{-2}	K'_6 = 0.0625 μm s^{-1} GPa^{-2}

Dependence of growth rate with temperature at constant pressure and with pressure at constant temperature is displayed in Fig. 5.5.a. and b.

By analyzing the graphs presented in Fig. 5.4.a and b representing the dependence of nucleation rate with temperature at constant pressure and with pressure and constant temperature, we concluded as follows:

- *At constant pressure, with temperature increase, the nucleation rate (number of diamond crystals) increases at maximum value, then decreases slowly to zero following an asymmetric curve;*
- *At constant temperature, with pressure increase, the nucleation rate (number of diamond crystals) increases continuously.*

Similarly, analysis of the graphs presented in Fig. 5.5.a and b, representing the dependence of growth rate with temperature at constant pressure, and with pressure and constant temperature, indicates that:

- *Under constant pressure and set time, growth rate (average size of diamond crystals) decreases with temperature increase;*
- *Under constant temperature and set time, growth rate (average size of diamond crystals) increases with temperature increase.*

Crystallographic phase analysis of graphite-cobalt specimens reacted under HP-HT conditions

In an attempt to understand metal catalyst contribution to graphite-diamond transformation, X-ray diffraction analysis was performed on graphite-cobalt specimens reacted to different - pressure-temperature conditions. In addition, X-ray diffraction analysis was performed on material resulted after reacted graphite-cobalt specimens were chemically (dissolution of cobalt in diluted nitric acid) and physically (separation of diamond from graphite in bromoform) processed.

Experimental conditions and the results of phase analysis by X-ray diffraction are included in Tab. 5.6. X-ray diffraction data reveal that diamond, as new phase, was detected in graphite-cobalt specimens reacted at pressure-temperature conditions at which diamond is thermodynamically stable (i.e. P =5.6 GPa; T = 1450 °C). On the other hand, if the graphite-cobalt specimen was reacted under pressure-temperature conditions at which diamond is not thermodynamically stable (i.e. P = 4.8 GPa; T = 1350 °C), only the initial phases, graphite and cobalt, were detected.

Following cobalt removal and diamond separation from graphite-cobalt specimens reacted under same pressure-temperature conditions, X-ray diffraction shows that untransformed carbon remains in initial phase – graphite.

To conclude, the results of phase analysis clearly indicate that following the treatment of graphite-cobalt catalyst specimens, at high pressure and high temperature, the only structural transformation is graphite-diamond transformation. No other phases and/or intermediary compounds, as precursors to graphite-diamond transformation, were identified.

6. Characterization of physical and chemical properties of synthesized diamond crystals

Nature and content of metallic impurities in synthesized diamond crystals

Analysis of the nature and content of impurities in diamond crystals synthesized under different pressure-temperature conditions was done by X-ray diffraction and inductively coupled plasma (ICP) spectroscopy. For this scope, diamond crystals of size 120/140 mesh (average size of approx. 115 μm) were chosen. Nature and concentration of residual cobalt impurities of diamond crystals synthesized under different pressure-temperature conditions are presented in Tab. 6.1. Cobalt content in diamond crystals synthesized at 5.6 GPa and temperatures ranging from 1375 °C to 1525°C is graphically displayed in Fig. 6.1.

As mentioned above, due to thermal inertia of the diamond synthesis capsule, in Tab. 6.1, reaction time has lower values at lower temperatures, and reaches the set value at higher temperatures.
Since only spectroscopic purity graphite and cobalt were used in all high pressure-high temperature experiments, synthesized diamond crystals contain only cobalt as metallic impurity in detectable concentrations. Data represented in Tab. 6.1, show clearly that cobalt is present in diamond crystals in f.c.c. phase; high temperature *beta* phase of cobalt. This result is in accordance with experimental results obtained by Wong (37). Using extended X-ray absorption fine structure (EXAFS) technique, coupled with conventional transmission electron microscopy (TEM), X-ray diffraction and X-ray fluorescence techniques, Wang confirmed that metallic impurities Ni, Co, Fe are present as f.c.c. crystallographic phases in the crystalline structure of diamond. Furthermore, using X-ray diffraction and metallographic techniques to analyze basic structural characteristics of phases presented as impurities in diamond crystals synthesized in carbon-nickel and carbon-cobalt systems, Vavrda (38), concluded that these impurities are not carbides, but primary solid solutions of carbon in the metal catalyst.
Data presented in Tab. 6.1 and Fig. 6.2 a. and b, also show a strong correlation between cobalt concentration in diamond crystals synthesized at different pressure-temperature conditions, specifically:

- *At constant pressure, P = 5.6 GPa, concentration of cobalt impurities decreases with temperature;*
- *At constant temperature, T = 1475 °C, concentration of cobalt impurities increases with pressure.*

Morpho-structural characteristics of synthesized diamond crystals

Due to pressure and temperature gradients inside the capsule, synthesized diamond crystals do not share same particle size and crystal characteristics (morphology, topography, crystal growth defects, color and transparency, etc.).
To study the morpho-structural characteristics of synthesized diamond the following pressure-temperature-time conditions were used: P = 5.2 GPa – 6.2 GPa; T = 1375 °C - 1525 °C, t = 600 sec. Taking into consideration that diamond crystals sharing same size range do not share same physical characteristics, such as crystal morphology (growth habit), surface topography (roughness) and color and transparency, the morpho-structural characterization of synthesized diamond crystals was made by defining the following main crystal types:

- Type A: crystals with cubic morphology featuring well developed (100) faces, heavily included, gray to black colors, opaque crystals;
- Type B: crystals with cubic morphology featuring well developed (100) faces and less developed (111) faces, deep yellow to light yellow colors, translucent and transparent crystals;
- Type C: crystals with cub-octahedron morphology featuring well developed (111) faces and less developed (100) faces, light yellow color, transparent crystals;
- Type D: crystals with no identifiable morphology, irregularly shaped, yellowish gray to gray colors, translucent crystals.

Since, in a sample of diamond crystals two or more types of diamond crystals and even atypically diamond crystals are present, in our analysis we singled out the predominant crystal type (highest weight percentage crystals). The results of morpho-structural analysis of diamond crystals synthesized under different pressure-temperature conditions are presented in Tab. 6.2.

Dependence of morpho-structural characteristics of diamond crystals with pressure and temperature can be summarized as follows:

- *At constant pressure with temperature increase, crystal growth habit evolves from predominantly cube with well developed (100) faces – type A crystals – to predominantly cub-octahedron with well developed (100) faces and less developed (111) faces – type B crystals – to predominantly cub-octahedron with well developed (111) faces and less developed (100) faces – type C crystals;*
- *At constant temperature with pressure increase, crystal growth habit evolves from predominantly cube with well developed (100) faces – type A crystals – to predominantly cub-octahedron with well developed (100) faces and less developed (111) faces – type B crystals – to predominantly crystals with no identifiable morphology, exceedingly irregularly shaped crystal – type D crystals.*

Mechanical strength of synthesized diamond crystals

Mechanical strength is controlled by intrinsic properties of diamond crystals. To a great extent, size, morphology, concentration of crystal growth defects (CGD) and residual metal catalyst (RMC) impurities are responsible for mechanical strength of diamond crystals.
Mechanical strength of diamond crystals was determined based on "on size" toughness test technique; toughness index being calculated as follows:

$$TI = (M / M_t) * 100 (\%) \qquad (6.1)$$

Toughness index was determined on diamond crystals of 50/60 mesh size or 250 - 297 μm (average size of approx. 274 μm), synthesized under different pressure-temperature conditions.
Toughness index of diamond crystals synthesized at constant pressure and different temperatures is included in Tab. 6.3 and graphically displayed in Fig. 6.3.
Toughness index of diamond crystals synthesized at constant temperature and different pressures is included in Tab. 6.4 and graphically displayed in Fig. 6.4.

Toughness index data indicate a strong dependence of mechanical strength of diamond crystals with pressure-temperature synthesis conditions, as follows:

- *At constant pressure, mechanical strength of diamond crystals increases with temperature;*
- *At constant temperature, mechanical strength of diamond crystals decreases with pressure.*

7. Conclusions

*** Formatting Issue - Best Available Paper/Graphic ***

- *A theoretical approach to catalytic diamond synthesis, by way of a mathematical model, has been proposed with the purpose of understanding the mechanism of the catalytic transformation of graphite to diamond in the binary Carbon-Metal (C-Me) systems with eutectic, at high pressure and high temperature (HP-HT);*

- *The mathematical model developed provides a representation of the HP-HT catalytic transformation of graphite to diamond in binary carbon-metal systems with eutectic from both thermodynamic and kinetic standpoint;*

- *The mathematical model developed, describes reasonably accurate the process of high pressure-high temperature catalytic transformation of graphite to diamond in binary carbon-metal systems with eutectic and has been reasonably validated, by comparing calculated data with from literature available for nickel-carbon system, and with experimental data acquired through diamond synthesis experiments in carbon-cobalt system, as follows:*
    - *calculated vs. experimental data from literature for carbon-nickel system at different pressures: 0 GPa, 2.6 GPa, 5.2 GPa, 5.4 GPa and 6.0 GPa, presented in Tab. 3.4;*
    - *calculated vs. experimental minimum and maximum temperatures required for diamond synthesis at P = 5.6 GPa presented in Tab. 5.2;*
    - *nucleation rate of diamond calculated as a function of temperature, at constant pressure, presented in Fig. 3.6, with experimental data acquired for dependence of nucleation rate of diamond with temperature, at constant pressure, presented in Tab. 5.4 and Fig. 5.4.a;*
    - *nucleation rate of diamond calculated as a function of pressure, at constant temperature, presented in Fig. 3.7, with experimental data acquired for dependence of nucleation rate of diamond with pressure, at constant temperature, presented in Tab. 5.4 and Fig. 5.4.b;*
    - *growth rate of diamond calculated as a function of temperature, at constant pressure, presented in Fig. 3.8, with experimental data acquired for dependence of growth rate of diamond with temperature, at constant pressure, presented in Tab. 5.5 and Fig. 5.5.a;*
    - *growth rate of diamond calculated as a function of pressure, at constant temperature, presented in Fig. 3.9, with experimental data acquired for dependence of growth rate of diamond with pressure, at constant temperature presented in Tab. 5.5 and Fig. 5.5.b.*

- *For a given graphite-metal catalyst system and under pressure-temperature conditions for which diamond is thermodynamically stable, intrinsic properties of diamond crystals are determined by nucleation and growth rates, which are controlled via pressure-temperature process parameters, thus:*
    - *subject to low nucleation and growth rates, well developed (cuboctahedron), large size diamond crystals with low level of crystal growth defects (CGD),*

residual metal catalyst (RMC) impurities and high mechanical strength could be synthesized;

- o *subject to high nucleation and growth rates, poorly developed (irregular shaped), small size diamond crystals with high level of crystal growth defects (CGD), residual metal catalyst (RMC) impurities and low mechanical strength could be synthesized*

- *Understanding the mechanism that governs the HP-HT catalytic transformation of graphite to diamond, represents a valuable approach toward an efficient control of the diamond synthesis processes, thus enabling synthesis of diamond crystals, whose physical and chemical properties are controlled by way of preassigned pressure-temperature synthesis parameters.*

References:

1. F. P. Bundy, J. Chem. Phys. 38(1980) 6930
2. F.P. Bundy, H.T, Hall, H.M. Strong, R.H. Wentorf, Nature 176 (1955), 51
3. H.P. Bovenkerk, F.P. Bundy, H.T. Hall, H.M. Strong, R.H. Wentorf, Nature 184 (1959) 1094
4. F.P. Bundy, H. P. Bovenkerk, H. M. Strong, R. H. Wentorf, J. Chem. Phys. 35 (1961) 383
5. A. A. Jardini, J. E. Tydings, S. B. Litvin, Am. Mineralogist 45 (1960) 27
6. H. Liander, E. Lundblad, Arkiv. Kemi. 16 (1960) 139
7. A. A. Jardini, J. E. Tydings, S. B. Litvin, Am. Mineralogist 47 (1962) 1393
8. R. H. Wentorf, H. P. Bovenkerk, J. Chem. Phys., 36 (1962) 1987
9. C. M. Higgins, P. Cannon, Nature, 194 (1962) 829
10. H. M. Strong, J. Chem. Phys., 39 (1963) 2057
11. L. F. Vereschaghin, et. al., Dokl. Akad. Nauk. 162 (1965) 1027
12. 47. M. Wakatsuki, Japan J. Appl. Phys., 5 (1966) 337
13. H. J. Rodewald, Chinia 15 (1961) 251
14. Yu. A. Litvin, Izv. Akad. Nauk. SSSR Neorg. Mat., 4 (1968) 175
15. P. Cannon, E.T. Collins, in "Reactivity of Solids" ed. G. H. Swab, Elsevier, Amsterdam 1965
16. R. H. Wentorf, Ber, Bunsenges. Phys. Chem. 70 (1966) 975
17. V. P. Butuzov et all., Rest Krist., 9 (1972) 69
18. L. F. Vereschaghin, et all., Dokl. Akad. Nauk. 192 (1970) 768
19. L. F. Vereschaghin, et all., Sovrem. Prob. Fiz. Kim. 1 (1968) 173
20. G. B. Bokii, A. I. Volkov, Kristalografiya 14 (no. 1) (1969)
21. A. A. Jukov, Jurnal Fiziceskoi Himii 41 (1967) 185
22. V. A. Ianciuk, Sverthverdie Materiali 5 (1980) 16
23. L.S. Palatnik, L. L. Gladkin, Dokl. Akad. Nauk SSSR (1971) 81
24. L.S. Palatnik, s.a. Fiz Tverd. Tela (1973) 2029
25. E.T. Peters, L. Kaufman, J.J. Ryan, "AFCRL Report 65 – 168"; Contract No. AF19 (628) 2466, february 1965

26. S. K. Dickinson, "AFCRL Report 70.0620", Physical Research Papers No. 434, November 1970, Air Force Cambridge Research Laboratories, Bedford, MA., USA
27. H. M. Strong, R. E. Hanneman, J. Chem. Phys. Vol. 46 No 9 (1967) 3668
28. R. Berman, F. Simon, Z. Elekrokem. 59 (1955) 333
29. M. Hansen, K. Anderco, "Constitution of Binary Alloys" (Mc. Graw Hill, N.Y. 1958)
30. H. M. Strong, Acta Metalurgica, 12 (1964) 1411
31. F. P. Bundy, J. Geophys. Res. 85 (1980) 6930
32. R. Berman, Thermal Properties in "Physical Properties of Diamond" ed. R. Berman (Clarendon Press, Oxford, 1965)
33. G. Munke, Diamond Research (1974) 7
34. I. Benea et all., *"High pressure apparatus for the synthesis of superhard materials"* RO Patent Nr. 90871/1986
35. V. E. Bean, S. Akimoto, P. M. Bell, S. Block, W. B. Holtzapel, M. H. Manghani, M. F. Nicol, S. M. Stishov, in "Research in High Pressure Science and Technology", Proc. Of 10th AIRAPT Conf. 1985, ed N. J. Trappeniers, et all. (North Holland, Amsterdam 1986) 52
36. E. V. Clougherty, L Kaufman, ASME Publication Paper No 62 – WA -258, 1963
37. J. Wong, E. F. Koch, C. I. Hejna, M.F. Garbausas, J. Appl. Phys., 58 (1985) 3388
38. T Vavrda, Sverhtverdie Materiali, 3 (1981) 27

*** Formatting Issue - Best Available Paper/Graphic ***

Tables & Figures:

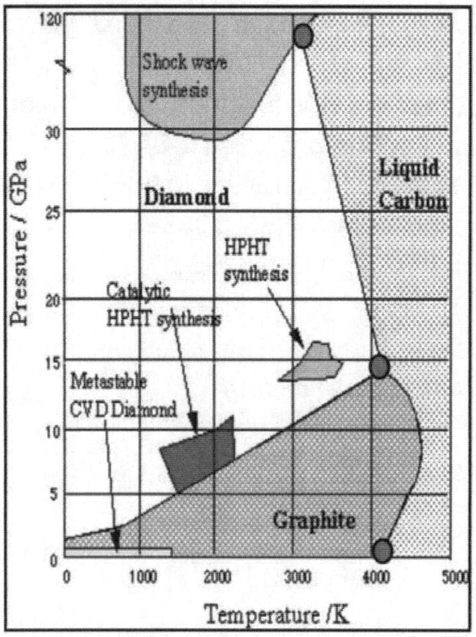

<u>Diamond-Liquid equilibrium line:</u>
diamond/metal/liquid triple point (hypothetic)
T ~ 2,800 K; P ~ 50 GPa
graphite/diamond/liquid triple point
T = 4,100 ± 100 K; P = 11 ± 1.0 GPa
<u>Graphite-Liquid equilibrium line:</u>
graphite/diamond/liquid triple point
P = 4,100 ± 100 K; P = 11 ± 1.0 Gpa
graphite/liquid/vapor triple point:
P = 4,020 ± 50 K; P = 0.011 – 0.014 GPa
max. point: T ~ 4,600 K; P ~ 6.5 GPa
<u>Graphite-Diamond equilibrium line:</u>
P (GPa) = 0.6865 + 0.00265 T (K) (Berman-Simon)
P (GPa) = 1.94 + 0.0025 T (°C) (Kennedy)

Fig. 1.1 Phase diagram of Carbon (Bundy 1980)

Tab. 3.1 – Thermodynamic data for S-L transformation of pure components C, Ni and Co

Transformation	ΔH (J/mol)	ΔS (J/mol*K)	ΔV (cm^3/mol)	Reference
G - L	90,376.50	-22.594	--	(25, 26)
D - L	89,121.30	-27.364	--	(25, 26)
Ni^{cfc} - L	17,614.60	-10.209	0.337	(25, 26)
Co^{cfc} - L	16,192.00	-9,163	0.263	(26)

Tab. 3.2 – Experimental data used for the calculation of interaction parameters L and constant K

***** Formatting Issue - Best Available Paper/Graphic *****

System	Equilibrium line	Point	$X*10^{-2}$ (at%)	T (K)	P(GPa)	Reference
Ni - C	G + L/ L	eutectic	0.088	1587	10^{-4}	(29)
		melting	0.200	2500	10^{-4}	(29)
		eutectic	0.137	1661	5.4	(27)
		eutectic	0.150	1600	6.7	(30)
	D + L / L	eutectic	0.135	1665	5.3	(29)
		eutectic	0.134	1667	5.4	(29)
C	G / L	melting	1.000	4000	6.5	(31)
		melting	1.000	4200	12.5	(31)
	D / L	melting	1.000	3350	6.5	(32)
		melting	1.000	4200	12.5	(31)

Tab. 3.3 – Calculated L and K values for solid/liquid equilibrium in C-Ni system

System	Equilibrium line	$E_G^L = X_{Me} * X_C((L_1^{(0)} - T*L_2^{(0)} + P(L_1^{(i)} + L_2^{(i)}))$				$V_C^{G,D/L} = K_1^{(i)} + P*K_2^{(i)}$	
		$L_1^{(0)}$ (J/mol)	$L_2^{(0)}$ (J/mol)	$L_1^{(i)}$ (J/mol)	$L_2^{(i)}$ (J/mol)	$K_1^{(i)}$ (cm^3/mol)	$K_2^{(i)}$ (cm^3/mol)
Ni - C	Ni^{cfc} + L / L (I = 1)	-72721.35	-28.786	-1462.30	309.61	--	--
	G + L / L (I = 2)	-72721.35	-28.786	-8222.30	443.08	5.40	-0.4032
	D + L / L (I = 3)	-72721.35	-28.786	-387.23	271.96	2.52	-0.0369

Tab. 3.4 - Comparison between calculated data and experimental data available for C-Ni system at high pressure

Point	Calculated data			Experimental data			
	$X*10^{-2}$ (at%)	T (K)	P(GPa)	$X*10^{-2}$ (at%)	T (K)	P(GPa)	Reference
Ni-G eutectic	0.088	1586.4	10^{-4}	0.091-0.112	1580-1591	10^{-4}	(29)
	0.112	1621.0	2.6	0.110	1637	2.6	(27)
	0.137	1660.9	5.4	0.137	1661	5.4	(27)
	0.143	1669.6	6.0	0.144	1670	6.0	(30)
Ni-D eutectic	0.134	1666.1	5.4	0.134	1667	5.4	(27)
G/D/L equilibrium	0.143	1729.6	5.4	0.150	1728	5.4	(27)
Ni(cfc)/G/L equilibrium	0.135	1658.1	5.2	0.135	1665	5.25	(27)

Tab. 3.5 – Experimental data used for the calculation of interaction parameters L and constant K

*** Formatting Issue - Best Available Paper/Graphic ***

System	Equilibrium line	Point	$X*10^{-2}$ (at%)	T (K)	P(GPa)	Reference
Co - C	G + L/ L	eutectic	0.116	1547	10^{-4}	(29)
		melting	0.200	2250	10^{-4}	(29)
		eutectic*	0.158	1616	5.4	calculated by similarity with Ni-C system
		eutectic*	0.172	1636	6.7	
	D + L / L	eutectic*	0.159	1665	5.0	
		eutectic*	0.158	1667	5.4	
C	G / L	melting	1.000	4000	6.5	31
		melting	1.000	4200	12.5	31
	D / L	melting	1.000	3350	6.5	32
		melting	1.000	4200	12.5	31

Tab. 3.6 – Calculated L and K values for solid/liquid equilibrium in C-Co system

System	Equilibrium line	$E_G^L = X_{Me} * X_C((L_1^{(0)} - T*L_2^{(0)} + P(L_1^{(i)} + L_2^{(i)}))$				$V_C^{G,D/L} = K_1^{(i)} + P*K_2^{(i)}$	
		$L_1^{(0)}$ (J/mol)	$L_2^{(0)}$ (J/mol)	$L_1^{(i)}$ (J/mol)	$L_2^{(i)}$ (J/mol)	$K_1^{(i)}$ (cm^3/mol)	$K_2^{(i)}$ (cm^3/mol)
Co - C	Co^{cfc} + L / L (i = 1)	69,870.75	-22.258	-1067.73	309.61	--	--
	G + L / L (i = 2)	69,870.75	-22.258	-7929.51	443.08	5.40	-0.4032
	D + L / L (i = 3)	69,870.75	-22.258	-3306.19	271.96	2.52	-0.0369

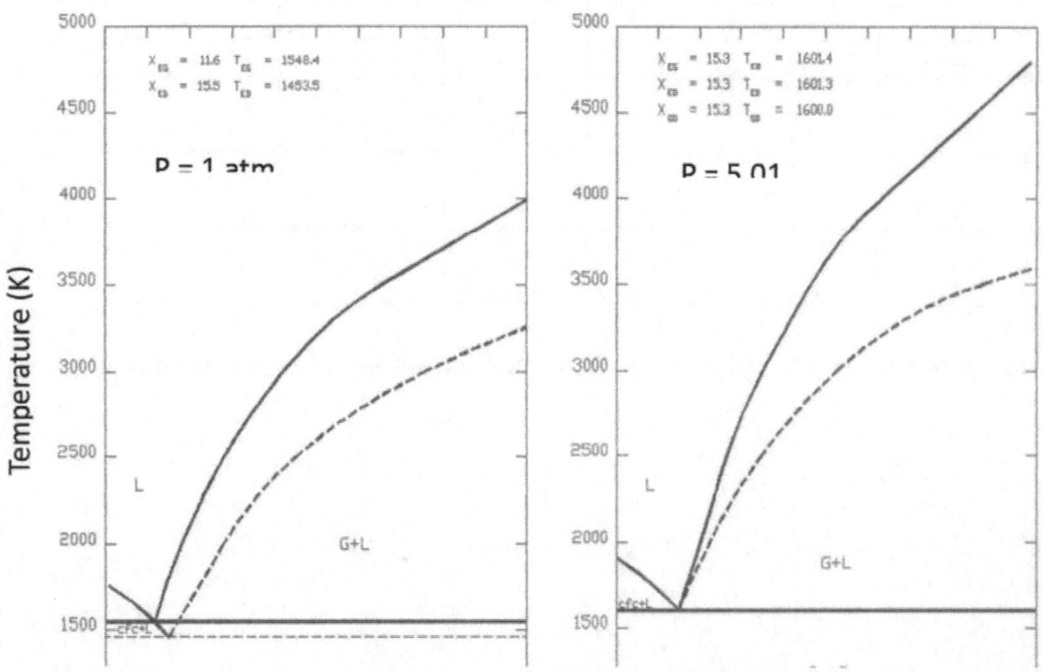

*** Formatting Issue - Best Available Paper/Graphic ***

Fig. 3.1 – Solid/liquid equilibrium phase diagrams of C-Co system calculated at 1 atm and 5.01 GPa

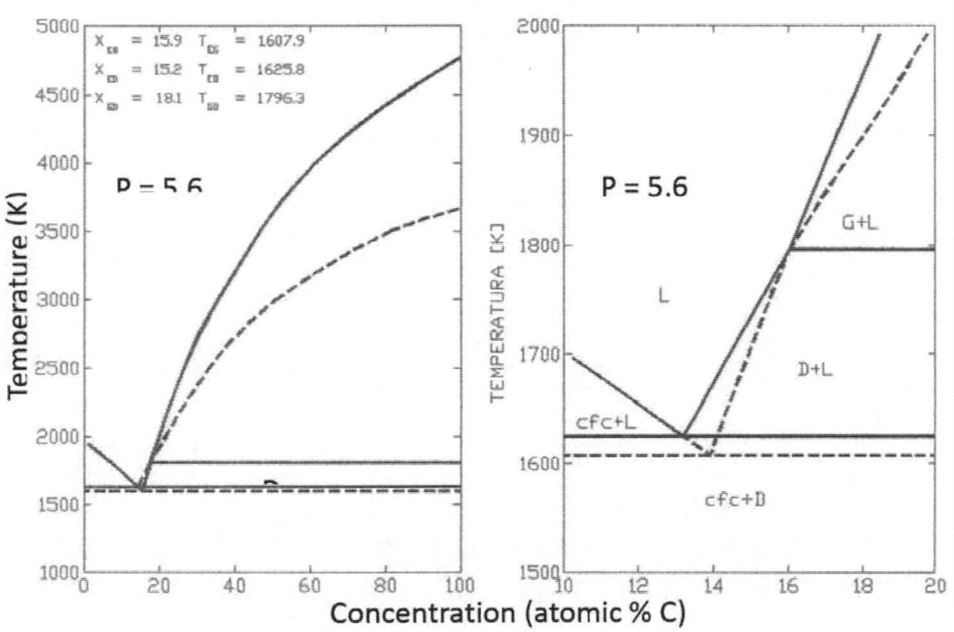

Fig. 3.2 – Solid/liquid equilibrium phase diagrams of C-Co system calculated at 5.6 GPa

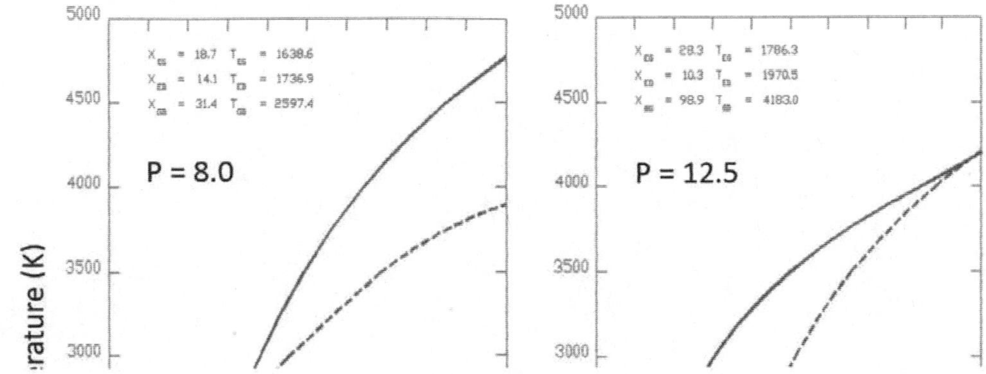

*** Formatting Issue - Best Available Paper/Graphic ***

Fig. 3.3 – Solid/liquid equilibrium phase diagrams of C-Co system calculated at 8.0 GPa & 12.5 GPa

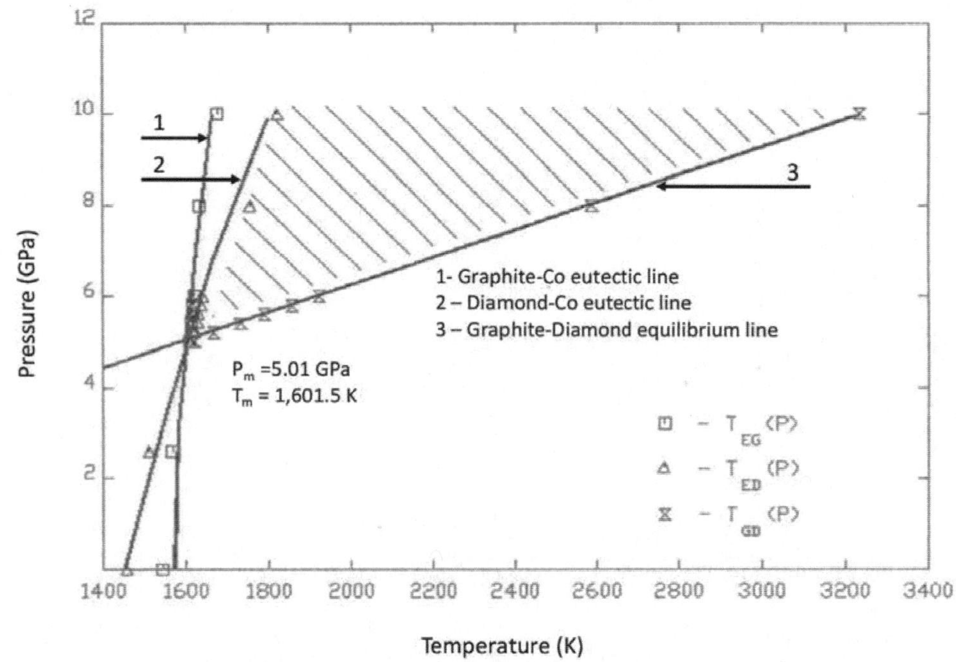

Fig. 3.4 – Pressure dependence of temperature of cobalt-graphite eutectic, cobalt-diamond eutectic and graphite/diamond/liquid equilibrium

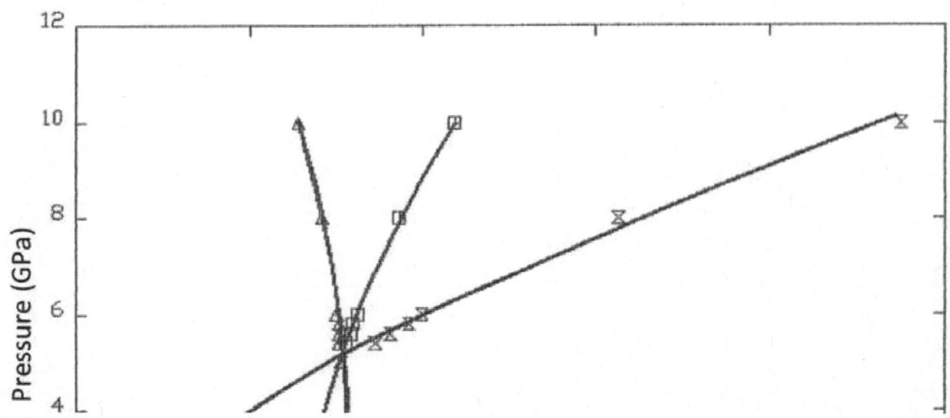

*** Formatting Issue - Best Available Paper/Graphic ***

Fig. 3.5 – Dependence of carbon concentration with pressure for cobalt-graphite eutectic, cobalt-diamond eutectic and graphite/diamond/liquid equilibrium

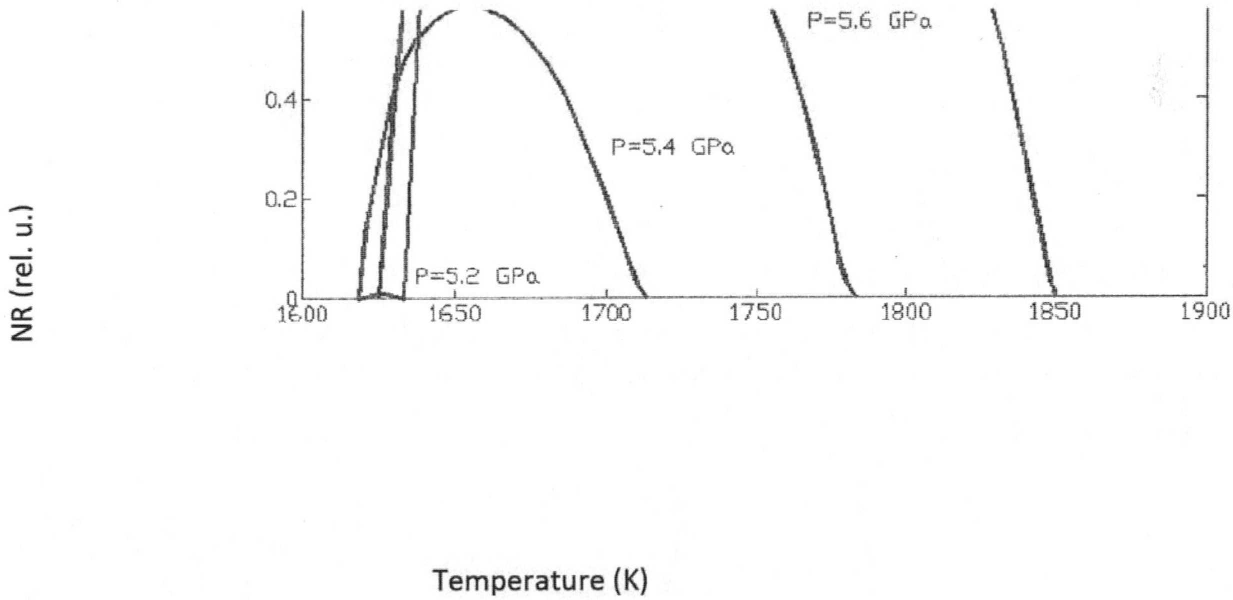

Fig. 3.6 – Dependence of nucleation rate of diamond with temperature at constant pressure

*** Formatting Issue - Best Available Paper/Graphic ***

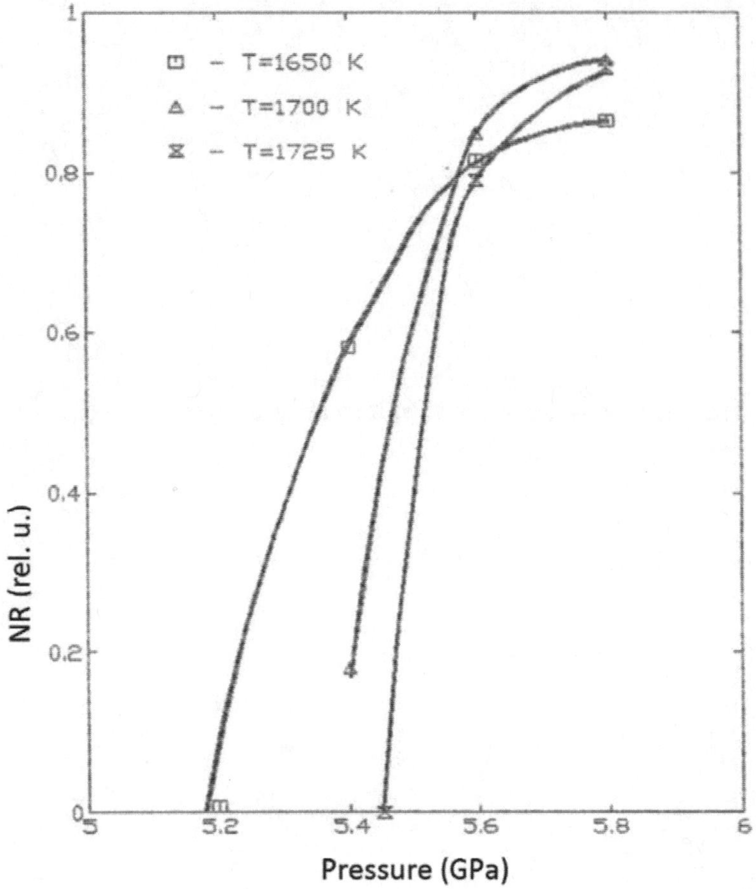

Fig. 3.7 – Dependence of nucleation rate of diamond with pressure at constant temperature

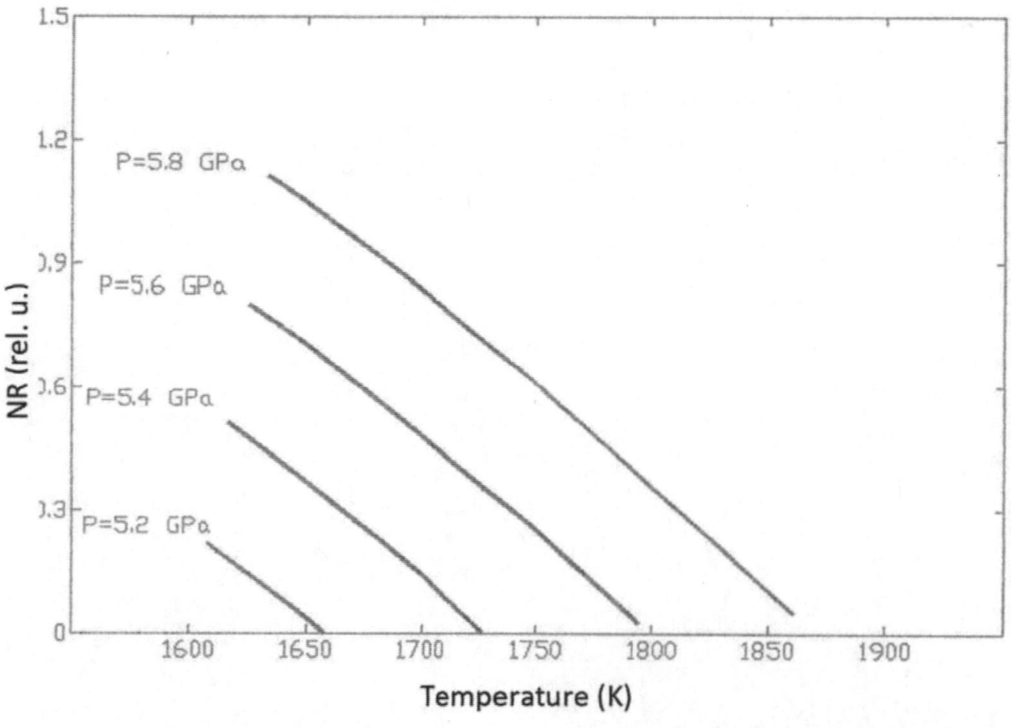

***** Formatting Issue - Best Available Paper/Graphic *****

Fig. 3.8 – Dependence of growth rate of diamond with temperature at constant pressure

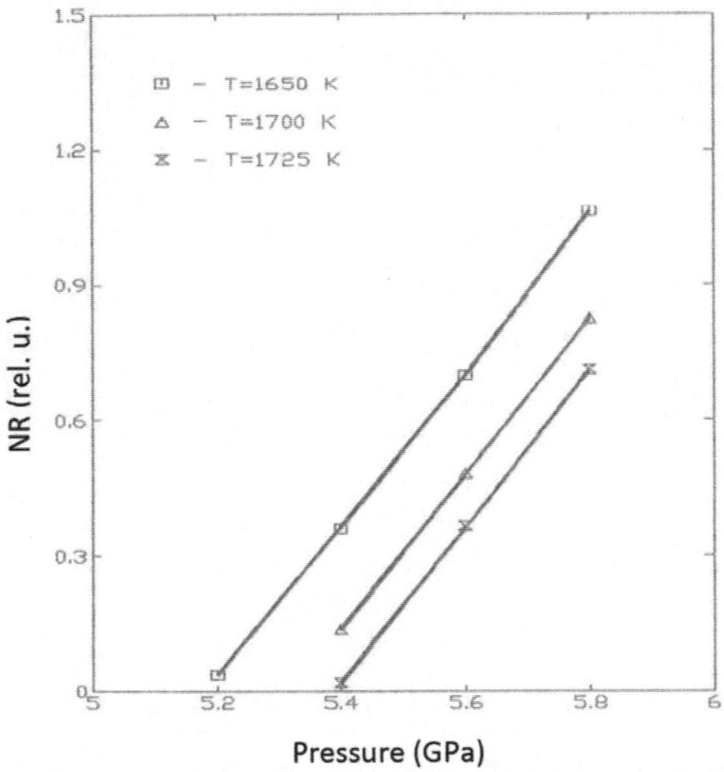

Fig. 3.9 – Dependence of growth rate of diamond with pressure at constant temperature

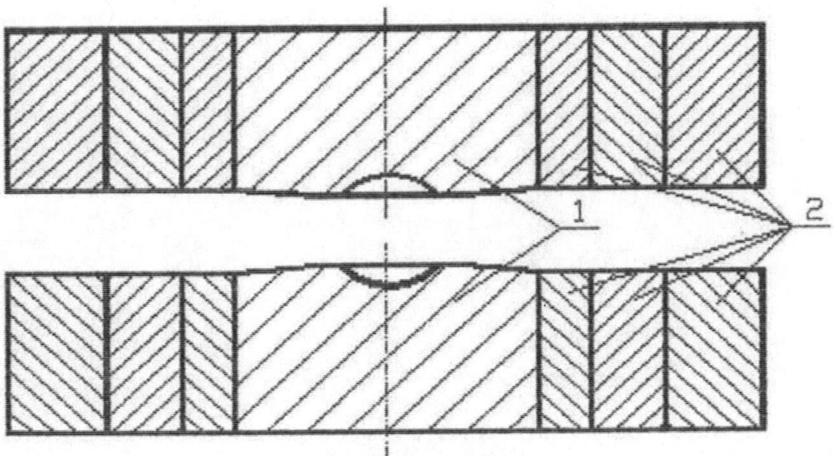

1 – Carbide anvils
2 – Steel rings

Fig. 4.1.a – High pressure apparatus assembly

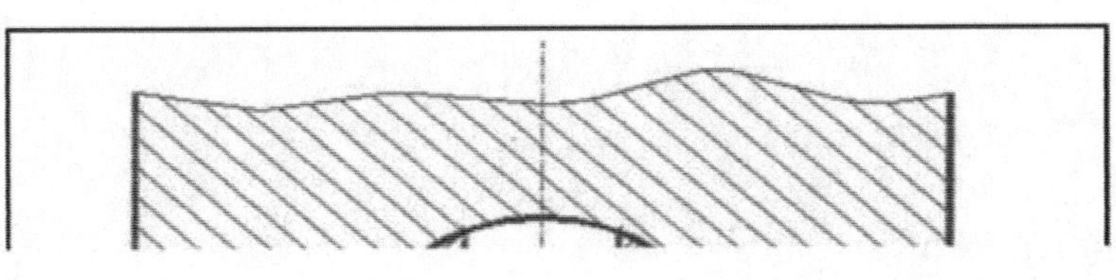

*** Formatting Issue - Best Available Paper/Graphic ***

Fig. 4.1.b – Gasket assembly

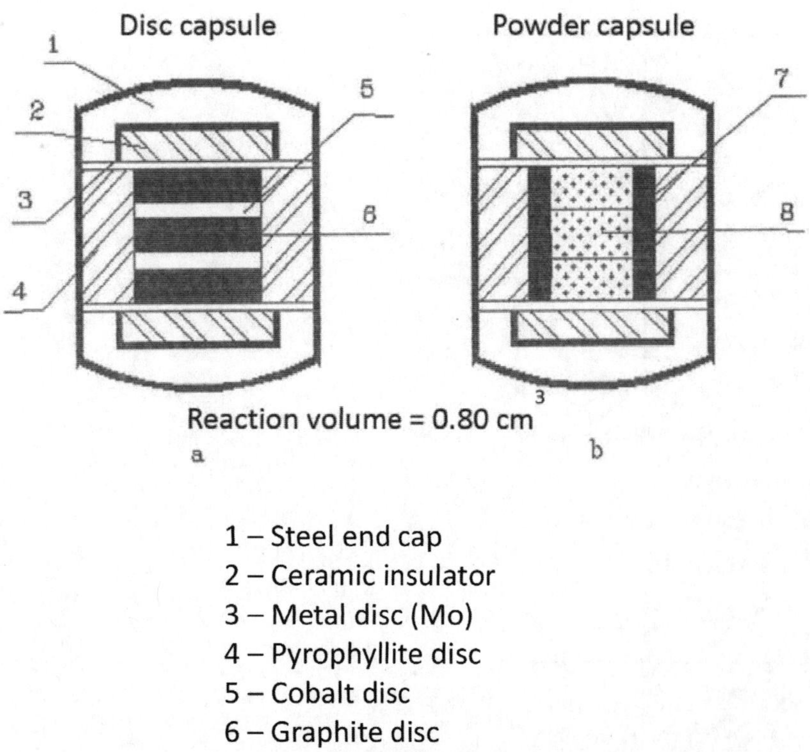

1 – Steel end cap
2 – Ceramic insulator
3 – Metal disc (Mo)
4 – Pyrophyllite disc
5 – Cobalt disc
6 – Graphite disc
7 – Graphite cylinder
8 – Graphite-Cobalt powder mi x

*** Formatting Issue - Best Available Paper/Graphic ***

Fig. 4.2 – Capsule assembly

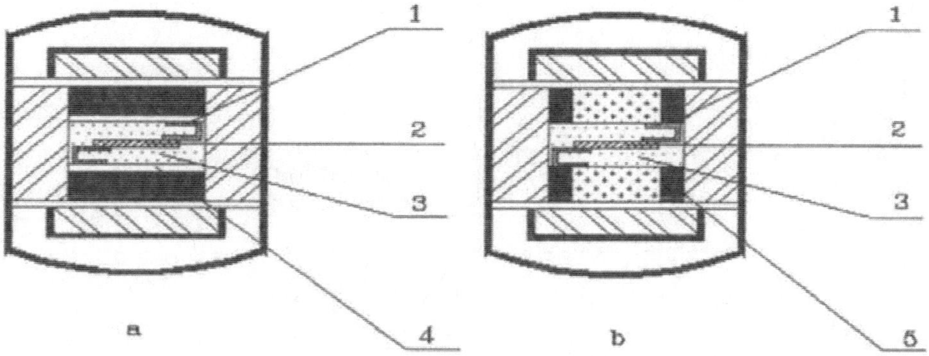

Disc capsule

Powder capsule

1 – Electric contact (Cu)
2 – Pressure sensor (Bi; Tl; Ba)
3 – NaCl media
4 – Metal disc (Mo)
5 - Graphite heater

Fig. 4.3 – Experimental setup for pressure calibration of both direct heating (a) and indirect heating (b)

Tab. 4.1 – Pressure values recommended for fixed pressure points on the international practical pressure scale

Pressure (GPa)	2.550 ± 0.006	3.68 ± 0.03	5.5 ± 0.1	7.7 ± 0.2	9.4 ± 0.3	12.3 ± 0.5	13.4 ± 0.6
Transition	Bi I-II	Tl II-III	Ba I-II	Bi III-V	Sn	high-Ba	Pb
Reference	(35)						

Tab. 4.2 – Experimental data acquired in pressure calibration experiments

		Direct heating capsule				Indirect heating capsule			
Transition	Pressure (GPa)	F (MN)	s (MN)	I (%)	E (%)	F (MN)	s (MN)	I (%)	E (%)
Bi I-II	2.55	2.21	0.04	1.8	92.3	2.18	0.04	1.8	93.5
Tl II-III	3.68	3.59	0.07	1.9	81.5	3.63	0.08	2.2	81.1
Ba I-II	5.50	5.67	0.13	2.2	76.4	5.64	0.14	2.4	78
Bi III-V	7.70	8.10	0.18	2.2	76.0	9.17	0.22	2.4	67.2

***** Formatting Issue - Best Available Paper/Graphic *****

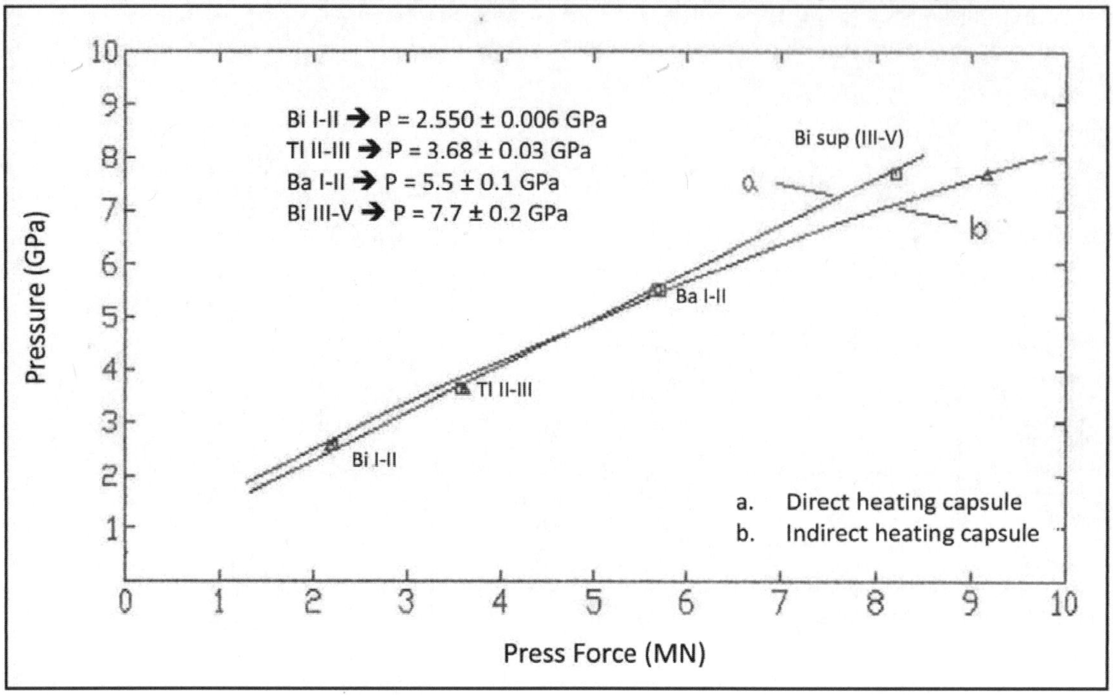

Fig. 4.4 – Pressure calibration curves for direct heating capsule (a) and indirect heating capsule

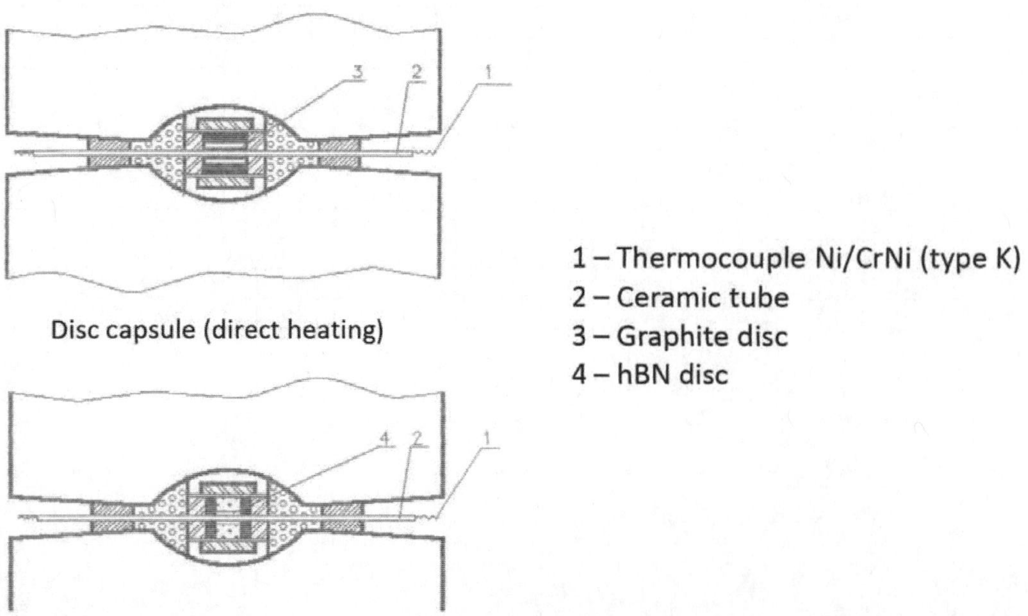

*** Formatting Issue - Best Available Paper/Graphic ***

Fig. 4.5 – Experimental setup for temperature calibration of both direct heating (a) and indirect heating (b)

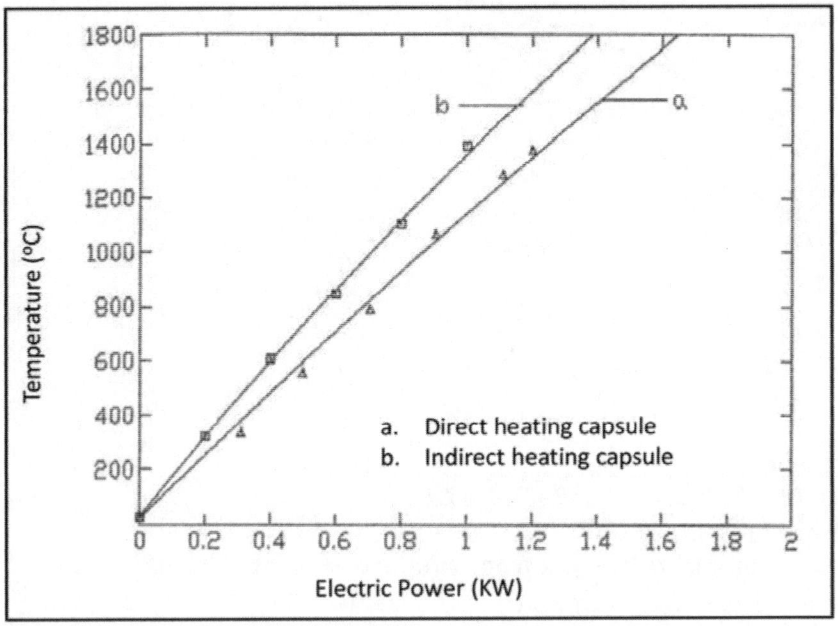

Fig. 4.6 – Temperature calibration curves for direct heating capsule (a) and indirect heating capsule

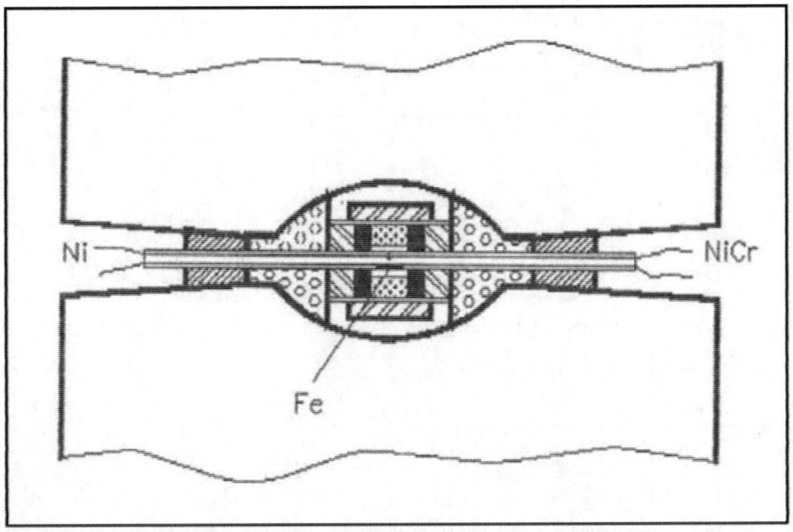

Fig. 4.7 – Experimental setup for detection of αFe – γFe transition at high pressure

*** Formatting Issue - Best Available Paper/Graphic ***

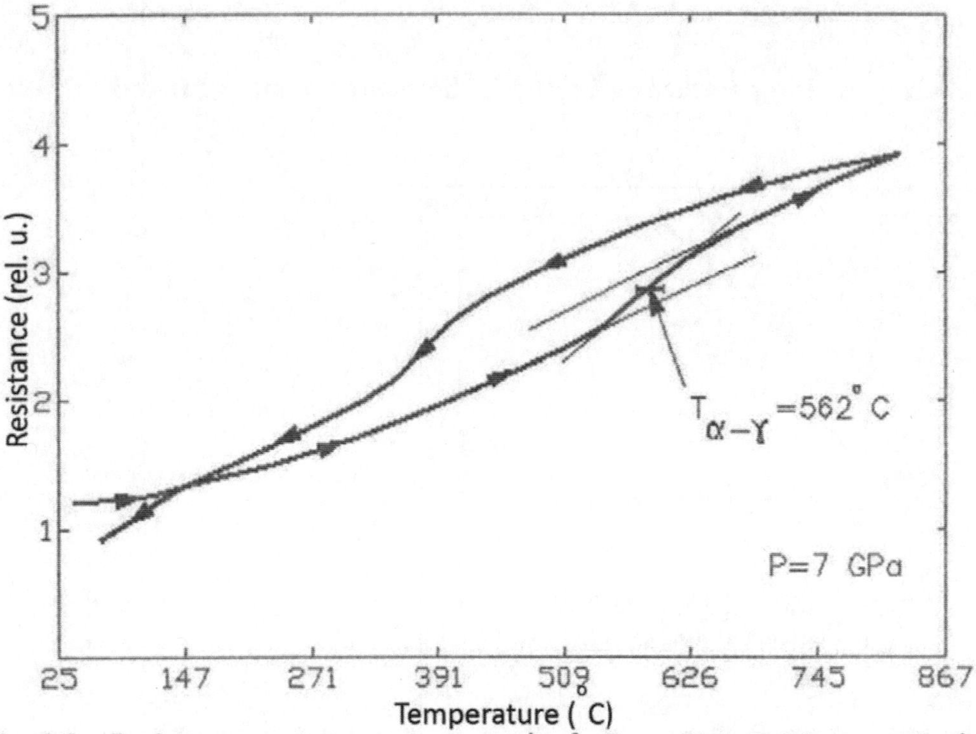

Fig. 4.8 – Resistance vs. temperature graph of αFe – γFe transition acquired at 7.0 GPa

Tab. 4.3 – Experimental data experimental vs. literature data (36) for αFe – γFe transition at high pressure

Fe^{α} → Fe^{γ} equilibrium (P,T)										
P(GPa)		0	2.5	3.5	3.6	5.5	7.0	7.5	7.7	0
	This work	910	747	n.a.	702	632	562	n.a.	545	522
T(oC)	Literature	910	735	705	n.a.	620	575	553	n.a.	540

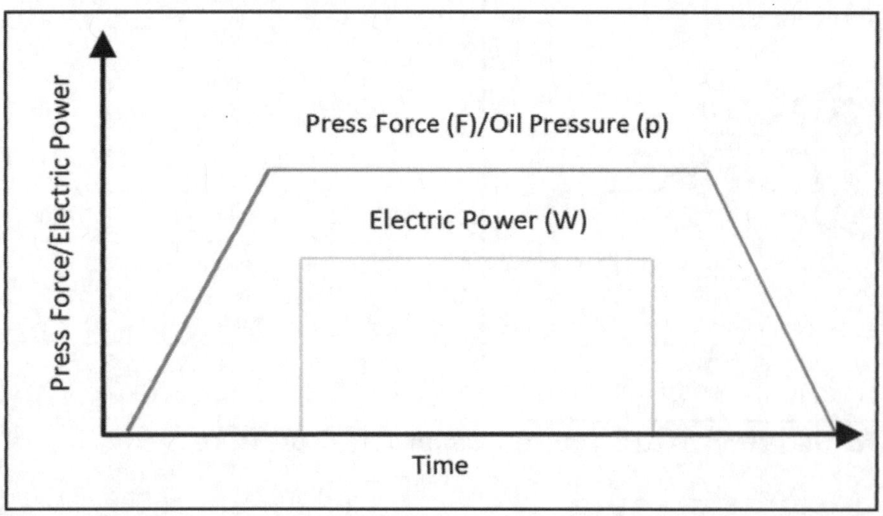

***** Formatting Issue - Best Available Paper/Graphic *****

Fig. 5.1 – Press force-electric power-time diagram used for diamond synthesis experiments

Tab. 5.1 – Experimental results for diamond formation in the carbon-cobalt system

P (GPa)	6.0	6.0	6.0	6.0	6.0	6.0	6.0	6.0	6.0	6.0	6.0	6.0	6.0	6.0
T (oC)	1300	1325	1350	1375	1400	1425	1475	1525	1575	1625	1650	1575	1700	1725
D/no-D	○	○	○	●	●	●	●	●	●	●	●	●	○	○

P (GPa)	5.8	5.8	5.8	5.8	5.8	5.8	5.8	5.8	5.8	5.8		
T (oC)	1300	1325	1350	1375	1400	1450	1500	1550	1575	1600	1625	1650
D/no-D	○	○	○	●	●	●	●	●	●	●	○	○

P (GPa)	5.6	5.6	5.6	5.6	5.6	5.6	5.6	5.6	5.6	5.6
T (oC)	1300	1325	1350	1375	1400	1450	1500	1525	1550	1575
D/no-D	○	○	●	●	●	●	●	●	○	○

P (GPa)	5.4	5.4	5.4	5.4	5.4	5.4	5.4	5.4	5.4
T (oC)	1300	1325	1350	1375	1400	1425	1450	1475	1500
D/no-D	○	○	●	●	●	●	●	○	○

P (GPa)	5.2	5.2	5.2	5.2	5.2	5.2	5.2
T (oC)	1300	1325	1350	1375	1400	1425	1450
D/no-D	○	○	●	●	●	○	○

P (GPa)	5.0	5.0	5.0	5.0
T (oC)	1300	1325	1350	1375
D/no-D	○	○	○	○

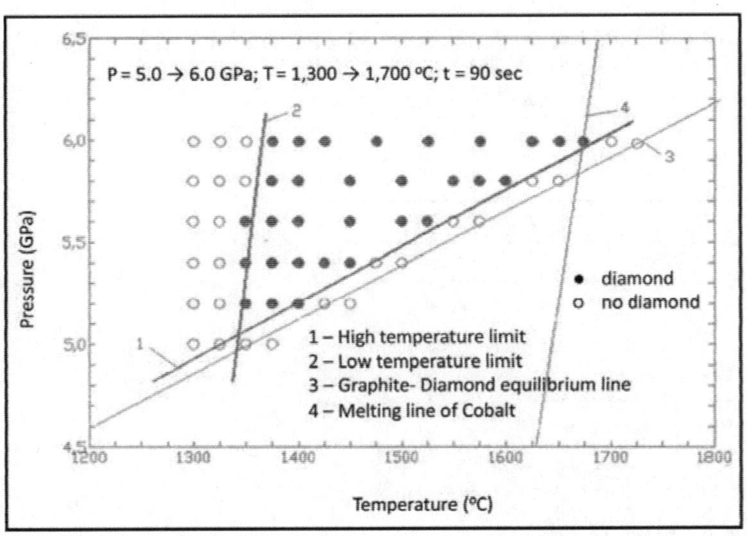

*** Formatting Issue - Best Available Paper/Graphic ***

Fig. 5.2 – Graphic representation of pressure-temperature conditions for diamond synthesis in C-Co system

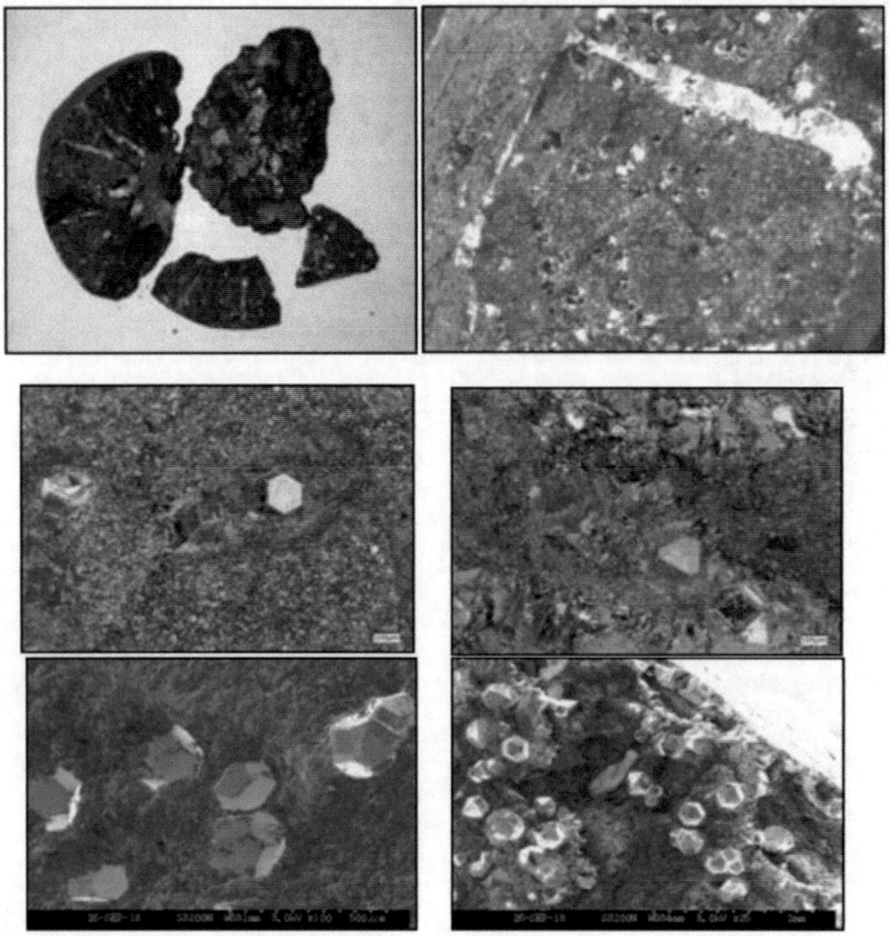

Fig. 5.3 – Optical and scanning electron microscopy images of synthesized diamond crystals

Tab. 5.2 – Experimental vs. calculated pressure-temperature conditions for diamond synthesis at P = 5.6 GPa

Experimental data (this work)			Calculated data (this work)		
P (GPa)		T °C	T °C	T K	
	T_{min}	1350	1353	1626	T_{ED}
P = 5.6	T_{max}	1525	1523	1796	T_{GD}

Tab. 5.3 – Average particle diameter per size grade and approx. number of particles per carat

ANSI	FEPA	Sieve Opening	Avg. Diameter	Number
325/400	D46	44-37	40.5	1,130,000
270/325	D54	53-44	48.5	660,000

*** Formatting Issue - Best Available Paper/Graphic ***

230/270	D64	63-53	58.0	384,000
200/230	D76	74-63	68.5	252,000
170/200	D91	88-74	81.0	140,000
140/170	D107	105-88	96.5	83,400
120/140	D126	125-105	115.0	49,400
100/120	D151	149-125	137.0	20,820
80/100	D181	177-149	163.0	17,140
60/80	D252	250-177	213.5	10,400
50/60	D301	297-250	273.5	3,240
40/50	D427	420-297	358.5	2,080

Tab. 5.4 – Experimental results and total number of particles (nucleation rate)

P	T	t	Md	Particle Size Distribution (ct)											N
(GPa)	(°C)	(sec)	(ct)	D252	D181	D151	D126	D107	D91	D76	D64	D54	D46	< D46	
5.6	**1325**	**300**	0												**0.00**
5.6	**1350**	**300**	11.9	2.2	2.2	1.9	1.6	1.1	1.0	0.5	0.5	0.3	0.3	0.3	**$1.95*10^6$**
5.6	**1375**	**300**	13.6	3.3	2.7	2.0	1.6	1.0	0.9	0.4	0.5	0.3	0.3	0.7	**$2.61*10^6$**
5.6	**1425**	**300**	13.9	3.4	2.9	2.5	0.9	1.4	1.1	0.4	0.4	0.3	0.3	0.3	**$1.98*10^6$**
5.6	**1475**	**300**	9.5	2.0	2.1	1.7	1.3	0.7	0.6	0.2	0.3	0.2	0.1	0.3	**$1.33*10^6$**
5.6	**1500**	**300**	7.3	1.7	1.7	1.4	1.0	0.5	0.4	0.1	0.2	0.1	0.1	0.2	**$0.80*10^6$**
5.6	**1525**	**300**	3.7	0.4	0.7	0.9	0.7	0.4	0.3	0.1	0.1	0.1	0.0	0.1	**$0.47*10^6$**
5.6	**1550**	**300**	0												**0.00**

P	T	t	Md	Particle Size Distribution (ct)											N
(GPa)	(°C)	(sec)	(ct)	D252	D181	D151	D126	D107	D91	D76	D64	D54	D46	< D46	
5.4	**1475**	**300**	0.0												**0.00**
5.6	**1475**	**300**	18.4	7.5	2.9	2.3	1.9	1.3	1.0	0.4	0.4	0.3	0.2	0.2	**$0.79*10^6$**
5.8	**1475**	**300**	27.1	12.3	3.9	3.0	2.6	1.7	1.6	0.6	0.6	0.4	0.2	0.3	**$2.21*10^6$**
6.0	**1475**	**300**	31.7	12.7	4.5	3.7	3.2	2.1	2.1	0.8	0.9	0.6	0.5	0.7	**$3.80*10^6$**
6.2	**1475**	**300**	36.7	12.5	5.8	4.6	4.1	2.8	2.7	1.1	1.2	0.8	0.6	0.7	**$4.46*10^6$**

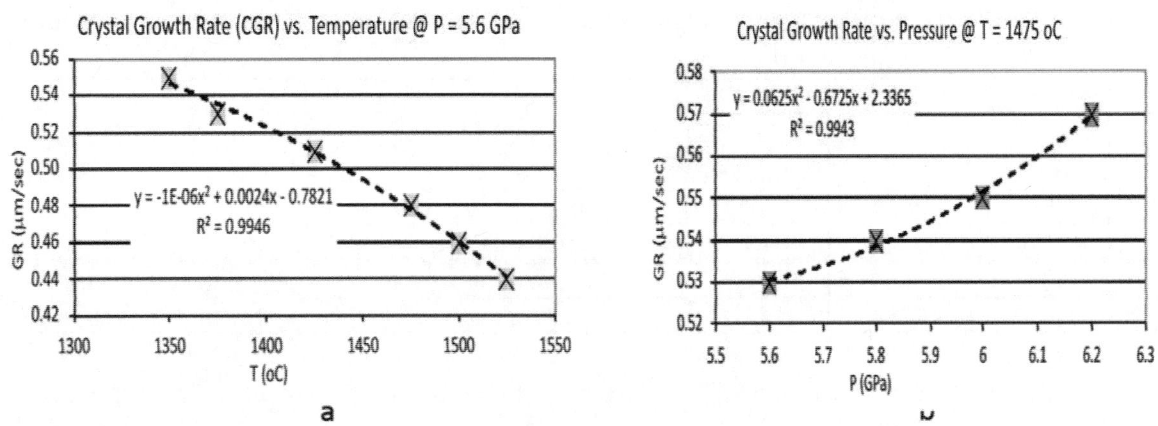

Fig. 5.4. a and b – Dependence of diamond nucleation rate with temperature at constant pressure and with pressure at constant temperature

*** Formatting Issue - Best Available Paper/Graphic ***

Tab. 5.5 – Experimental results and growth rate of diamond crystals

P	T	t	Md	Particle Size Distribution (ct)											d-avg	GR
(GPa)	(°C)	(sec)	(ct)	D252	D181	D151	D126	D107	D91	D76	D64	D54	D46	< D46	(μ)	(μ/sec)
5.6	1350	260	11.8	2.6	2.1	1.8	1.6	1.1	0.9	0.4	0.4	0.3	0.2	0.3	142.75	0.55
5.6	1375	270	13.1	3.1	2.6	2.0	1.6	1.0	0.9	0.4	0.4	0.2	0.2	0.6	144.34	0.53
5.6	1425	280	14.8	3.3	2.9	2.5	2.2	1.3	1.0	0.3	0.4	0.3	0.3	0.3	142.14	0.51
5.6	1475	300	9.2	1.9	2.0	1.7	1.3	0.7	0.5	0.2	0.3	0.2	0.2	0.3	143.36	0.48
5.6	1500	300	6.4	1.3	1.6	1.3	0.9	0.5	0.4	0.1	0.1	0.1	0.1	0.1	138.58	0.46
5.6	1525	300	3.6	0.4	0.7	0.9	0.7	0.4	0.3	0.1	0.1	0.0	0.0	0.1	131.32	0.44
5.6	1550	300	0.0												0.00	0.00

P	T	t	Md	Particle Size Distribution (ct)											d-avg	GR
(GPa)	(°C)	(sec)	(ct)	D252	D181	D151	D126	D107	D91	D76	D64	D54	D46	< D46	(μ)	(μ/sec)
5.6	1475	300	11.2	4.6	1.8	1.4	1.2	0.8	0.7	0.2	0.2	0.1	0.1	0.1	159.15	0.53
5.8	1475	300	26.7	12.2	4.0	2.9	2.6	1.6	1.6	0.5	0.5	0.3	0.3	0.3	163.16	0.54
6.0	1475	300	31.1	13.1	4.4	3.5	2.9	2.0	2.0	0.9	0.9	0.6	0.3	0.6	166.11	0.55
6.2	1475	300	35.3	14.2	5.0	4.0	3.4	2.5	2.5	0.9	0.9	0.6	0.6	0.6	170.41	0.57

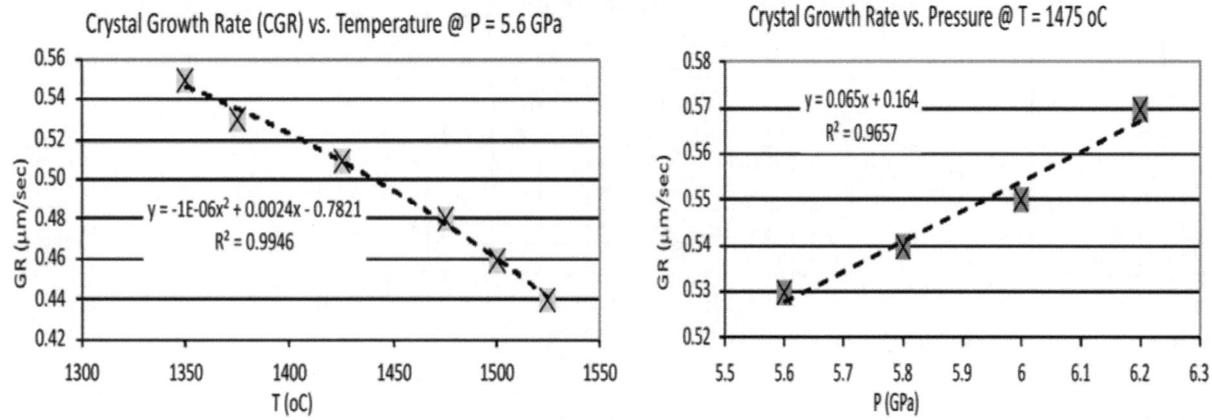

a b

Fig. 5.5. a and b – Dependence of growth rate of diamond with temperature at constant pressure and with pressure at constant temperature

Table 5.6 – X-ray diffraction phase analysis results on graphite-cobalt specimens reacted under different pressure-temperature conditions

Exp. Conditions				XRD phase analysis	Remarks
G-Co mass ratio	P (GPa)	T (°C)	t (sec)		
G/Co = 1/1	5.6	1450	300	Graphite; Diamond; Co (c.f.c.)	
G/Co = 1/1	4.8	1350	300	Graphite; Co (c.f.c.)	
G/Co = 1/1	5.6	1450	300	Graphite (hexagonal)	after Co removal from reacted specimen
G/Co = 1/1	4.8	1350	300	Graphite (hexagonal)	

Tab. 6.1 – Nature and content of Co impurity in diamond crystals synthesized under different pressure-temperature conditions

*** Formatting Issue - Best Available Paper/Graphic ***

P (GPa)	T (°C)	t (sec)	Co Phase	Co wt %
5.6	1375	260	fcc-Co	1.47
5.6	1425	270	fcc-Co	0.74
5.6	1475	280	fcc-Co	0.33
5.2	1375	300	fcc-Co	2.28
5.6	1525	300	fcc-Co	0.22
6.0	1475	300	fcc-Co	0.54

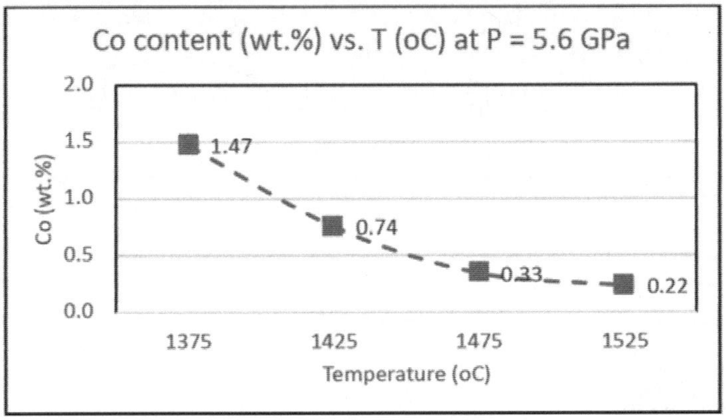

Fig. 6.1 – Dependence of Co content (%) with temperature at constant pressure

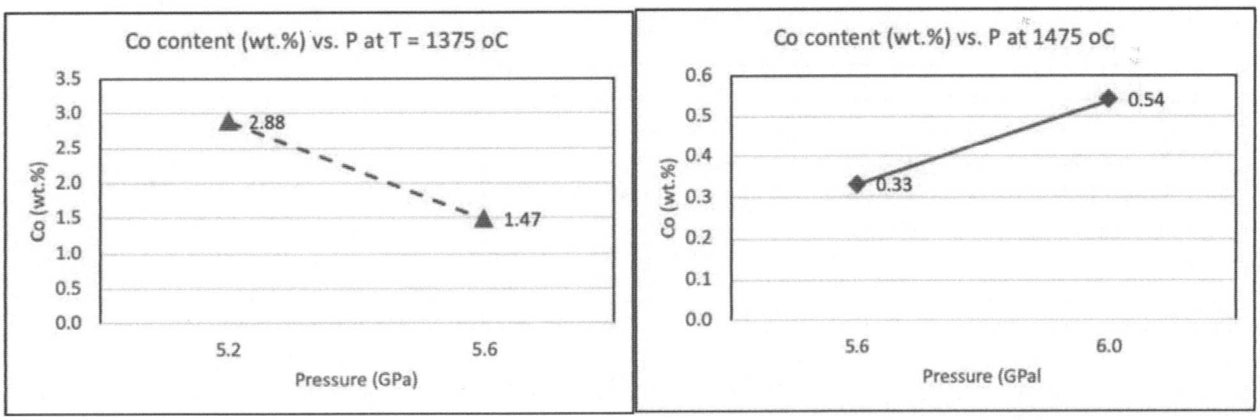

a b

Fig. 6.2 – Dependence of Co content (%) with temperature at constant pressure (a), and with pressure at constant temperature (b)

Tab. 6.2 – Morpho-structural characteristics of synthesized diamond crystals under different pressure-temperature conditions

*** Formatting Issue - Best Available Paper/Graphic ***

P (GPa)	T (°C)	t (sec)	Size (μm)	Morpho-structural characteristics of diamond crystals
5.6	1375	600	297-250	A-type crystal: majority B-type crystals: small number
5.6	1425	600	250-297	A-type crystals: large number B-type crystal: large number; C-type: relativ small number
5.6	1475	600	250-297	B-type crystals: approx. half C-type crystals: approx hallf
5.6	1525	600	250-297	C-type crystals: majority B-type crystals: reduced number
5.2	1375	600	250-297	Stratified grown black crystals; graphite layer alternating with diamond layers; A-type crystals: v. small numbers
6.0	1475	600	250-297	D-type crystals: large number multi-crystalline aggregates & macles: large number

Tab. 6.3 – Dependence of toughness index of synthesized diamond crystals with temperature at constant pressure (P = 5.6 GPa)

P (GPa)	T (°C)	t (sec)	Size	TI (%)
5.6	1375	600	250-297	13
5.6	1425	600	250-297	18
5.6	1475	600	250-297	29
5.6	1525	600	250-297	55

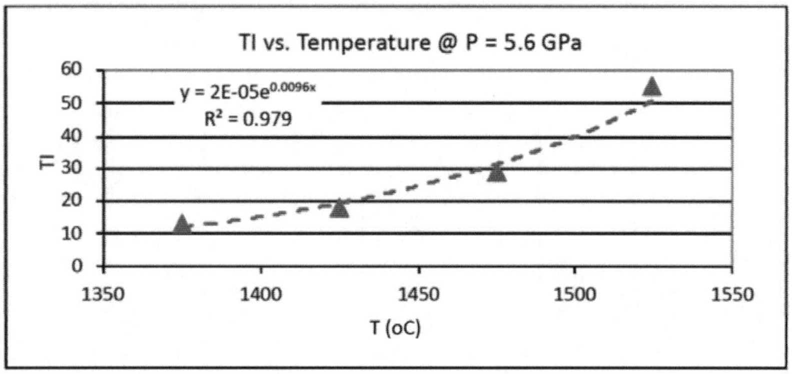

Fig. 6.3 – Dependence of toughness index of synthesized diamond crystals with temperature at constant pressure (P = 5.6 GPa)

Tab. 6.4 – Toughness index of synthesized diamond crystals with temperature at constant pressure

P (GPa)	T (°C)	t (sec)	Size	TI (%)
5.2	1475	600	250-297	
5.4	1475	600	250-297	65
5.6	1475	600	250-297	29
6.0	1475	600	250-297	15

*** Formatting Issue - Best Available Paper/Graphic ***

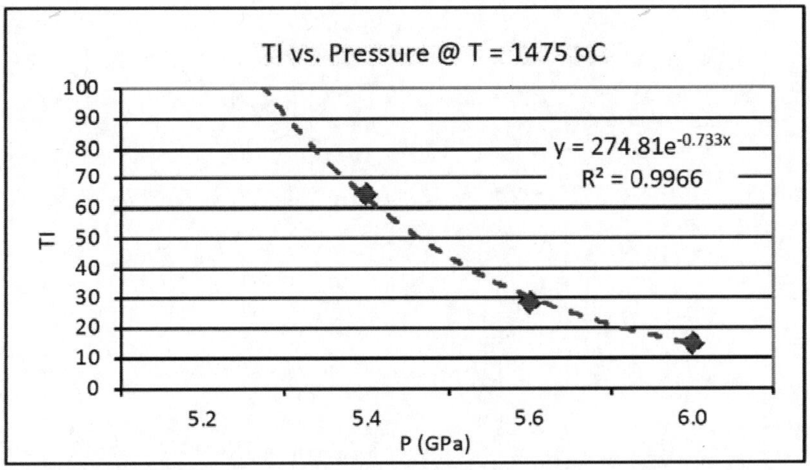

Fig. 6.4 – Dependence of toughness index of synthesized diamond crystals with pressure at constant temperature

AUTHOR INDEX